T0190344

Handbook of II-VI Semiconductor-Based Sensors and Radiation Detectors

Ghenadii Korotcenkov

Editor

Handbook of II-VI Semiconductor-Based Sensors and Radiation Detectors

Volume 2, Photodetectors

 Springer

Editor
Ghenadii Korotcenkov
Department of Physics and Engineering
Moldova State University
Chisinau, Moldova

ISBN 978-3-031-20512-5   ISBN 978-3-031-20510-1   (eBook)
https://doi.org/10.1007/978-3-031-20510-1

© The Editor(s) (if applicable) and The Author(s), under exclusive license to Springer Nature
Switzerland AG 2023
This work is subject to copyright. All rights are solely and exclusively licensed by the Publisher, whether
the whole or part of the material is concerned, specifically the rights of translation, reprinting, reuse of
illustrations, recitation, broadcasting, reproduction on microfilms or in any other physical way, and
transmission or information storage and retrieval, electronic adaptation, computer software, or by similar
or dissimilar methodology now known or hereafter developed.
The use of general descriptive names, registered names, trademarks, service marks, etc. in this publication
does not imply, even in the absence of a specific statement, that such names are exempt from the relevant
protective laws and regulations and therefore free for general use.
The publisher, the authors, and the editors are safe to assume that the advice and information in this book
are believed to be true and accurate at the date of publication. Neither the publisher nor the authors or the
editors give a warranty, expressed or implied, with respect to the material contained herein or for any
errors or omissions that may have been made. The publisher remains neutral with regard to jurisdictional
claims in published maps and institutional affiliations.

This Springer imprint is published by the registered company Springer Nature Switzerland AG
The registered company address is: Gewerbestrasse 11, 6330 Cham, Switzerland

e who are interested in exactly these compounds, we can recommend the Metal
les series which is published by Elsevier.

he aim of this three-volume book is to provide an updated account of the state
e art of multifunctional II-VI semiconductors, from fundamental sciences and
erial sciences to their applications as various sensors and radiation detectors,
, based on this knowledge, formulate new goals for further research. This book
vides interdisciplinary discussion of a wide range of topics, such as synthesis of
/I compounds, their deposition, processing, characterization, device fabrication,
testing. Topics of the recent remarkable progresses in application of nanoparti-
s, nanocomposites, and nanostructures consisting of II-VI semiconductors in
ious devices are also covered. Both experimental and theoretical approaches
re used for this analysis.

Currently, there exist books on II-VI semiconductors. However, some of them
re published too long ago and cannot reflect the current state of research in this
ea. Other published books focus on a limited number of topics, from which topics
lated to various sensor applications such as gas sensors, humidity sensors, and
osensors are almost completely excluded. When considering photodetectors, the
cus is also only on the analysis of IR photodetectors. Although sensors operating
the visible, ultraviolet, terahertz, and X-ray ranges also hold great promise for
pplications. With these books, we will try to close this gap.

Our three-volume book *Handbook of II-VI Semiconductor-Based Sensors and
adiation Detectors* is the first to cover both chemical sensors and biosensors and
ll types of photodetectors and radiation detectors based on II-VI semiconductors.
t contains a comprehensive and detailed analysis of all aspects of the application of
I-VI semiconductors in these devices. This makes these books very useful and
comfortable to use. Combining this information in three volumes, united by com-
non topics, should help readers in finding the necessary information on required
subject.

Chapters in *Handbook of II-VI Semiconductor-Based Sensors and Radiation
Detectors. Vol. 1: Materials and Technologies* describe the physical, chemical, and
electronic properties of II-VI compounds, which give rise to an increased interest in
these semiconductors. Technologies that are used in the development of various
devices based on II-VI connections are also discussed in detail in this volume.

*Handbook of II-VI Semiconductor-Based Sensors and Radiation Detectors. Vol.
2: Photodetectors* focuses on the consideration of all types of optical detectors,
including IR detectors, visible detectors, and UV detectors. This consideration
includes both the fundamentals of the operation of detectors and the peculiarities of
their manufacture and use. An analysis of new trends in development of II-VI
semiconductors-based photodetectors is also given.

*Handbook of II-VI Semiconductor-Based Sensors and Radiation Detectors. Vol.
3: Sensors, Biosensors and Radiation Detector* describes the use of II-VI compounds in other fields such as radiation detectors, gas sensors, humidity sensors,
optical sensors, and biosensors. The chapters in this volume provide a comprehensive overview of the manufacture, parameters, and applications of these devices.

Preface

Binary and ternary semiconductors of II-VI group (ZnS, ZnSe, Zn
CdTe, HgTe, HgS, HgSe, HgCdTe, CdZnTe, CdSSe, and HgZnTe) a
among researchers because of their remarkable physical and chem
which, as a group, are unique. II-VI compounds possess a very wic
electronic and optical properties. Most materials of group II-VI are se
with a direct band gap and high optical absorption and emission c
addition, binary II-VI compounds are easily miscible, providing a con
of properties. As results, the II-VI semiconductors possess band gap, v
wide range. Therefore, II-VI compounds can serve as efficient light e
as light diodes and lasers, solar cells, and radiation detectors operating
from IR to UV and X-ray. II-VI compound-based devices can also co
range. Besides common photovoltaic applications, II-VI semiconduct
potential candidates for a variety of electronic, electro-optical, sensing
electric devices. In particular, nanoparticles of II-VI semiconductors, su
tum dots, one-dimensional structures, and core-shells structures, can l
development of gas sensors, electrochemical sensors, and biosensors. T
conductors, when downsized to nanometer, have become the focus o
because of their tunable band structure, high extinction coefficient, poss
ple exciton generation, and unique electronic and transport properties. It
tant that II-VI semiconductors can be easily prepared in high quality
polycrystalline, and nanocrystalline films. The concentration of charge ca
also vary in II-VI semiconductors in wide range due to doping. Thus, tl
II-VI films represents an economical approach to the synthesis of semico
for various applications. It should be noted that the range of technical app
for II-VI compounds goes beyond the better-known semiconductors such a
and some of III-V compounds.

Formally, metal oxides such as CdO and ZnO also belong to II-VI comp
However, we will not cover them in this book. In recent years, these com
have been allocated to a separate group, "metal oxides," and many books hav
devoted to their discussion, in contrast to other II-VI compounds. In part

We believe that these books will enable the reader to understand the present status of II-VI semiconductors and their role in the development of new generation of photodetectors, sensors, and radiation detectors. I am very pleased that many well-known experts with extensive experience in the development and research of II-VI semiconductor sensors and radiation detectors were involved in the preparation of the chapters of these books.

The target audience for this series of books are scientists and researchers working or planning to work in the field of materials related to II-VI semiconductors, i.e., scientists and researchers whose activities are related to electronics, optoelectronics, chemical and bio sensors, electrical engineering, and biomedical applications. I believe this three-volume book may also be of interest to practicing engineers and project managers in industries and national laboratories who would like to develop II-VI semiconductor-based radiation sensors and detectors but do not know how to do it, and how to select the optimal II-VI semiconductor for specific applications. With numerous references to an extensive resource of recently published literature on the subject, these books can serve as an important and insightful source of valuable information, providing scientists and engineers with new ideas for understanding and improving existing II-VI semiconductor devices.

I believe that these books will be very useful for university students, doctoral students, and professors. The structure of these books offers the basis for courses in materials science, chemical engineering, electronics, optoelectronics, environmental control, chemical sensors, photodetectors, radiation detectors, biomedical applications, and many others. Graduate students may also find the book very useful in their research and understanding of the synthesis of II-VI semiconductors, study, and application of this multifunctional material in various devices. We are confident that all of them will find the information useful for their activities.

Finally, I thank all the authors who contributed to these books. I am grateful that they agreed to participate in this project and for their efforts to prepare these chapters. This project would not have been possible without their participation. I am also very grateful to Springer for the opportunity to publish this book with their help. I would like also to inform that my activity related to editing this book was funded by the State Program of the Republic of Moldova project 20.80009.5007.02.

I am also grateful to my family and wife, who always support me in all my endeavors.

Chisinau, Moldova   Ghenadii Korotcenkov

Contents

About the Editor

 **Ghenadii Korotcenkov** received his PhD in physics and technology of semiconductor materials and devices in 1976 and his Doctor of Science degree (doctor habilitate) in physics of semiconductors and dielectrics in 1990. He has more than 50-year experience as a teacher and scientific researcher. For a long time he was a leader of gas sensor group and manager of various national and international scientific and engineering projects carried out in the Laboratory of Micro- and Optoelectronics, Technical University of Moldova, Chisinau, Moldova. International foundations and programs such as the CRDF, the MRDA, the ICTP, the INTAS, the INCO-COPERNICUS, the COST, and NATO have supported his research. From 2007 to 2008, he carried out his research as an invited scientist at Korea Institute of Energy Research (Daejeon). Then, from 2008 to 2018, Dr. G. Korotcenkov was a research professor in the School of Materials Science and Engineering at Gwangju Institute of Science and Technology (GIST) in Korea. Currently, G. Korotcenkov is a chief scientific researcher at Moldova State University, Chisinau, Moldova.Scientists from the former Soviet Union know the results of G. Korotcenkov's research in the study of Schottky barriers, MOS structures, native oxides, and photoreceivers based on III–Vs compounds such as InP, GaP, AlGaAs, and InGaAs. His current research interests since 1995 include material sciences, focusing on metal oxide film deposition and characterization (In_2O_3, SnO_2, ZnO, TiO_2), surface science, thermoelectric conversion, and design of physical and chemical sensors, including thin film gas sensors.G. Korotcenkov is the author or editor of 45 books and special issues, including the 11-volume Chemical Sensors series published by Momentum Press; 15-volume *Chemical Sensors* series published by Harbin Institute of Technology Press, China; 3-volume *"Porous Silicon: From Formation to Application"* issue published by CRC Press; 2-volume *Handbook of Gas Sensor Materials* published by Springer; 3-volume *Handbook of Humidity Measurements* published by CRC Press; and 6 proceedings of the

international conferences published by Trans Tech Publ., Elsevier, and EDP Sciences. In addition, currently he is a series editor of Metal Oxides book series published by Elsevier. Since 2017, more than 35 volumes have been published within this series.G. Korotcenkov is the author and coauthor of more than 650 scientific publications, including 31 review papers, 38 book chapters, and more than 200 peer-reviewed articles published in scientific journals (h-factor=42 (Web of Science), h=44 (Scopus) and h=59 (Google scholar citation), 2022). He is the holder of 17 patents. He presented more than 250 reports at national and international conferences, including 17 invited talks. G. Korotcenkov, as a cochairman or member of program, scientific, and steering committees, has participated in the organization of more than 40 international scientific conferences. Dr. G. Korotcenkov is a member of editorial boards of five scientific international journals. His name and activities have been listed by many biographical publications including Who's Who. His research activities have been honored by the National Prize of the Republic of Moldova (2022), the Honorary Diploma of the Government of the Republic of Moldova (2020), an award of the Academy of Sciences of Moldova (2019), an award of the Supreme Council of Science and Advanced Technology of the Republic of Moldova (2004), the Prize of the Presidents of the Ukrainian, Belarus, and Moldovan Academies of Sciences (2003), Senior Research Excellence Award of the Technical University of Moldova (2001, 2003, 2005), the National Youth Prize of the Republic of Moldova in the field of science and technology (1980), among others. Some of his research results and published books have won awards at international exhibitions. G. Korotcenkov also received a fellowship from the International Research Exchange Board (IREX, United States, 1998), Brain Korea 21 Program (2008–2012), and BrainPool Program (Korea, 2007–2008 and 2015–2017). https://www.scopus.com/authid/detail.uri?authorId=6701490962   https://publons.com/researcher/1490013/ghenadii-korotcenkov/ https://scholar.google.com/citations?user=XR3RNhAAAAAJ&hl https://www.researchgate.net/profile/G_Korotcenkov

Contributors

Claire Abadie Sorbonne Université, CNRS, Institut des NanoSciences de Paris, INSP, Paris, France

M. Abdullah Nasiriya Nanotechnology Research Laboratory (NNRL), College of Science, University of Thi-Qar, Nasiriya, Iraq

Department of Physics, College of Science, University of Thi-Qar, Nasiriya, Iraq

Amin H. Al-Khursan Nasiriya Nanotechnology Research Laboratory (NNRL), College of Science, University of Thi-Qar, Nasiriya, Iraq

Baqer O. Al-Nashy Department of Physics, College of Science, University of Misan, Omarah, Iraq

Shonak Bansal Electronics and Communication Engineering Department, Chandigarh University, Gharuan, India

Alessio Bosio Department of Mathematical, Physical and Computer Sciences, University of Parma v.le delle Scienze, Parma, Italy

Mariarosa Cavallo Sorbonne Université, CNRS, Institut des NanoSciences de Paris, INSP, Paris, France

K. -W. A. Chee School of Electronic and Electrical Engineering, Kyungpook National University, Daegu, Republic of Korea

School of Electronics Engineering, College of IT Engineering, Kyungpook National University, Daegu, Republic of Korea

Tung Huu Dang Sorbonne Université, CNRS, Institut des NanoSciences de Paris, INSP, Paris, France

S. A. Dvoretsky Rzhanov Institute of Semiconductor Physics of the Siberian Branch of the RAS, Novosibirsk, Russia

S. M. Dzyadukh National Research Tomsk State University, Tomsk, Russia

Sema Ebrahimi Materials and Energy Research Center (MERC), Karaj, Iran

Light, Nanomaterials, Nanotechnologies (L2n) Laboratory, CNRS EMR7004, The University of Technology of Troyes, Troyes Cedex, France

Department of Physics and Mathematics, University of Hull, Cottingham Road, UK

G.W.Gray Centre for Advanced Materials, University of Hull, Cottingham Road, UK

D. I. Gorn National Research Tomsk State University, Tomsk, Russia

Charlie Gréboval Sorbonne Université, CNRS, Institut des NanoSciences de Paris, INSP, Paris, France

Jaker Hossain Solar Energy Laboratory, Department of Electrical and Electronic Engineering, University of Rajshahi, Rajshahi, Bangladesh

Satyabrata Jit Department of Electronics Engineering, Indian Institute of Technology (Banaras Hindu University), Varanasi, India

Tania Kalsi Nano-Materials and Device Lab, Department of Nanoscience and Materials, Central University of Jammu, Rahya-Suchani, Jammu & Kashmir, India

Adrien Khalili Sorbonne Université, CNRS, Institut des NanoSciences de Paris, INSP, Paris, France

Ghenadii Korotcenkov Department of Physics and Engineering, Moldova State University, Chisinau, Moldova

Abdul Kuddus Solar Energy Laboratory, Department of Electrical and Electronic Engineering, University of Rajshahi, Rajshahi, Bangladesh

Graduate School of Science and Engineering, Saitama University, Saitama, Japan

Hemant Kumar Department of Electronics and Communication Engineering, Jaypee Institute of Information Technology, Noida, India

Pragati Kumar Nano-Materials and Device Lab, Department of Nanoscience and Materials, Central University of Jammu, Rahya-Suchani, Jammu & Kashmir, India

Sandeep Kumar ICAR-Central Institute for Subtropical Horticulture, Lucknow, Uttar Pradesh, India

Emmanuel Lhuillier Sorbonne Université, CNRS, Institut des NanoSciences de Paris, INSP, Paris, France

Paweł Madejczyk Military University of Technology, Warsaw, Poland

N. N. Mikhailov Rzhanov Institute of Semiconductor Physics of the Siberian Branch of the RAS, Novosibirsk, Russia

Shaikh Khaled Mostaque Solar Energy Laboratory, Department of Electrical and Electronic Engineering, University of Rajshahi, Rajshahi, Bangladesh

M. Muthukumar ICAR-Central Institute for Subtropical Horticulture, Lucknow, Uttar Pradesh, India

Nima Naderi Materials and Energy Research Center (MERC), Karaj, Iran
Photonics Research Centre, University of Malaya, Kuala Lumpur, Malaysia

S. N. Nesmelov National Research Tomsk State University, Tomsk, Russia

Jiajia Ning Key Laboratory of Physics and Technology for Advanced Batteries, Ministry of Education, College of Physics, Jilin University, Changchun, China

Mingfa Peng School of Electronic and Information Engineering, Jiangsu Province Key Laboratory of Advanced Functional Materials, Changshu Institute of Technology, Changshu, Jiangsu, People's Republic of China

John C. Peterson The James Franck Institute, The University of Chicago, Chicago, IL, USA

Igor Pronin Department of Nano- and Microelectronics, Penza State University, Penza, Russia

Md. Ferdous Rahman Department of Electrical and Electronic Engineering, Begum Rokeya University, Rangpur, Bangladesh

Rada Savkina National University "Kyiv-Mohyla Academy", Kyiv, Ukraine
V. Lashkaryov Institute of Semiconductor Physics at NAS of Ukraine, Kyiv, Ukraine

Nupur Saxena Organisation for Science Innovations and Research, Garha Pachauri, India

Tetyana Semikina Lashkarev Institute of Semiconductor Physics, National Academy of Science of Ukraine, Kiev, Ukraine

G. Y. Sidorov Rzhanov Institute of Semiconductor Physics of the Siberian Branch of the RAS, Novosibirsk, Russia

Oleksii Smirnov V. Lashkaryov Institute of Semiconductor Physics at NAS of Ukraine, Kyiv, Ukraine

Xuhui Sun Institute of Functional Nano & Soft Materials (FUNSOM), and Jiangsu Key Laboratory for Carbon-based Functional Materials and Devices, Soochow University, Suzhou, Jiangsu, People's Republic of China

Victor V. Sysoev Yuri Gagarin State Technical University of Saratov, Saratov, Russia

A. V. Voitsekhovskii National Research Tomsk State University, Tomsk, Russia

M. V. Yakushev Rzhanov Institute of Semiconductor Physics of the Siberian Branch of the RAS, Novosibirsk, Russia

Benyamin Yarmand Materials and Energy Research Center (MERC), Karaj, Iran

Part I
IR Detectors Based on II–VI Semiconductors

Chapter 1
Introduction in IR Detectors

Ghenadii Korotcenkov

1.1 Introduction

Infrared (IR) range is the range of electromagnetic radiation from 0.78 to 1000 μm, which is divided into sub-ranges:

- near IR, NIR: 0.78–1 μm;
- short wavelength IR, SWIR: 1–3 μm;
- medium wavelength IR, MWIR: 3–6 μm;
- long wavelength IR, LWIR: 6–12 μm;
- very long wavelength IR, VLWIR: 12–20 μm;
- far infrared region (FIR): 20–1000 μm

In comparison with visible and ultraviolet rays, infrared radiation has small energy, for example 1.24 eV at $\lambda = 1$ μm, 0.12 eV at 10 μm, and ~0.01 eV at 100 μm.

For the first time the presence of infrared radiation was found in 1800 in the process of experiments conducted by the English astronomer William Herschel. A clearer understanding of IR radiation was obtained in 1900 through Plank's law (Eq. 1.1). According to Plank's law, every physical object spontaneously emits radiation in a wide range of wavelengths. The peak wavelength of the radiation corresponds to the equilibrium temperature of the object. Spectral radiation emittance, calculated according to Plank's law, is shown in Fig. 1.1. As is seen, the peak radiation of objects at room temperature (~300 K), is ~10 μm. The surface of the Sun, which has a temperature of ~6000 K, has a maximum radiation in the visible range, although radiation in the IR region is also present.

$$M(\lambda) = C_1 \lambda^{-5} \left[\exp\left(C_2 / \lambda \cdot T \right) - 1 \right]^{-1} \tag{1.1}$$

G. Korotcenkov (✉)
Department of Physics and Engineering, Moldova State University, Chisinau, Moldova
e-mail: ghkoro@yahoo.com

© The Author(s), under exclusive license to Springer Nature Switzerland AG 2023
G. Korotcenkov (ed.), *Handbook of II-VI Semiconductor-Based Sensors and Radiation Detectors*, https://doi.org/10.1007/978-3-031-20510-1_1

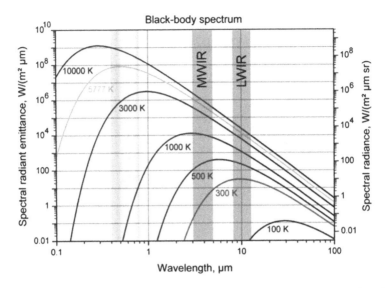

Fig. 1.1 Spectral emittance of objects at given equilibrium temperatures. (Reprinted from Karim and Andersson [10]. Published 2013 by IOP as open access)

Table 1.1 Main applications of infrared photodetectors

Community	Applications
Military	Reconnaissance, navigation, night vision, guided missiles.
Commercial	
Civil	Police, firemen, border post.
Environment	Pollution control, natural resources, energy savings, meteorology.
Industry	Maintenance, fabrication processes control, nondestructive tests, optical communications.
Medical	Thermography.
Science	
Astronomy	Observation of the universe in the infrared region.
Physics, chemistry	IR spectroscopy.

Where T – absolute temperature (K), C_1 – first radiation constant = $3.74 \cdot 10^4$ Wμm^4/cm^2, C_2 – second radiation constant = $1.44 \cdot 10^4$ μmK, λ – wavelength (μm).

It should be noted that the development of the IR technologies has intensified significantly only in the last 40–50 years. The main applications of the IR devices are shown in Table 1.1. During this period, significant advances have been made in the development of various IR photoreceivers (see Fig. 1.2) and as a result, it is currently impossible to imagine many extremely important applications without the use of IR detectors. IR detectors have become the basis of space surveillance systems, ballistic missile launch detection systems, non-contact temperature measurement, motion sensors, IR spectroscopy, night vision devices, warhead homing

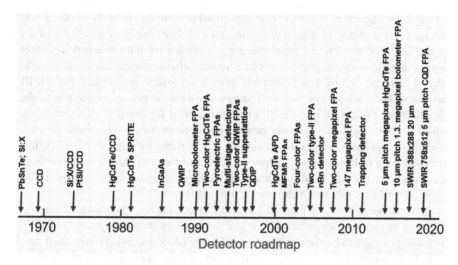

Fig. 1.2 History of the development of infrared detectors and systems. (Adapted from Rogalski et al. [21]. Published 2020 by PAS as open access)

systems, and holographic information recording and processing systems. IR sensors are widely used in astronomy, medicine, ultra-long range optical communication systems, rangefinders, meteorological exploration, climatology, laser search and visualization, etc. [4, 8, 12, 30]. It is the photodetectors, which in the overwhelming majority of cases determines such basic parameters of IR systems as range, sensitivity, spectral scope, noise immunity, resolution, dynamic range and other characteristics of the equipment.

1.2   IR Photodetectors

Two types of detectors are currently used to detect infrared radiation – photonic (cooled and uncooled) and thermal.

1.2.1   Thermal (Non-selective) IR Detectors

Thermal radiation detectors are non-selective devices, i.e. they have the same spectral characteristic over a wide range of the electromagnetic spectrum (up to hundreds of micrometers) [4, 23, 34]. The operation of thermal radiation detectors is based on the conversion of radiation energy into heat, and then into electrical energy. In bolometers and thermistors, an increase in the temperature of the receiver changes its electrical conductivity, in thermopiles, a thermo-EMF appears, in a pyroelectric receiver, the value of the surface charge changes, and in dielectric bolometers using ferroelectric capacitors (barium strontium titan) as sensing elements, its dielectric changes with a change in temperature constant and, therefore, the capacitance of the capacitor. Thermal detector manufacturing technologies have reached a certain degree of perfection and predetermined a number of advantages, due to which sensors of this type occupy a quantitatively dominant position in the market of the IR detectors. Their advantages are well known. In its most general form, this is the simplicity of the design and the absence of the need for cryogenic temperatures, which leads to significantly lower power consumption for power supply. In addition, there is practically no need for service. For pyroelectrics and thermopiles, this is the absence of both a power supply and the need for temperature compensation. As a result, the noise (1/f) of such devices is sharply reduced. Competitive price also plays a significant role when choosing a thermal sensor [23].

In the development of thermal radiation detectors, the maximum effect was achieved in recent years, when silicon micro-electromechanical systems (MEMS) technology began to be used in their manufacture [2, 3, 13, 34]. This made it possible to manufacture micro-sized elements of thermal sensors and provide them with good thermal isolation. In particular, the state-of-the-art technology has made it possible to fabricate monolithic silicon-based focal plane arrays (FPAs) with more than 10^5 pixels and a pixel size of 50 μm. When using vanadium oxide (VOx) as a sensitive material for MEMS microbolometers, fabricated using Si-based MEMS technology, a temperature noise equivalent (NETD) of 40–50 mK for an aperture number of 1 has been achieved, which is an undeniable progress [16, 29]. The achievement of such results has led to the fact that of all types of IR thermal detectors, uncooled MEMS-based detectors are of the greatest interest. Their appearance and structure are shown in Fig. 1.3. The interest in such devices is due to the fact that MEMS-based detector arrays combine the ability to form images of very good quality with the low cost and ease of use inherent in uncooled detectors. While their sensitivity does not come close to that of the best HgCdTe detector arrays, uncooled MEMS-based detector arrays provide a level of performance that most users are satisfied with. Currently MEMS-based detector arrays, including microbolometer

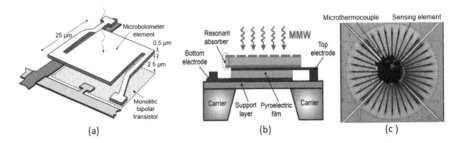

Fig. 1.3 (**a**) Modern 25 μm Microbolometer pixel (https://movitherm.com/), (**b**) Pyroelectric sensor. Sketch of the pyroelectric sensor with an integrated resonant absorber. (**c**) The CMOS-MEMS thermopile after drop-coating of graphene. ((**b**) Reprinted from Kuznetsov et al. [13]. Published 2018 by Nature as open access; (**c**) Reprinted from Chen and Chen [3]. Published 2020 by MDPI as open access)

Table 1.2 Thermal detectors

Mechanism	Comment	Advantages	Disadvantages
nn			
Mercury in glass	Thermometer	Simple	Low R Very slow
Gas in chamber	Golay cell	Available Simple	Very slow
Thermal coefficient of resistance			
Carbon resistor	First bolometer	Easy to model Easy to make	Need T_{LHe} (4 K)
Vanadium oxide (VO_x)	Microbolometer FPAs	Available Silicon processing Good stability	Low R
Amorphous Si	Microbolometer FPAs	Available Silicon processing	Less sensitive than VO_x
Thermal EMF (Seebeck effect)			
Thermocouple	One junction pair		Slow, low R
Thermopile	Many junction pairs MEMS thermopile	Available Simple Inexpensive	Slow, low R
Ferroelectric and pyroelectric			
Barium strontium titanate	First high volume AC coupled	Available High R Inexpensive	Needs chopper

Source: Reprinted with permission from Vincent et al. [33]. Copyright 2016: Wiley

arrays and MEMS-based thermoelectric (thermopile) arrays with acceptable parameters, are commercially available worldwide [6, 34]. Comparative characteristics of thermal detectors are given in Table 1.2.

However, thermal detectors also have disadvantages that turn out to be quite significant when choosing an IR sensor for work in applications, requiring high parameters, such as military equipment [29]. Such disadvantages are the structural and technological complexity of thermal insulation of many detector pixels from each other and from the substrate, their sensitivity to temperature fluctuations and vibrations with relatively small relative changes in the electrical characteristics of the material per one degree of change in the object temperature, the inertia of the spectral response and low uniformity of the image. For thermal detectors, there is also the problem of decreasing pixel size. For example, the pixel sensitivity of a microbolometer is highly dependent on the pixel area; therefore, there is a need to maximize both the optical absorption area and thermal insulation. Unfortunately, when the pixel size is reduced, this condition cannot be fulfilled, and therefore, when using conventional single-level micromachining processes, performance degradation is observed as the unit cell size is reduced below 40 μm. That is why, in most MEMS bolometers, the pixel size does not decrease below 50 μm. Due to the problems mentioned earlier, when using thermal detectors, there are often problems with obtaining a clear image at long distances [29].

1.2.2   Photonic Radiation Detectors

The main group of IR detectors in terms of various applications is the so-called photon or quantum radiation detectors [23]. Photonic radiation detectors ensure the conversion of the incident photon flux into an electrical signal due to the direct interaction of photons with the electronic subsystem of the receiver material (see Fig. 1.4). The sensitivity of the photon detector is proportional to the number of absorbed photons. Such a receiver is selective, i.e. it only reacts to photon of radiation with a certain frequency (wavelength). In other words, photonic detectors respond to photons whose energy exceeds certain threshold values, for example, the semiconductor band gap ("intrinsic" detectors).

In turn, photonic radiation detectors are divided into receivers with an external photoelectric effect (photomultipliers) and receivers with an internal photoelectric effect (photoresistors, photodiodes, phototransistors, etc.). In modern infrared systems, receivers with an internal photoelectric effect are most widely used. In receivers of this type, three main physical phenomena are used, caused by the effect of radiation on a semiconductor: the phenomenon of photoconductivity, photovoltaic and photoelectric effects. Since IR radiation has small energy (energy is inversely proportional to wavelength), these detectors are cooled down to cryogenic temperatures in order to increase infrared detection efficiency/sensitivity. Quantum detectors react very quickly to IR radiation (response time is order of µs), but they have response curves with a detectivity that varies greatly with wavelength.

Photoconductors The operation of a photoresistor (PR) or photoconductor (PC) is based on the change in the electrical conductivity of the sensitive layer during irradiation (read Chap. 2, Vol. 2). It is important to note that the first IR photonic receivers were photoconductive because of the simplicity of the technology, and the relative ease of achieving near-ideal infrared performance and excellent reliability. At the same time, it was found that the most sensitive PRs are also the most inertial. For a number of them, a direct relationship has been established between the PR

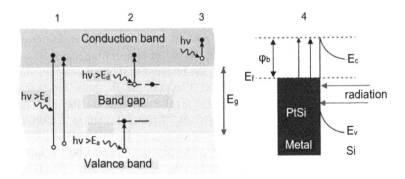

Fig. 1.4 Photonic mechanisms of excitation of the electron subsystem in photonic detectors: (1) intrinsic excitation, (2) impurity excitation, (3) absorption by free carriers, (4) excitation in Schottky diodes. (Idea from Kulchitsky et al. [12])

sensitivity threshold and the time constant τ. Often, the mobility of electrons and holes in a semiconductor is very different, as a result of which faster carriers can pass through the detector several times before the carriers recombine. This provides a gain mechanism. Operating temperature is another important factor. Decreasing the temperature of the sensitive layer expands the spectral range of its operation in the IR spectral region and increases its integral sensitivity. With cooling, the noise decreases, and, consequently, the detecting ability of the PR increases. In addition, upon cooling, the time constant and its dark resistance increase.

The advantages of PR are small size and weight, lower supply voltage compared to photoemission detectors and the ability to work in a wide spectral range. The PRs usually have a very high integral sensitivity and low power consumption not exceeding several watts. The disadvantages of photodetectors of this class include increased inertia, a significant dependence of characteristics and parameters on temperature, a small linear zone of the energy characteristic, and the dependence of the output signal on the illumination area of the sensitive layer. Currently, the main materials used for the manufacture of IR photoresistors are CdS, PbS, PbSe, InSb, Ge:Au and HgCdTe. Photo detectors on CdS, PbS, PbSe, InSb, and HgCdTe are intrinsic, and Ge:Au-based devices are extrinsic IR photoconductive detectors.

Photodiodes A photodiode (PD) is usually called a semiconductor radiation detector based on the use of one-way conductivity of a p-n junction, upon illumination of which an electromotive force (EMF) is formed (photovoltaic mode), or in the presence of a voltage source in the photodiode circuit, its reverse current changes (photodiode mode). Currently, materials for the manufacture of IR photodiodes are mainly Ge, Si, InSb, as well as ternary compounds such as InGaAs and HgCdTe. The features of the operation of HgCdTe-based photodiodes are described in [1] and the Chaps. 2, 3, 4, 5, 6, 7, and 8. There are many methods for manufacturing IR PDs, each of which has its own strengths and weaknesses, depending on the specific design of the photodetector and its purpose. For HgCdTe, this information can be found in [17, 19] and Chap. 15, Vol. 1.

The spectral sensitivity of PDs changes when switching from photovoltaic mode to photodiode mode. This sensitivity also depends on the temperature of the semiconductor material used for PDs fabrication. With decreasing temperature, the spectral characteristic and its maximum shift to the short-wavelength region, and the dark current as well as the noise level also decreases. Therefore, IR detectors, especially those designed to work in the LWIR spectral range, are forced to operate at low temperatures, down to cryogenic temperatures. In many cases, the temperature of operation is a critical parameter, because the cooling system often dominates the size, weight and reliability of the detector system.

The time constant of the photodiode τ_{PD} largely depends on the method of its manufacture and the size of the photosensitive area. For PDs, the τ_{PD} value is usually close to 10^{-5} s; for diffusion p-n junction with small areas, τ_{PD} can reach 10^{-6} s. In special PDs with a small thickness, $\tau_{PD} \sim 10^{-10}$ s can be achieved. Comparatively large dark currents when conventional photodiodes are switched on in the photodiode mode make it impossible to use them for measuring low light flows. In this case, it is necessary to work in the photovoltaic mode, in which the detecting ability of the system is determined not by the low noise of the receiver, but by noises of its electronic circuit or subsequent electronic links.

To improve the sensitivity for detecting infrared radiation, avalanche photodiodes (APDs) are often used [16]. Avalanche photodiodes include areas of high electrical field. Carrier multiplication is carried out by transferring sufficient kinetic energy to the carrier to create an additional electron-hole pair by impact ionization. There is always some excess noise associated with multiplication, but this can be minimized with designs that allow one carrier to be multiplied. The ideal APD is an inexpensive device with low dark noise, wide spectral and frequency response, and a gain from 1 to 10^6 or more. The characteristics of avalanche diodes based on HgCdTe are discussed in the Chap. 3, Vol. 2.

In addition to the above devices, it is possible to note the Schottky barrier photodiodes belonging to the group of photoemissive detectors (see Fig. 1.4). They are characterized by a relatively simple manufacturing technology and their parameters are close to those of p-i-n PDs. The most popular Schottky-barrier IR detector is the PtSi-p-Si detector, which can be used for IR detection in the 3–5 μm spectral range [11]. Radiation is transmitted through the p-type silicon and is absorbed in the metal PtSi (Platinum silicide) layer, producing hot holes which are then emitted over the potential barrier into the silicon, leaving the silicide negatively charged. This fundamental difference in the detection mechanism underlies the unique properties of Schottky IR detectors, including their exceptional spatial uniformity and their modified Fowler spectral response. The Schottky barrier height of the PtSi detector is ~0.22 eV, corresponding to a cutoff wavelength of ~5.6 μm. Due to the Fowler dependence, the quantum efficiency (QE) of the PtSi detector in the 3–5 μm MWIR regime is relatively low [14]. The main advantage of the Schottky-barrier detectors is that they can be fabricated as monolithic arrays in a standard silicon process [23].

Photodiodes formed on the basis of heterostructures are also present in a large number on the market of IR detectors [5]. Some options for using heterostructures in the development of IR detectors are described in Chaps. 5 and 6, Vol. 2.

Phototransistors A phototransistor (PT) is a semiconductor device with photocurrent amplification properties with two p-n junctions, in which there is a directional movement of current carriers. Phototransistors have a high quantum efficiency (about 100). However, the presence of the second p-n junction leads to a significant increase in noise. Therefore, it is often preferable to use photodiodes with an additional stage of the signal amplifier, the noise of which affects the detection ability of the device less than the noise arising in the phototransistor. The disadvantages of phototransistors also include: significant instability of parameters and characteristics over time when the ambient temperature changes; and lower detection ability than photodetectors. It should be noted that some PTs have a "blind spot" in the center of the sensitive layer due to the shading of a part of the base by the emitter. Therefore, when using them, it is necessary to distribute the flow over the entire photosensitive surface of the PT.

Table 1.3 Infrared detectors used in single detector assemblies

Detector	(μm)	Common applications, and comments
Si (PV)	0.1–1.1	Optical communication, fire sensing, light and laser power measurement, photon counting
InGaAs (PV)	0.7–1.8	Optical communications, FTIR, gas detection, light and laser power measurement, tunable diode laser spectroscopy (TDLS), moisture analyzers. Replaces Ge (faster)
InAs (PV)	0.9–3.5	FTIR, non-contact temperature measurement, laser monitoring, gas analyzers, spectrophotometry.
InSb (PC, PV)	1.0–5.5	FTIR, spectrophotometry, thermometry, remote sensing, gas analysis
InAsSb (PV)	3.3–11	Gas measurement (CH_4, CO_2, CO, NH_3, O_3, etc.), flame monitoring (CO_2 resonance radiation), radiation thermometry
PbS (PC)	1–3.6	Non-contact temperature measurement, spark detecting, flame control, moisture measurement, spectrophotometry. These detectors are used especially when large-area detectors are required because they are significantly less expensive than comparable III-V (InGaAs) detectors.
PbSe (PC)	1–5.8	Gas analysis, laser power measurement, medical CO_2 detection, non-contact temperature measurement, flame detection, fire detection, moisture monitoring.
HgCdTe (PC, PV)	1.0–16	FTIR, industrial process control, heat-seeking guidance, laser warning receiver, laser monitoring, temperature monitoring, gas detection, remote sensing. Используется вместо PbSe and PbS where high detectivity is needed, low bias voltage, selective peak wavelength response, and fast response times.
Thermopile	0.1–100	Fire sensing, intrusion, laser power, temperature measurement, gas detection.
Pyroelectric (piezoelectric)	0.1–100	Fire sensing, motion detection, laser power, temperature measurement. Most sensitive of thermal detectors, sensitive to vibration
Bolometer and microbolometer	0.1–100	FTIR, astronomy
Golay cell and microgolay cell	0.1–100	FTIR

FTIR Fourier-transform infrared spectroscopy, *NDIR* nondispersive infrared spectroscopy, *PC* photoconductive, *PV* photovoltaic

As for the most common commercial and science applications of IR detectors, they are listed in Table 1.3. A description of some of these applications can be found in [33].

1.2.3 IR Photodetectors Array

A certain share in the market of modern IR photodetectors is occupied by photodetector (PD) arrays developed for thermal imaging equipment [12] and for astronomical application [8]. The main distinctions of photodetector arrays from single photodetectors are listed in the Table 1.4.

Table 1.4 Comparison of two general IR detector configurations

eature	Single detector assembly	PDs array
Detectors (pixels)	A few: 1, 2, 4, perhaps 16	Thousands, millions
Equipment and software	Can be simple	Specialized
Output and responsivity measurement	AC (usually)	DC (usually)
Record data for every element	Yes	Yes
Report data for every element	Yes – tabular	No, or graphical only
Report array statistics	Optional	Essential
Figures of merit	Basic	Basic + some unique
Relative cost	Low	Expensive
Quantity per year	Many	Few
Electronics	Discrete or small ASICs	ROIC
Primary applications	Various nonimaging	Imaging

Source: Data extracted from Vincent et al. [33]
ASICs application-specific ICs, *ROIC* a readout- integrated circuit

A focal plane array (FPA) of photodetectors is created by combining separate, usually identical, photosensitive elements on a single chip. The individual detectors in an array are often referred to as pixels, short for picture elements; pixel is generally thought of as the smallest single component of a digital image. However, the process of developing an integrated detector array is much more complicated than manufacturing a separate detector element (read the Chap. 4, Vol. 2). A fundamental limitation in the design of detector arrays is that light easily hits adjacent pixels in the array, resulting in false counts or crosstalk. There are approaches that can be taken to mitigate this limitation, but they add additional complexity to production. In addition, the manufacturing process of the array becomes even more complex due to the requirement to maintain low leakage currents in individual pixels, which makes the manufacturing process even more cumbersome.

It is important to note that the first HgCdTe-based photodetector arrays were made on the basis of photoconductive detectors. HgCdTe photoconductive detectors have been in routine production since the early 1980s and are often called first-generation detectors. They have been designed for spectral ranges of 8–12 μm and 3–5 μm. Detectors operated at 80 K for LWIR and at 80–200 K for MWIR spectral range. The reliable performance, low levels of defects and easily understood physics has led to a long product life for photoconductive arrays. However, the size of the matrix in such detectors was limited, and the first-generation thermal imaging systems had to use sophisticated optics to scan the infrared image over the matrix to construct the scene. Therefore, photoconductive HgCdTe detectors have been very successful in producing arrays of up to a few hundred elements for use in first-generation thermal imaging systems. The limitations of photoconductive detectors emerged when the need arose for very large focal plane arrays. In addition, the low impedance of the photoconductor made it unsuitable for injecting into silicon charge transfer devices or field effect transistors; hence, each element required a lead-out

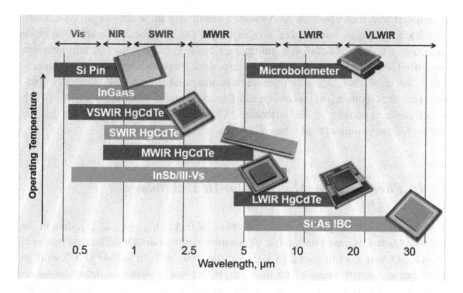

Fig. 1.5 Materials used for development of large format FPAs. (https://www.raytheon.com)

through the vacuum encapsulation to an off-focal plane amplifier. All this significantly limited the capabilities of matrices based on photoconductive detectors. In this regard, the use of photoconductive detectors in the development of photodetectors arrays has now been limited. Currently, photodiodes and heterostructures of various types are the basis of such photodetectors arrays. For a more detailed discussion of photodetectors arrays, see Chap. 4, Vol. 2.

The market for high performance thermal imaging cameras based on FPAs is currently split between two material systems: InSb and HgCdTe [12, 31]. FPAs present on the market are shown on Fig. 1.5. InSb has a large share of the market of cooled thermal imaging cameras because the first FPAs were originally obtained from this material. At the same time, HgCdTe has several key advantages that could make it the material of choice for the development of third generation thermal imaging cameras. HgCdTe can operate at much higher temperatures ($T_{oper}^{max} < 200K$), than InSb (T_{oper}^{max} <100–120 K), and therefore MCT-based cameras can be smaller and consume less power. For FPAs designed to operate in the SWIR range, other materials can also be used, mainly InGaAs (https://www.flir.com). However, InGaAs / InP-based FPAs can have high noise in the spectral range above 1.7 μm due to defects caused by the lattice mismatch between InGaAs and InP, which is used as substrate for growing InGaAs epitaxial layers.

As for astronomical applications, in addition to InSb and HgCdTe, they also use Si:As-based FPAs [8, 31]. The fabrication of FPAs on the base of Si:As does not present any particular difficulties, as standard technologies are used and large, high quality silicon wafers are available. In addition, there are no problems with thermal expansion mismatch between the FPA and the multiplexer. As a result, the best samples of matrices based on Si:As with a sensitivity up to 28–40 μm have more than 2 k × 2 k pixels with a pixel size of 18 μm [32]. For Si:As devices, the more challenging task is to develop a multiplexer capable of operating at temperatures around 10 K required to achieve low detector dark currents. However, this temperature is outside the range in which standard commercial silicon CMOS devices operate. The development of special multiplexers requires additional costs, which sometimes compensate for the advantages of Si:As. This is due to the fact that the market for astronomical Si:As-based devices is very small [8].

1.2.4 Photosensitive Materials for IR Technology

If we turn to the history of the development of IR technology [4, 22, 23], then in the early stages it was based on the use of complex semiconductors. The first practical IR detector was manufactured in 1933 on the basis of lead sulfide (PbS) with an operating sensitivity range of up to 3 μm. In the late 1940s, this detection wavelength was increased to 5 μm through the use of lead selenide (PbSe) and lead telluride (PbTe). In the 1950s, research began on semiconductor compounds from III-V, IV-IV and II-VI groups, which led to the emergence of new photosensitive materials suitable for the development of various IR photodetectors. InSb, $InAs_{1-x}Sb_x$, $Pb_{1-x}Sn_xTe$ (LTT), InGaAs and HgCdTe (MCT) were one of them. A large number of other combinations of compounds were also tested, but in most cases, there were problems with crystal growth or there were limitations on doping, which made them unsuitable for the manufacture of IR devices.

It should be noted that in the late 1960s – early 1970s, PbSnTe was actively developed in parallel with HgCdTe [7]. Unlike HgCdTe, PbSnTe of the required quality was easy to grow, and therefore high-quality IR photodiodes on their basis for the LWIR spectral range were demonstrated relatively quickly. However, in the late 1970s, work on PbSnTe-based detectors was discontinued [1]. There were two reasons for this: a high dielectric constant and a large mismatch in the temperature coefficient of expansion (TCE) of PbSnTe with TCE of the Si. The scanning-based IR imaging systems required relatively fast response times so that the scanned image was not blurred. The high dielectric constant did not allow the development of devices with the low capacitance required for fast operation. Large TCE, led to the destruction of indium bonds in the hybrid structure between the silicon-based reader and the PbSnTe-based detector array during repeated thermal cycling from room temperature to cryogenic operating temperature.

Concerning current IR technology (www.hamamatsu.com; www.teledynejudson.com; www.photonicsolutions.co.uk; https://vigo.com.pl), it is mainly based on InSb, InGaAs and HgCdTe [17]. One should note that the use of HgCdTe allowed to cover a wider range of wavelengths of infrared radiation compared to other materials. Depending on the composition of $Hg_{1-x}Cd_xTe$, photodetectors based on this material can cover the entire spectral IR range from SWIR to VLWIR (see Fig. 1.6). In addition, HgCdTe has nearly ideal properties for the development of electronic

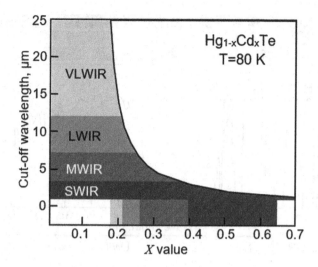

Fig. 1.6 The cut-off wavelength for $Hg_{1-x}Cd_xTe$ as a function of composition, x

avalanche photodiodes (e-APDs) for the infrared region (read the Chap. 3, Vol. 2). The high mobility ratio of HgCdTe contributes to the creation of conditions under which single-carrier multiplication with low noise figures take place. Such IR detectors are effective when used in photon-starved applications, such as long-range imaging and astronomy [1]. Other advantages of HgCdTe can be found in the Chap. 4, Vol. 1.

Currently, HgCdTe-based detectors are mainly used for spectral ranges of 1–2.5 μm, 3–5 μm and 8–14 μm. InSb is more suitable for the spectral range of 3–5 μm, while InGaAs is more suitable for the spectral range of 0.4–2.3 μm [9, 17]. The spectral areas of their application are shown in Figs. 1.5 and 1.7. However, even until now, inexpensive polycrystalline thin-film photoconductive detectors based on PbS and PbSe remain the preferred choice for many applications in the spectral range of 1–3 μm and 3–5 μm. It should be noted that in recent decades there has also been interest in such compounds as $Hg_{1-x}Mn_xTe$, $Hg_{1-x}Zn_xTe$ and $Hg_{1-x}Cd_xSe$ as potential alternatives to HgCdTe for infrared detectors. Photoconductive and photovoltaic detectors have been reported using these materials, but the devices have not yet reached technology perfection like $Hg_{1-x}Cd_xTe$ [26, 27].

We must also not forget about extrinsic (photon energies smaller than the bandgap) photoconductive detectors, which were developed in the early 1950s. Research has shown that using extrinsic photoconductive response from germanium, doped by copper, mercury, zinc, and gold allows developing IR receivers capable to work in the 8–14 μm (LWIR) and 14–30 μm (VLWIR) spectral range. However, they must operate at very low temperatures (30–50 K) to achieve similar performance to intrinsic detectors, and they sacrifice quantum efficiency to avoid the need for thick detectors.

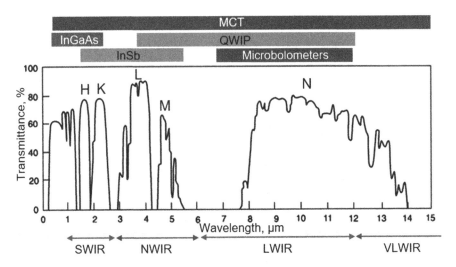

Fig. 1.7 Basic semiconductor materials for cooled thermal imaging arrays and operating spectral ranges. This Figure also shows the specific absorption bands of the atmosphere. (Data extracted from Rogalski [22, 23] and Piotrowski and Rogalski [17])

There are also Si-based intrinsic and extrinsic IR photoconductive detectors. Silicon intrinsic IR detectors are usually designed and fabricated as Schottky-barrier detectors (PtSi detector). Such detectors can be used for detection in the range of 3–5 μm. As for longer wavelengths, in this area can be used extrinsic silicon IR photodetectors, such as Si:As or Si:Sb [20]. The developed devices are designed to operate in the range of 5–40 μm. However, the sensitivity range can be extended up to approximately 300 μm. As well as Ge-based IR PDs, operating temperatures of Si-based extrinsic detectors lies in the cryogenic temperature range (8–30 K). A detailed review of a bulk Si and Ge IR detectors can be found in [28].

It is important to note that in recent years there has been a significant increase in interest in extrinsic IR photoconductive detectors based on germanium and especially silicon [20]. This interest is due to the creation of multi-element FPAs for use in space and land infrared astronomy, as well as on spacecraft for various purposes. Advances in Si-based FPAs technology, the creation of low-noise semiconductor preamplifiers and deep-cooled multiplexers, as well as unique designs of extrinsic silicon IR PDs and deep-cooling equipment have achieved record-breaking detectivity, close to the emission limit even under exceedingly low backgrounds in space [15].

1.2.5 Comparison of Thermal and Photonic Infrared Detectors

Comparing the relative response of photon and thermal detectors as a function of wavelength with either a vertical scale of W^{-1} or photon^{-1}, then as seen in Fig. 1.8, photon detectors show a linear increase in response on the dependences, recalculated

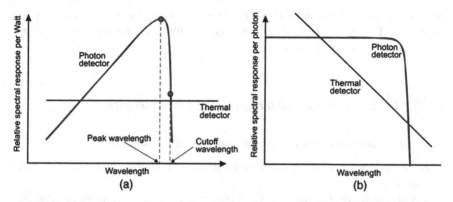

Fig. 1.8 Relative spectral response for a photon and thermal detector for (**a**) constant incident radiant power and (**b**) photon flux, respectively. (Reprinted with permission from Rogalski et al. [24]. Copyright 2000: SPIE)

per unit power of the incident radiation, until the cutoff wavelength is reached. An increase in response is associated with an increase in the number of photons to achieve the same power with increasing wavelength. The cutoff wavelength is determined by the detector material used. At the same time, thermal detectors tend to be spectrally flat in this case. This is due to the fact that their response is proportional to the energy absorbed, which, when recalculated per unit of incident radiation power, does not depend on the wavelength. If the sensor response is recalculated per photon, then the photon detectors in this case are usually flat, and the thermal sensors have a linearly decreasing response.

Compared to thermal detectors, cooled photonic sensors have a NETD of 10–20 mK, and this indicator practically does not change over a wide range of integration times (5–7 ms and more). The detectivity of photonic detectors is about two orders of magnitude higher than that of thermal detectors. This allows large aperture ratios to be used when designing thermal imaging cameras [29]. In addition, photon-type detectors have better signal-to-noise ratio and faster response time. For many applications, thermal detectors may not provide the required response time. Consequently, heat detectors are not suitable for infrared thermal imaging cameras that use higher frame rates and multispectral performance. At the same time, photon detectors require deep cryogenic cooling, and this leads to a complication of the device, an increase in its geometric dimensions, weight and high power consumption. Only in the last decade, due to progress in the development of photodetectors, it was possible to increase the operating temperatures from 77 to 150 K. In addition, if the service life of the photodetector array of the photon detector itself can be determined for decades, then the service life of the cooler does not exceed 30 thousand operating hours (the best models), after which the cooler must be replaced. All this leads to an increase in the cost of sensors of this type in comparison with thermal detectors, as well as to an increase in the cost of their operation. Nevertheless, the price criterion, as a rule, is not decisive in applications such as military and astronomy, where the technical characteristics of photodetectors come first, and not their price. Therefore, often in such applications, photonic detectors dominate, as devices with the best parameters.

1.2.6 Parameters Characterizing IR Photodetectors

The main parameters characterizing IR detectors are the followings:

- *Dark current* – current measured in the absence of radiation under the operating conditions.
- *Photo sensitivity (Responsibility)* – is the output voltage (or output current) per watt of incident energy when noise is ignored.
- *Quantum efficiency* – Quantum efficiency (QE) represents the percentage of photo-generated electron-hole pairs to the number of incoming photons.
- *Integral sensitivity* – sensitivity to non-monochromatic radiation of a given spectral range.
- *Spectral sensitivity* – dependence of the sensitivity on the radiation wavelength.
- *Current sensitivity* – the sensitivity of the photodetector, in which the measured electrical value is the photocurrent, and the voltage sensitivity is the sensitivity when the measuring value is the voltage at the output of the radiation detector.
- *Noise of the radiation detector* – the output chaotic signal at the output of the photodetector measured in the absence of radiation. Noise does not allow registering arbitrarily small signals that become imperceptible against its background, i.e. restrict the limiting capabilities of the device.
- *Noise Equivalent Power (NEP)* – NEP is a measure of flux in order to generate a signal that is equal to the noise level.
- *Noise Equivalent Temperature Difference (NETD)* – The minimum temperature difference, producing a signal level equal to the noise of the detector, is defined as NETD. NETD is an important figure of merit, which can be considered in order to estimate thermal detection performance.
- *Detectivity (D*)* – the value of the minimum radiation flux, which produces a signal at the output equal to the intrinsic noise. It is inversely proportional to the square root of the area of the radiation receiver. It is measured in 1/W.
- *Specific Detectivity* – The detectivity multiplied by the square root of the product of a frequency bandwidth of 1 Hz and an area of 1 cm². Measured in $cm^2Hz^{1/2}/W$.
- *Response time, or time constant* τ – the time required to establish a signal at the output corresponding to the input action. The time constant τ determines the cutoff frequency of modulation of the signal at the input of the photodetector.
- *Working temperature* – the maximum temperature of the sensor and the environment at which the sensor is able to perform its functions correctly.
- *Photodetector resistance* – for photoresistors, the dark resistance measured in the absence of radiation is considered as a photodetector resistance. For photodiodes, the value of the differential resistance R_D is usually given, which is equal

to the ratio of small increments of the signal voltage to the photocurrent under the given operating conditions.

Most of the important parameters of photonic photodetectors are controlled by the noise level. The nature of this noise and approaches to reduce it are given in Table 1.5.

As for the spectral detectivity of various photodetectors, this data for a number of commercially available IR detectors are shown in Fig. 1.9. The temperatures at which these parameters were determined are also indicated there. It is seen that in many cases spectral detectivity approaches the theoretical limit values.

1.2.7 The Role of the Atmosphere in IR Technology

All IR detectors, except for sensors used in space work in the Earth's atmosphere, which has its own specific absorption bands (see Fig. 1.7). The absorption bands of water vapor with a center of 6.3 μm and of carbon dioxide with centers of 2.7 and 15 μm limit the transmission of radiation by the atmosphere in the wavelength range of 2–20 μm, determining the position of the so-called atmospheric transparency windows: 3–5 and 8–12 μm (Fig. 1.2). According to the established classification, these transparency windows correspond to the mid- (MWIR) and long-wave (LWIR) ranges of the IR spectrum. In the international photometric system, the position of the transparency windows is standardized corresponding to the wavelength ($\lambda \pm \Delta\lambda$):

- in the visible and early near infrared range: B, V, R, J – up to 1.2 μm;
- H- range – (1.6 ± 0.1) μm;
- K- range – (2.2 ± 0.3) μm;

Table 1.5 Limits to photonic detector performance

Limits	Noise origin	How to reduce?
Fundamental	Background radiation noise	Spatial and spectral filtering
	Signal photon noise	Cannot be reduced
	Heterodyne photon noise	Cannot be reduced
Less fundamental	Auger thermal generation	Selection of semiconductors, non-equilibrium depletion
	Internal radiative generation	Design of the detector
	Radiative generation from adjacent elements	Design of the detector
Technological	Shockley-read generation	Elimination of Shockley-read centers
	Thermal generation at surfaces, interfaces and contacts	Improved surface treatment and technology
	Low frequency noise	Zero bias operation, improved technology
	Amplifier noise	Improved electronic interface

Source: Reprinted from Piotrowski [18]. Published 2004 by PAS as open access

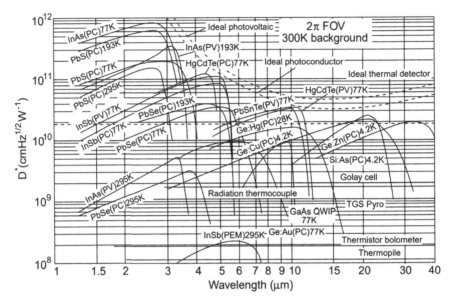

Fig. 1.9 Comparison of the D* of various commercially available infrared detectors when operated at the indicated temperature. Chopping frequency is 1000 Hz for all detectors except the thermopile, thermocouple, thermistor bolometer, Golay cell and pyroelectric detector. For these detectors a chopping frequency is 10 Hz. Each detector is assumed to view a hemispherical surround (2π field of view) at a temperature of 300 K. Theoretical curves for the background-limited D* (dashed lines) for ideal photovoltaic and photoconductive detectors and thermal detectors are also shown. *PC* photoconductive detector, *PV* photovoltaic detector. (Reprinted from Rogalski [25]. Published 2000 by MYU K.K. as open access)

- L- range – (3.6 ± 0.45) μm;
- M- range – (4.6 ± 0.5) μm;
- N- range – (10 ± 2) μm;
- Q- range – (20 ± 0.4) μm.

Between the transparency bands there are bands of complete absorption of IR radiation by the atmosphere, mainly by carbon dioxide CO_2 (2.6–2.9; 4.2–4.4 μm) and water vapor H_2O (5–8 μm). This means that when measured in these spectral ranges, the Earth's atmosphere absorbs most of the radiation coming from the object. Thus, in order to detect the IR signal, one must use the so-called atmospheric transparency windows. These spectral ranges have their advantages in different applications. Infrared spectral range 0.78–3 μm (NIR and SWIR) is used in fiber-optic communication lines, devices for external observation of objects and equipment for chemical analysis. SWIR ends at 2.5 μm and differs from medium and long wavelengths (MWIR and LWIR) that objects in this spectral range generate extremely little intrinsic thermal radiation. Observation in the SWIR range is carried out due to reflected light, that is, illumination from some powerful source is required, so this range is in many ways similar to the visible range. All wavelengths from 2 to 5 μm

(SWIR and MWIR) are used in pyrometers and gas analyzers that monitor the level of contamination in a various gas environment. It is also believed that the spectral range of 3–5 μm (MWIR) is more suitable for systems that capture images of objects with a high intrinsic temperature or in applications where contrast is required higher than sensitivity. Compared to MWIR, the LWIR range is more suitable for operation in smoky and dusty conditions. It was also found that measurements in the LWIR range are not affected by solar reflections and sea surface gloss. It has been experimentally established that in the LWIR range, the interference from the radiation of the inhomogeneity of the sky, re-reflected from the rough sea surface, is about 10 times less than in MWIR range. This range of 8–15 μm (LWIR), popular for special applications, is also used wherever it is necessary to see and recognize any objects in the fog.

Acknowledgments This research was funded by the State Program of the Republic of Moldova, project 20.80009.5007.02.

References

1. Baker IM (2017) II-VI narrow bandgap semiconductors: optoelectronics. In: Kasap S, Capper P (eds) Springer handbook of electronic and photonic materials. Springer, pp 867–896
2. Bao A, Lei C, Mao H, Li R, Guan Y (2019) Study on a high performance MEMS infrared thermopile detector. Micromachines 10:877
3. Chen S-J, Chen B (2020) Research on a CMOS-MEMS infrared sensor with reduced graphene oxide. Sensors 20:4007
4. Corsi C (2010) History highlights and future trends of infrared sensors. J Mod Opt 57(18):1663–1686
5. Ghimire H, Jayaweera PVV, Somvanshi D, Lao Y, Unil Perera AG (2020) Recent progress on extended wavelength and split-off band heterostructure infrared detectors. Micromachines 11:547
6. Graf A, Arndt M, Sauer M, Gerlach G (2007) Review of micromachined thermopiles for infrared detection. Meas Sci Technol 18(7):R59–R75
7. Harman TC, Melngailis J (1974) Narrow gap semiconductors. In: Wolfe R (ed) Applied solid state science. Academic, New York
8. Hodapp KW, Hall DNB (2006) Introduction to detectors: possible status in 2010–2020. In: Whitelock P, Dennefeld M, Leibundgut B (eds) Proceedings IAU symposium, No. 232. International Astronomical Union, pp 40–51
9. Joshi AM, Ban VS, Mason S, Lange MJ, Kosonocky WF (1992) 512 and 1024 element linear InGaAs detector arrays for near-infrared (1–3 μm) environmental sensing. Proc SPIE 1735:287–295
10. Karim A, Andersson JY (2013) Infrared detectors: advances, challenges and new technologies. IOP Conf Ser: Mater Sci Eng 51:012001
11. Kimata M (2001) Metal silicide Schottky infrared detector arrays. In: Capper P, Elliott CT (eds) Infrared detectors and emitters: materials and devices. Kluwer Academic Publishers, Boston, pp 77–98
12. Kulchitsky NA, Naumov AV, Startsev VV (2020) Infrared focal plane array detectors: "post pandemic" development trends. Part 1. Photonics Russia 14(3):234–244. (In Russian)
13. Kuznetsov SA, Paulish AG, Navarro-Cía M, Arzhannikov AV (2018) Selective pyroelectric detection of millimetre waves using ultra-thin metasurface absorbers. Sci Rep 6:21079

14. Lin TL, Park JS, George T, Jones EW, Fathauer RW, Maserjian J (1993) Long-wavelength PtSi infrared detectors fabricated by incorporating a p^+ doping spike grown by molecular beam epitaxy. Appl Phys Lett 62(25):3318–3320

15. McCreight CR, McKelvey ME, Goebel JH, Anderson GM, Lee JH (1986) Detector arrays for low-background space infrared astronomy. Laser Focus/Electro-Optics 22:128–133

16. Norton PR (2006) Third-generation sensors for night vision. Opto-Electron Rev 14:283–296

17. Piotrowski J, Rogalski A (2007) High-operating-temperature infrared photodetectors. SPIE, Bellingham

18. Piotrowski J (2004) Uncooled operation of IR photodetectors. Opto-Electron Rev 12(1):111–122

19. Reine MB (2001) HgCdTe photodiodes for IR detection: a review. Proc SPIE 4288:266–277

20. Rieke GH (2007) Infrared detector arrays for astronomy. Annu Rev Astron Astrophys 45:77–115

21. Rogalski A, Kopytko M, Martyniuk P (2020) 2D material infrared and terahertz detectors: status and outlook. Opto-Electron Rev 28:107–154

22. Rogalski A (2012) History of infrared detectors. Opto-Electron Rev 20:279–308

23. Rogalski A (2003) Infrared detectors: status and trends. Prog Quant Electron 27:59–210

24. Rogalski A, Adamiec K, Rutkowski J (2000) Narrow-gap semiconductor photodiodes. SPIE-The International Society for Optical Engineering, Bellingham

25. Rogalski A (2000) Infrared detectors at the beginning of the next millennium. Sens Mater 12(5):233–288

26. Rogalski A (1991) $Hg_{1-x}Mn_xTe$ as a new infrared detector material. Infrared Phys 31:117–166

27. Rogalski A (1989) $Hg_{1-x}Zn_xTe$ as a potential infrared detector material. Prog Quant Electron 13:299–253

28. Sclar N (1984) Properties of doped silicon and germanium infrared detectors. Prog Quant Electron 9:149–257

29. Smuk S, Kochanov Y, Petroshenko MP, Solomitskii D (2014) IRnova long-wavelength infrared sensors based on quantum wells. Komponenti Tehnologia 1:20–25. (In Russian)

30. Sobrino JA, Del Frate F, Drusch M, Jiménez-Muñoz JC, Manunta P, Regan A (2016) Review of thermal infrared applications and requirements for future high-resolution sensors. IEEE Trans Geosci Remote Sens 54:2963–2972

31. Starr B, Mears L, Fulk C, Getty J, Beuville E, Boe R et al (2016) RVS large format arrays for astronomy. Proc SPIE 9915:99152X

32. Tan CL, Mohseni H (2018) Emerging technologies for high performance infrared detectors. Nano 7(1):169–197

33. Vincent JD, Hodges SE, Vampola J, Stegall M, Pierce G (2016) Fundamentals of infrared and visible detector operation and testing, 2nd edn. Wiley, Hoboken

34. Xu D, Wang Y, Xiong B, Li T (2017) MEMS-based thermoelectric infrared sensors: a review. Front Mech Eng 12(4):557–566

Chapter 2
Photoconductive and Photovoltaic IR Detectors

Rada Savkina and Oleksii Smirnov

2.1 Introduction to Photoconductive and Photovoltaic IR Detectors on II-VI Semiconductors

The photoconductivity and photovoltaic effect-based devices are the most widely exploited photon detectors of the infrared (IR) radiation. As we already know from the previous chapters, photon detectors have significant advantages over other technologies in the field of detecting IR radiation such as fast response, high sensitivity, and wavelength selectivity. A few excellent scientific works have been published on IR detectors and most of them have been devoted to HgCdTe. HgCdTe is a variable-gap II-VI semiconductor, which is used most often in the production of IR photon detectors since it able to cover a wide wavelength range from 1 to 25 μm by controlling the Hg content. Prof. Antoni Rogalski, one of the world's leading researches in the field of IR optoelectronics, on the pages of his one of the last reviews concluded that: *"HgCdTe ... has triggered the rapid development of the three "detector generations" considered for military and civilian applications ..."* [45]. In the opinion of Antoni Rogalski, unique position of this material is conditioned by *"...composition-dependent tailorable energy band gap over the entire 1–30 μm range, large optical coefficients that enable high quantum efficiency, and favorable inherent recombination mechanisms that lead to long carrier lifetime and high operating temperature"* [42].

R. Savkina (✉)
National University "Kyiv-Mohyla Academy", Kyiv, Ukraine

V. Lashkaryov Institute of Semiconductor Physics at NAS of Ukraine, Kyiv, Ukraine
e-mail: rk.savkina@ukma.edu.ua; savkina@nas.gov.ua; rada.k.savkina@gmail.com

O. Smirnov
V. Lashkaryov Institute of Semiconductor Physics at NAS of Ukraine, Kyiv, Ukraine
e-mail: alex_tenet@isp.kiev.ua

© The Author(s), under exclusive license to Springer Nature Switzerland AG 2023
G. Korotcenkov (ed.), *Handbook of II-VI Semiconductor-Based Sensors and Radiation Detectors*, https://doi.org/10.1007/978-3-031-20510-1_2

This chapter provides data about photoconductive and photovoltaic infrared detectors manufactured from HgCdTe, as well as from the alternative ternary alloy systems, such as HgZnTe and HgMnTe. Their design, performance, advantage, and disadvantages are evaluated and compared. Infrared photon detectors operating in the middle (3–5 μm) and long wavelength (8–14 μm) infrared spectral range require cryogenic cooling to achieve useful performance. Background-limited performance is typically not achieved without significant cooling of the IR photon detectors. At the same time, there are known some concepts of the high operating temperature photon detection proposed and implemented to improve the performance of IR photodetectors near room temperature, which will also be described in this chapter.

2.2 Hg-Based Materials for IR Photon Detectors

Narrow-gap mercury-cadmium-telluride technologies are well developed now, and today this material is one of the basic semiconductors for photon detectors from near IR (wavelength $\lambda \sim 1.5$ μm) to long IR ($\lambda \sim 20$ μm) and is used in large-scale arrays with silicon CMOS readouts. We will not focus here on the properties of HgCdTe. The advantages of these compounds are discussed in the Chap. 4, Vol. 1. We only point out that there are several important features denoted in [37] that make HgCdTe an excellent material for infrared detectors:

– HgCdTe can be made n-type or p-type by several relatively convenient methods, at carrier concentrations required for most high-performance photodiode architectures. There exists a clear trend the use of extrinsic acceptor doping at photodiods manufacturing to avoid the strong Shockley-Read recombination associated with the Hg vacancy.
– Device-quality material can be grown by a variety of methods – both liquid phase epitaxy and vapor phase epitaxy methods such as MBE and MOVPE.
– The lattice mismatch between HgTe and CdTe is small, approximately 0.3%. This allows epitaxial growth of high-quality HgCdTe films on IR-transparent CdTe or nearly-lattice-matched IR-transparent CdZnTe substrates, with dislocation densities in the low-10^5 cm^{-2} range. IR-transparent substrates that are less costly, are available in much larger areas, and are more rugged than CdTe and CdZnTe, such as sapphire and silicon, can be used for epitaxial growth of HgCdTe films with dislocation densities that are acceptably low (mid-10^6 cm^{-2}) for many important photodiode applications.
– The thermal expansion coefficient of HgCdTe sufficiently closed to that of silicon as well as the relatively low dielectric constant ($\varepsilon_s = 18 \times \varepsilon_0$) of HgCdTe have opened important technological possibilities. This is important for fast response in laser pulse detectors where small RC time constants, for suppressing preamplifier noise below the detector noise and for technologically viable hybrid arrangements of HgCdTe detector arrays.

Table 2.1 The main physical properties for HgZnTe and HgMnTe solid solutions

	Lattice constant $a(x)$, (nm) at 300 K	Density $\rho(x)$ (g/cm³) at 300 K	The composition of the solid solution
$Hg_{1-x}Zn_xTe$	0.6461–0.0361x	8.05–2.41x	$0.10 \leq x \leq 0.40$
$Hg_{1-x}Mn_xTe$	0.6461–0.0121x	8.12–3.37x	$0.08 \leq x \leq 0.30$
Energy gap E_g (eV)			
$Hg_{1-x}Zn_xTe$	$-0.3 + 0.0324x^{0.5} + 2.731x\text{-}0.629x^2 + 0.533x^3 + 5.3 \times 10^{-4}\, T(1\text{-}0.76x^{0.5}\text{-}1.29x)$		
$Hg_{1-x}Mn_xTe$	$-0.253 + 3.466x + 4.9 \times 10^{-4}Tx\text{-}2.55 \times 10^{-3}\, T$		
Intrinsic carrier concentration, n_i (cm⁻³)			
$Hg_{1-x}Zn_xTe$	$(3.067 + 11.37x + 6.548\cdot10^{-3}\, T\text{-}3.633\cdot10^{-2}Tx) \times 10^{14}\cdot E_g^{3/4}\cdot T^{3/2}\exp(-5802E_g/T)$		
$Hg_{1-x}Mn_xTe$	$(4.615\text{-}1.59x + 2.64 \times 10^{-3}\, T\text{-}1.7 \times 10^{-2}Tx + 34.15x^2) \times 10^{14}\cdot E_g^{3/4}\cdot T^{3/2}\exp(-5802\, E_g/T)$		

	Momentum matrix element P (eV cm)	Spin-orbit splitting energy $\varDelta$ (eV)
$Hg_{1-x}Zn_xTe$	8.5×10^{-8}	1.0
$Hg_{1-x}Mn_xTe$	$(8.35\text{-}7.94x) \times 10^{-8}$	1.08

	Effective masses	Mobilities (cm²/Vs)
$Hg_{1-x}Zn_xTe$	$m_e^*/m = 5.7 \times 10^{-16}\cdot E_g/P^2$ $m_h^*/m = 0.6$	$\mu_e = 9 \times 10^8 b/T^{2a}$ $a = (0.14/x)^{0.6}$ $b = (0.14/x)^{7.5}$ $\mu_h = \mu_e/100$
$Hg_{1-x}Mn_xTe$	$m_e^*/m = 5.7 \times 10^{-16}\cdot E_g/P^2$ $m_h^*/m = 0.5$	$\mu_e = 9 \times 10^8 b/T^{2a}$ $a = (0.095/x)^{0.6}$ $b = (0.095/x)^{7.5}$ $\mu_h = \mu_e/100$

	Static dielectric constant ε	High frequency dielectric constant ε_∞
$Hg_{1-x}Zn_xTe$	$20.206\text{-}15.153x + 6.5909x^2\text{-}0.951826x^3$	$13.2\text{-}19.1916x + 19.496x^2\text{-}6.458x^3$
$Hg_{1-x}Mn_xTe$	$20.5\text{-}32.6x + 25.1x^2$	$15.2\text{-}28.8x + 28.2x^2$

Source: Data extracted from Ref. [43]

Among the narrow-gap II-VI semiconductors, HgZnTe and HgMnTe have been studied as potential alternatives to HgCdTe solid solution. The main physical properties for both ternary alloys are presented in the Table 2.1. Data was obtained from the Ref. [43]. Both HgZnTe and HgMnTe exhibit compositional-dependent optical and transport properties like HgCdTe material with the same energy gap. Some physical properties of alternative alloys indicate on structural advantage in comparison with HgCdTe [39, 44]. For example, introducing ZnTe in HgTe decreases statistically the ionicity of the bond improving the stability of the HgZnTe alloy. Interdiffusion coefficient is about 10 times lower in HgZnTe than in HgCdTe. In the range of temperatures typical for IR detectors operation (77 K), the spin-independent properties of HgMnTe are practically identical to the properties of HgCdTe, discussed exhaustively in the literature. But the question of HgMnTe lattice stability is rather ambiguous.

The first HgZnTe photoconductive detectors were fabricated by Z. Nowak and M.E. Ejsmont in the early 1970s (see Ref. in Rogalski [39]). Then, it was shown that

$Hg_{0.885}Zn_{0.15}Te$ can be used as a material for high-quality ambient-temperature 10.6 μm photoconductors with detectivity around 10^8 cm $Hz^{1/2}$ W^{-1} [35]. The research group at Santa Barbara Research Center has developed very long wavelength HgZnTe photoconductive detectors with $\lambda = 17$ μm at temperature ≥ 65 K with peak detectivity $D^* = 8 \times 10^{10}$ cm $Hz^{1/2}$ W^{-1} obtained for the best wafers [32]. The best quality n^+-p HgZnTe photodiodes were manufactured by ion implantation technique. The measurements reveal comparable value of R_0A for both HgZnTe and HgCdTe photodiodes at 77 K in 1980s [2].

In comparison with HgCdTe, HgZnTe detectors are easier to prepare due to their relatively high hardness. The maximum of microhardness for HgZnTe is more than twice that one for HgCdTe (see Fig. 2.1). Moreover, HgZnTe is the material that is more resistant to dislocation formation and plastic deformation than HgCdTe. It was found that fixed charges at the anodic oxide-HgZnTe interface are lower than that of HgCdTe and is around 2×10^{10} cm^{-2} at 90 K.

Based on HgMnTe, mainly p-n junction photodiodes are created. Good quality p-n HgMnTe and HgCdMnTe junctions manufactured by annealing in Hg-saturated atmospheres of as-grown, p-type samples were made with the detectivities in the 3- to 5- and 8- to 12-μm spectral ranges closed to the background limit [6]. Other type of detector – avalanche photodiodes, was developed on the base HgMnTe and HgCdMnTe compounds. The R_0A product for such detector was 2.62×10^2 Ω cm^2, which is equivalent to a detectivity value of 1.9×10^{11} cm $Hz^{1/2}$ W^{-1} at 300 K [46].

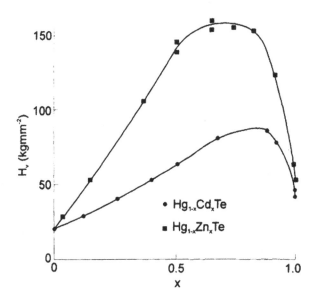

Fig. 2.1 Microhardness H versus composition x of HgZnTe and HgCdTe. (Reprinted with permission from Ref. [48]. Copyright 1986: Elsevier)

At the same time, neither HgZnTe nor HgMnTe have ever been systematically explored in the device context [43]. The prime reason for this is a serious problems encountered in these crystals' growth including:

- the essential separations between the liquids and solidus curves resulting in the high segregation coefficients;
- a weak variation of the growth temperature causes a large composition variation;
- high mercury pressure over the melt both in the case of HgZnTe and HgMnTe is an unfavorable condition for the growth of homogeneous bulk crystals.

2.3 Photoconductive and Photovoltaic IR Detectors: Design, Performance, Advantage, and Disadvantages

2.3.1 Photonic Mechanism of Detection

Photonic mechanism of detection (see Fig. 2.2) consists in direct conversion of incident photons into conducting electrons either bound to lattice atoms (intrinsic absorption) or to impurity atoms (extrinsic, impurity absorption) or with free electrons within a material. A key difference between intrinsic and extrinsic detectors is that extrinsic detectors require much cooling to achieve high sensitivity at a given spectral response cutoff in comparison with intrinsic detectors. According to the classification presented in [21], there are two main types of photon detectors that work with the majority and the minority charge carriers. If the dominant carrier is the majority carrier, then the sensing is photoconductive (PC) in nature. For minority carrier devices, both photoconductive and photovoltaic (PV) modes of detection can be utilized. Historical perspective on PC and PV detectors in HgCdTe was reviewed in [19], and the theory of photoconductors and a conventional photodiode can be found in monographs [41, 43].

Fig. 2.2 Fundamental optical excitation processes in semiconductors: A intrinsic absorption, B extrinsic absorption, C free carriers' absorption, C_b conduction band, V_b valence band

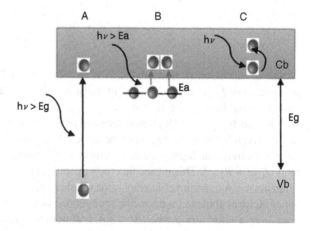

2.3.2 Photonic Detector Characterization

The following parameters are typically used to characterize and compare the perfor-mance of different materials systems and device architectures: responsivity (R_v), noise equivalent power (NEP), and detectivity (D) [41]. The responsivity is a func-tion of the wavelength of the incident radiation. The responsivity of IR detector is defined as the ratio of the electrical output signal (voltage V_s, or current I_s) to an input signal power in the form of a known photon flux (P_λ):

$$R_v = \frac{V_s}{P_\lambda} \text{ or } R_i = \frac{I_s}{P_\lambda} \tag{2.1}$$

The unit of responsivity is volts per watt (V/W) or amperes per watt (A/W).

The noise equivalent power is the signal level that produce a signal-to-noise ratio of 1. It can be written in terms of responsivity:

$$NEP = \frac{V_n}{R_v} = \frac{I_n}{R_i} \tag{2.2}$$

where V_n and I_n are a noise voltage and noise current respectively.

The unit of NEP is watt. If NEP is quoted for a fixed reference bandwidth, which is often assumed to be 1 Hz, this "NEP per unit bandwidth" has a unit off watts per square root hertz (W/Hz$^{1/2}$).

The detectivity D is the reciprocal of NEP:

$$D = \frac{1}{NEP} \tag{2.3}$$

Both NEP and detectivity are functions of electrical bandwidth (Δf) and detector area (A_d), so a normalized detectivity D^* suggested in [15, 16] is defined as

$$D^* = \frac{\left(A_d \Delta f\right)^{1/2}}{NEP} \tag{2.4}$$

and is expressed in unit cm Hz$^{1/2}$ W^{-1}, which recently is called "Jones." The importance of D^* is that this figure of merit permits comparison of detectors of the same type but having different areas.

The detectivity D^* of IR photodetector is limited by generation and recombination rates G and R in the active region of the device A_d. For a given wavelength and operat-ing temperature, the highest device performance can be obtained by maximizing the ratio $\eta/[(G + R)t]^{1/2}$. This means that high quantum efficiency η can be obtained with a thin device. A quantum efficiency is the number of photocarriers generated per num-ber of incident photons for a specific semiconductor with reflectivity r and is given by $\eta = (1-r)\cdot(1-exp(-\alpha x))$, where $0 \le \eta \le 1$. If both the generation and recombination rates and electrical and optical area of IR detector are equal, the detectivity of an opti-mized photodetector is limited by thermal processes and it can be expressed as

$$D^* = 0.31 \frac{\lambda}{hc} k \left(\frac{\alpha}{G} \right)^{1/2} \tag{2.5}$$

where $1 \leq k \leq 2$ and is dependent on the contribution of recombination and back-side reflection. The ratio of the absorption coefficient to the thermal generation rate, α/G, is the fundamental figure of merit of any material intended for infrared photodetection.

An important factor influencing the performance of an IR detector is unwanted fluctuations in the measured signal defined as noise. All detectors are limited in the minimum radiation power that they can detect, by noise that is determined by the fluctuation of the charge carriers generated both thermally within the detector (the internal noise) and by the incident background flux of IR radiation that is absorbed by the detector material (the radiation noise). The ultimate performance of infrared detectors is reached when the internal noise is low compared to the radiation noise. The practical operating limit for most infrared detectors is not the signal fluctuation limit but the background fluctuation limit, also known as the background-limited infrared photodetector (BLIP) limit. The expression for BLIP detectivity can be written as

$$D^*_{BLIP} = \frac{\lambda}{hc} \left(\frac{\eta}{2\Phi_B} \right)^{1/2} \tag{2.6}$$

where Φ_B is the total background photon flux density reaching the detector:

$$\Phi_B = \sin^2 \left(\frac{\theta}{2} \right) \int_{\lambda_c}^{0} \frac{2\pi c \, d\lambda}{\lambda^4 \left(\exp\left(\frac{hc}{\lambda k T_B} \right) - 1 \right)}, \tag{2.7}$$

$$\sin^2 \left(\frac{\theta}{2} \right) = \frac{\Phi_B(\theta)}{\Phi_B(2\pi)} \tag{2.8}$$

where the integrand is Planck's photon emittance at temperature T_B, θ is a field of view angle. The background-limited detectivity D^* relative to 2π FOV becomes

$$\frac{1}{\sin\left(\frac{\theta}{2} \right)} = \frac{D^*_{BLIP}(\theta)}{D^*_{BLIP}(2\pi)} \tag{2.9}$$

Equation (2.9) holds for photovoltaic detectors. Photoconductive detectors have a lower D^*_{BLIP} by a factor of $\sqrt{2}$:

$$D^*_{BLIP} = \frac{\lambda}{2hc}\left(\frac{\eta}{\Phi_B}\right)^{1/2} \tag{2.10}$$

The following paragraphs will describe the main properties of the PC and PV IR detectors. The theory of photoconductive and photovoltaic IR detectors can be found in monographs [41, 43].

2.3.3 Photoconductive IR Detectors

Photoconductive IR detector is a type of photodetector that are based on semiconductor materials whose conductivity increases under the absorption of incident photon flux density $\Phi_s(\lambda)$ resulted in non-equilibrium charge carriers' generation. Excitation of electrons into the conduction band or holes into the valence band occurs from another band or impurity states within the band (impurity-bound states in energy gap, quantum wells, or quantum dots). Alternative term for photoconductive detector is photoresistor, light-dependent resistor and photocell [31]. The photoconductive effect involves applying a bias voltage across a uniform piece of detector material to generate a photocurrent proportional to the photoexcited electron concentration (see Fig. 2.3). For low-resistance material, PC detector is usually operated in a constant current circuit. For high-resistance photoconductors, a constant voltage circuit is preferred, and the signal is detected as a change in current in the bias circuit. Herewith, the basic value describing intrinsic or extrinsic photoconductivity is a short-circuit photocurrent:

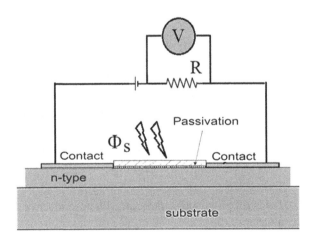

Fig. 2.3 Schematic view of unit cells for PC detector

$$I_{ph} = Aq\eta\Phi_s g, \tag{2.11}$$

where $A = wl$ is a detector area, g is a PC gain represented the number of electrons flowing through the electrical circuit per photon absorbed [26], η is a quantum efficiency, Φ_s – is the incident photon flux density. The value of PC gain depends upon relation between the drift length $L_d = v_d\tau$ and interelectrode spacing, l.

The basic requirements for high photoconductive responsivity at a given wavelength λ follow from the expression for the voltage responsivity of PC detector:

$$R_v = \frac{V_s}{P_\lambda} = \frac{\eta}{At}\frac{\lambda\tau}{hc}\frac{V_b}{n_0} \tag{2.12}$$

where the absorbed monochromatic power $P_\lambda = \Phi_s Ah\nu$. In other word, one must have high quantum efficiency η, long excess carrier lifetime τ, the smallest possible piece of crystal, low thermal equilibrium carrier concentration n_0, and the highest possible bias voltage V_b. Current gain of photoconductors is their advantage, which leads to higher responsivity than is possible, for example, with non-avalanching photovoltaic detectors. However, serious problem of photoconductors operated at low temperature is nonuniformity of detector element due to recombination mechanisms at the electrical contacts and its dependence on electrical bias.

The frequency dependent responsivity can be determined by the equation

$$R_v = \frac{\eta}{At}\frac{\lambda\tau_{ef}}{hc}\frac{V_b}{n_0}\frac{1}{\left(1+\omega^2\tau_{ef}^2\right)^{1/2}} \tag{2.13}$$

where τ_{ef} is the effective carrier lifetime.

However, there are two limits on applied bias voltage and, respectively, voltage responsivity increase – Joule heating of the detector element and sweep-out of minority carriers. It should be noted that for typical detector sizes of 50×50 μm^2 and long lifetimes of excess carriers (1μs in 8–14 μm devices at 77 K and 10 μs in 3–5 μm devices at higher temperatures), such parameters as contacts, drift, and diffusion have a significant effect on the PC detector performance (read Section 9.1.1.1 in Rogalski [41]).

The total noise voltage of a photoconductor is

$$V_{noise}^2 = V_j^2 + V_{GR}^2 + V_{1/f}^2 \tag{2.14}$$

where terms in expression are the fundamental types of internal noise sources usually operative in photoconductive detectors: Johnson–Nyquist (sometime called thermal) noise (V_j), generation–recombination (G–R) noise (V_{GR}) and the third form of noise, not amenable to exact analysis, is $1/f$ noise ($V_{1/f}$).

Johnson–Nyquist noise is associated with the finite resistance R of the device and is caused by random movements of charge carriers whose temperature is bigger than 0 K. Their thermal energy increases as temperature increases. Thermal noise occurs in the absence of external bias as a fluctuating voltage or current depending upon the method of measurement. The root mean square of Johnson–Nyquist noise voltage in the bandwidth Δf is given as

$$V_J^2 = 4kTR\Delta f \tag{2.15}$$

where k is the Boltzmann constant. This noise has "white" character.

At finite bias currents, the carrier density fluctuations cause resistance variations, which are observed as noise exceeding Johnson–Nyquist noise. This type of excess noise in photoconductive detectors is referred to as G–R noise. The G–R noise is due to the random generation of free charge carriers by the crystal vibrations and their subsequent random recombination. It leads to conductivity changes that will be reflected as fluctuations in current flow through the crystal. The $(G$–$R)$ noise voltage for equilibrium conditions is equal

$$V_{gr}^2 = 2(G + R)lwt(Rqg)^2 \Delta f \tag{2.16}$$

where G and R are the volume generation and recombination rates, $g = \tau/t_t$ is a photoconductive gain which can be defined as a ratio of free carrier lifetime, τ, to transit time, t_t, between the ohmic contacts, and l, w, t are the geometric dimensions of the detector.

Finally, $1/f$ noise is associated with the presence of potential barriers at the contacts, interior, or surface of the detector. Its reduction to an acceptable level depends greatly on the processes of preparing the contacts and surfaces. The two most current models for the explanation of $1/f$ noise are assuming the fluctuations in the mobility of free charge carriers (Hooge's model) [14] and in the free carrier density (McWhorter's model) [49].

2.3.4   Photovoltaic IR Detector

A photovoltaic effect occurs in structures with built-in potential barriers. The most widely used PV detector is the *p-n junction photodiode* (see Fig. 2.4a), where a strong internal electric field exists across the junction even in the absence of radiation. When a photoexcited electron-hole pair are injected optically into the vicinity of such barriers, the electron and hole are separated by the space-charge field causing a change in voltage across the open-circuit cell or a current to flow in the short-circuited case. The role of the built-in electric field is to cause the charge carriers of opposite sign to move in opposite directions depending upon the external circuit. Minority carriers are readily accelerated to become majority carriers on the other side. This way a photocurrent is generated that shifts the current-voltage characteristic in the direction of negative or reverse current, as shown in Fig. 2.4b.

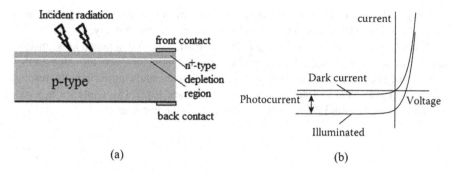

Fig. 2.4 (a) Schematic of typical *p-n* junction photodiode and (b) its current-voltage characteristic for the illuminated and nonilluminated diode

The current-voltage (*I-V*) relationship for photodiode in which the dark current is only due to diffusion current, is given by the well-known Shockley diode equation:

$$J_{dark}(V) = J_s\left(\exp\frac{qV}{kT} - 1\right)$$ (2.17)

With illumination, the total current density in the *p-n* junction becomes:

$$J(V,\Phi) = J_{dark}(V) - J_{ph}(\Phi)$$ (2.18)

where the dark current density, J_{dark}, depends on V and the photocurrent density, $J_{ph} = \eta q A \Phi$, depends on the photon flux density Φ. Here J_s is the saturation current.

As it was noted above, the theory of a conventional photodiode can be found in monographs [41, 43]. Here we will discuss only the main points. For a photodiode, the important parameters are open-circuit voltage and short-circuit current. The open-circuit voltage produced by the accumulation of electrons and holes on the two sides of the junction can be obtained by multiplying the short-circuit current by

$R = \left(\dfrac{\partial I}{\partial V}\right)^{-1}_{V=V_b}$, where V_b is the bias voltage. In many applications the photodiode is operated at zero-bias voltage $V_b = 0$ which corresponds to the resistance R_0. The $R_0 A$ ($\Omega \cdot cm^2$) product is a parameter that makes it possible to inform the order of magnitude of the dark current and makes it possible to evaluate the quality of the photodetector i.e., this is a figure of merit of a PV detector. This product for ideal diffusion-limited diode is given by [41]:

$$(R_0 A)_D = \frac{kT}{qJ_s}$$ (2.19)

$$J_s = (kT)^{1/2} n_i^2 q^{1/2}\left[\frac{1}{p}\left(\frac{\mu_e}{\tau_e}\right)^{1/2} + \frac{1}{n}\left(\frac{\mu_h}{\tau_h}\right)^{1/2}\right]$$ (2.20)

where p and n are the hole and electron majority carrier concentrations, τ_e and τ_h the electron and hole lifetimes in the p- and n-type regions, respectively.

Long [26] and Tredwell and Long [47] showed that the fundamental noise mechanisms in HgCdTe photodiodes and HgCdTe photoconductors were the same. In the case of the photodiode at zero-bias voltage $V_b = 0$ in thermal equilibrium, the resulting noise is identical to Johnson –Nyquist noise (see Eq. 2.15), while for a diode exposed to background flux density Φ_b this expression becomes

$$V_J^2 = \left(4kT + 2q^2\eta AR_0\Phi_b\right)R_0\Delta f$$

(2.21)

In the absence of a background-generated current, the current noise is equal to the Johnson noise ($4kTR_0\Delta f$) at zero bias, and it tends to the usual expression for shot noise ($2qI_D\Delta f$) for voltages greater than a few kT in either direction [41]. Due to the absence of recombination noise, the limiting $p–n$ junction's noise level can ideally be $\sqrt{2}$ times lower than that of the photoconductor.

Detectivity in the case of the photodiode can be determined as

$$D^* = \frac{\eta\lambda q}{hc}\left(\frac{4kT}{R_0 A} + 2q^2\eta\Phi_B\right)^{-1/2}$$

(2.22)

Here in the case of background-limited performance and $\dfrac{4kT}{R_0 A} \ll 2q^2\eta\Phi_B$, we obtain Eq. (2.10), but in the case of t thermal noise limited performance and $\dfrac{4kT}{R_0 A} \gg 2q^2\eta\Phi_B$, Eq. (2.22) converts to

$$D^* = \frac{\eta\lambda q}{2hc}\left(\frac{R_0 A}{kT}\right)^{1/2}$$

(2.23)

It should be noted that for real diodes, the process of carrier transfer is much more complicated, and to determine the quality of photodiodes, it is necessary to consider such phenomena as generation–recombination within the depletion region, tunneling through the depletion region, surface effects, impact ionization, and space-charge limited current. For photodiode device structures, the critical region is the p-n junction, which can be formed using different approaches, including, ion implantation, Hg diffusion, impurity diffusion, and type conversion. The various HgCdTe-based $p–n$ junction photodiode architectures are collected and described in review [37] and are presented in Table 2.2.

Over the years, several kinds of PV IR devices have been developed, including homojunction and heterojunction photodiodes and *metal-insulator-semiconductor* (MIS) photo-capacitors. Among them, p-n junction PV diodes have become the mainstream technology. In general, there are two groups: one is an n-on-p junction diode, and the other is a p-on-n junction diode, where the latter device has been demonstrated to have lower dark current and thus higher operating temperature. Regarding the fabrication technology, there are generally two classes of p-n

Table 2.2 Cross section views of various photodiode architectures

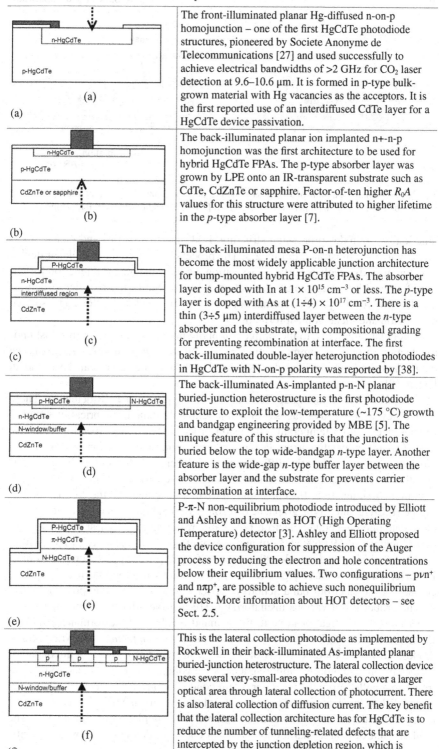

(a)	The front-illuminated planar Hg-diffused n-on-p homojunction – one of the first HgCdTe photodiode structures, pioneered by Societe Anonyme de Telecommunications [27] and used successfully to achieve electrical bandwidths of >2 GHz for CO_2 laser detection at 9.6–10.6 μm. It is formed in p-type bulk-grown material with Hg vacancies as the acceptors. It is the first reported use of an interdiffused CdTe layer for a HgCdTe device passivation.
(b)	The back-illuminated planar ion implanted n+-n-p homojunction was the first architecture to be used for hybrid HgCdTe FPAs. The p-type absorber layer was grown by LPE onto an IR-transparent substrate such as CdTe, CdZnTe or sapphire. Factor-of-ten higher R_0A values for this structure were attributed to higher lifetime in the *p*-type absorber layer [7].
(c)	The back-illuminated mesa P-on-n heterojunction has become the most widely applicable junction architecture for bump-mounted hybrid HgCdTe FPAs. The absorber layer is doped with In at 1×10^{15} cm^{-3} or less. The *p*-type layer is doped with As at $(1 \div 4) \times 10^{17}$ cm^{-3}. There is a thin $(3 \div 5$ μm$)$ interdiffused layer between the *n*-type absorber and the substrate, with compositional grading for preventing recombination at interface. The first back-illuminated double-layer heterojunction photodiodes in HgCdTe with N-on-p polarity was reported by [38].
(d)	The back-illuminated As-implanted p-n-N planar buried-junction heterostructure is the first photodiode structure to exploit the low-temperature (~175 °C) growth and bandgap engineering provided by MBE [5]. The unique feature of this structure is that the junction is buried below the top wide-bandgap *n*-type layer. Another feature is the wide-gap *n*-type buffer layer between the absorber layer and the substrate for prevents carrier recombination at interface.
(e)	P-π-N non-equilibrium photodiode introduced by Elliott and Ashley and known as HOT (High Operating Temperature) detector [3]. Ashley and Elliott proposed the device configuration for suppression of the Auger process by reducing the electron and hole concentrations below their equilibrium values. Two configurations – pνn+ and nπp+, are possible to achieve such nonequilibrium devices. More information about HOT detectors – see Sect. 2.5.
(f)	This is the lateral collection photodiode as implemented by Rockwell in their back-illuminated As-implanted planar buried-junction heterostructure. The lateral collection device uses several very-small-area photodiodes to cover a larger optical area through lateral collection of photocurrent. There is also lateral collection of diffusion current. The key benefit that the lateral collection architecture has for HgCdTe is to reduce the number of tunneling-related defects that are intercepted by the junction depletion region, which is particularly important for lower temperature operation, e.g., 40 K for LWIR, and for very large optical areas.

Source: Data extracted from Ref. [37]

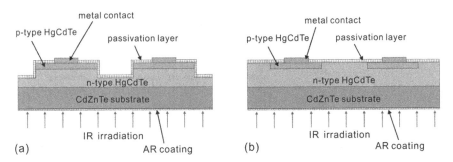

Fig. 2.5 Schematic diagram of (**a**) mesa structure and (**b**) planar structure of p-on-n HgCdTe photodiodes. (Reprinted with permission from Ref. [25]. Copyright 2015: AIP Publishing)

junction photodiodes processing structure used: mesa structures and planar structures, which are illustrated in Fig. 2.5. The processing procedure of a planar device structure usually involves surface passivation, window etching, *p-n* junction formation, followed by metal contact deposition, whereas that of mesa device structures usually involves *p-n* junction formation, mesa isolation etching, followed by surface passivation and metal contact formation [25].

A popular alternative to the simple *p-n* photodiodes especially for ultrafast photodetection in optical communication, measurement, and sampling systems is the *p-i-n photodiode*. A detailed discussion of such photodetectors can be found in Donati [8].

Schottky barrier photodiodes have also found application as IR detectors. These devices reveal some advantages over *p–n* junction photodiodes: fabrication simplicity, absence of high temperature diffusion processes, and high speed of response [41]. This is a majority carrier device with the thermionic emission process which much more efficient than the diffusion process. Therefore, for a given built-in voltage, the saturation current in a Schottky diode is several orders of magnitude higher than in the *p–n* junction. Recent studies have demonstrated that the combination of *n*-type HgCdTe and *p*-type graphene can construct a Schottky junction to effectively dissociate photogenerated electron-hole pairs, resulting in a highest external quantum efficiency of 69.06% at 3 μm and maximum high responsivity of 2.6 A/W in visible to mid-infrared wavelength region (0.5÷5.2) μm [53]. A heterostructure of Bi-Layer Graphene-CdTe-HgCdTe for MWIR photodetector that has an Ohmic contact for the electrons but a Schottky barrier for the holes was proposed by [10].

The properties of various insulator/HgCdTe interfaces have attracted numerous experimental investigations due to the demand for the high-performance HgCdTe IR detectors, such as PV detectors or *metal-insulator-semiconductor* (MIS) photodiodes. MIS photodiode consists of a metal gate separated from a semiconductor surface by an insulator. By applying a negative voltage to the metal electrode, electrons are repelled from the insulator – semiconductor interface, creating a depletion region. When incident photons create hole–electron pairs, the minority carriers drift a way to the depletion region and the volume of the depletion region shrinks. The total amount of charge that a photogate can collect is defined as its well capacity.

The general theory of MIS devices as applied to HgCdTe is reviewed in [20]. The passivation effect of ZnS [30], Al_2O_3 [54], CdTe [51] and other materials deposition on a HgCdTe MIS detector were investigated.

2.4 High Operation Temperature IR Detector

Usually, IR detectors are used mainly in special areas, so the main task is to optimize their sensitivity, spatial and time resolution. To achieve good signal-to-noise performance and a very fast response, the photon IR detectors require cryogenic cooling. To obtain high performance HgCdTe-based detectors, temperatures typically $80 \div 200$ K are required. Significant cooling is needed to reduce noise and leakage currents resulting from the thermal generation and recombination (G-R) processes near room temperature. The cooling adds substantial cost and bulk to the IR sensor, thus, in general, the photon IR sensors are relatively complex, costly, and not highly portable. At the same time affordability is ignored. Today, commercial and government industries (medicine, coercion and rescue services, transportation, etc.) are showing increasing interest in "affordable" IR receivers. The civil market forms requirements to the price, the sizes, convenience in use and accordingly roughens such parameters, as sensitivity, equivalent noise of a difference of temperatures, and inertia of receivers.

The detectivity of an optimized IR detector (see Eq. 2.5) is depended on the ratio of the absorption coefficient to the thermal generation $\alpha/_G$, i.e., is limited by the processes in the device workspace. As we know (see, for example, [19]), the most important generation and recombination mechanisms in HgCdTe: Shockley-Read-Hall (SRH), radiative and Auger. Auger processes become more important and degrade the performances of HgCdTe-based IR devices as the energy band gap is decreased and/or the temperature is increased. The Auger mechanism imposes fundamental limitations to the LWIR HgCdTe detector performance. The Shockley-Read-Hall (SRH), recombination mechanism involves the recombination of electron-hole pairs via defect levels within the energy bandgap of the material. Therefore, it is not an intrinsic process and its effect on detector noise and external quantum efficiency can be suppressed as HgCdTe processing improves. To reduce the dark current and increase the detector operating temperature, it is essential to suppress these defect-related dark current mechanisms, in which *passivation* plays a critical role. Successes of IR devices performance associated with passivation are reviewed in [25].

There are several strategies of improving IR photon detectors high-temperature performance such as the suppression of radiative recombination using photon recycling [22], the suppression of both radiative and Auger recombination with carrier depletion [3, 4], the suppression of Auger recombination using band structure engineering in strained layer superlattices – both for III-V [13] and for II-VI [17] compounds. All three effects provide significant enhancements in detectivities. Development of the high operating temperatures (HOT) photodetectors based on

II-VI compounds were specifically focused on HgCdTe solid solutions [36, 42]. Many new concepts have been implemented and tested to improve their performance [18, 29, 45].

2.4.1 Ways to Improve Detector's Performance Without Cooling

Key improvements of IR technology through innovations in material growth and device design have made it possible to create IR devices that operate at room temperature. Next, we will talk about *barrier structures, Auger suppressed structures* and *photon trapping structures.*

A new concept of infrared detector named the *nBn* detector has been proposed by Maimon and Wicks [28]. Figure 2.6 shows examples of the barrier IR HOT detectors and their bandgap diagrams, that eliminate the cooling requirements of photodetectors operating in the MWIR range. *nBn* detector is designed to reduce the dark current (G-R current originating within the depletion layer associated with Shockley–Read–Hall process) and noise without impeding the photocurrent (signal). The barrier serves also to reduce the surface leakage current. The *n*-type semiconductor on one side of the barrier constitutes a contact layer for biasing the device, while the *n*-type narrow-bandgap semiconductor on the other side of the barrier is a photon-absorbing layer whose thickness should be comparable to the absorption length of light in the device, typically several microns. The barrier should be located near the minority-carrier collector and away from the region of optical absorption. *nBn* – detector somewhat resembles a typical *p–n* photodiode, except that the junction (space-charge region) is replaced by an electron-blocking unipolar barrier (*B*), and that the *p*-contact is replaced by an *n*-contact. The structure can filter out majority carriers while collecting minority carriers, similarly to a photodiode. It can be stated that the *nBn* design is a hybrid between a PC and PV photodetector.

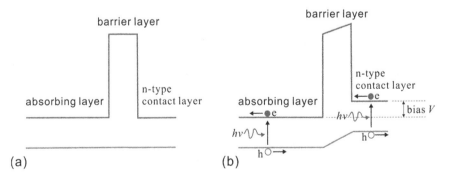

Fig. 2.6 Schematic energy band diagram of an ideal nBn detector under (**a**) zero bias and (**b**) illumination and low reverse bias V. (Reprinted with permission from Ref. [25]. Copyright 2015: AIP Publishing)

The results obtained by [1] indicate that the composition, doping, and thickness of the barrier layer in MWIR HgCdTe nBn detectors can be optimized to yield performance levels comparable with those of ideal HgCdTe *p-n* photodiodes. It is also shown that introduction of an additional barrier at the back contact layer of the detector structure (nBnn+) leads to substantial suppression of the Auger G-R mechanism. This results in an order-of-magnitude reduction in the dark current level compared with conventional nBn or p-n junction-based detectors, thus enabling background-limited detector operation above 200 K. The idea to optimize the nBn structure by an asymmetric barrier - abrupt on the contact side to efficiently block the majority carriers, and gradual on the absorption layer side to plane down the remaining potential barrier for the collected photocarriers, is discribed in [12].

The *Auger suppressed structure* also provides a significant reduction in dark current at high operating temperatures. This device structure first proposed by Ashley and Elliott [3] is formed as the lightly doped absorber located between two layers of high doped and high bandgap. The absorber region of the device is near intrinsic and is π−type (lightly doped *p*-type) or ν−type (lightly doped *n*-type). In the case of high reverse bias, the device design ($P^+/\nu/N^+$) reduces the carrier concentration in the central absorber layers well below thermal equilibrium and, thus, suppresses Auger processes and effectively reduces the dark current. It has been shown that the reduction in the dark current is around 12 times at temperatures above 200 K. Numerical simulations also shows that the dark current is significantly smaller in Auger suppressed structure compared with standard double layer planar heterostructure (DLPH) ones, and this leads to improved detectivities for a wide range of temperatures [50]. A MWIR (6 μm cutoff) DLPH detector and a HOT detector with the same cutoff will operate with $D^* = 5 \times 10^{10}$ cm Hz$^{1/2}$ W^{-1} detectivity at ~120 K and ~170 K respectively. By implementing an Auger suppressed architecture, an improvement of 50 K in the operating temperature opens the possibility of thermoelectrically (TE) cooled high performance MWIR IR photon detectors.

There are other ways to improve the performance of IR detectors associated with size effect of reducing the amount of thermal generation as well as optical optimization [25, 33]. Thermally generated dark current is directly proportional to the material volume of the detector. *Photon trapping technology* is an important approach to reducing dark current and enhancing the detector operating temperature. Photonic crystal structures can be used to effectively enhance the light absorption in detectors with small material volume. To reduce dark current without dropping quantum efficiency, it was proposed the photonic crystal structure which is a HgCdTe-based MWIR [52] and LWIR IR detector [25].

One approach to reducing the material volume of the detector is to reduce the thickness of the absorption layer. For example, it was found that reducing the thickness of the absorption region of HgCdTe photodiodes operating at 200 K [23] is good solution to improve the response speed of a detector. On the other hand, for standard HgCdTe detectors, most of the light is absorbed in the absorption layer

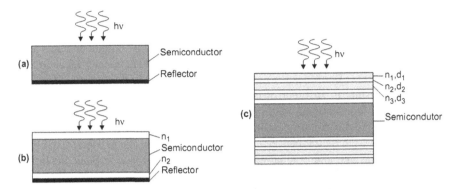

Fig. 2.7 Schematic structure of interference enhanced photodetectors: (**a**) the simplest structure, interference occurs between the waves reflected at the rear and the front surface of semiconductor (**b**) structure immersed between two dielectric layers supplied with backside reflector, and (**c**) structure immersed between two photonic crystals. (Reprinted with permission from Ref. [33]. Copyright 2004: Elsevier)

around 10 μm, leading to a high quantum efficiency. However, devices with too thin absorption layer (to less than 1 μm) would suffer from poor quantum efficiency and reduced responsivity. Therefore, maintaining high light absorption in a thin HgCdTe layer becomes the critical point.

A possible way to improve the performance of IR detector is to reduce its thickness that can be achieved by using multiple pass of radiation with a backside reflector. Various proposed optical resonator structures are shown in Fig. 2.7. More efficient light absorption can be achieved by utilizing interference phenomena to setup a resonant cavity within the detector by using two dielectric layers on the front and rear surface of the detector, the schematic structure of which is shown in Fig. 2.7b. An important limitation of the optical cavity application is that the gain in quantum efficiency can be achieved only in narrow spectral regions. Another limitation comes from the fact that efficient optical resonance occurs only for near-perpendicular incidence and is less effective for oblique incidence.

An increase in apparent "optical" size of the detector using a various type of optical concentrators such as optical cones, conical fibers, and other types of optical concentrators also leads to an increase in the performance of IR photodetector. More efficient enhancement can be achieved by placing the absorber in optical resonant cavities (see Ref. [18]).

In addition to classical photoresistors and photodiodes, three other types of IR detectors can operate at near 300 K: magnetic concentration detectors, photoelectromagnetic (or PEM) detectors and Dember effect detectors [33].

2.4.2   *Photoelectromagnetic Effect IR Detectors*

The *photoelectromagnetic* effect is caused by an in-depth diffusion of photogene-rated carriers whose trajectories are deflected in a magnetic field. The driving force for diffusion is the gradient of carrier concentration that is caused by the non-uniform absorption of radiation in the different layers of bulk semiconductor. Theoretical analysis indicates that the maximum voltage responsivities of PEM detectors can be reached in relatively strong magnetic fields $B \sim \mu_e^{-1}$ for samples with high resistance. At room temperature the radiation is almost uniformly absorbed within the diffusion length in HgCdTe material since the ambipolar diffusion length in narrow gap semiconductors is several μm while the absorption of radiation is relatively weak ($1/\alpha \approx 10$ μm). In such cases, a low recombination velocity at the front surface and a high recombination velocity at the back surface is necessary for a good PEM detector response.

The best at the time of 2004, uncooled HgCdTe 10.6 μm PEM detectors exhib-ited experimental voltage responsivity of about 0.1 V/W (width of 1 mm) and detec-tivities of about 1×10^7 cm Hz$^{1/2}$ W^{-1}, which is by a factor of ~3 below the predicted ultimate value 3.4×10^7 cmHz$^{1/2}$ W^{-1}, [33]. Gaziyev et al. [11] have reported that the specific detectivity D^* of manufactured on the basis of HgCdTe ($x = 0,2$) monocrys-tals PEM detector for MWIR region with a maximum responsivity near 6 μm was about $(0.8 \div 0.9)$ 10^8 cmHz$^{1/2}$ W^{-1} on frequency of 1200 Hz.

As of today, company VIGO System S.A. has developed uncooled HgCdTe-based photovoltaic IR detectors based on PEM effect in the semiconductor with peak detectivity on the level 1.6×10^8 cmHz$^{1/2}$ W^{-1} (see Table 2.3). The devices are designed for the maximum performance at 10.6 μm and especially useful as a large active area device to detect CW (Continuous Wave) and low frequency modulated radiation. These devices are mounted in specialized packages with incorporated magnetic circuit inside. Zinc selenide anti-reflection coated (wZnSeAR) window prevents unwanted interference effects and protects against pollution. Spectral detectivity of HgCdTe-based PEM detectors presented in Table 2.3 is shown in Fig. 2.8.

Today the measured performance of PEM detectors is the same as that of the PC detectors for the same spectral range (read Sect. 2.5). In contrast to photoconduc-tors, PEM detectors do not require electrical biasing. The frequency characteristics

Table 2.3 2.0–12.0 μm HgCdTe ambient temperature PEM detectors with ZnSe window. Optimum wavelength $\lambda_{opt} = 10.6$ μm

HgCdTe-based PEM detector 20 °C	Optical area, A_0 (mm × mm)	Resistance, R (Ω)	Current responsivity-optical area length product, R_i (A·mm/W)	Peak Detectivity, D^a (cm Hz$^{1/2}$ W^{-1})	Time constant, t (nsec)
PEM series	1 × 1	≥40	≥0.002	≥2.0 × 10^7	≤1.2
PEMI[a] series	2 × 2	40–100	≥0.01	≥1.6 × 10^8	

Source: Data extracted from https://vigo.com.pl/
[a]Optically immersed

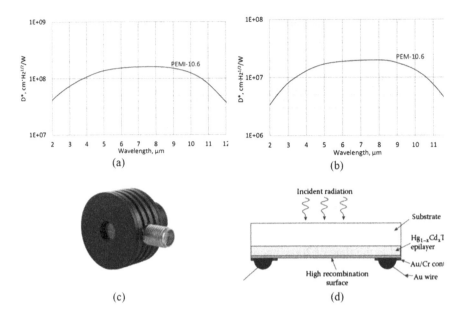

Fig. 2.8 Spectral detectivity of HgCdTe-based PEM detector (**a**) with optical immersion and (**b**) without optical immersion; (**c**) detector image and (**d**) cross section of a back side illuminated element of PEM detector. (Reprinted from https://vigo.com.pl/)

of a PEM detector is flat over a wide frequency range, starting from constant current. This is due to lack of low frequency noise and a very short response time [33]. PEM detectors have been also fabricated with other HgTe-based ternary compound such as HgZnTe and HgMnTe [34].

2.4.3 Magnetoconcentration IR Detectors

Unlike the PEM detector, the device based on the *magnetoconcentration* effect is biased with an electric field perpendicular to the magnetic field. As a result, the Lorentz force directs current carriers to the surface of high surface recombination velocity, the concentration of electrons and holes in the bulk of the material is diminished and thus the regions become depleted. This leads to reduction of thermally generated noise. At the same time, such devices exhibited a large low-frequency noise when biased to achieve a sufficient depletion of semiconductor that have been severe obstacles to their widespread applications. Improvement of performance was observed at frequency >100 kHz.

2.4.4 Dember Effect IR Detectors

Detectors based on the *Dember* effect have also been proposed [9]. Dember effect is known as an effect of photodiffusion – the appearance of the potential difference in the direction of radiation absorption and bulk photodiffusion. Two conditions are required for the photovoltage generation: the distribution of photogenerated carriers should be nonuniform and the diffusion coefficients of electrons and holes must be different. The best performance is achievable for a device with thickness of the order of a diffusion length, with a low surface recombination velocity and a low reflection coefficient at the illuminated front surface and with a large recombination velocity and a large reflection coefficient at the nonilluminated back surface (Fig. 2.9b). Since the device is not biased and the noise voltage is determined by the Johnson–Nyquist thermal noise. Detectivities as high as $\approx 2.4 \times 10^8$ cmHz$^{1/2}$ W^{-1} and of $\approx 2.2 \times 10^9$ cmHz$^{1/2}$ W^{-1} are predicted for optimized 10.6 µm devices at 300 and 200 K, respectively (see Fig. 2.9a).

2.5 PC and PV IR Detectors Manufacturing

The first report about photoconductive effect in HgCdTe was published by Lawson et al. in 1959 [24]. Historical review of the further development of HgCdTe-based photoconductors is summarized in [40] and in monograph *Infrared Detectors* [41]. Infrared photon detector technology has reached a high level of sophistication. The results of study in this area made it possible to move to the stage of device manufacturing.

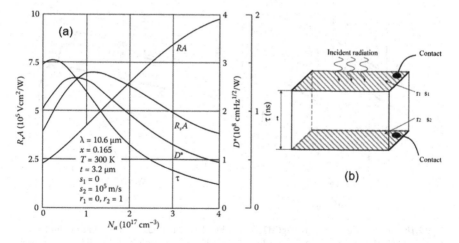

Fig. 2.9 The calculated normalized responsivity (R_vA), detectivity (D^*) and bulk recombination time (τ) of uncooled 10.6-µm HgCdTe Dember detector as a function of acceptor concentration. (Reprinted with permission from Ref. [9]. Copyright 1991: Elsevier)

Table 2.4 HgCdTe photoconductive detectors produced by Teledyne Judson Technologies

HgCdTe-based PC detector series J15n[*]	Active Size (square), mm	Cutoff Wavelength, μm	Peak Wavelength, μm	Peak D*@ 10KHz, cm Hz$^{1/2}$ W^{-1}	Time constant, t, μsec
n = 12	0.25...4	>12	11 ± 1	$5 \times 10^{10}...1.5 \times 10^{10}$	0.15÷0.5
n = 14	0.1...2	>13.5	13	$5 \times 10^{10}...2.5 \times 10^{10}$	0.5
n = 16	0.1...2	>16.6	14	$4 \times 10^{10}...2 \times 10^{10}$	0.3
n = 22	0.25...2	>22	16	$1 \times 10^{10}...6 \times 10^{9}$	0.1
n = 24	1	>24	20	5×10^{9}	0.1

[*]J15 Series detectors are designed for operation in the 2–26 μm wavelength region and for cryogenic operation at 77 °K.

Thermoelectrically cooled HgCdTe-based PC detector	Active size (square), mm	Cutoff wavelength, μm	Peak wavelength, μm	Peak D*@ 10KHz, cm Hz$^{1/2}$ W^{-1}	Time constant, t, μsec
Series J15TE2 −40 °C	0.25...1	4.0 ± 0.25	~4	4×10^{10}	5
	0.25...1	4.5 ± 0.25	~4.4	2.5×10^{10}	3
	0.25...1	5.0 ± 0.25	~4.8	1.5×10^{10}	2
Series J15TE3 −65 °C	0.1...1	>5	~4.8	3×10^{10}	2
	0.25...1	–	10.6	2×10^{8}	1...5
Series J15TE4 −80 °C	0.1...1	>5	~4.8	6×10^{10}	2
	0.25...1	–	10.6	6×10^{8}	1...5

Source: Data extracted from www.teledynejudson.com

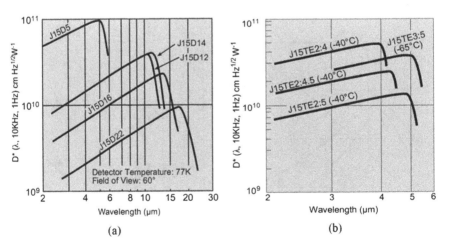

Fig. 2.10 Spectral detectivity of HgCdTe-based PC detector series J15n (**a**) and thermoelectrically cooled HgCdTe-based PC detector series J15TE (**b**) produced by Teledyne Judson Technologies. (Reprinted from www.teledynejudson.com)

Teledyne Judson Technologies (TJT) is a global leader for IR sensors from the Visible to VLWIR spectrum from 1969 year. This company produces the background limited PC detectors with state-of-the-art performance (see Table 2.4) for thermal imaging, CO_2 laser detection, industrial process control, FTIR spectroscopy, missile guidance, and night vision. J15 Series PC HgCdTe detectors is composed of a thin layer (10–20 μm) of HgCdTe with metalized contact pads defining the active area. Spectral detectivity of HgCdTe-based PC detector series J15n is shown in Fig. 2.10a. Its value reaches 10^{11} cm $Hz^{1/2}$ W^{-1} for MWIR spectral region and is at the level of 10^{10} cm $Hz^{1/2}$ W^{-1} and higher for the LWIR spectral region, i.e., these detectors work at the BLIP level. These detectors are low impedance devices,

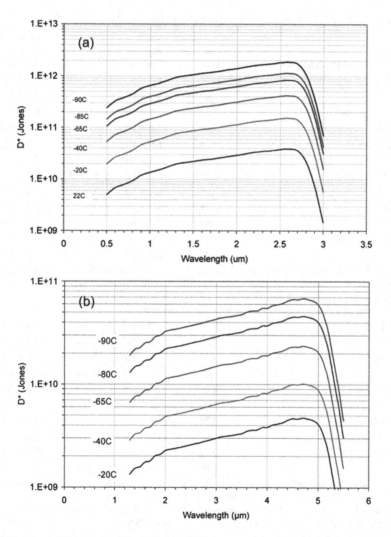

Fig. 2.11 Spectral detectivity of HgCdTe-based PV detector series J19 with 2.8 μm cutoff (**a**) and 5 μm cutoff (**b**) produced by Teledyne Judson Technologies. (Reprinted from www.teledyne-judson.com)

typically 10–150 ohms, and require a low voltage noise preamplifier. It should be noted that the series of thermoelectrically cooled HgCdTe PC detectors designed for industrial and military applications demonstrates high detectivity at the level $(1 \div 5) \times 10^{10}$ cm Hz$^{1/2}$ W^{-1} in the 2–5 µm wavelength region without liquid nitrogen cooling (see Fig. 2.10b).

J19 Series PV HgCdTe detectors are high-quality photodiodes for use in the 500 nm to 2.8 µm and 500 nm to 5.0 µm spectral ranges (see Fig. 2.11a, b). Unlike the photoconductors commonly used in the 500 nm to 5.0 µm region, HgCdTe photodiodes operate in the photovoltaic mode and do not require a bias current for operation. According to the manufacturer this makes J19 detectors the better choice

Table 2.5 HgCdTe photovoltaic detectors produced by Teledyne Judson Technologies

HgCdTe-based PV detector series J19	Active size diameter, mm	Shunt impedance, Ohm	Dark current@ −0.1 V, A	Peak D*@ 1 KHz, cm Hz$^{1/2}$ W^{-1}
J19:2.8-18C (*↓)	0.25	1.5×10^4	2.0×10^{-6}	2.8×10^{10}
22 °C	1.00	1.5×10^3	2.0×10^{-5}	3.5×10^{10}
J19TE1:2.8-66C	0.25	2.0×10^5	1.0×10^{-7}	1.1×10^{11}
−20 °C	1.00	2.0×10^4	1.0×10^{-6}	1.4×10^{11}
J19TE2:2.8-66C	0.25	1.5×10^6	2.0×10^{-8}	2.9×10^{11}
−40 °C	1.00	1.5×10^5	2.0×10^{-7}	3.7×10^{11}
J19TE3:2.8-66C	0.25	8.0×10^6	5.0×10^{-9}	5.9×10^{11}
−65 °C	1.00	8.0×10^5	5.0×10^{-8}	7.4×10^{11}
J19TE4:2.8-3CN	0.25	3.2×10^7	3.0×10^{-9}	8.0×10^{11}
−85 °C	1.00	3.2×10^6	3.0×10^{-8}	1.0×10^{12}
J19TE4:2.8-3VN	0.25	6.4×10^7	2.0×10^{-9}	8.6×10^{11}
−90 °C	1.00	6.4×10^6	2.0×10^{-8}	1.1×10^{12}

* Series detectors are high-quality HgCdTe photodiodes for use in the 500 nm to 2.8 µm spectral range. Data for detectors with active size 1 mm; 50% Cutoff Wavelength is 2.8 µm; Peak Wavelength is 2.6 µm; Peak Responsivity is 1.3 A/W. FOV is 180°, 60° FOV at −90 °C. Maximum reverse bias voltage for all detectors is 0.2 V.

J19TE1:5-66C (**↓)	0.25	4.0×10^2	5.0×10^{-5}	5.6×10^9
−20 °C	1.00	4.0×10^1	5.0×10^{-4}	4.7×10^9
J19TE2:5-66C	0.25	1.0×10^3	2.0×10^{-5}	1.1×10^{10}
−40 °C	1.00	1.0×10^2	2.0×10^{-4}	1.0×10^{10}
J19TE3:5-66C	0.25	3.2×10^3	6.0×10^{-5}	2.2×10^{10}
−65 °C	1.00	3.2×10^2	6.0×10^{-4}	2.4×10^{10}
J19TE4:5-3CN	0.25	7.2×10^3	3.0×10^{-6}	3.8×10^{10}
−80 °C	1.00	7.2×10^2	3.0×10^{-5}	4.6×10^{10}
J19TE4:5-3VN	0.25	1.2×10^4	2.0×10^{-6}	5.4×10^{10}
−90 °C	1.00	1.2×10^3	2.0×10^{-5}	6.8×10^{10}

** Series detectors are high-quality HgCdTe photodiodes for use in the 500 nm to 5.0 µm spectral range. Data for detectors with active size 1 mm; 50% Cutoff Wavelength is 5.0 µm; Peak Wavelength is 4.5 µm; Peak Responsivity is 1.0–2.2 A/W. FOV is 180°, 45° FOV at −90 °C. Maximum reverse bias voltage for all detectors is 0.5 V.

Source: Data extracted from www.teledynejudson.com

for DC and low-frequency applications, as it does not exhibit the low frequency or $1/f$ noise characteristic of the PbS, PbSe and HgCdTe photoconductors. The J19TE detectors are mounted on thermoelectric coolers (TEC) where one-stage (TE1) is used for −20 °C operation and subsequent stages (TE2, TE3, TE4) are used to achieve lower temperatures (see Table 2.5). Cooling an HgCdTe photodiode reduces noise and improves detectivity. Cooling also increases shunt resistance HgCdTe photodiodes and improve their response at longer wavelengths with a reduction in temperature. J19TE detectors offer superior pulse response for applications in monitoring and detecting high-speed pulsed lasers.

Another company, VIGO System S.A., has developed a unique technology for manufacturing instruments for quick and convenient detection of 1–16 μm infrared radiation. IR detectors operate in ambient temperature or are cooled with simple and inexpensive thermoelectric coolers (see Table 2.6). PC series features uncooled IR

Table 2.6 HgCdTe photoconductive detectors produced by VIGO System S.A.

HgCdTe-based PC series*/PCI series** 20 °C	Active size (square), mm²	Resistance R, Ω	Peak wavelength, μm	Peak D*, cm Hz$^{1/2}$ W^{-1}	Time constant, t, nsec
PC-5/PCI-5	From	≤1200	5.0	≥1.5 × 10⁹/6.0 × 10⁹	≤5000
PC-6/PCI-6	0.05 × 0.05	≤600	6.0	≥7.0 × 10⁸/2.5 × 10⁹	≤500
PC-9/PCI-9	to 4 × 4	≤300	9.0	≥1.0 × 10⁸/5 × 10⁸	≤10
PC-10.6/PCI-10.6		≤120	10.6	≥1.9 × 10⁷/1.0 × 10⁸	≤3

*1.0–12.0 μm HgCdTe ambient temperature photoconductive detectors.
**1.0–12.0 μm HgCdTe ambient temperature, optically immersed photoconductive detectors

Thermoelectrically cooled HgCdTe-based PC detector	Active size (square), mm	Resistance R, Ω	Peak wavelength, μm	Peak D*, cm Hz$^{1/2}$ W^{-1}	Time constant, t, nsec
PC-2TE series* 230 °K	From	≤1200	5.0	≥2.0 × 10¹⁰	≤20,000
	0.05 × 0.05	≤800	6.0	≥6.0 × 10⁹	≤4000
	to 2 × 2	≤400	9.0	≥9.0 × 10⁸	≤40
		≤300	10.6	≥4.0 × 10⁸	≤10
		≤200	12.0	≥1.0 × 10⁸	≤3
		≤150	13.0	≥4.0 × 10⁷	≤2
PC-3TE series ** 210 °K	From	≤400	9.0	≥1.5 × 10⁹	≤60
	0.05 × 0.05	≤300	10.6	≥4.5 × 10⁸	≤20
	to 2 × 2	≤300	12.0	≥1.8 × 10⁸	≤5
		≤300	13.0	≥1.2 × 10⁸	≤4
PC-4TE series *** 195 °K	From	≤500	9.0	≥1.9 × 10⁹	≤80
	0.05 × 0.05	≤400	10.6	≥5.0 × 10⁸	≤30
	to 2 × 2	≤400	12.0	≥4.0 × 10⁸	≤7
		≤400	13.0	≥2.0 × 10⁸	≤6
		≤300	14.0	≥1.0 × 10⁸	≤5

*1.0–14.0 μm HgCdTe two-stage thermoelectrically cooled PC detectors
**1.0–15.0 μm HgCdTe three-stage thermoelectrically cooled PC detectors
***1–16 μm HgCdTe four-stage thermoelectrically cooled PC detectors

Source: Data extracted from https://vigo.com.pl/

PC detectors based on sophisticated HgCdTe heterostructures for the best performance and stability. The devices should operate in optimum bias voltage and current readout mode. Performance at low frequencies is reduced due to $1/f$ noise. The $1/f$ noise corner frequency increases with the cut-off wavelength.

By 2003, VIGO had been manufacturing detectors with the use of Isothermal Vapour Phase Epitaxy (ISOVPE) of HgCdTe. It did not require costly devices, but on the other hand it did not enable fabrication of complicated semiconductor structures, like ones in which the band gap could be increased or decreased in consecutive layers of a heterostructure. The possibility of free spatial shaping of the band gap and of doping level adjustment was ensured by a Metal Organic Chemical Vapour Deposition (MOCVD) reactor commissioned in 2003.

Optical immersion technique was used to improve parameters of VIGO System detectors (PCI series, see Table 2.6). Optical immersion means the use of a certain type of lens, which is an integral part of an IR detector. The lens enables collecting an amount of optical radiation falling on the device larger than the one that could be collected only due to the physical area of the device. As a result, the detector picks up more usable signal – as much as a larger device – while retaining a smaller area. A detector with an immersion lens is best suited for operation with low power of optical signal, which means the applications where the highest detectivity of the detector is required. The use of optical immersion enables improving the signal-to-noise ratio almost 11 times without any need for additional customer's interference with the measurement system. The degree of detector parameters improvement depends on the material from which the immersion lens is made. In the VIGO System detectors comprising a GaAs substrate and an integrated immersion lens made of the same material, the refractive index of the lens is equal to 3.3. That means the detectivity is improved 3.3 times in a detector with a hemispherical lens, and nearly 11 times in a detector with a hyperhemispherical lens.

Figure 2.12 shows the performance of the PC HgCdTe devices produced by VIGO System S.A. Without optical immersion MWIR photovoltaic detectors are sub−BLIP devices with performance close to the generation−recombination limit, but well−designed optically immersed devices approach BLIP limit when thermoelectrically cooled with 2−stage Peltier coolers. Situation is less favorable for >8−μm LWIR photovoltaic detectors; they show detectivities below the BLIP limit by an order of magnitude. Typically, the devices are used at zero bias.

HgCdTe ambient temperature PV detectors with anti-fringing technology (PV-5-AF series) applied for gas detection, monitoring and analysis (CO, HF, NH_3, C_2H_2, CH_4, C_2H_6, HCl, H_2CO, SO_2, CO_2, N_2O) are presented in the Table 2.7. To make these detectors immune to unwanted optical fringing effects, VIGO developed anti-fringing technology (internal modification of substrate's surface) and successfully applied it. Characteristics of the uncooled IR PV multiple junction detectors (PVM series) designed for the maximum performance at 10.6 μm and applied for the laser power monitoring are also shown in the Table 2.7.

Figure 2.13 shows peak detectivity D* of IR PV HgCdTe-based detectors produced by VIGO System S.A. without (1) and with (2) optical immersion. Reverse bias may significantly increase response speed and dynamic range. It also results in

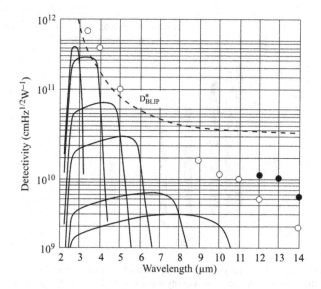

Fig. 2.12 Typical spectral detectivity of HgCdTe detectors with 2−stage TE coolers (solid lines). The best experimental data (white dots) are measured for detectors with FOV equal 36°. BLIP detectivity is calculated for FOV = 2π. Black dots are measured for detectors with 4−stage TE cooler. (Reprinted from Ref. [29]. Published 2013 by Polish Academy of Sciences (PAN) as open access)

Table 2.7 HgCdTe ambient temperature and minor cooling photovoltaic detectors produced by VIGO System S.A.

HgCdTe-based PV detectors 20 °C	Active size (square), mm^2	Resistance R, Ω	Peak wavelength, μm	Peak D*, cm Hz$^{1/2}$ W^{-1}	Time constant, t, nsec
PV-5-AF1 × 1-TO39-NW-90	1 × 1	~8	4.4 ± 0.2	~1.45 × 10^9	≤570
PV-5-AF0.1 × 0.1-TO39-NW-90	0.1 × 0.1	~265		~3.55 × 10^9	≤177
PVM-10.6-1 × 1-TO39-NW-90	1 × 1	≥30	8.5 ± 1.5	≥2.0 × 10^7	≤1.5
PVM-2TE-10.6-1 × 1-TO8-wZnSeAR-70 (230 °K)		≥90	8.5 ± 2.0	≥2.0 × 10^8	≤4

Source: Data extracted from https://vigo.com.pl/

improved performance at high frequencies, but 1/f noise that appears in biased devices may reduce performance at low frequencies. The value of the detectivity increases both with low cooling and in optically immersed PV devices. Moreover, the combination of these methods allowed VIGO System S.A to achieve the peak parameters which are comparable with detectivity of the PV devices obtained by Teledyne Judson Technologies (see Table 2.5 and Fig. 2.11).

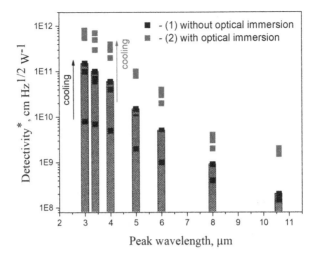

Fig. 2.13 Peak detectivity of IR photovoltaic detectors based on sophisticated HgCdTe hetero-structures for the best performance and stability without (1) and with (2) optical immersion. The arrow indicates the direction of the operating temperature decreasing of the detector – from 20 °C (lower D^* value) to 195 °K (four-stage thermoelectric cooler). (Data extracted from https://vigo. com.pl/)

References

1. Akhavan ND, Jolley G, Umana-Membreno GA, Antoszewski J, Faraone L (2015) Theoretical study of midwave infrared HgCdTe nBn detectors operating at elevated temperatures. J Electron Mater 44:3044–3055
2. Ameurlaine J, Rousseau A, Nguyen-Duy T, Triboulet R (1988) (HgZn)Te infrared photovoltaic detectors. Proc SPIE 0929:14–20
3. Ashley T, Elliott CT (1985) Non-equilibrium devices for infrared detection. Electron Lett 21(10):451–452
4. Ashley T, Elliott CT, Harker AT (1986) Non-equilibrium modes of operation for infrared detectors. Infrared Phys 26(5):303–315
5. Bajaj J (2000) State-of-the-art HgCdTe infrared devices. Proc SPIE 3948:42–54
6. Becla P (1986) Infrared photovoltaic detectors utilizing $Hg_{1-x}Mn_xTe$ and $Hg_{1-x-y}Cd_xMn_yTe$ alloys. J Vac Sci Technol A 4(4):2014–2018
7. Destefanis G, Chamonal JP (1993) Large improvement in HgCdTe photovoltaic detector performance at LETI. J Electron Mater 22:1027
8. Donati S (2021) Photodetectors: devices, circuits and applications, 2nd edn. Wiley-IEEE Press, New York
9. Djuric Z, Piotrowski J (1991) Dember IR photodetectors. Solid State Electron 34:265–269
10. Ganguly S, Tonni FF, Ahmed SZ, Ghuman P, Babu S, Dhar NK et al (2021) Dissipative quantum transport study of a bi-layer graphene-CdTe-HgCdTe heterostructure for MWIR photodetector. In: 2021 IEEE research and applications of photonics in defense conference (RAPID). IEEE, pp 1–2
11. Gaziyev FN, Nasibov IA, Ibragimov TI, Huseynov EK (2005) HgCdTe based PEM detector for middle range of IR spectrum. Proc SPIE 5834:123–132
12. Gravrand O, Boulard F, Ferron A et al (2015) A new nBn IR detection concept using HgCdTe material. J Electron Mater 44:3069–3075

13. Grein CH, Young PM, Flatté ME, Ehrenreich H (1995) Long wavelength InAs/InGaSb infrared detectors: optimization of carrier lifetimes. J Appl Phys 78(12):7143–7152
14. Hooge FN (1969) 1/f noise is no surface effect. Phys Lett 29A:123–140
15. Jones RC (1952) Performance of detectors for visible and infrared radiation. In: Morton L (ed) Advances in electronics, vol 5. Academic, New York, pp 27–30
16. Jones RC (1959) Phenomenological description of the response and detecting ability of radiation detectors. Proc IRE 47:1495–1502
17. Jung HS, Grein CH, Becker CR (2003) Superlattices for very long wavelength infrared detectors. In: Materials for infrared detectors III, vol 5209. International Society for Optics and Photonics, pp 90–98
18. Kalinowski P, Mikołajczyk J, Piotrowski A, Piotrowski J (2019) Recent advances in manufacturing of miniaturized uncooled IR detection modules. Semicond Sci Technol 34:033002
19. Kasap S, Willoughby A, Capper P, Garland J (2011) Mercury cadmium telluride: growth, properties and applications. Wiley
20. Kinch MA (1981) Metal-insulator-semiconductor infrared detectors. Semicond Semimet 18:313–378
21. Kinch MA (2007) Fundamentals of infrared detector materials, vol 76. SPIE Press, Bellingham/Washington, DC
22. Kopytko M, Jóźwikowski K, Martyniuk P, Rogalski A (2019) Photon recycling effect in small pixel pin HgCdTe long wavelength infrared photodiodes. Infrared Phys Technol 97:38–42
23. Kopytko M, Martyniuk P, Madejczyk P, Jóźwikowski K, Rutkowski J (2018) High frequency response of LWIR HgCdTe photodiodes operated under zero-bias mode. Opt Quant Electron 50(2):1–12
24. Lawson WD, Nielson S, Putley EH, Young AS (1959) Preparation and properties of HgTe and mixed crystals of HgTe-CdTe. J Phys Chem Solids 9:325–329
25. Lei W, Antoszewski J, Faraone L (2015) Progress, challenges, and opportunities for HgCdTe infrared materials and detectors. Appl Phys Rev 2(4):041303
26. Long D (1980) Photovoltaic and photoconductive infrared detectors. In: Keys RJ (ed) Optical and infrared detectors, Part of the book series "topics in applied physics", vol 19. Springer, Berlin/New York, pp 101–147
27. Maille JHP, Salaville A (1979) Semiconductor Devices. U.S. Patent 4,132,999, January 2
28. Maimon S, Wicks GW (2006) nBn detector, an infrared detector with reduced dark current and higher operating temperature. Appl Phys Lett 89:151109
29. Martyniuk P, Rogalski A (2013) HOT infrared photodetectors. Opto-Electron Rev 21(2):240–258
30. Meena VS, Mehata MS (2021) Investigation of grown ZnS film on HgCdTe substrate for passivation of infrared photodetector. Thin Solid Films 731:138751
31. Paschotta R (2008) Photoconductive detectors. In: Encyclopedia of laser physics and technology, 1st edn. Wiley-VCH. https://www.rp-photonics.com/photoconductive_detectors.html
32. Patten EA, Kalisher MH, Chapman GR, Fulton JM, Huang CY, Norton PR et al (1991) HgZnTe for very long wavelength infrared applications. J Vac Sci Technol B 9(3):1746–1751
33. Piotrowski J, Rogalski A (2004) Uncooled long wavelength infrared photon detectors. Infrared Phys Technol 46:115–131
34. Piotrowski J, Rogalski A (2007) High-operating temperature infrared photodetectors. SPIE Press, Bellingham
35. Piotrowski J, Adamiec K, Maciak A, Nowak Z (1989). ZnHgTe as a material for ambient temperature 10.6 μm photodetectors. Appl Phys Lett 54(2):143–144
36. Piotrowski J, Pawluczyk J, Piotrowski A, Gawron W, Romanis M, Kłos K (2010) Uncooled MWIR and LWIR photodetectors in Poland. Opto-Electron Rev 18:318–327
37. Reine MB (2001) HgCdTe photodiodes for IR detection: a review. Proc SPIE 4288:266–277
38. Riley KJ, Lockwood AH (1980) HgCdTe hybrid focal-plane arrays. Proc SPIE 217:206
39. Rogalski A (1992) New ternaiy alloy systems for infrared detectors Proc SPIE 1845:52–60

40. Rogalski A (2005) HgCdTe infrared detector material: history, status and outlook. Rep Prog Phys 68:2267–2336. https://doi.org/10.1088/0034-4885/68/10/R01
41. Rogalski A (2010) Infrared detectors, 2nd edn. CRC Press, Boca Raton, p 898
42. Rogalski A (2011) Recent progress in infrared detector technologies. Infrared Phys Technol 54(3):136–154
43. Rogalski A (2019) Infrared and terahertz detectors, 3nd edn. CRC Press, Boca Raton, p 1044
44. Rogalski A, Adamiec K, Rutkowski J (2000) Narrow-gap semiconductor photodiodes. Vol. 77. SPIE Press: Chap. 8 HgZnTe and HgMnTe Photodiodes, pp. 337–360
45. Rogalski A, Martyniuk P, Kopytko M, Hu W (2021) Trends in performance limits of the HOT infrared photodetectors. Appl Sci 11:501
46. Shin SH, Pasko JG, Lo DS, Tennant WE, Anderson JR, Gorsk M et al (1986) $Hg_{1-x-y}Mn_xCd_yTe$ alloys for $1.3-1.8$ µm photodiode applications. MRS Online Proc Library 89:267–274
47. Tredwell TJ, Long D (1977) Detection of long wavelength infrared moderate temperatures. Final report, NASA Lyndon B. Johnson Space Center Contract NAS9-14180, 5s (NASA Accession No. N78-13876)
48. Triboulet R, Lasbley A, Toulouse B, Granger R (1986) Growth and characterization of bulk HgZnTe crystals. J Cryst Growth 79(1–3):695–700
49. Van der Ziel A (1959) Fluctuation phenomena in semiconductors. Butterworths, London
50. Velicu S, Grein CH, Emelie PY, Itsuno A, Phillips JD, Wijewarnasuriya P (2010) Non-cryogenic operation of HgCdTe infrared detectors. Quant Sens Nanophotonic Devices VII 7608:760820
51. Wang X, He K, Chen X, Li Y, Lin C, Zhang Q et al (2020) Effect of annealing on the electro-physical properties of CdTe/HgCdTe passivation interface by the capacitance–voltage characteristics of the metal–insulator–semiconductor structures. AIP Adv 10(10):105102
52. Wehner JGA, Smith EPG, Venzor GM, Smith KD, Ramirez AM, Kolasa BP et al (2011) HgCdTe photon trapping structure for broadband mid-wavelength infrared absorption. J Electron Mater 40(8):1840–1846
53. You C, Deng W, Liu M, Zhou P, An B, Wang B et al (2021) Design and performance study of hybrid graphene/HgCdTe mid-infrared photodetector. IEEE Sensors J 21(23):26708–26715
54. Zhang P, Ye ZH, Sun CH, Chen YY, Zhang TN, Chen X et al (2016) Passivation effect of atomic layer deposition of Al_2O_3 film on HgCdTe infrared detectors. J Electron Mater 45(9):4716–4720

Chapter 3
II–VI Compound Semiconductor Avalanche Photodiodes for the Infrared Spectral Region: Opportunities and Challenges

K. -W. A. Chee

3.1 Introduction

Avalanche photodiodes (APDs) are highly sensitive semiconductor devices that harness the photoelectric effect to convert optical energy into electrical energy. Inherently, the APDs deliver a high current gain, and hence high sensitivity, upon the application of a reverse bias voltage, and they enjoy a distinctly higher signal-to-noise ratio (SNR) compared to that of conventional PIN photodiodes, as well as a short response time, and low dark current. Nevertheless, APDs invariably require a higher operating voltage. The primordial phenomenon for the internal gain is called impact ionization, which increases the photocurrent flow in response to an incident optical power. Under a sufficiently strong electric field, the highly energetic charge carriers generated by impact ionization may trigger cascades of ionization events akin to avalanching. Figure 3.1 clarifies such a process of avalanche multiplication via impact ionization. The electrons and holes will drift in opposite directions in the electric field. Further, the avalanche gain of the APD will undesirably amplify the dark current and the shot noise due to the electrical junction(s); in fact, the stochastic nature of the APD gain is liable for the excess noise to the photocurrent output. Hence, compared to the classical PIN photodiode, the APDs tend to generate a higher level of noise; the noise current is generally amplified by a factor that is at least comparable to the avalanche gain, thereby affecting the SNR.

As demonstrated by McIntyre [2], the total spectral shot noise current, $<i_{shot}>$ can be analytically described depending on the mean avalanche gain as follows:

K. undefined.-W. A. Chee (✉)
School of Electronic and Electrical Engineering, Kyungpook National University, Daegu, Republic of Korea

School of Electronics Engineering, College of IT Engineering, Kyungpook National University, Daegu, Republic of Korea
e-mail: kwac2@cantab.net

© The Author(s), under exclusive license to Springer Nature Switzerland AG 2023
G. Korotcenkov (ed.), *Handbook of II-VI Semiconductor-Based Sensors and Radiation Detectors*, https://doi.org/10.1007/978-3-031-20510-1_3

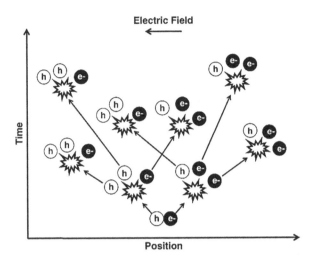

Fig. 3.1 Symbolic process of three-particle generation via impact ionization and avalanching through consecutive ionization events. (Reprinted with permission from Ref. [1]. Copyright 2020: Elsevier B.V)

$$\left\langle i_{shot}^{2} \right\rangle = 2q\left[I_{dark,s} + <M>^{2} \left(I_{photo} + I_{dark,b} \right) F(M) \right] \Delta f \qquad (3.1)$$

through $F(M) = kM + \left(2 - \dfrac{1}{M} \right)(1-k)$ or $F(M) = \dfrac{M}{k} + \left(2 - \dfrac{1}{M} \right)\left(1 - \dfrac{1}{k} \right)$ in the

pure electron or hole injection limit, respectively; the main noise source is assignable to the statistical Poissonian fluctuations of the impact ionization, which is a random avalanche process that engenders an excess noise factor, $F(M)$, where M is the current gain and k is the ratio of the hole to electron ionization coefficients (β/α). I_{photo}, $I_{dark,b}$ and $I_{dark,s}$ are the photocurrent, bulk leakage current and surface leakage current, respectively, and Δf is the system bandwidth. Since impact ionization occurs for hot carriers under the influence of an adequately strong electric field, the coefficients α and β are governed by the scattering processes due to the electrons and holes, and the electric field. Under high-field conditions, the mobile charge carriers gain kinetic energy at a faster rate. Hence, α and β increase with the electric field, so that k tends to unity.

Another performance consideration for APDs is their bandwidth. Emmons [3] showed that the bandwidth is determined by the transit time of the charge carriers across the avalanche region to attain the high gain. The higher the required gain, the lower the bandwidth. Hence, a trade-off relationship in conflicting design requirements arises although this limitation can be mitigated by a lower k-value. When the k-value is lowered, both the excess noise and bandwidth performance can be enhanced. Analogously, the ionization coefficients of the electrons and holes in the multiplication region are major determinants of the excess noise and gain-bandwidth product. Infrared (IR) APDs fabricated in elemental group IV semiconductors such as Si or Ge, and certain III–V compound semiconductors, such as InGaAs, nP,

InAlAs, AlGaAs, or AlGaAsP, exhibit high excess noise at high gain because of significant initiation by both carrier types in the avalanche process proceeding from the high scattering rates under a high reverse bias. Indeed, Si APDs generally harbor lower levels of noise than several of the III–V counterparts [4]; for example, $F(M)$ is between 2 to 3, and 4 to 5, for Si and III–Vs, respectively. In terms of the optimal SNR, since the optimum gain depends partially on $F(M)$, the former characteristically ranges from 50 to 1000 for Si APDs but plummets to 10 through 40 for Ge and InGaAs APDs. Early efforts in HgCdTe APD development, which featured hole-dominated avalanching, *i.e.*, hole-injected APDs (or h-APDs), centred on applications in short-wave IR (SWIR) optical fiber communications [5–9]. Nevertheless, the linear-mode APD array has its genesis in the elucidation of the physics of avalanche multiplication in HgCdTe toward noiseless, extraordinarily high gain. Since the seminal report by Kinch et al. in 2004 on HgCdTe electron-initiated APDs (e-APDs) [10] – avalanching is dominated by one carrier type, *i.e.*, electrons – coupled with step change advances in the control over the heteroepitaxial growth of multilayer structures, considerable interest has been roused in exploiting virtually noise-free gain exceeding 1,000 at room-temperature operation not only in the SWIR band, but in the mid-wave IR (MWIR) and long-wave IR (LWIR) bands.

Only matched by InAs or AlInAsSb, HgCdTe is the only compound semiconductor so far reported that is being exploited with attractive avalanche gain characteristics that are overwhelmingly dominated by electrons, attainable at the appropriate alloy composition [11]. Compared with other materials—including Si, Ge, InGaAsP/InP, and GaSb/AlAsSb [12, 13]—HgCdTe can bestow low excess noise, short response time, and high intrinsic avalanche gain in SWIR, MWIR, and LWIR applications [8, 14–17]. Therefore, HgCdTe APDs are suited for IR wave front sensing, advanced laser rangefinders, light detection and ranging (LiDAR), astronomy, gated viewing in SWIR, 3D-imaging radars, long-haul optical fiber communications, photon counting, *etc.* [18–23]. Although molecular beam epitaxy (MBE)-grown lattice-matched InAlAs/InGaAs heterostructures have led to an impressive gain maximum of 2,000 (at 235 K), their optimal noise performance constrains the gain to sub-20 with an $F(M)$ value of 2.3 at room temperature [24]. Instead, the liquid-phase epitaxy (LPE)-grown HgCdTe e-APDs exhibited a virtually noise-free gain of up to 150 (at 196 K), thus fulfilling the noise and gain performance specifications for thresholded linear-mode photon counting that would otherwise necessitate multi-stage upscaling of the InAlAs/InGaAs APD to satisfy the equivalent performance standards [25]. Given that the MBE-grown HgCdTe IR APDs have already established a market presence underscored by their assuring reputation for ultralow dark current levels, there is enormous interest in increasing the sensitivity and readout rates of focal plane arrays (FPAs) in third-generation (3G) technology whereby the high avalanche gain means reduction in the exposure times of cameras. The present chapter surveys the principles, design, fabrication, engineering, and technologies, as well as the opportunities and challenges related to II–VI compound semiconductor APDs for the IR spectral region.

3.2   Background and Roadmap

Several materials and architectures have been explored for adaptation in photodetector technologies [26]. Figure 3.2 provides the technology roadmap in the exploitation of IR materials over the previous decades [27]. Modern IR detector technology progressed rapidly post-World War 2. The first practical IR detectors used PbS, providing sensitivity up to ~3µm. Meanwhile, high-performance PbSe and PbTe detectors were also developed. Particularly, since the Bell Labs demonstration of the first operating transistor in 1947, InSb, HgCdTe, and Si modern IR detectors were designed. Moreover, PbSe, PbTe, and InSb detectors extended the spectral sensitivity to 3–5 µm, of the MWIR regime. The direct bandgap ternary compound semiconductors, HgCdTe and PbSnTe led to the first advanced progress of photodetector technologies. Nevertheless, the greater dielectric constant of PbSnTe limited its high-frequency performance, and the high thermal expansion coefficient constrained its integration in CMOS processes.

The first-generation systems included Si and HgCdTe technologies, which were based on individual linear elements. These early designs were without multiplexing functions embedded in the focal plane. The combination of MBE and advanced photolithographic processes became the cornerstone of the semiconductor industry that enabled rapid FPA development. Charge coupled device (CCD) imagers were launched in the 1970s by AT&T Bell Labs., which subsequently ushered in a new era of second-generation FPAs integrated with focal plane readout integrated

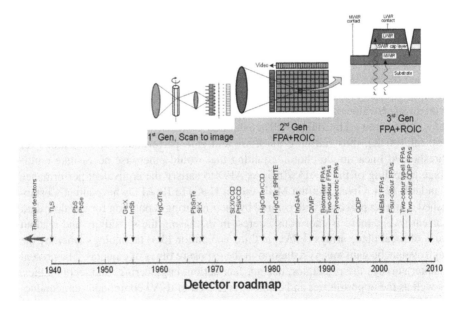

Fig. 3.2 Infrared detector and systems roadmap. (Reprinted with permission from Ref. [28]. Copyright 2009: AIP Publishing)

circuits (ROICs). The ROICs multiplex the signals, enact signal processing, and read out the array. The second-generation systems adopted monolithic and hybrid FPA systems incorporating multiplexing ROICs; signal multiplexing could be performed from a very large photodetector array. At present, the 3G FPAs, which comprise several orders of magnitude more pixels and superior on-chip properties compared to that of the second generation, are under active development.

The PIN photodiodes have been very successfully exploited in high-energy particle physics experiments [29–32]. Nevertheless, despite the use of a state-of-the-art amplifier necessary because of the lack of an internal gain, the noise is significant, composed of several hundred electron fluctuations. The APDs retain an exquisite internal gain that enables a high SNR, but $F(M)$ limits the optimal current gain. Since the beginning of the present millennium, the Geiger-mode APD was developed, allowing detection of individual photons. This high-sensitivity detector technology has evolved since the first Si single-photon detectors, as depicted in Fig. 3.3, namely, the planar APD (Fig. 3.3a) developed by RCA company [33], and the reach-through APD (Fig. 3.3b) of Shockley Semiconductor Laboratory [34], in the 1960s.

3.3 Alloy Composition and Technology

The addition of elemental Cd opens the bandgap of the Hg-based binary semimetal HgTe to operate in the IR wavelength range. Hence, the II–VI ternary compound, HgCdTe, is a highly versatile material system that has become an industry standard in the fabrication of IR detectors for applications encompassing the spectral region from 2.2 to 9.7 μm. By carefully adjusting the Hg:Cd fraction, the spectral range connoting wavelengths from 1 to 30μm [27, 36] can be controlled via the tunable bandgap between 0 and 1.5 eV of the $Hg_{1-x}Cd_xTe$ alloy to match the atmospheric windows of the solar and IR spectra for passive thermal imaging and sensing. For

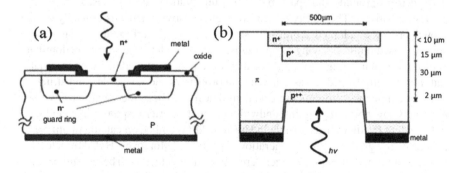

Fig. 3.3 The earliest single-photon APDs. (**a**) Planar APD from RCA company and (**b**) Reach-through APD from Shockley Semiconductor Laboratory. (Reproduced with permission from Ref. [35]. Copyright 2006: Elsevier B.V)

variable alloy compositions from x = 0.1 to 0.7, the hole to electron impact ionization ratio is highly favorable for a low $F(M)$ value. Corresponding to the energy band structure, the considerable asymmetry between the effective masses of conduction band electrons and valence band heavy holes leads to a highly unbalanced ionization coefficient for the holes and electrons. The threshold impact ionization energy for the electron is marginally above the bandgap, but that for the hole surpasses twice this value. Further, the electron mobility is two orders of magnitude greater than the hole mobility in HgCdTe [37–41], implying wildly variable scattering rates of holes and electrons. A high avalanche gain can therefore be achieved alongside a low $F(M)$ value as the multiplication process is initiated by a single charge carrier type; electron injection for lower x-values (x < 0.6), $i.e.$, k < 1, or hole injection for $0.6 \leq x \leq 0.7$, $i.e.$, k $\geq$ 1 [39, 42, 49], desirable for the SWIR to LWIR [11, 43–48] or only SWIR [5–7, 9, 49–51] spectral bands, respectively. The dominant carrier multiplication process in the HgCdTe APD can be limited to one type of charge carrier, such as the focus of the majority of the work involving low Cd content, $i.e.$, x < 0.6 for e-APDs [10]. Moreover, a nearly constant exponential gain increase in the range as high as >100 has been characterized, in tandem with an approximately constant SNR [11, 52–54]. Not only the gain values are generally well above 1,000 but the bandwidth is not sensitive to the gain [39]. Even at a low reverse bias, HgCdTe e-APDs exhibit high gain and low excess noise besides a THz-scale bandwidth [55]. An $F(M)$ close to unity (1.1–1.4) is achievable in this material [56]. Conversely, h-APDs tolerate a lower gain (< 100) and a higher $F(M)$ at the equivalent reverse bias condition.

The high internal gain, high quantum efficiency and sensitivity, low excess noise, and short response time of HgCdTe APDs cater to applications in the SWIR spanning between 1 and 3 μm, the MWIR from 3.2 to 5 μm, and the LWIR from 7.5 to 14 μm [8, 14–17, 46, 52, 57]. The SWIR is harnessed in astronomy and in active imaging when illuminated by IR sources ($e.g.$, 1.55-μm Nd:YAG lasers); and the MWIR and LWIR regimes are exploited in passive thermal IR regimes- the latter usually involves imaging in conditions clouded with fog and smoke. The largest market opportunities for environmental and planetary science applications are derived from ground- and space-based instrumentation that emphasize on single-photon sensitivity. The ability to achieve high sensitivity depends crucially on the growth processes adopted. At present, several state-of-the-art thermal imaging sensors are based on InSb having 5.5-μm sensitivity. InSb detectors currently dominate cooled thermal imaging systems since they were first developed based on this material. Importantly, HgCdTe has fundamental properties and favorable band structure features conducive to producing excellent detectors. High linear gain at low bias, extremely low noise, high gain-bandwidth product, shorter response time, and gain independent bandwidth achievable in HgCdTe APDs embody a catalog of attributes suited for integration in next generation (3G) FPAs. Moreover, HgCdTe detectors can operate at much higher temperatures than InSb detectors. These features have an edge over that of the III–V counterparts, and HgCdTe-based cameras can be made more portable with high power efficiencies.

3.4 General Architecture and Operation

Figure 3.4 reveals the basic architecture of the APD, composed of a moderately doped p-absorber and a lightly doped n-avalanche region. As the reverse bias increases, the junction depletion extends further into the lightly doped *n*-region with an electric field rising to a critical level that instigates avalanche multiplication. In

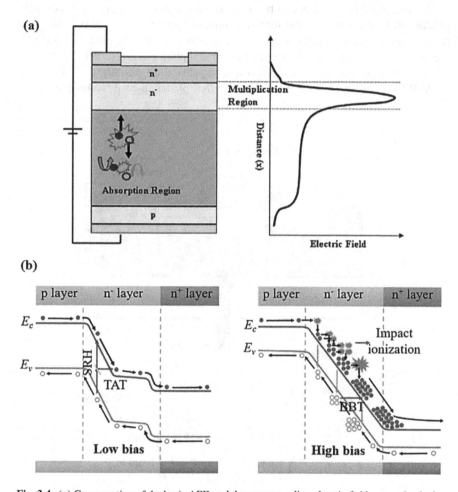

Fig. 3.4 (**a**) Cross-section of the basic APD and the corresponding electric field across the device structure under a reverse bias. The maximum electric field strength occurs in the multiplication region, which entails the lightly doped *n*-type region. (Reprinted with permission from Ref. [19]. Copyright 2011: Elsevier Ltd.). (**b**) Energy band diagram at a low (left) or high (right) reverse voltage. The Shockley-Read-Hall (SRH) and trap-assisted tunneling (TAT) mechanisms at a low reverse voltage, and the band-to-band tunneling (BBT) and avalanche mechanisms at a high reverse voltage, are depicted for the dark current generation. (Reprinted from Ref. [58]. Published 2021 by Nature Portfolio as open access under the terms of the Creative Commons Attribution 4.0 International License)

the abutting p-absorber, the electric field separates the photogenerated electron-hole pairs, with one charge carrier type drifting towards the avalanche region. The photocurrent output is amplified via impact ionization by the injected carriers in the avalanche layer. Typically, the APD is designed to maximize the minority carrier injection from the absorber layer into the multiplication region to optimize for low excess noise. The avalanche region is a crucial determinant of the avalanche gain, excess noise, and gain-bandwidth product.

Practical APD designs need to be optimized for high gain, high gain-bandwidth product, low excess noise, short response time, and high temperature stability. The two examples of architectures commonly being fabricated are the planar n-on-p and the annular n-on-p high density-vertical integrated photodiode (HDVIP). The basic cross-sectional structures of a back-illuminated planar n-on-p e-APD and the n-on-p HDVIP are exemplified in Fig. 3.5. In the planar n-on-p technology, the

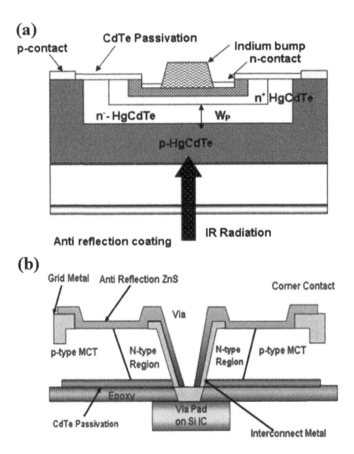

Fig. 3.5 Cross-section of the (**a**) back-illuminated planar n-on-p or (**b**) front-illuminated n-on-p HDVIP architecture. (Reprinted with permission from Ref. [19]. Copyright 2011: Elsevier Ltd.)

electron-hole pairs generated in the p-absorber upon optical irradiation will diffuse/drift into the depletion region (or avalanche region) within the n⁻-region. As the reverse bias increases, the depletion region extends toward the device surface. Meanwhile, the avalanche gain, and thus, sensitivity, increases with the electric field. Hence, the avalanche region is governed by the n-layer thickness. Conversely, in the HDVIP technology, the depletion region can extend beyond the n-region, into the p-absorber, upon increasing the reverse bias up to the critical field for the onset of avalanche multiplication. The electric field is uniform across the depletion region, leading to a homogeneous avalanche gain and dark current distribution, thereby averting an adverse impact on sensitivity [55].

The APD may be operated in the linear mode or Geiger mode. In the linear mode, the bias is applied near the breakdown voltage, so that the rate at which the charge carriers are collected at the device terminals outstrips the rate at which the electron-hole pairs are generated, thus resulting in an avalanche photocurrent decay. With a finite gain, the individual photons can be distinguished. Therefore, linear-mode APDs can be leveraged in photon counting applications. In the Geiger mode, an extraordinarily high non-linear gain is used to sense single-photon signals. The bias is applied above the breakdown voltage (pragmatically known as overbias), so that the multiplication rate dominates the collection rate. In this on-state, the photocurrent initially increases exponentially under the photo-induced breakdown, and the electrical junction becomes conductive following an optical input, so that even a single-photon input can result in strong electrical current pulses. In theory, the gain is infinite if the bias is held above the breakdown voltage. Within a certain finite period of time, the electrons and holes accumulate, respectively, at the n- and p-type edges of the depletion region, thus generating a built-in electric field opposing the externally applied field. The APD remains in the on-state until the applied bias is reduced to quench the device, turning off the electrical conduction and thwarting damage to the junction. As the applied bias is held low to quench the device, the mid-gap states release the trapped carriers. After a certain amount of elapsed time, the overbias is restored but there is a low probability of after-pulsing (breakdown due to the charge carriers released from the deep levels). Contrary to that in the linear mode, the electrical response is governed by the bias circuit instead of the optical signal power; this mode of operation is unsuitable for photon counting applications. Especially pertinent to HgCdTe APDs that display highly temperature-sensitive performance, cryogenic cooling employing liquid nitrogen and rescaled junction areas are usually indispensable to minimize the dark current in the case of the narrow bandgap property [59], and especially if there are less stringent demands on the purity of the semiconductor alloy. Geiger APDs are deployed to sense faint optical sources in applications where a short duty cycle and limited portability are tolerable.

3.5   Fabrication and Processing

It is well-known that HgCdTe APDs may encounter high fabrication costs and bottlenecks in mass production that stem from the weak Hg-Te bonding. The challenge in growing HgCdTe layers is primarily attributable to the high vapor pressure of Hg; the homogenous composition may be non-trivial to achieve in bulk crystals and epilayers. In industry, there is a systemic transition over the recent three to four decades away from the use of bulk grown HgCdTe materials toward epitaxially grown structures based on the metal-organic vapor phase epitaxy (MOVPE), LPE, and MBE techniques, to allow highly controllable lattice-matched growth of compound semiconductors [47, 60–66]. Particularly, Leonardo MW Ltd. (formerly SELEX Galileo Infrared Ltd.) [64, 67] and Vigo Systems SA adopt MOVPE-growth and CEA-Leti configures an in-house LPE or MBE system [68, 69] for the fabrication of their 3G IR FPAs. The epitaxial growth techniques shun the high-pressure processes imperative for bulk growth. Furthermore, in view of the high degree of control afforded by epitaxial growth, superior APD performance can be achieved via the enhanced crystal quality, and the uniformity of the alloy composition, doping, and layer thicknesses. The absorber, junction, and avalanche regions can be independently optimized in separate optical absorption, charge transport, and multiplication (SACM) structures. Advanced APD heterostructures can be easily realized through device localization and engineering (e.g., bandgap, quantum well, or impact ionization engineering, etc.), enabling high-sensitivity e-APDs, multispectral APD functions with high spectral resolution capabilities, and large format arrays consisting of highly sensitive pixels offering high spatial resolution. The drive toward lower manufacturing costs relies on the use of large-area growth techniques and lower-cost substrate materials.

MOVPE is a large-area growth technique that had earlier enjoyed successes in their application to a range of III–V compound semiconductor technologies [70]. For example, MOVPE-GaN light emitting diodes became commercially available by the mid-1990s. II–VI semiconductors having wide or narrow bandgaps are also grown by MOVPE [71]. The growth temperature for MOVPE is generally higher than that for MBE, but being lower than that for LPE, cannot attain the crystalline quality of the latter. Nevertheless, bandgap engineering throughout the device can be performed with much greater ease. Foundational for the MOVPE growth technique is the interdiffused multilayer process: when depositing alternating thin layers of CdTe and HgTe grown under their individually optimized conditions (not exceeding a combined thickness of 200–250 nm), the binary compounds interdiffuse to form uniform HgCdTe crystals within a relatively short amount of time owing to their high diffusion coefficients. A short annealing procedure at the growth temperature at the end of the diffusion process fully homogenizes the HgCdTe layer, yielding a composition contingent on the relative thickness of the individual CdTe and HgTe layers. Central to the flexibility and versatility of the interdiffused multilayer process is the notion that the flux of each binary compound can be individually controlled and optimized for dopant incorporation, and the total thickness of the crystalline HgCdTe layer depends on the growth rate and growth duration.

Through the interdiffused multilayer process, the heterolayer doping and thickness can be completely optimized. The intrinsic defects commonly considered in the grown HgCdTe layers include the Hg vacancies, Hg interstitials, Te antisites, and several other defect complexes. Under normal MOVPE growth conditions, the amount of metal vacancies depends on the process parameters. The substrate orientation is a dominant factor in the nucleation growth, dopant incorporation efficiency, and growth rates, because of the germane variety of step energies involved [72]. On (100) substrates, the growth rates are slowest, whereas (111)B substrates grant faster growth rates and suppress the macrodefect density. In the former case, the substrates are slightly misoriented from the (100) plane to mitigate strain-induced defect formation. In the latter, the twin boundaries present in the grown layers act like donors [73], and As [74] (I [63]) dopant incorporation is substantially stymied (enhanced); the most popularly adopted dopants in MOVPE-HgCdTe layers are As (p-type) and I (n-type). The (211)B [75] and (552)B [76] substrates have also been investigated as alternatives. For CdZnTe substrates, the (211)B orientation is favored for device-quality fabrication, by courtesy of the ease in suppressing the formation of macrodefects; the (211)B orientation also serves as the preferred growth direction for MBE. CdZnTe substrates typify the traditional option for the lattice-matched growth of HgCdTe. Even so, there are several drawbacks. The per unit area substrate cost is up to two orders of magnitude greater compared to the alternative GaAs. In the preferred <111> orientation for bulk HgCdTe growth, the uniform alloy composition control crucial in a LWIR architecture cannot be attained; Zn segregation along the growth axis leads to uneven Zn concentrations within (100) substrates. Given the higher MOVPE growth temperature (compared to that in MBE or LPE), outgassing is enhanced of residual Cu impurities from Te precipitates in the substrates that accumulate in the grown HgCdTe layer leading to fluctuations in the background doping (Cu acts as an active dopant); hence there is a lack of reproducibility control of the low n-type doping common in several device architectures.

Attractive substrates for MOVPE growth are Si and GaAs, which require the introduction of buffer layers to alleviate the lattice mismatch-induced defect generation, and autodoping effects. On account of the low-cost opportunity, large-area wafer availability, and thermal matching properties to the ROICs, Si represents an outstanding substrate candidate. However, challenges include the requisite to maintain the preferred (100) orientation as well as the necessity of an in-situ process to rid the thermal oxide. Solutions have included using Ge [77], GaAs [78], or ZnTe/CdTe [79] interfacial layers, or utilizing a high-temperature oxide desorption annealing process [80]. Also readily available in large wafer sizes (up to 150 mm diameter) and nominally more expensive than Si, GaAs has been considered as the most important substrate for low-density, device-quality MOVPE-HgCdTe [75, 81]. For example, highly affordable high-performance LWIR FPAs have been fabricated using GaAs substrates for MOVPE-grown HgCdTe materials bump bonded to ROICs [82]. The lattice mismatch encountered on GaAs substrates (14.6%) is smaller than that on Si substrates (19.3%), hence the use of GaAs substrates is a priori expected to lead to higher FPA performance by virtue of the correlated superior crystal quality obtainable for the grown HgCdTe material. Typically, the CdTe seed/buffer layer thickness is optimized for the crystalline quality of the grown HgCdTe as well as to minimize the rate of Ga out-diffusion from the substrate that may induce autodoping effects [75, 81]. In-situ annealing techniques have also been incorporated into the growth reactor procedure to remove Hg vacancies in order that the intentionally introduced dopants during growth are activated and dominate the electrical properties of the fabricated devices [83]. Alternatively, an ultrathin layer of Cd is introduced intervening the interdiffused multilayer process to generate the excess metal to counter the Hg vacancies [84].

In IR FPAs, the size and density of hillocks (macrodefects) will determine the cosmetic (defective pixel clusters) quality. To enhance the surface morphology and crystal quality of the grown HgCdTe layers, Giess et al. [85] successfully reduced the pyramidal hillock density to <10 cm^{-2} at the MOVPE-HgCdTe/GaAs (100) interface by a priori treatment of the substrate using an alkali metal containing solution, such as aqueous KOH solution [86]. The adsorbed alkali metal atoms (e.g., Na or K) on the substrate facilitates the homogeneous nucleation of the CdTe layer on GaAs [75, 86]. Suh et al. [87] reduced the surface hillock density to 90 cm^{-2} by simply occluding the Hg bath with a lid in front of the heating susceptor (see Fig. 3.6) to suppress the pre-reaction kinetics of the Cd precursor with Hg and facilitate homogeneous nucleation of HgCdTe on the GaAs (100) substrate. The ever-challenging issue of reducing the hillock density to <10^3 cm^{-2} [73, 88] by means of the foregoing approaches ensure best-in-class crystallinity of the grown HgCdTe material for 3G FPA applications. Nevertheless, the thermal mismatch between the GaAs substrate and Si ROICs remains a technological quandary to be addressed for large-area FPA manufacture. On the contrary, Si substrates are compatible with CMOS processes and so they are desirable for integration with ROICs. Temperature cycling reliability is also expected to be superior as a consequence of the excellent mechanical integrity and thermal capability.

For MBE-HgCdTe, alternative substrates are GaAs [89], Ge [90], and Si [91].

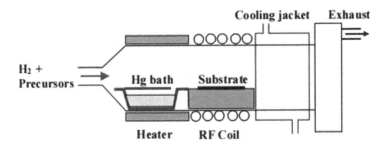

Fig. 3.6 Schematic illustrating the MOVPE reactor within which a Hg bath is occluded with a lid before the susceptor. (Reprinted with permission from Ref. [87]. Copyright 2002: Elsevier B.V)

He et al. [92] compared the MBE-HgCdTe on GaAs (211) B and Si (211) substrates. A buffer layer of CdTe was grown on the substrate prior to the HgCdTe nucleation to alleviate the lattice mismatch. Figure 3.7a uncovers the extent of lattice tilt for epilayers grown on the (211) substrate, thereby indicating that the tilt stems intrinsically from the misfit strain and is linearly related to the relative difference between the in-plane lattice constants. To reduce the density of threading dislocations, the CdTe buffer layer was grown on the substrates misoriented toward the [111] direction by 1.0°. Figure 3.7b shows that the X-ray double-crystal rocking curve full-width at half-maximum (FWHM) value reduces with the tilt angle between the plane of the CdTe buffer epilayer and Si (211) substrate from 4.2° to 2.75°, and then becomes largely constant for smaller tilt angles; the FWHM value remains largely unchanged for the CdTe buffer epilayer grown on the GaAs (211) substrate. These results imply that the misfit strain can be mitigated using misoriented Si substrates corresponding to the angle of lattice tilt. Figure 3.8 compares the crystal quality of the CdTe buffer epilayer deposited on misoriented Si (211) with that on GaAs (211). Indeed, the FWHM values of the X-ray rocking curves or etch pit density (EPD) due to CdTe on the misoriented Si substrate became comparable with that on the GaAs substrate. Hence, the crystal quality in these two disparate cases is essentially identical. As the CdTe buffer layer thickness increases, the EPD, and thus, the threading dislocation density, decreases, owing to the strain mitigation effects.

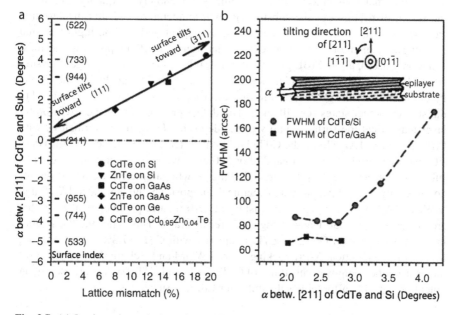

Fig. 3.7 (**a**) Lattice mismatch dependence of the tilt angle between the CdTe/ZnTe (211) and (211) substrate. (**b**) FWHM of X-ray rocking curves of CdTe on Si/GaAs as a function of tilt angle between the CdTe (211) and Si (211) substrate. (Reprinted with permission from Ref. [92]. Copyright 2007: Elsevier B.V)

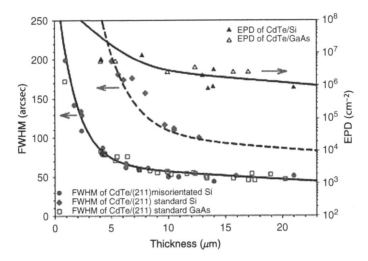

Fig. 3.8 CdTe buffer epilayer thickness dependence of etch pit density (EPD) and the FWHM of X-ray rocking curves for the various substrates. (Reprinted with permission from Ref. [92]. Copyright 2007: Elsevier B.V)

Large-area FPAs need to have excellent lateral homogeneity of the deposited active layer materials in terms of alloy composition and crystal quality. Control of the lateral distribution of the substrate growth temperature and flux dispensation from the sources with rotation mechanisms, coupled with in-line high-precision mapping and monitoring techniques (Fourier-transform IR mapping is the most commonly used), allow MBE growth of uniform HgCdTe layers on large-area substrates [93, 94]. The layer thickness can be characterized by analyzing the Fabry-Perot interference fringes. Moreover, the detector performance is directly related to the surface defect states (or hillocks). The defect propagation in the CdTe buffer layer may affect the surface quality of the deposited HgCdTe. Nevertheless, with MBE-grown 3" HgCdTe on the CdTe buffer layer, the surface defect (≥ 2 μm) density was measured to be less than 300 or 500 cm^{-2} when involving the GaAs or Si substrate, respectively [92]. Provided that the growth parameters are optimized for homogeneity of crystal quality and chemical composition, the fabricated LWIR and MWIR FPA based on 3" epilayers displayed performance parameters at 80 K that are on par with that fabricated using CdZnTe substrates. This is despite the smaller lattice mismatch with GaAs of ZnTe compared with CdTe (7.8% versus 14.6%). Moreover, the identical X-ray rocking curve FWHM and etch pit density measurements from the composite substrates of CdTe/GaAs and CdTe/Si indicate their great potential as highly affordable substitutes for CdZnTe substrates.

3.6   Device Concept Design and Engineering

There are several strategies to combat the electronic noise that blights the performance of optical transceivers incorporating IR APD technology. Low excess noise and high quantum efficiency are integral conditions for background-limited performance determined by photon statistics [95]. The multiplication layer can be fabricated in wide bandgap material to forestall, e.g., trap-assisted tunneling, to reduce the overall leakage current. Indeed, the surface leakage currents and 1/f noise conduce to the dark current [96], which can in turn be detrimental to the SNR performance of the APD. Concurrently, the absorption layer thickness can be reduced to control the thermally induced leakage current. High-carrier concentration layers can envelope the absorption layer to suppress the 1/f noise. Significantly, the contact layers can be doped heavily to reduce the sheet resistance undergirding 1/f noise. Graded bandgap structures can be fabricated in the absorption layer (bandgap engineering) to enhance the field-assisted transport of charge carriers into the multiplication region to elevate the quantum efficiency.

The temperature stability of the APD performance is an important design and technology consideration for high-reliability systems. New-generation FPAs need to operate at higher temperatures so as to benefit from (1) reduced cooler power dissipated from the Stirling cooling engine and (2) an increased operating span as well as (3) an improved portability design. Pillans et al. [97] demonstrated using high-operating-temperature (HOT) MWIR HgCdTe APDs, high-quality imaging at temperatures as high as 210 K. Compared to the other material systems utilized in the fabrication of HOT LWIR photodetectors, only the use of the ternary HgCdTe alloy is appropriate, affording a low doping concentration (10^{13} cm^{-3}) and a high Shockley-Real-Hall carrier lifetime (>1 ms) [98]. Nevertheless, LWIR HgCdTe APDs are known to exhibit a temperature-dependent dark current that can degrade the SNR performance. Consequently, the device concept of unipolar barrier layer engineering was introduced into the HgCdTe detector [99–103]. The dark current was shown to be effectively curtailed by the barrier layer engineering, thus forging new potential in HOT HgCdTe devices [104–110]. Charge carrier diffusion in the absorption region can be inhibited and the Auger noise contribution can be suppressed so that the overall dark current performance is enhanced at high temperature [111–113]. To accommodate suitable amplification properties at high reverse bias, He et al. [114] developed a pBp structure, consisting of a p-type contact layer and a graded barrier layer between the absorber and the substrate (see Fig. 3.9), which significantly lowered the dark current without penalty to the avalanche gain at high-temperature operation.

Figure 3.10 portrays the bias voltage and temperature dependence of the dark current characteristics. The dark current increases monotonically with temperature at low bias voltages. Nevertheless, a high bias voltage is required to raise the intrinsic gain. In the pBp structure, the dark current exhibits a negative temperature coefficient at high bias voltages, therefore favoring high-temperature operation without compromising the high avalanche gain.

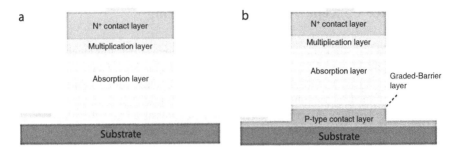

Fig. 3.9 Comparison of the (**a**) conventional APD and (**b**) pBp APD structure. (Reproduced from Ref. [114]. Published 2020 by the Optica Publishing Group as open access)

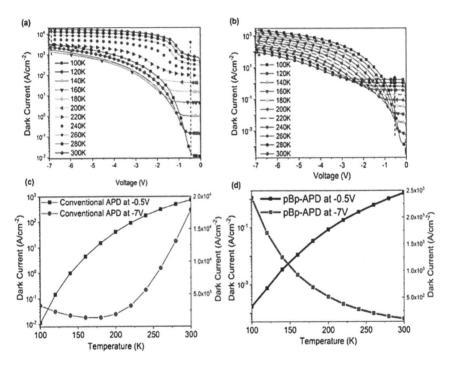

Fig. 3.10 Variation of dark current as a function of bias voltage and temperature as a parameter in the (**a**) conventional HgCdTe APD and (**b**) HgCdTe pBp APD. The dark current versus temperature characteristics at – 0.5 and –7 V are shown in the (**c**) conventional HgCdTe APD and (**d**) HgCdTe pBp APD. (Adapted from Ref. [114]. Published 2020 by the Optica Publishing Group as open access)

Figure 3.11 compares the carrier concentration distributions, and electric field/ potential profiles in the devices at a high operating bias of – 7 V. Within the multiplication layer of the pBp structure, the high electric field separates the photogenerated electron-hole pairs and promotes(suppresses) electron(hole)-initiated impact ionization. Electron depletion in the absorption layer significantly enhances the

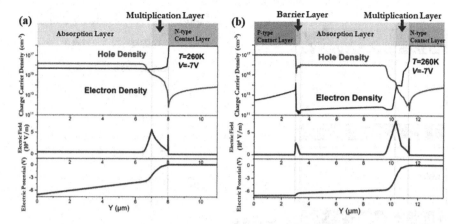

Fig. 3.11 Carrier concentration distributions, and electric field and electric potential profiles in the (**a**) conventional APD and (**b**) pBp APD structures. (Adapted from Ref. [114]. Published 2020 by the Optica Publishing Group as open access)

dark current performance, and the high operating bias has no appreciable effect on the unipolar blocking capability of the barrier layer. The barrier layer hinders the diffusion of electrons into the absorption layer yet facilitates charge injection into the multiplication layer, and therefore, the dark current is suppressed under a high reverse voltage.

Using additionally complex multi-heterojunction architectures, dual-waveband detectors (DWBs) have been developed that allow concurrent readout of optical flux from discordant IR spectral bands, *e.g.*, MWIR and LWIR [115–117]. Both wavebands are heterogeneously incorporated into a single detector, thereby providing the ability to toggle between the two as required. The cost advantage is significant compared to that of two separate image sensors. Fused imaging can be performed to delineate the features of importance via spectral variances. A dual-band IR metalorganic chemical vapor deposition (MOCVD)-structure studied by Kopytko et al. [115] is shown in Fig. 3.12 depicting that of the N^+-n-P^+-p-N^+ (capitalized letters refer to wide bandgap and the lower-case letters denote the absorption regions) structure, whereby the wide bandgap p^+-material As-doped to 5×10^{17} cm^{-3} is sandwiched between lightly doped narrow bandgap absorber materials. The abutting n- and p-layers having the appropriately matched bandgaps are the MWIR (3 to 4.2 µm) and LWIR (4.2 to 5.2 µm) absorption regions. The absorption layers are stacked vertically in a fashion permitting transmission of the IR flux through the MWIR layer (MW1) before the LWIR layer (MW2). The MW1 active area has a Cd molar ratio of 0.32 and is non-intentionally doped, with a background impurity concentration of 10^{15} cm^{-3} featuring n-type conductivity. The MW2 active layer has a Cd molar ratio of 0.28 and is As-doped to 5×10^{15} cm^{-3}. The electrical contacts are formed to the wide bandgap layers I-doped to 2×10^{17} cm^{-3}. The selected band is determined by the applied bias polarity.

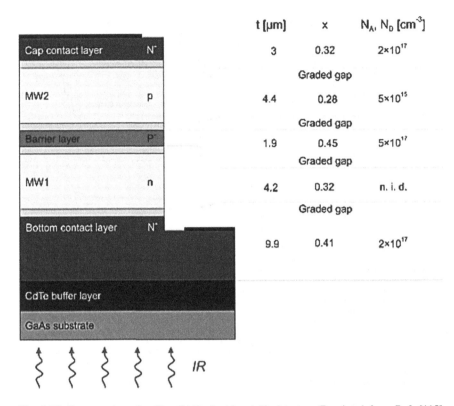

		t [μm]	x	N_A, N_D [cm^{-3}]
Cap contact layer	N$^+$	3	0.32	2×10^{17}
			Graded gap	
MW2	p	4.4	0.28	5×10^{15}
			Graded gap	
Barrier layer	P$^-$	1.9	0.45	5×10^{17}
			Graded gap	
MW1	n	4.2	0.32	n. i. d.
			Graded gap	
Bottom contact layer	N$^+$	9.9	0.41	2×10^{17}
CdTe buffer layer				
GaAs substrate				

IR

Fig. 3.12 Cross-section of a Hg$_{1-x}$Cd$_x$Te dual-band IR detector. (Reprinted from Ref. [115]. Published 2019 by Springer as open access under the terms of the Creative Commons Attribution 4.0 International (CC BY) License)

The response time of e-APDs has also been carefully studied and engineered. For excellent response time characteristics, the following conditions must be met: (1) reduced diffusion time of the minority carrier to the multiplication region, (2) minimal hole and electron transit time across the avalanche layer, and (3) small RC time constant arising from the junction capacitance and series resistance of the absorber. Perrais et al. [54, 118] reported a record-level gain of 2,800 and gain-bandwidth product of 1.1 THz from the planar MWIR HgCdTe p-APD [54, 118], and an avalanche gain of 3,500 and gain-bandwidth product of 2.1 THz from the front-illuminated planar MWIR HgCdTe e-APD [119]. Besides eliminating preferential avalanche breakdown at the junction edge and minimizing the number of bulk microplasmas [3], Li et al. [120] demonstrated that with an effective guard ring design, the dark current, *F(M)*, and mean square noise of the MWIR HgCdTe e-APD can be reduced while maintaining a high bandwidth and a high-temperature performance stability.

3.7   Concluding Remarks and Outlook

We have surveyed the highly versatile HgCdTe material system employed in industry-standard IR APDs suited for state-of-the-art photon counting or 3G FPAs. The ability to fine-tune the alloy composition of the compound semiconductor through advanced heteroepitaxial techniques possessing a very high degree of control of the thin epilayers and defect propagation has led to bandgap engineering and impact ionization localization opportunities that enable a high avalanche gain, a high temperature stability performance, low dark current, short response time, and low excess noise. The basic architecture, and fabrication and processing technologies, have been discussed. The physics and operating principles of the HgCdTe APD have been addressed, revealing innovative design approaches to attain record-level photoelectric performance in SWIR, MWIR, and LWIR applications. Notably, since Cd destabilises the weak Hg-Te bond, the use of Zn and Mn as candidate substitutes has been explored in MBE-HgZnTe and MBE-HgMnTe semiconductor alloys, respectively, with the former exhibiting Hall mobilities comparable with that of the best of MBE-HgCdTe crystals on GaAs (100) substrates [121]. While noting certain challenges in fabrication, performance and cost, insights provided in this work are anticipated to invigorate interest in the adoption and engineering of II–VI compound semiconductor device concepts for advanced photodetection capabilities.

Acknowledgements This research was supported by the Kyungpook National University Research Fund, 2021, the Kyungpook National University Excellent New Research Fund, the Dongil Culture and Scholarship Foundation for academic research in 2022, and the Brain Korea (BK) 21 FOUR project funded by the Ministry of Korea (4199990113966) and the National Research Foundation of Korea.

References

1. Huntington AS (2020) 1 – Types of avalanche photodiode. In: Huntington AS (ed) InGaAs avalanche photodiodes for ranging and lidar. Woodhead Publishing, Sawston, pp 1–92
2. McIntyre R (1966) Multiplication noise in uniform avalanche diodes. IEEE Trans Electron Dev ED-13(1):164–168
3. Emmons R (1967) Avalanche-photodiode frequency response. J Appl Phys 38(9):3705–3714
4. Marshall AR, David JP, Tan CH (2010) Impact ionization in InAs electron avalanche photodiodes. IEEE Trans Electron Dev 57(10):2631–2638
5. Shin SH, Pasko JG, Law HD, Cheung DT (1982) 1.22-μm HgCdTe/CdTe avalanche photodiodes. Appl Phys Lett 40(11):965–967
6. Alabedra R, Orsal B, Lecoy G, Pichard G, Meslage J, Fragnon P (1985) An $Hg_{0.3}Cd_{0.7}Te$ avalanche photodiode for optical-fiber transmission systems at $\lambda = 1.3$ μm. IEEE Trans Electron Dev 32(7):1302–1306
7. Nguyen Duy T, Meslage J, Pichard G (1985) CMT: the material for fiber optical communication devices. J Cryst Growth 72(1):490–495

8. de Lyon T, Baumgratz B, Chapman G, Gordon E, Hunter A, Jack M et al (1999) Epitaxial growth of HgCdTe 1.55-μm avalanche photodiodes by molecular beam epitaxy. In: Proc. SPIE 3629, 256-267. https://doi.org/10.1117/12.344562
9. Ve'rie' C, Raymond F, Besson J, Nguyen Duy T (1982) Bandgap spin-orbit splitting resonance effects in $Hg_{1-x}Cd_xTe$ alloys. J Cryst Growth 59(1):342–346
10. Kinch MA, Beck JD, Wan CF, Ma F, Campbell J (2004) HgCdTe electron avalanche photodiodes. J Electron Mater 33(6):630–639
11. Beck JD, Wan C-F, Kinch MA, Robinson J, Mitra P, Scritchfield R et al (2004) The HgCdTe electron avalanche photodiode. In: Proc. SPIE, 5564, 44-53. https://doi.org/10.1117/12.565142
12. Song H-Z (2018) Chapter 9: Avalanche photodiode focal plane arrays and their application to laser detection and ranging. In: Chee KWA (ed) Advances in photodetectors-research and applications. IntechOpen, pp 145–168
13. Dehzangi A, Li J, Razeghi M (2021) Low noise short wavelength infrared avalanche photodetector using sb-based strained layer superlattice. Photonics 8(5):148
14. Campbell JC (2021) Evolution of low-noise avalanche photodetectors. IEEE J Sel Top Quant Electron 28(2):3800911
15. Rothman J (2018) Physics and limitations of HgCdTe APDs: a review. J Electron Mater 47(10):5657–5665
16. Singh A, Pal R (2017) Infrared avalanche photodiode detectors. Defence Sci J 67(2):159
17. Reine M, Marciniec J, Wong K, Parodos T, Mullarkey J, Lamarre P et al (2008) Characterization of HgCdTe MWIR back-illuminated electron-initiated avalanche photodiodes. J Electron Mater 37(9):1376–1386
18. Pal R (2017) Infrared technologies for defence systems. Def Sci J 67(2):133–134
19. Singh A, Srivastav V, Pal R (2011) HgCdTe avalanche photodiodes: a review. Opt Laser Technol 43(7):1358–1370
20. Rothman J, Foubert K, Mollard L, Péré-Laperne N, Salvetti F, Kerlain A, Reibel Y (2014) HgCdTe avalanche photodiodes: Application for infra-red detection. In: Proceedings of 11th International Workshop on Low Temperature Electronics (WOLTE). 07-09 July 2014, Grenoble. https://doi.org/10.1109/WOLTE.2014.6881011
21. Rogalski A (2003) Infrared detectors: status and trends. Prog Quantum Electron 27(2):59–210
22. de Borniol E, Rothman J, Lefoul X (2016) Time resolved infrared detection with HgCdTe avalanche photodiodes. In: OSA Technical Digest of Lasers Congress 2016 (ASSL, LSC, LAC). 30 October–3 November 2016,Boston, LW4B.2. https://doi.org/10.1364/LSC.2016.LW4B.2
23. Rothman J, de Borniol E, Abergel J, Lasfargues G, Delacourt B, Dumas A et al (2017) HgCdTe APDs for low-photon number IR detection. In: OSA technical Digest (online) (Optica Publishing Group), paper MM8C.5. https://doi.org/10.1364/MICS.2016.MM8C.5
24. Huntington AS, Compton MA, Williams GM (2007) Linear-mode single-photon APD detectors. In: Proc. SPIE 6771, 67710Q. https://doi.org/10.1117/12.751925
25. Williams GM, Compton M, Ramirez DA, Hayat MM, Huntington AS (2013) Multi-gain-stage InGaAs avalanche photodiode with enhanced gain and reduced excess noise. IEEE J Electron Dev Soc 1(2):54–65
26. Chee KWA (2018) Introductory Chapter: Photodetectors. In: Chee KWA (ed) Advances in photodetectors – research and applications, vol 3-8. IntechOpen, London
27. Rogalski A (2011) Recent progress in infrared detector technologies. Infrared Phys Technol 54(3):136–154
28. Rogalski A, Antoszewski J, Faraone L (2009) Third-generation infrared photodetector arrays. J Appl Physics 105(9):091101
29. Simon A, Kalinka G, Jakšić M, Pastuović Ž, Novák M, Kiss ÁZ (2007) Investigation of radiation damage in a Si PIN photodiode for particle detection. Nucl Instrum Methods Phys Res, Sect B 260(1):304–308
30. Ahmad I, Betts RR, Happ T, Henderson DJ, Wolfs FLH, Wuosmaa AH (1990) Nuclear spectroscopy with Si PIN diode detectors at room temperature. Nuclear Instruments and Methods

in Physics Research Section A: Accelerators, Spectrometers, Detectors and Associated Equipment 299(1):201–204

31. Andreani L, Bontempi M, Rossi PL, Rignanese LP, Zuffa M, Baldazzi G (2014) Comparison between a silicon PIN diode and a CsI(Tl) coupled to a silicon PIN diode for dosimetric purpose in radiology. Nuclear Instruments and Methods in Physics Research Section A: Accelerators, Spectrometers, Detectors and Associated Equipment 762:11–15

32. Pantazis J, Huber A, Okun P, Squillante MR, Waer P, Entine G (1994) New, high performance nuclear spectroscopy system using Si-PIN diodes and CdTe detectors. IEEE Trans Nucl Sci 41(4):1004–1008

33. McIntyre RJ (1961) Theory of microplasma instability in silicon. J Appl Phys 32(6):983–995

34. Haitz RH (1964) Model for the electrical behavior of a microplasma. J Appl Phys 35(5):1370–1376

35. Renker D (2006) Geiger-mode avalanche photodiodes, history, properties and problems. Nuclear Instruments and Methods in Physics Research Section A: Accelerators, Spectrometers, Detectors and Associated Equipment 567(1):48–56

36. Norton P (2002) HgCdTe infrared detectors. Opto-electron Rev 10:159–174

37. Ralph SH, Neil TG, Jean G, Janet EH, Andrew G, David CH et al (2005) Photomultiplication with low excess noise factor in MWIR to optical fiber compatible wavelengths in cooled HgCdTe mesa diodes. In: Proc. SPIE 5783, 92540P. https://doi.org/10.1117/12.603386

38. Ma F, Li X, Campbell JC, Beck JD, Wan C-F, Kinch MA (2003) Monte Carlo simulations of $Hg_{0.7}Cd_{0.3}Te$ avalanche photodiodes and resonance phenomenon in the multiplication noise. Appl Physics Lett 83(4):785–787

39. Beck JD, Wan C-F, Kinch MA, Robinson JE (2001) MWIR HgCdTe avalanche photodiodes. In: Proc. SPIE 4454, 188-196. https://doi.org/10.1117/12.448174

40. Yoo SD, Kwack KD (1997) Theoretical calculation of electron mobility in HgCdTe. J Appl Phys 81(2):719–725

41. Murthy OVSN, Venkataraman V, Sharma RK, Vurgaftman I, Meyer JR (2009) Multicarrier conduction and Boltzmann transport analysis of heavy hole mobility in HgCdTe near room temperature. J Appl Physics 106(11):113708

42. Orsal B, Alabedra R, Valenza M, Pichard G, Meslage J (1985) Impact ionization rates for electrons and holes in Hg0.3Cd0.7Te in avalanche photodiodes for optical fiber transmission systems at $\lambda = 1.3$ μm. J Crystal Growth 72:496–503. https://doi.org/10.1016/0022-0248(85)90197-6

43. Han X, Guo H, Yang L, Zhu L, Yang D, Xie H et al (2022) Dark current and noise analysis for long-wavelength infrared HgCdTe avalanche photodiodes. Infrared Phys Technol 123:104108

44. Rothman J, Foubert K, Lasfargues G, Largeron C, Zayer I, Sodnik Z et al (2014) High operating temperature SWIR HgCdTe APDs for remote sensing. In: Proc. SPIE 9254, 92540P. https://doi.org/10.1117/12.2069486

45. Zhu L, Guo H, Deng Z, Yang L, Huang J, Yang D et al (2022) Temperature-Dependent Characteristics of HgCdTe Mid-Wave Infrared E-Avalanche Photodiode. IEEE J Selected Topics Quantum Electron 28(2: Optical Detectors):1–9

46. Reine MB, Marciniec JW, Wong KK, Parodos T, Mullarkey JD, Lamarre PA et al (2007) HgCdTe MWIR Back-illuminated electron-initiated avalanche photodiode arrays. J Electron Mater 36(8):1059–1067

47. Singh A, Shukla AK, Pal R (2015) HgCdTe e-avalanche photodiode detector arrays. AIP Adv 5(8):087172

48. Sieck A, Benecke M, Eich D, Oelmaier R, Wendler J, Figgemeier H (2018) Short-wave infrared HgCdTe electron avalanche photodiodes for gated viewing. J Electron Mater 47(10):5705–5714

49. Orsal B, Alabedra R, Valenza M, Lecoy GP, Meslage J, Boisrobert CY (1988) $Hg_{0.4}Cd_{0.6}Te_{1.55}$-μm avalanche photodiode noise analysis in the vicinity of resonant impact ionization connected with the spin-orbit split-off band. IEEE Trans Electron Dev 35(1):101–107

50. Orsal B, Alabedra R, Maatougui A, Flachet JC (1991) $Hg_{0.56}Cd_{0.44}Te$ 1.6- to 2.5-μm avalanche photodiode and noise study far from resonant impact ionization. IEEE Trans Electron Dev 38(8):1748–1756

51. Royer M, Brossat T, Fragnon P, Meslage J, Pichard G, Duy TNG (1983) Détecteurs HgCdTe pour télécommunication par fibres optiques. Annales des Télécommunications 38(1):62–72

52. Kerlain A, Bonnouvrier G, Rubaldo L, Decaens G, Reibel Y, Abraham P et al (2012) Performance of mid-wave infrared HgCdTe e-avalanche photodiodes. J Electron Mater 41(10):2943–2948

53. Asbrock J, Bailey S, Baley D, Boisvert J, Chapman G, Crawford G et al (2008) Ultra-High sensitivity APD based 3D LADAR sensors: linear mode photon counting LADAR camera for the Ultra-Sensitive Detector program. In: Proc. SPIE 6940, 69402O. https://doi.org/10.1117/12.783940

54. Perrais G, Gravrand O, Baylet J, Destefanis GL, Rothman J (2007) Gain and dark current characteristics of planar HgCdTe avalanche photo diodes. J Electron Mater 36:963–970

55. Rothman J, Perrais G, Destefanis G, Baylet J, Castelein P, Chamonal JP (2007) High performance characteristics in pin MW HgCdTe e-APDs. In: Proc. SPIE 6542, 654219. https://doi.org/10.1117/12.723465

56. Rothman J, Lasfargues G, Delacourt B, Dumas A, Gibert F, Bardoux A, Boutillier M (2017) HgCdTe APDS for time resolved space applications. CEAS Space J 9:507–516

57. G. Finger, I. Baker, M. Downing, D. Alvarez, D. Ives, L. Mehrgan, et al. (2017) Development of HgCdTe large format MBE arrays and noise-free high speed MOVPE EAPD arrays for ground based NIR astronomy. in Proc. SPIE 10563, 1056311. DOI: https://doi.org/10.1117/12.2304270

58. Chen J, Chen J, Li X, He J, Yang L, Wang J et al (2021) High-performance HgCdTe avalanche photodetector enabled with suppression of band-to-band tunneling effect in mid-wavelength infrared. npj Quantum Mater 6(1):103

59. Zhou T, Chee KWA (2019) Chapter 7: Overcoming the Bandwidth-Quantum Efficiency Trade-Off in Conventional Photodetectors. In: Chee KWA (ed) Advances in photodetectors – research and applications. IntechOpen, London, United Kingdom, pp 115–126

60. Michael DJ, James FA, Anderson C, Steven LB, George C, Gordon E et al (2001) Advances in linear and area HgCdTe APD arrays for eyesafe LADAR sensors. In: Proc. SPIE 4454, 198-211. https://doi.org/10.1117/12.448175

61. Michael J, Jim A, Steven B, Diane B, George C, Gina C et al (2007) MBE based HgCdTe APDs and 3D LADAR sensors. In: Proc. SPIE 6542, 65421A. https://doi.org/10.1117/12.724347

62. Mallik S, Hultquist K, Ghosh S, Velicu S, Hyeson J (2005) MBE grown mid-infrared HgCdTe avalanche photodiodes on Si substrates. In: 63rd Device Research Conference Digest, 20-22 June 2005, Santa Barbara, CA, USA, pp 75-76. https://doi.org/10.1109/DRC.2005.1553062

63. Mitra P, Case FC, Reine MB, Starr R, Weiler MH (1997) Doping in MOVPE of HgCdTe: orientation effects and growth of high performance IR photodiodes. J Cryst Growth 170(1):542–548

64. Hall DNB, Baker IM, Finger G (2016) Towards the next generation of L-APD MOVPE HgCdTe arrays: beyond the SAPHIRA 320 × 256. In: Proc. SPIE 9915, 99150O. https://doi.org/10.1117/12.2234370

65. Baker IM, Maxey C, Hipwood LG, Weller HJ, Thorne P (2012) Developments in MOVPE HgCdTe arrays for passive and active infrared imaging. In: Proc. SPIE 8542, 85421A. https://doi.org/10.1117/12.981850

66. Maxey CD, Capper P, Baker IM (2019) Chapter 9: MOVPE growth of cadmium mercury telluride and applications. In: Irvine S, Capper P (eds) Metalorganic Vapor Phase Epitaxy (MOVPE). Wiley, New Jersey, pp 293–324

67. Ian B, Chris M, Les H, Keith B (2016) Leonardo (formerly Selex ES) infrared sensors for astronomy: present and future. In: Proc. SPIE 9915, 991505. https://doi.org/10.1117/12.2231079

68. Johan R, Pierre B, Julie A, Sylvain G, Gilles L, Lydie M et al (2018) HgCdTe APDs detector developments at CEA/Leti for atmospheric lidar and free space optical communications. In: Proc. SPIE 11180, 111803S. https://doi.org/10.1117/12.2536055
69. Ferret P, Zanatta JP, Hamelin R, Cremer S, Million A, Wolny M, Destefanis G (2000) Status of the MBE technology at leti LIR for the manufacturing of HgCdTe focal plane arrays. J Electron Mater 29(6):641–647
70. Behet M, Hövel R, Kohl A, Küsters AM, Opitz B, Heime K (1996) MOVPE growth of III–V compounds for optoelectronic and electronic applications. Microelectron J 27(4):297–334
71. Mullin JB, Cole-Hamilton DJ, Irvine SJC, Hails JE, Giess J, Gough JS (1990) MOVPE of narrow and wide gap II–VI compounds. J Cryst Growth 101(1):1–13
72. Snyder DW, Mahajan S, Ko EI, Sides PJ (1991) Effect of substrate misorientation on surface morphology of homoepitaxial CdTe films grown by organometallic vapor phase epitaxy. Appl Phys Lett 58(8):848–850
73. Capper P, Maxey CD, Whiffin PAC, Easton BC (1989) Substrate orientation effects in $Cd_xHg_{1-x}Te$ grown by MOVPE. J Cryst Growth 96(3):519–532
74. Švob L, Chèze I, Lusson A, Ballutaud D, Rommeluère JF, Marfaing Y (1998) Crystallographic orientation dependence of As incorporation in MOVPE-grown CdTe and corresponding acceptor electrical state activation. J Cryst Growth 184-185:459–464
75. Gawron W, Madejczyk P, Kłos K, Rutkowski J, Piotrowski A, Rogalski A, Mróz W (2009) Surface smoothness improvement of HgCdTe layers grown by MOCVD. Bullet Polish Acad Sci Technical Sci 57(2):139–146
76. Mitra P, Case FC, Glass HL, Speziale VM, Flint JP, Tobin SP, Norton PW (2001) HgCdTe growth on (552) oriented CdZnTe by metalorganic vapor phase epitaxy. J Electron Mater 30(6):779–784
77. Wang W-S, Bhat I (1995) Growth of high quality CdTe and ZnTe on Si substrates using organometallic vapor phase epitaxy. J Electron Mater 24(5):451–455
78. Jones CL, Hipwood LG, Shaw CJ, Price JP, Catchpole RA, Ordish M et al (2006) High performance MW and LW IRFPAs made from HgCdTe grown by MOVPE. In: Proc. SPIE 6206, 620610. https://doi.org/10.1117/12.667610
79. Hails JE, Keir AM, Graham A, Williams GM, Giess J (2007) Influence of the silicon substrate on defect formation in MCT grown on II-VI buffered Si using a combined molecular beam epitaxy/metal organic vapor phase epitaxy technique. J Electron Mater 36(8):864–870
80. Maruyama K, Nishino H, Okamoto T, Murakami S, Saito T, Nishijima Y et al (1996) Growth of (111) HgCdTe on (100) Si by MOVPE using metalorganic tellurium adsorption and annealing. J Electron Mater 25(8):1353–1357
81. Nishino H, Murakami S, Saito T, Nishijima Y, Takigawa H (1995) Dislocation profiles in HgCdTe(100) on GaAs(100) grown by metalorganic chemical vapor deposition. J Electron Mater 24(5):533–537
82. Hipwood LG, Jones CL, Walker D, Shaw CJ, Abbott P, Catchpole RA et al (2007) Affordable high-performance LW IRFPAs made from HgCdTe grown by MOVPE. In: Proc. SPIE 6542, 65420I. https://doi.org/10.1117/12.720647
83. Madejczyk P, Piotrowski A, Gawron W, Kłos K, Pawluczyk J, Rutkowski J et al (2005) Growth and properties of MOCVD HgCdTe epilayers on GaAs substrates. Opto-electron Rev 13(3):239–251
84. Piotrowski A, Kłos K (2007) Metal-organic chemical vapor deposition of $Hg_{1-x}Cd_xTe$ fully doped heterostructures without postgrowth anneal for uncooled MWIR and LWIR detectors. J Electron Mater 36(8):1052–1058
85. Giess J, Hails JE, Graham AP, Blackmore GW, Houlton MR, Newey J et al (1995) The role of surface adsorbates in the metalorganic vapor phase epitaxial growth of (Hg,Cd)Te onto (100) GaAs Substrates. J Electron Mater 24:1149–1153
86. Suh S-H, Song J-H, Moon S-W (1996) Metalorganic vapor phase epitaxial growth of hillock free (100) sol HgCdTeGaAs with good electrical properties. J Cryst Growth 159(1–4):1132–1135

87. Suh S-H, Kim J-S, Kim HJ, Song J-H (2002) Control of hillock formation during MOVPE growth of HgCdTe by suppressing the pre-reaction of the Cd precursor with Hg. J Cryst Growth 236(1):119–124

88. Irvine SJC, Giess J, Gough JS, Blackmore GW, Royle A, Mullin JB et al (1986) The potential for abrupt interfaces in $Cd_xHg_{1-x}Te$ using thermal and photo-MOVPE. J Cryst Growth 77(1–3):437–451

89. He L, Yang J, Wang S, Wu Y, Fang W (1999) Recent progress in molecular beam epitaxy of HgCdTe. Adv Mater 11(13):1115–1118

90. Zanatta JP, Ferret P, Theret G, Million A, Wolny M, Chamonal JP, Destefanis G (1998) Heteroepitaxy of HgCdTe (211)B on Ge substrates by molecular beam epitaxy for infrared detectors. J Electron Mater 27(6):542–545

91. De Lyon TJ, Rajavel RD, Jensen JE, Wu OK, Johnson SM, Cockrum CA, Venzor GM (1996) Heteroepitaxy of HgCdTe(112) infrared detector structures on Si(112) substrates by molecular-beam epitaxy. J Electron Mater 25(8):1341–1346

92. He L, Chen L, Wu Y, Fu XL, Wang YZ, Wu J et al (2007) MBE HgCdTe on Si and GaAs substrates. J Cryst Growth 301-302:268–272

93. Nosho BZ, Roth JA, Jensen JE, Pham L (2005) Lateral uniformity in HgCdTe layers grown by molecular beam epitaxy. J Electron Mater 34(6):779–785

94. Faurie JP, Million A, Boch R, Tissot JL (1983) Latest developments in the growth of $Cd_xHg_{1-x}Te$ and CdTe–HgTe superlattices by molecular beam epitaxy. J Vac Sci Technol A 1(3):1593–1597

95. Gordon NT, Baker IM (2001) Assessment of infrared materials and devices. In: Capper P, Elliott CT (eds) Infrared detectors and emitters: materials and devices. Springer US, Boston, pp 23–42

96. Bae SH, Lee SJ, Kim YH, Lee HC, Kim CK (2000) Analysis of 1/f noise in LWIR HgCdTe photodiodes. J Electron Mater 29(6):877–882

97. Pillans L, Ash RM, Hipwood L, Knowles P (2012) MWIR mercury cadmium telluride detectors for high operating temperatures. In: Proc. SPIE 8353, 83532W. https://doi.org/10.1117/12.919015

98. Kopytko M, Rogalski A (2022) New insights into the ultimate performance of HgCdTe photodiodes. Sensors Actuators A Phys 339:113511

99. Martyniuk P, Rogalski A (2015) MWIR barrier detectors versus HgCdTe photodiodes. Infrared Phys Technol 70:125–128

100. Maimon S, Wicks GW (2006) nBn detector, an infrared detector with reduced dark current and higher operating temperature. Appl Phys Letters 89(15):151109

101. Itsuno AM, Phillips JD, Velicu S (2011) Design and modeling of HgCdTe nBn detectors. J Electron Mater 40(8):1624–1629

102. Uzgur F, Kocaman S (2019) Barrier engineering for HgCdTe unipolar detectors on alternative substrates. Infrared Phys Technol 97:123–128

103. Itsuno AM, Phillips JD, Velicu S (2012) Design of an Auger-suppressed unipolar HgCdTe NBvN photodetector. J Electron Mater 41(10):2886–2892

104. Kopytko M, Kębłowski A, Gawron W, Madejczyk P, Kowalewski A, Jóźwikowski K (2013) High-operating temperature MWIR nBn HgCdTe detector grown by MOCVD 21(4):402–405

105. Kopytko M, Kębłowski A, Gawron W, Pusz W (2016) LWIR HgCdTe barrier photodiode with Auger-suppression. Semicond Sci Technol 31(3):035025

106. Kopytko M, Jóźwikowski K (2013) Numerical analysis of current–voltage characteristics of LWIR nBn and p-on-n HgCdTe photodetectors. J Electron Mater 42(11):3211–3216

107. Kopytko M, Wróbel J, Jóźwikowski K, Rogalski A, Antoszewski J, Akhavan ND et al (2015) Engineering the bandgap of unipolar HgCdTe-based nBn infrared photodetectors. J Electron Mater 44(1):158–166

108. Kopytko M, Jóźwikowski K (2015) Generation-recombination effect in MWIR HgCdTe barrier detectors for high-temperature operation. IEEE Trans Electron Dev 62(7):2278–2284

109. Akhavan ND, Umana-Membreno GA, Gu R, Antoszewski J, Faraone L (2018) Delta doping in HgCdTe-based unipolar barrier photodetectors. IEEE Transa Electron Dev 65(10):4340–4345
110. Madejczyk P, Gawron W, Martyniuk P, Keblowski A, Pusz W, Pawluczyk J et al (2017) Engineering steps for optimizing high temperature LWIR HgCdTe photodiodes. Infrared Phys Technol 81:276–281
111. Rogalski A, Kopytko M, Martyniuk P (2018) Performance prediction of p-i-n HgCdTe long-wavelength infrared HOT photodiodes. Appl Opt 57(18):D11–D19
112. Vallone M, Goano M, Bertazzi F, Ghione G, Palmieri A, Hanna S et al (2019) Reducing inter-pixel crosstalk in HgCdTe detectors. Optical and Quantum Electronics 52(1):25
113. Vallone M, Goano M, Bertazzi F, Ghione G, Hanna S, Eich D et al (2020) Constraints and performance trade-offs in Auger-suppressed HgCdTe focal plane arrays. Appl Opt 59(17):E1–E8
114. He J, Li Q, Wang P, Wang F, Gu Y, Shen C et al (2020) Design of a bandgap-engineered barrier-blocking HOT HgCdTe long-wavelength infrared avalanche photodiode. Opt Express 28(22):33556–33563
115. Kopytko M, Gawron W, Kębłowski A, Stępień D, Martyniuk P, Jóźwikowski K (2019) Numerical analysis of HgCdTe dual-band infrared detector. Opt Quantum Electron 51(3):62
116. Fatih U, Serdar K (2019) A dual-band HgCdTe nBn infrared detector design. In: Proc. SPIE 11129, 1112903. https://doi.org/10.1117/12.2529240
117. Philippe T, Gérard D, Philippe B, Jacques B, Olivier G, Johan R (2008) Advanced HgCdTe technologies and dual-band developments. In: Proc. SPIE 6940, 69402P. https://doi.org/10.1117/12.779902
118. Perrais G, Rothman J, Destefanis G, Chamonal J-P (2008) Impulse response time measurements in $Hg_{0.7}Cd_{0.3}Te$ MWIR avalanche photodiodes. J Electron Mater 37(9):1261–1273
119. Perrais G, Derelle S, Mollard L, Chamonal J-P, Destefanis G, Vincent G et al (2009) Study of the transit-time limitations of the impulse response in mid-wave infrared HgCdTe avalanche photodiodes. J Electron Mater 38(8):1790–1799
120. Li Q, Wang F, Wang P, Zhang L, He J, Chen L et al (2020) Enhanced performance of HgCdTe midwavelength infrared electron avalanche photodetectors with guard ring designs. IEEE Trans Electron Dev 67(2):542–546
121. Faurie JP, Reno J, Sivananthan S, Sou IK, Chu X, Boukerche M, Wijewarnasuriya PS (1986) Molecular beam epitaxial growth and characterization of HgCdTe, HgZnTe, and HgMnTe on GaAs(100). J Vac Sci Technol A 4(4):2067–2071

Chapter 4
IR Detectors Array

Ghenadii Korotcenkov

4.1 Introduction

In recent years, the pace of development of thermal imaging technology has significantly accelerated [20, 52, 60, 64, 65]. Since its inception, the market for infrared (IR) thermal imaging technology has grown primarily due to its military applications. Thermal imaging devices using a matrix of thermal and photon detectors allow combat operations in poor visibility conditions, detect hidden objects and give target designation. Thermal imaging cameras in the form of binoculars, optical sights, guidance and homing systems are widely used in the army, navy and aviation, vision systems for ground combat robots and perimeter security systems. Today, the military sector still provides the market with some growth, but the priorities for its development have changed (see Fig. 4.1a). At present the main growth in the market is provided by the sectors of civil and medical thermography, security and fire surveillance, as well as applications related to ensuring the safety of navigation, control and protection of water and coastal objects, astrophysics, space and science imaging [9, 10, 23, 24, 32, 35, 45–47, 73]. Such devices are used for rescue operations during extinguishing fires, rescuing people in smoke conditions, identifying the most dangerous areas with high temperatures. Devices using thermal imaging cameras allow observation in poor visibility conditions and detecting people with high temperatures in a crowd. According to the forecast of Maxtech International (USA), the market for civil and military infrared systems, amounting to $10.5 billion in 2017, will exceed $17 billion in 2023 and should exceed $20 billion in 2025.

In the development of thermal imaging cameras, various approaches can be used. In particular, such systems can be developed on the basis of (a) single photodetectors and two-dimensional (line and frame) scanning using a scanning

G. Korotcenkov (✉)
Department of Physics and Engineering, Moldova State University, Chisinau, Moldova
e-mail: ghkoro@yahoo.com

© The Author(s), under exclusive license to Springer Nature Switzerland AG 2023 79
G. Korotcenkov (ed.), *Handbook of II-VI Semiconductor-Based Sensors and Radiation Detectors*, https://doi.org/10.1007/978-3-031-20510-1_4

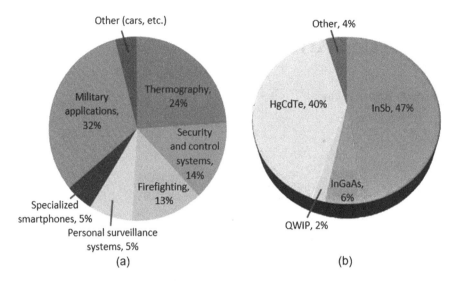

Fig. 4.1 (a) Areas of use of thermal imaging (Yole Development 2018, http://www.yole.fr/); (b) Relative shares of sales of photonic FPAs on different materials (Maxtech International, https://maxtech-intl.com/)

optical-mechanical system, (b) line of photodetectors and a simplified frame scan, or (c) grouped several lines (with time delay and accumulation) and a low-speed sweep system. The latter include vacuum devices with electronic scanning of the receiving target. However, the presence of optical-mechanical scanning systems leads to a significant complication of the equipment, a decrease in its manufacturability and reliability. When observing fast moving objects, there is also great difficulty in creating a high-speed mechanical scanner. Electron beam scanning has the best performance. But the large dimensions of cathode-ray tubes and the need for high-voltage power supply, an incandescent cathode, deflecting and focusing systems do not allow the creation of portable transmitting video cameras. Therefore, in recent years, the main direction in the development of such systems is the use of "simultaneously looking" – focal-plane, two-dimensional solid-state multi-element radiation detectors or Focal-Plane Array (FPA), the operation of which does not require the use of optical-mechanical scanning systems. All other things being equal, such thermal imaging cameras outperform scanning systems in terms of the combination of such parameters as reliability, sensitivity, signal-to-noise ratio, image quality, speed, dynamic range, and spatial resolution. Other advantages of such devices are low weight, dimensions and power consumption, noiseless operation, the ability to communicate with modern computers, video and TV equipment, and digital image processing in real time. At present, developments have reached such a level that large-format matrices (with the number of pixels $10^3 \times 10^3$ and more) are actually already technical analogs of the retina in the organs of vision. For example, modern technology allows the manufacture of IR-FPAs, which can have up to 10^8 IR detectors, which corresponds to the number of sensitive receptors in the human eye (~$2 \cdot 10^8$). This provides not only a significant expansion of the spectral range of vision, but also the implementation of optoelectronic systems with the maximum achievable values of the threshold sensitivity, inertia and information capacity.

4.1.1   Materials and Types of IR Detectors

IR-FPAs can be made on the basis of various materials that determine such charac- teristics of thermal imaging cameras as the operating spectral range, temperature sensitivity, the need for cooling the array and the speed of the system [20]. There are two main types of infrared detectors used in thermal imaging cameras: photon detectors and thermal detectors. Thermal detectors include microbolometric and pyroelectric detectors. Table 4.1 shows the materials on the basis of which various types of IR photodetectors arrays are developed.

The most important advantage of bolometric infrared detectors is the ability to operate without cooling (at temperatures of about 300 K), while most photon detec- tors operate at cryogenic temperatures (usually at least 77 K). In addition, the cost of FPAs based on uncooled bolometers in industrial production is two orders of magnitude lower than the cost of photonic arrays [69]. However, they are noticeably inferior in other parameters to photonic FPAs.

Table 4.1 Materials used in IR-FPAs of various types

Types of IR detectors	Material of photodetector array	Spectral range, μm
Photonic	Lead chalcogenide (PbS, PbSe);	1.5–6
	Compounds mercury-cadmium-tellurium HgCdTe (MCT);	1–20
	Indium Antimonide (InSb);	3–5
	Platinum silicide – silicon Schottky barrier structures (PtSi-Si);	3–5
	Doped silicon (Si:X) and germanium (Ge:X);	1.5–40
	Multilayer structures with quantum wells based on GaAs /AlGaAs (QWIP detectors);	8–14
	Compound indium-gallium-arsenide (InGaAs)	0.3–2.3
Thermal microbolometric	Vanadium oxides V_xO;	8–14
	Polycrystalline and amorphous silicon	8–14
Thermal pyroelectric	Lead zirconate;	8–12
	Barium-strontium niobate and titanate;	
	Triglycine sulfate	

4.1.2 Photonic IR FPAs and Basic Materials Used to Develop Them

4.1.2.1 Materials Used in the Development of Photonic IR FPAs

Currently, the main materials used in the development of photon-type IR FPAs are HgCdTe (MCT), InSb, InGaAs, InAs/GaSb and GaAs/AlGaAs. InAs/GaSb heterostructures are used to fabricate type-II superlattices (T2SLs), while GaAs/AlGaAs are used to fabricate quantum well heterostructures (type-I superlattices, T1SLs). We should also not forget the quantum dots of various materials, based on which in recent years it has also been possible to develop efficient IR photodetectors, the so-called quantum dot IR photodetector (QDIP) (Chap. 7, Vol. 2).

Quantum well IR photodetectors (QWIP) on the base of GaAs/AlGaAs heterostructures have a narrow spectral region and a relatively low quantum efficiency (less than 10%), therefore, they require a longer signal accumulation time than devices based on InSb and HgCdTe (quantum efficiency is about 90%). However, the spectral characteristics of QWIP detectors can be flexibly adjusted by changing the width and height of the quantum wells formed by the GaAs and AlGaAs layers. In addition, the ability to use commercial manufacturing processes developed for III-V compounds significantly reduces the cost of the ongoing development. One-color IR photodetector arrays based on quantum-dimensional structures with formats 320×256 and 640×480 are already produced by companies such as Sander, Lookheed Martin and QWIP Technologies, and these FPAs are already being used in a number of commercial applications. This is proof that the technology of growing quantum-dimensional structures (QDS) has become effective.

Interest in FPAs development based on InAs/GaSb type-II superlattices (T2SLs) has emerged in recent years [59]. There are two primary motivations for development of InAs/GaSb T2SLs. The first is related to the existing problems of reproducibly fabricating high-operability HgCdTe focal plane arrays (FPAs) at reasonable cost. The second motivation is theoretical predictions of lower Auger recombination for T2SL detectors compared to HgCdTe, which can be translated into a fundamental advantage of T2SL over HgCdTe in terms of lower dark current, provided that other parameters such as Shockley-Read-Hall (SRH) lifetime are equal. At present InAs/GaSb T2SL photodetectors offer similar performance to HgCdTe at an equivalent cut-off wavelength, but with a sizeable penalty in operating temperature, due to the inherent difference in SRH lifetimes. It is predicted that since the future infrared (IR) systems will be based on the room temperature operation of depletion-current limited arrays with pixel densities that are fully consistent with background- and diffraction-limited performance due to the system optics, the material system with long SRH lifetime will be required. Since T2SLs are very much resisted in attempts to improve its SRH lifetime, currently the only material that meets this requirement is HgCdTe [59]. However, despite the fact that the modern version of the T2SLs technology is as yet in its infancy, the rapid progress made in T2SLs over the past few years has shown great promise for this material for a future development of IR technologies [19, 41].

As for FPAs based on QDIP, they are still under development and therefore cannot yet compete with FPAs based on MCT [9, 27]. Achieving acceptable results requires significant investment and fundamental research. In connection with the above, FPAs made of mercury-cadmium-tellurium (MCT) and indium antimonide (InSb) are currently mainly used for high-sensitivity and long-range thermal imaging cameras. In terms of their parameters, devices based on these materials are close to theoretical characteristics. It is expected that HgCdTe and InSb will remain the main materials for IR technology for at least the next 10-15 years. At present, approximately an equal number of devices are produced based on these compounds, and this ratio will remain the same. However, in the field of military applications where high sensitivity and speed are required, MCT still dominates [9]. HgCdTe is currently the most prevalent material system used in high performance infrared detectors due to the following reasons [60–62, 64]:

- the tailorable energy band gap covers the entire spectral range of 1–30 μm, which includes important spectral ranges such as 1–2.5 μm, 3–5 μm, and 8–14 μm,
- large optical coefficients that enable high quantum efficiency, and
- favorable inherent recombination mechanisms that lead to high operating temperature.

4.1.2.2 Photonic IR FPAs

Modern hybrid multi-element IR photodetectors usually consist of two parts: a multi-element photosensitive structure and a silicon multiplexer, connected to each other by group cold welding using indium bumps (Fig. 4.2). A linear or matrix multiplexer or readout integrated circuits (ROIC) is an integrated circuit that provides the required electrical modes of operation of photosensitive elements, reads electrical signals obtained as a result of photoelectric conversion of incident IR radiation by photodetector, performs preprocessing, and analog-digital conversion of the photo signal, and ultimately largely determines the quality of the resulting image [44]. Examples of ROICs development by Sofradir and LETI one can find in [6, 26, 29]. One of these ROICs is shown in Fig. 4.3.

Usually, the composition of image signal generators also includes shift registers, clock generators, analog-to-digital converters, blocks for precision correction of pixel inhomogeneity, blocks for replacing photosignals from non-working elements, blocks for subtracting the background signal component. They may include electronic devices that provide high-frequency reading of image fragments, as well as devices for integrating two or more colors (if photosensitive elements have sensitivity in two or more spectral ranges), devices for encoding images into conventional colors, etc. The combination of the photosensitive element and the microelectronic path in a single housing led to a decrease in size, to a decrease in parasitic capacitances and to improved shielding from external interference, that is, to an improvement in the threshold and operational characteristics of optoelectronic equipment. In addition, such a combination made it possible to ensure the transportability of the output signal from photodetectors and to significantly simplify its

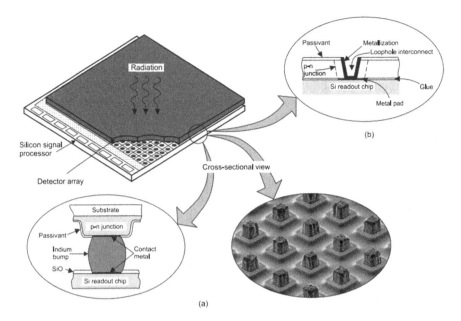

Fig. 4.2 Hybrid IR FPA with independently optimized signal detection and readout: (**a, b**) Cross-section of the photodiodes during hybridization using (**a**) indium bump technique and (**b**) loophole technique. (Reprinted from Ref. [61]. Published 2008 by De Gruyter as open access)

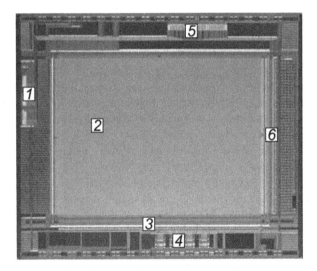

Fig. 4.3 The photo of 384 × 288 silicon readout integrated circuit with 25 μm pixel pitch for MWIR and LWIR HgCdTe based FPA: 1 – digital-to-analog converters (DACs) block, 2 – array of pixel cells, 3 – column followers and decoder, 4 – preamplifiers and output buffers, 5 – digital block, 6 – row decoder. (Reprinted from Ref. [83]. Published 2015 by IOP as open access)

processing [32]. Silicon multiplexer chips are subject to more stringent uniformity requirements than conventional integrated circuit; and meeting these requirements are often beyond the scope of standard factory manufacturing processes for making very large scale integrated circuits. By the type of construction, photonic detectors arrays can be created both on the basis of complementary metal-oxide-semiconductor (CMOS) switches [33], and on the basis of a charge-coupled devices (CCD). CCD technology is used for not very large scale arrays and their technology is more complicated than the CMOS production line. In comparison with CCDs, the MOS multiplexers exhibit important advantages. CMOS technology provides a higher realizable capacity of the storage MOS capacitors, which is important at radiation wavelengths above 5 micrometers, at which the background illumination is high. In addition, tighter packing, higher and more uniform electrical parameters at 77 K, lower control voltages, and greater dynamic range are achieved [64]. Leading world firms such as SOFRADIR (www.sofradir.com), Indigo Systems Corporation (www.indigosystems.com), etc., have developed and are currently serially producing a series of linear (4 × 288, 6 × 460, 1 × 512) and matrix (128 × 128, 320 × 240, 320 × 256, 384 × 288, 640 × 512) silicon multiplexers for long-range (LWIR), medium (MWIR) and near infrared (NIR) spectral ranges. Thus hybridized assembly of the FPA and ROIC is glued to the surface of the carrier sapphire substrate from the side of the silicon microcircuit. On the contact pads of the substrate, leads are wired for reading photosignals, supplying power and control signals. Thermal sensors are installed on the same substrate and a cooled diaphragm is attached. The appearance of such assemblies is shown in Fig. 4.4a, b. Image shown in Fig. 4.4c demonstrates the resolution capability of such FPAs with pixel size smaller than 15 μm [49].

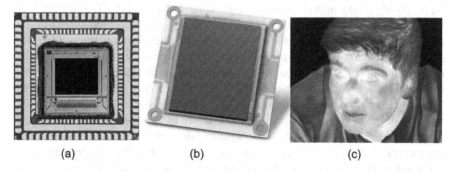

(a)   (b)   (c)

Fig. 4.4 (a) A standard SAPHIRA detector (https://www.leonardocompany.com/en/products/saphira-1). The dark rectangle in the center is the HgCdTe APD array, and the temperature sensor is visible in the upper-left corner. In operation it is mounted to a 68-pin leadless chip carrier. HgCdTe APD array hybridized to a complementary metal-oxide-semiconductor (CMOS) ROIC; (b) HgCdTe 1280 × 1024 SCORPIO MW detector with pixel pitch of 15 μm (www.Sofradir.com); (c) Image taken with SELEX SuperHawk 1280 × 1024 MWIR HgCdTe FPA module with pitch of 8 μm (https://www.leonardocompany-us.com/). This image was obtained from a liquid argon cooled FPA in the test room at Selex ES. (Reprinted with permission from Ref. [49]. Copyright 2015: SPIE)

4.2 Photovoltaic HgCdTe-Based FPAs

4.2.1 Technological Developments

Photovoltaic HgCdTe FPAs are based on both *p*-type and *n*-type materials. The highest quality structures are grown on $Cd_{1-y}Zn_yTe$ (CZT) (y = 0.03–0.05) substrates, matched to MCT in the crystal lattice constant, which makes it possible to obtain structures with a small amount of dislocation ($<10^5$ cm^{-2}). The CZT substrate growth performed by vertical gradient freeze (VGF) bulk method. Current boules grown at LETI and Sofradir are above 110 mm in diameters, while maintaining the very high crystalline quality required for MCT growth, low dislocation density (in the low 10^4 cm^{-2}) and low precipitate concentration [11]. Usually crystalline orientation is set to (111) in order to perform the liquid phase epitaxy of the MCT layer onto this substrate. The Zn content in CdZnTe is optimized to guarantee lattice matching with the targeted MCT composition. After dicing into 7 × 7 cm wafers, the substrate surface has to be prepared for MCT growth using liquid phase epitaxy (PLE), molecular beam epitaxy (MBE) [50] or metalorganic vapour phase epitaxy (MOVPE) methods [10].

The complex and multistage technology for producing MCTs includes deep purification of the starting Cd, Hg and Te, the synthesis of precursors, and the growth of HgCdTe epitaxial layers by various methods. One of the important parameters of MCT layers designed for FPAs is the uniformity of the composition over the area. Moreover, for photodetectors operating in the range of 8–12 μm, the change in the composition of $Hg_{1-x}Cd_xTe$ films should not exceed $\Delta x = 0.0002$ by 10 mm. At present, the main industrial method for the manufacture of epitaxial layers in the world's leading companies producing IR photodetectors arrays is the method of liquid-phase epitaxy on a cadmium-zinc-tellurium compound (CZT) substrate. The advantages of this method are the relatively low cost and high productivity of the equipment, automatic additional cleaning of the surface at the initial stage of growth, additional cleaning from impurities in the growth process, and uniformity of the composition over the area. In addition, photodetectors fabricated using liquid epitaxy exhibited higher efficiency and lower dark currents. However, large-area CdZnTe substrates remain expensive products with poorly reproducible characteristics. Besides that, the use of single-crystal CdZnTe as substrates, due to the limited diameter of the grown single crystals, significantly limits the possibilities of creating megapixel photodetectors. As is known, megapixel IR FPAs require large-area MCT plates with a given high lateral compositional homogeneity. All this together with the relatively low yield due to cluster defects and blinking pixels, makes the cost of the technology using CdZnTe substrates prohibitive. In this regard, technologies are being developed everywhere that make it possible to exclude the use of CdZnTe substrates when growing large area HgCdTe epitaxial layers [3, 40]. Studies have shown that this problem can be solved by using low-temperature molecular beam epitaxy (MBE) of HgCdTe films. The experiment showed that MBE is the most developed, flexible method for obtaining such a material in the form of hetero-epitaxial structures. The built-in ellipsometric equipment allows high-precision control of the composition of the epitaxial MCT film and its changes during growth. This opens up wide possibilities for optimizing the design of heterostructures, which makes it possible to simplify the technology of fabricating IR FPAs with extremely high parameters. Another important advantage of the MBE method for growing MCTs is the use of cheap silicon wafers. This means that HgCdTe growth on large-area Si substrates is to enable larger array formats and potentially reduced FPA cost compared to FPA fabricated using smaller and more expensive CdZnTe substrates.

For cooled hybrid large format IR FPAs, there is a reliability issue in thermal cycling from room temperature to liquid nitrogen temperature. This problem is simplified precisely when using silicon substrates for growing MCTs. But it must be admitted that the large difference in chemical composition and crystal lattice parameters mismatch between MCT and Si makes the problem of fabricating FPAs based on MCT/Si structures with suitable parameters into an extremely difficult task [45, 46]. Only the optimization of the processes of pre-epitaxial preparation of the silicon substrate surface and the conditions for the formation of the CdTe/ZnTe buffer layer [50], as well as the improvement of MCT growth processes and the use of post-growth processes made it possible to obtain heteroepitaxial MCT structures by the MBE method with acceptable parameters; without interphase boundaries, with a uniform distribution over the surface of morphological V-defects and etching pits with densities less than 10^3 cm^2 and 10^6 cm^2, respectively [12, 17]. This method made it possible to grow homogeneous MCT heterostructures on silicon substrates with a diameter of more than 100 mm. Using MBE system, researchers produced epitaxial HgCdTe layers on (211) Si substrates with very low macro defect density and uniform Cd composition across the epitaxial wafers. These HgCdTe/Si composite wafers have shown growth defect densities less than 10 defects /cm^2, approximately 100 times better than can be achieved on CdZnTe substrates, due to the better crystalline quality of the starting substrate [8]. Comparative characteristics of the processes of growing epitaxial MCT layers on CdZnTe and Si substrates are given in the Table 4.2. As you can see, Si-based substrate technology for growing HgCdTe epitaxial layers has a number of significant advantages that can significantly reduce the cost of FPAs based on HgCdTe.

It is important to note that in the fabrication of devices based on epitaxial HgCdTe films grown on Si substrates, the same etching, passivation, and metallization schemes can be used as in the fabrication of detectors based on HgCdTe /CdZnTe structures. Wherein, Bangs et al. [8] have shown that manufacturing processes for detectors throughout the area 150 mm HgCdTe/Si wafers usually produced high performing detector pixels from edge to edge of the photolithographic limits across the wafer, offering 5 times the printable area as compared with 6 × 6 cm CdZnTe substrates. Large-format (2 K × 2 K) MWIR FPAs fabricated using large area HgCdTe layers grown on 6-inch diameter (211) silicon substrates demonstrated NEDT operability better than 99.9%. Moreover, SWIR and MWIR detector performance characteristic for devices fabricated on HgCdTe/Si substrates are comparable to those established for devices fabricated on HgCdTe/CdZnTe wafers. HgCdTe

Table 4.2 Advantages of Si-based substrate technology for HgCdTe material development

Parameter	Bulk CdZnTe	Si
Maximum size	7×7 cm^2	150 мм in diameter
Maximum area	$\sim$50 cm^2	$\sim$180 cm^2
Scalability	No	Yes
Cost	$220/cm^2	$\sim$$1/cm^2
Thermal match to Si ROIC	No	Yes
Robustness	Brittle	Hard
Lattice match to MCT	Yes	No
Surface	Smooth	Smooth
Orientation available	(111)	(112)
Substrate quality (dislocations)	<10,000 cm^2	<100 cm^2
Impurities	Low	Extremely low

Source: Data extracted from [12] and [20]

devices fabricated on both types of substrates have demonstrated very low dark current, high quantum efficiency and full spectral band fill factor characteristic of HgCdTe [8, 15].

As for the very technology of manufacturing photodetectors that form a matrix of photodetectors, this technology does not differ from the technology for manufacturing discrete photodetectors described earlier in [40] and Chap. 15, Vol. 1. To create p-n junctions, the method of ion implantation of B$^+$ or As$^+$, depending on the type of conductivity of the main absorbing layer, with a dose of $(2–3)\cdot10^{13}$ cm^{-2} is usually used. In this case, the main mechanism for the formation of the n-type doped region in p-HgCdTe is when boron ions knock out mercury atoms, which pass into interstices, becoming electrically active defects responsible for the formation of the n-region of the p-n junction.

A HgCdTe avalanche photodiodes (APDs) are also used to make the focal plane array [68, 75]. It is believed that HgCdTe avalanche photodiodes (APDs) are particularly suitable for low flux detection. Indeed, the possibility to pre-amplify the incoming photonic signal into the photodiode itself (with no major degradation of the signal to noise ratio [53] is very interesting when the detection performance is limited by the ROIC noise as it is the case in the very low flux detection for astronomy [23, 24]. As for the configuration of the photodetectors themselves that form the array, they can be made both in planar design and in the form of mesa structures. However, the best results were obtained using mesa diode technology and MCT films grown by Metal Organic Vapour Phase Epitaxy (MOVPE) [1]. Advantages of this technology include high optical efficiency, near perfect modulation transfer function (MTF) and very low dark current. Photodiodes in FPAs usually operate in the back-side illumination mode.

Depending on the tasks being solved, epitaxial structures can have both single-layer and multilayer structures. So, Gu et al. [25] when developing their FPAs used the structure shown in Fig. 4.5. An x = 0.4 Auger suppressing barrier is grown between the substrate and absorption layer to reduce Auger injection and thus

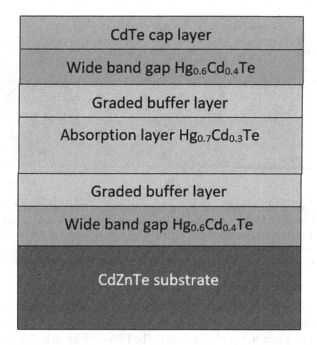

Fig. 4.5 A multilayer structure of HgCdTe material designed and grown for FPAs operated in 3–5 μm spectral range. (Data extracted from Ref. [25])

deduce the dark current. The x = 0.3 $Hg_{1-x}Cd_xTe$ was grown as the main absorption layer to ensure that the device is operating with a 5 μm cut off wavelength at 77 K (mid wavelength infrared, MWIR). The x = 0.4/0.3 double layer heterostructure was designed to reduce the generation-recombination dark current. Both sides of the absorption layer are buffered with a graded layer from x = 0.3 to x = 0.4 to improve the material quality.

When solving some problems, p-n junctions can be formed directly during the deposition of epitaxial HgCdTe layers. So, when developing an array of dual-band detectors, the structures shown in Fig. 4.6a were used. Dual-band detectors extend the single-color process by simply growing a second n-type absorbing layer on top of the p-type layer to form two back-to-back photodiode junctions, which is referred to as an n-p-n or triple-layer heterojunction (TLHJ) device structure. A triple-layer n-p-n heterojunction was grown by molecular-beam epitaxy (MBE) on 100 mm (211) Si wafers with ZnTe and CdTe buffer layers [50]. The MWIR/LWIR dual band epitaxial structures had low macro defect densities (<300 cm^{-2}). Inductively coupled plasma etched detector arrays with 640 × 480 dual band pixels (20 μm) were mated to dual-band readout integrated circuits (ROICs) to produce FPAs. The measured 80 K cutoff wavelengths were 5.5 μm for MWIR and 9.4 μm for LWIR, respectively. The dual-band FPA architecture achieves nearly simultaneous detection of two spectral bands while still being producible for pixel dimensions between 30 μm and 20 μm, or even smaller. The shorter-wavelength (band 1) radiation is

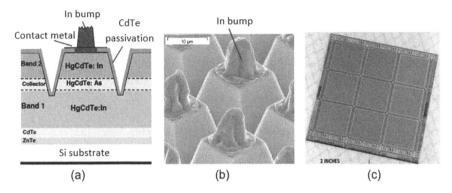

Fig. 4.6 (a) Cross-section of single-mesa dual-band detector architecture applied to HgCdTe on Si; (b) SEM image of 20-micron-unit-cell dual-band detectors; (c) A 6 cm × 6 cm molecular-beam epitaxy (MBE) triple-layer heterojunction (TLHJ) HgCdTe dual-band L/MWIR detector wafer with a 512 × 512 30-μm-unit-cell detector array product die layout, with test structure detector die along the upper and lower portions of the wafer. (Reprinted with permission from Ref. [71]. Copyright 2011: Springer)

detected via the p-on-n junction, while the longer-wavelength (band 2) radiation passes through the shorter-cutoff n-type absorbing layer and can be detected via the n-on-p junction nearer to the pixel contact. The FPAs exhibited high pixel operability in each band, with noise equivalent differential temperature (NEDT) operability of 99.98% for the MWIR band and 99.6% for the LWIR band at 84 K [50].

To form a multi-element system of contacts, as a rule, multilayer metallization based on Cr, Au, In and Ni with different layer thicknesses is used. The intermediate nickel layer gives the contacts the necessary mechanical strength and stability, while the indium layer provides ductility and elasticity. The last layer of indium with a thickness of 3–5 micrometers is thermally processed in a vacuum. As a result, indium contacts are formed in the form of balls 8–10 micrometers in height, protruding above the crystal surface as it is shown in Fig. 4.6b. Indium columnar microcontacts up to 10–12 micrometers in height can be formed by other technological methods. Indium columnar microcontacts are required for cold welding of two crystals, a photodetectors array and a multiplexer. To do this, two crystals are placed in parallel one above the other, oriented in accordance with the circuit topology, and with the help of a specially selected value of the mechanical load interconnected via indium bumps using flip-chip bonding [32]. The combination of many thousands of indium microcontacts of the photodiode array crystal and the silicon crystal of the microcircuit can be carried out using an infrared microscope with visualization of the abutting microcontacts through the silicon substrate. This method is the most common in the manufacture of hybrid FPAs. However, it is important to note that in addition to the indium bump technique a hybrid FPAs detectors and multiplexers can also be fabricated using loophole interconnection [7]. In this case, the detector and the multiplexer chips are glued together to form a single chip before detector fabrication. The photovoltaic detector is formed by ion implantation and loopholes are drilled by ion milling, and electrical interconnection between each detector and its corresponding input circuit is made through a small hole formed in each detector. The loophole interconnection technology offers more stable mechanical and thermal features than flip-chip hybrid architecture. Despite the great laboriousness of this process, the loophole interconnection technology offers more stable mechanical and thermal features than flip-chip hybrid architecture. This is the approach used by Strong et al. [74] when developing FPAs with extremely small pixel sizes.

One of the most important parameters of matrix photodetectors is the magnitude of the inter-element coupling. Strong coupling can result in significant blurring of the thermal image. Three types of interconnection are possible: optical, electrical and photoelectric. The optical coupling is determined by the quality of the optical path of the thermal imaging device. Photovoltaic is associated with the diffusion of photogenerated carriers in the common semiconductor layer of the photodiode array. The voltage drop determines the electrical coupling when the current flows from the p-n junctions to the inactive contacts, as well as the interconnection between the input channels of the cooled silicon microcircuit. It is known that the properties of IR MCT-based photodiodes substantially depend on the chemical structure and electronic properties of the surface. Therefore, during the formation of the matrix, as well as during the manufacture of individual photodetectors, the passivation of the MCT surface is carried out, which helps to reduce surface currents and prevents the degradation of structures. As a passivating coating, CdTe and ZnS layers up to 100 nm thick are usually used [4]. These layers are formed using technological parameters that exclude prolonged temperature treatments. The latter is necessary to preserve the chemical composition of MCT films, and hence their properties. For example, these layers can be deposited by thermal evaporation method.

It should be noted that the FPAs manufacturing technology is constantly being improved. For example, Teledyne has developed a process for removal the CdZnTe substrate material, and substrate removal is now standard for all NIR, SWIR and MWIR focal plane arrays made by Teledyne [36]. Substrate removal involves stripping all of the CdZnTe substrate after hybridization to leave a layer of HgCdTe only 7 to 10 μm thick. After removing the substrate, an anti-reflective (AR) coating is applied to the HgCdTe. Experiment has shown that removing the substrate has a number of advantages, including the following:

- Increase in sensitivity to visible light, down to 380 nm, with a significant increase in quantum efficiency below 1.3 μm (see Fig. 4.7a);
- Elimination of fluorescence from cosmic rays absorbed in the CdZnTe substrate, which is very important for low light level space applications;
- Elimination of Fabry-Perot fringes that can occur in the substrate with narrow band illumination, such as in spectrometers.

Teledyne has also developed a process to fabricate lightly doped HgCdTe detectors that can be fully depleted with minimal (1–2 volt) reverse bias [36]. A completely depleted detector has removed most of the free electrons, which suppresses the dark Auger current signal to the point where the "dark current" dominates due to the

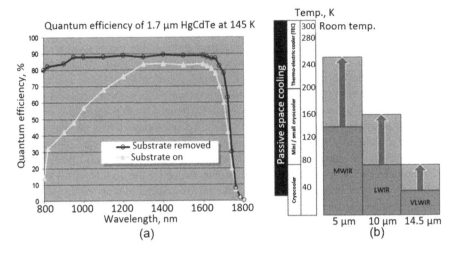

Fig. 4.7 (a) Quantum Efficiency of a substrate removed HgCdTe detector with 1.7 μm cutoff; (b) Increase in operating temperature from the use of fully depleted HgCdTe. (Reprinted from Ref. [36]. Published 2019 by SPIE as open access)

background radiation seen by the detector. Depending on the cut-off wavelength and operating temperature, the background radiation is 10–100 times lower than the dark current. This offers a great advantage for space applications, allowing operation at much higher temperatures with passive cooling or smaller cryo-coolers. This means that when developing FPAs you can use less expensive cooling options with longer cooler operating lifetime. Figure 4.7b shows the increase in operating temperature that can be obtained with the use of fully depleted HgCdTe detectors.

4.2.2 Photonic Cooled Detectors

Unfortunately, in order to achieve the desired result in terms of sensitivity and resolution, deep cryogenic cooling is required for the operation of devices with photon detectors in the IR region, especially in LWIR and VLWIR spectral regions. All objects in the infrared region of the spectrum are "self-luminous" if their temperature is above absolute zero, so the IR receivers themselves can "glow" in the range of their sensitivity (3–5 and 8–14 micrometers), making it difficult to detect weak radiation coming from outside. Therefore, in order to increase the detecting ability, it is necessary to extinguish the intrinsic radiation of the sensitive element, adjacent diaphragms and other elements of the device. This is achieved by cooling the photodetector to temperatures at which the self-radiation noise (dark current) becomes negligible. In addition, cooling the receiver prevents excessive heating of sensitive elements with low heat capacity and ensures the stability of the functional properties of semiconductor elements.

For deep (cryogenic) cooling of the FPAs (T = 75–80 K), liquid nitrogen or a cooling machine operating in a closed Split-Stirling cycle are used. For not deep cooling (T = 150–250 K) or thermal stabilization of the operation of an uncooled photodetector array, a thermoelectric cooling system based on Peltier elements is used. A cooling machines operating on a closed Split-Stirling cycle have low power consumption and dimensions, which allow the cooling element to be placed inside a thermal imaging camera. Cooling to operating temperature occurs in 5–8 min, which takes about 3 watts of power.

A typical design of a cooled photon detectors array is shown in Fig. 4.8. A hybrid photodetector unit, including an array of photosensitive elements and a silicon read-out integrated circuit, is mounted in a vacuum case. Cooling of FPAs is provided by a microcryogenic cooling system (MCS) integrated with the FPA housing.

4.2.3 Performances of MWIR and LWIR FPAs

Most thermal imaging cameras operate in the 3–5 μm (MWIR) and 8–14 μm (LWIR) spectral ranges, which correspond to the atmospheric transparency windows in the infrared region. Therefore, we will consider their parameters. Among the characteristics of FPAs, the main ones are:

– temperature sensitivity NETD (Noise Equivalent Temperature Difference)
– temperature difference equivalent to noise;
– the number of pixels that make up the matrix;
– image acquisition speed;
– the need for cooling the matrix.

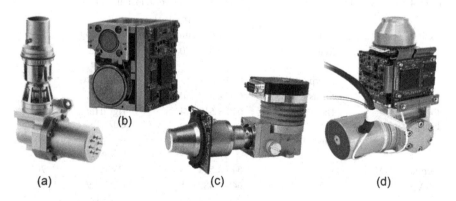

Fig. 4.8 Appearance of FAPs developed by various firms: (**a**) Scorpio MW (France) (https://www.lynred.com/products/scorpio-mw-product-range); (**b**) CD640-12 MW-μ (USA) (https://www.leonardodrs.com/); (**c**) ASTROH-640KPT15A810 (Russia) (https://astrohn.ru/product/astron-640krt15a810/); (**d**) PELICAN DLW (Israel) (https://www.scd.co.il/products/pelican-d-lw/)

The best QWIPs have NETD below 10 mK, typical −20 mK, medium −35 mK. QWIP-based FPAs are available in 256 × 256, 320 × 240, 320 × 256, 640 × 512 matrix formats [18]. For devices based on MCT detectors, NETD sensitivity: for the best models −10 mK, typical −15 mK, medium −20 mK; the resolution of the commercially available FPAs is up to 640 × 512 pixels.

Long-wave infrared (LWIR) operation requires cooling to 80 K, while medium-wave infrared (MWIR) operation often requires cooling to 120 K. The update rate for HgCdTe or InSb matrices usually ranges from 100 to 400 Hz; for FPAs based on QWIPs at full resolution, this frequency is in the range of 50–250 Hz.

For multi-pixel FPAs, aside from the large format, very high level of performance is of great importance in quantum efficiency (QE), dark current and noise [30]. While the noise is mainly addressed by the ROIC input stage, QE and dark current are highly dependent on the photodiode structure. To obtain a high QE, it is necessary to optimize the generation and collection of photo carriers: for this purpose, a high pixel fill factor as well as a sufficiently thick absorbing layer are required. Concerning dark current, at the low operating temperatures considered here, the dark currents are most likely depletion currents associated with SRH recombination defects in the space charge region of the photodiode [23, 24]. The mitigation of such depletion currents implies the use of the best quality materials as well as the best narrow gap surface passivation.

4.2.3.1 Cooled FPAs for the Spectral Range of 8–12 μm

Cooled FPAs for the spectral range of 8–12 μm are presented in the Table 4.3. They are serially produced by the world's leading manufacturers and are widely represented on the world market of FPAs. The main formats are 320 × 256 and 640 × 512

Table 4.3 FPAs for the spectral range of 8–12 μm of various world manufacturers

Country, firm	Brand	Format	Pixel size, μm	Technology	T_{oper}, K
France, Sofradir	Mars L	320 × 256	30	HgCdTe/ CdZnTe	80
	Scorpio LW	640 × 512	15	HgCdTe/Ge	80
	Sirius LW	640 × 512	20	QWIP	73
England, Finmeccanica	Harier LW	640 × 512	24	HgCdTe/ CdZnTe	80
	Hawk LW	640 × 512	16	HgCdTe/ CdZnTe	80
Germany, AIM	–	640 × 512	15	QWIP	70
USA, DRS	CD640-12-B	640 × 480	15	HgCdTe/ CdZnTe	80
China, GST	C615M	640 × 512	15	HgCdTe	80
Russia, Astron	A-640KRT15A810	640 × 512	15	HgCdTe/Si	78

Source: Data extracted from [73] and [54]

pixels. Megapixel FPAs in this spectral range are mainly manufactured on the basis of MCTs and QWIPs. FPAs with a spectral range of 8–12 micrometers provide the best temperature sensitivity and noise immunity in smoky and dusty conditions. In this spectral range, there is a maximum of the intrinsic thermal radiation of bodies at a temperature of 300 K. For example, the human body, by virtue of being at a temperature of ~300 K, emits radiation that peaks around 10 μm. According to Wien's law of displacement, the maximum of the intrinsic thermal radiation of bodies when they are heated shifts to the short-wave region. This is why the spectral range of 1–5 μm is better suited for detecting more heated bodies.

4.2.3.2 Cooled Photodetectors Array for the Spectral Range of 3–5 μm

Cooled FPAs for the spectral range of 3–5 μm serially produced by leading companies are presented in the Table 4.4. It can be seen that InSb and MCT are the main materials used for the development of such devices. It is important to note that for MCT-based FPAs, the operating temperature can be increased to 110–120 K without degrading performance, ensuring operation in modes limited by background radiation [75]. For FPAs based on InSb, such an increase in operating temperature is impossible. This is due both to different coefficients of thermal expansion of the band gap (Eg) in MCT and InSb (for HgCdTe dEg/dT > 0, and for InSb dEg/dT < 0), and the initially shorter-wavelength boundary of the photosensitivity of HgCdTe-based photodetectors, developed for the region of 3–5 μm. The red border of the photosensitivity of $Hd_{1-x}Cd_xTe(x = 0.3)$-based photodetectors is about 5 μm, and based on InSb −5.6 μm at T = 80 K. The main matrix format for FPAs developed for

Table 4.4 FPAs for the spectral range of 3–5 μm of various world manufacturers

Country, firm	Brand	Format	Pixel size, μm	Technology	T_{oper}, K
France, Lynred	Jupiter MW	1280 × 1024	15	HgCdTe	80
France, Sofradir	Scorpio MW	640 × 512	15	HgCdTe	80
	Jupiter MW	1280 × 1024	15	HgCdTe	80
Israel, SCD	Pelican MW	640 × 512	15	InSb	80
	Black bird	1920 × 1536	10	InSb	80
	Hawk MW	640 × 512	16	HgCdTe/CdZnTe	80
	Hawk HD	1280 × 1024	8	HgCdTe/CdZnTe	80
Germany, AIM	HiPIR 1280 M	1280 × 1024	15	HgCdTe	80
USA, FLIR	Neutrino	640 × 512	15	InSb	80
USA, DRS	CD640-12-M	640 × 480	12	HgCdTe	80
Korea, i3System	640-15 K-8	640 × 512	15	InSb	80
China, GST	C615M	640 × 512	15	T2SL	80
Russia, Orion	–	640 × 512	15	InSb	80
Russia, Sapfir	–	320 × 256	30	InSb	80

Source: Data extracted from [73] and [54]

3–5 µm range is 640 × 512 pixels. However, a number of companies makes the transition to 1280 × 1024 and even 1920 × 1080 and 1920 × 1536 formats [79].

The factors that favor the use of IR FPAs developed for the 3–5 µm spectral range are a greater contrast, more favorable weather conditions, when with an increase in the water vapor content in the atmosphere, the transmission in the 3–5 µm region decreases more slowly than in the region of 8–14 µm. As a result, greater transparency of the atmosphere is achieved in high humidity conditions, and better resolution due to lower optical diffraction in this spectral range. More information on the parameters of commercially available FPAs can be found on the websites of the companies developing these FPAs (http://www.raytheon.com; http://teledynesi. com/imaging; http://www.sofradir.com; http://www.aim-ir.com; http://www.scd. co.il; http://www.flir.com; https://astrohn.ru; https://www.leonardodrs.com; https:// www.lynred.com; https://www.gst-ir.net; https://orio-ir.ru/en/; https://www.gst-ir.net).

4.3 Trends in FPAs Development

4.3.1 Pixel Size Reduction

The main trend at present is to reduce the weight, size and power consumption of photoelectronic modules. Decreasing the pixel size and increasing the format is a general trend for almost all world developers and manufacturers of IC FPAs (see Figs. 4.9 and 4.10a) [3, 9, 48]. AIM Infrared Modules (Germany), BAE Systems (USA), Brandywine photonics LLC (USA), CalSensors Inc. (USA), EGIDE USA (USA), China Germanium Co. Ltd. (China), FLIR Systems (USA), SCD (Israel), Raytheon Vision Systems (USA), RICOR (Israel), Selex ES (UK), Thales Cryogenics (France), Lynred (France), Spectrolab Inc. (USA) and others work in this direction. By decreasing the pixel size, it is possible to increase their overall number in a given die size, leading to a higher resolution and opening the way for new applications such as persistent surveillance. In particular, a decrease in the pitch and an increase in the format leads to a significant increase in the range of object recognition. On the other hand, maintaining the same number of pixels while decreasing the pixel size results in a much smaller crystal size, which allows detectors to be manufactured with smaller dimensions, weight, power, and cost (SWaP-C). Reducing the pixel dimensions can also reduce the size of the optics to support lower SWaP-C at the system level [14, 49, 57]. Achieving these goals will naturally require an improvement in the technology for growing high-quality epitaxial MCT layers with reduced defectiveness.

640 × 512 pixels format of FPAs with a step of 15 µm is currently the main format of FPAs and, apparently, in terms of price-quality ratio, it will remain so for the next 5–10 years. As a rule, photodetectors forming the array have a mesa structure shown in Fig. 4.10b. Such structure helps to reduce the physical area of the p-n

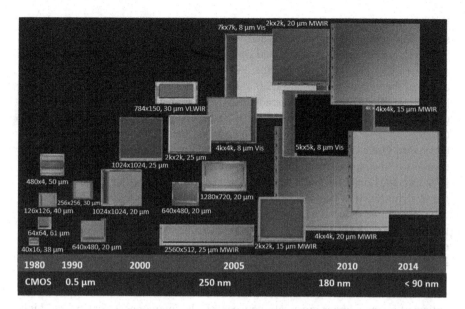

Fig. 4.9 Focal plane array size progression. Illustrates the timeline and progression of increasing array format and size with corresponding reduction in pixel size. (https://www.raytheon.com) (Reprinted with permission from Ref. [72]. Copyright 2016: SPIE)

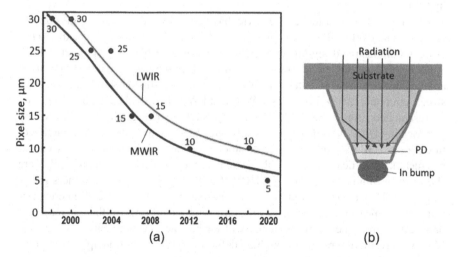

Fig. 4.10 (a) Dynamics of pixel size reduction for CdHgTe-based APDs. Data extracted from [45]; (b) Schematic diagram of Selex ES mesa pixel. (Adapted with permission from Ref. [49]. Copyright 2015: SPIE)

junction, and hence to reduce the dark current. As it is seen in Fig. 4.10b, despite its small size, optical capture remains across the full pixel area and photons, entering the pixel, are effectively trapped and optically concentrated into the junction and absorber [49]. However, the leading FPAs developing companies have already achieved the megapixel format (1280 × 1024 elements) as commercially available [21, 49]. SOFRADIR (France) already in 2015 offered 10 μm pitch HgCdTe MWIR diode array [51]. Sofradir is introducing a 10 μm pixel pitch detector in either XGA (1024 × 768) or HD720 (1280 × 720) formats, named Daphnis. Selex ES Ltd. (Leonardo, Great Britain) has developed a technology for manufacturing SuperHawk matrices of megapixel format in the mid-wave infrared range with a step of 8 μm [49]. Lynred (France) is going to offer matrices with even smaller pitch, 5–7 μm, in the near future. It is predicted that the pixel size of long-wave HgCdTe infrared detectors can be reduced to 5 μm, while that of mid-wave HgCdTe infrared detectors can be reduced to 3 μm. Currently, the prices for such photodetectors arrays are quite high, which does not allow to FPAs developers to make a massive transition to it. However, many predict that such a transition will take place by 2025 [45, 46]. Although, it must be admitted that this will not be easy to achieve. This task presents serious challenges associated with detector processing technologies, lower optical efficiency, larger cross-talk between pixels, interconnect density, and high charge capacity. According to estimates made by Dhar et al. [20], to achieve high sensitivity for LWIR FPAs with 5 μm pixels (for example <30 mK) it is required a large amount of integrated charge to be placed in very small unit cells. For a 5 μm planar unit cell, the charge capacity in standard ROIC technology is less than one million electrons, whereas 8 to 12 million electrons are required for good sensitivity [20].

In spite of these challenges, Teledyne has demonstrated recently small 40 × 40 test HgCdTe LWIR FPAs with pixel size of 5 μm [77]. Interestingly, that these devices demonstrated significantly better performance than was predicted [78]. Authors attribute this effect to lower doping and thinner active layer in optimized architecture resulting in reduction of Auger-limited diffusion dark currents. Full size, 5 μm pitch 1280 × 720 pixel MWIR and LWIR FPAs have also been reported by researchers from DRS Technologies [5]. Such a small pitch was achieved by downscaling DRS' high-density vertically integrated photodiode (HDVIP) architecture (Fig. 4.10b) with interconnects to silicon readout chip obtained by dry etched and metalized vertical vias. The collection efficiency, cross-talk, and operability are shown to be similar to larger pitch HDVIP FPAs reported by these authors previously [74]. They also observed lower dark current levels (factor of 2–4) in comparison to those measured in larger, 12 μm and 15 μm pitch FPAs. So there is reason to believe that these new developments will enable the development of lower cost FPAs with resolution ranging from high definition (HD) resolution up to many millions of pixels.

4.3.2   Quantum-Dimensional Structures

Another important direction in the development of photodetectors arrays is the development of technologies based on the formation of quantum-dimensional struc-tures [58] (see also Chaps. 5, 7 and 8, Vol. 2). At present, such photodetectors arrays are manufactured using wide-gap III-V semiconductor-based heterostructures. However, due to quantum-size effects, these structures have sensitivity in the middle or long-wave infrared spectral ranges.

Currently, the most developed quantum- dimensional photodetectors are photo-resistors based on structures with multiple quantum wells (QWIP), which, depending on the material used, form type-1 (GaAs /AlGaAs), type-II (InAs / GaSb and InAs/GaInSb) and type-III (HgTe/CdTe) superlattices. Such quantum-dimensional structures also include barrier superlattices and structures based on quantum dots. A description of these structures and photodetectors based on them can be found in the Chaps. 5 and 6, Vol. 2. The listed quantum-dimension-based photodetectors, including those in the form of photodetectors array, are already commercially available. According to published forecasts [32], in 10–15 years, IR FPAs, including dual-range FPAs, with ultimate sensitivity will be available in formats up to 1 K-1 K (or $10^3 \times 10^3$ pixels) and more.

4.3.3   Monolithic and Multispectral FPAs

Development of monolithic IR photodetectors [37, 55, 66, 80, 81] can also be attrib-uted to the promising direction of FPAs development. Currently, IR photodetectors arrays are mainly manufactured using hybrid technology, where an MCT-based photodiode array and a silicon readout integrated circuit (multiplexer) are formed separately, and then connected element by element using indium bump technique. However, the hybrid technology has limitations on the number of pixels and the resistance of the FPA formed to mechanical and thermal influences. Monolithic IR photodetectors, in which photosensitive elements are formed in MCT layers grown in the cells of a silicon multiplexer, make it possible to get rid of the above disad-vantages [38, 63, 82]. Unfortunately, no significant success has yet been achieved in this direction. However, studies carried out in recent years [13, 16, 28] have shown that using MCT quantum dots to fabricate photon detectors forming FPAs on the silicon ROIC surface can facilitate this task. In particular, the QD layers are amor-phous what permits fabrication of devices directly onto ROIC substrates, as shown in Fig. 4.11 with no restrictions on pixel or array size and with a short cycle of production. In addition, the monolithic integration of QD-based detectors into ROIC does not require any hybridization steps. The individual pixels are determined by the area of the metal contact pads located on the top of the ROIC surface. Methods of wet chemistry can be used to synthesize colloidal nanoparticles. This so-called top-surface photodetector offers a 100% fill factor and is compatible with

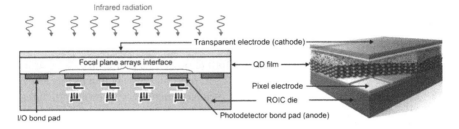

Fig. 4.11 IR monolithic array structure based on CQDs. (Reprinted from Ref. [58]. Published 2021 by MDPI as open access)

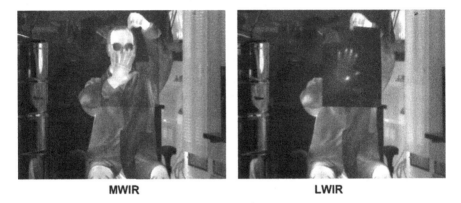

Fig. 4.12 An example of an image taken at T = 78 K for an army DBFM 640 × 480 M/LWIR FPA (# 7586704) in MWIR and LWIR spectral ranges with a field of view of f/5 and a frame rate of 30 Hz. (Reprinted with permission from Ref. [56]. Copyright 2005: SPIE)

post-processing at the top of complementary metal-oxide semiconductor (CMOS) electronics. It is expected that the successful implementation of this new class of infrared technology may contribute to low-cost CMOS cameras for a wide variety of applications [58]. It should be noted that these expectations are met. The first SWIR cameras based on QD thin-film photodiodes, monolithically fabricated on silicon ROICs, have already been released [34]. The Acuros camera has resolution 1920 × 1080 (2.1 megapixels, 15-μm pixel pitch) and uses 0.4 to 1.7 μm broadband spectral response.

Development of two-color and multi-spectral focal plane arrays is another promising direction in the development of thermal imaging technology based on MCT [3, 9]. Almost all leading firms [50, 71] actively develop two-color and multi-spectral FPAs. Their use in optoelectronic systems increases the likelihood of target detection and recognition. For example, Fig. 4.12 demonstrates the capabilities of such a FPAs developed by Raytheon Vision Systems [56]. In this imagery, the subject is holding a sheet of plastic that transmits in the MWIR band but absorbs in the LWIR band. The difference in the images demonstrates the multi-spectral capability of these FPAs. Thus, high information content and compactness of devices are the driving forces of the development of this direction. It is believed that in the near future, dual-spectral IR FPAs will become commercially available.

4.3.4   High Operation Temperature FPAs

Increasing the operating temperatures of FPAs and the development of microcryo-genic cooling systems for "high-operation temperature" FPAs operated at interme-diate temperatures (HOT: ~150–300 K), can also be attributed to important directions in the development of IR technologies in the near future [3, 31, 51, 58, 73]. If the operating temperature of the FPAs increases without degrading image quality, then smaller coolers can be used and the SWaP and system cost can be reduced. How does this, along with pixel size reduction, affect the size of FPAs, developed and fabricated by DRS Network and Imaging Systems [57], shown in Fig. 4.13.

For example, Shkedy et al. [67] reported that an increase in FPA operating tem-perature from 77 to 150 K reduced the cooler power consumption from 25 W to 7.5 W. A more significant advantage can be obtained if the operating temperature is increased to >200 K, where an inexpensive thermoelectric cooler can be imple-mented. Typically, the increase in operating temperatures of the MWIR and LWIR infrared detectors try to resolve via reducing dark current, reducing the rate of heat

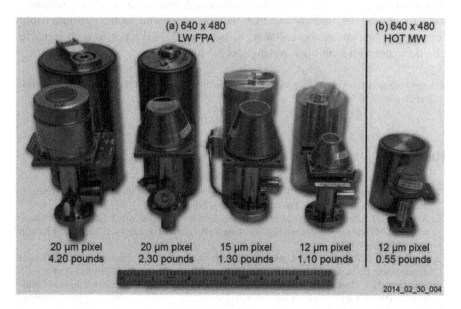

Fig. 4.13 (**a**) Progressive decrease in a DRS production 640 × 480 FPA LW package as a function of pixel pitch, and (**b**) the reduction in package size for high operating temperatures (HOT). (Reprinted with permission from Ref. [57]. Copyright 2014: SPIE)

release in the active region and minimizing the active volume of the detector without reducing the quantum efficiency [3, 9, 22, 42, 49, 58, 70]. According to Kinch [39], in order to achieve this goal, the doping concentration below 5×10^{13} cm^{-3} is required. However, despite the sixty-year history of the development of HgCdTe-based photodetectors, its ultimate limit of HOT performance, which lies in the range of 140–300 K, has not been reached. In most publications, FPAs operating temperatures do not exceed 110–120 K [31, 51] and only a few designs allow FPAs to operate at higher temperatures [67, 79]. For example, Thorne et al. [79] reported that their MWIR FPAs (640 × 512 array format, 16 µm pixel pitch) can operate at 155 K. To achieve such operating temperature, they used possibilities of the MOVPE process. Using mesa diode heterostructures (see Fig. 4.10b), the volume of the absorber was restricted to limit the volume for diffusion current generation. In addition, the ability to dope the MCT absorber of p-type ensured low Auger recombination, and the ability to locate the junction in the region with a higher bandgap reduced thermal generation within the depletion layer. The distance between the p-n junction and the absorber, as well as the band gap were also monitored. Even more successes in this direction are expected in the coming years. For example, Kinch [39] predicted that large area ultra-small pixel diffraction-limited and background-limited photon detecting MW and LW HgCdTe FPAs operating at room temperature will be available within the next ten years. The basis for such predictions is the long Shockley-Read-Hall (SRP) carrier lifetime in HgCdTe, which makes this material a great candidate for FPAs operated at 300 K. Simulations carried out by Akhavan et al. [2] have shown that the new IR detector architecture based on a unipolar nBn structure can also help reduce the dark current and increase the operating temperature. The results obtained showed that photodetectors with such a structure could operate at temperatures above 200 K. Kopytko et al. [43] reported such HgCdTe nBn detectors with a cut-off wavelength of 3.5 µm.

Acknowledgments This research was funded by the State Program of the Republic of Moldova, project 20.80009.5007.02.

References

1. Abbott P, Thorne PM, Arthurs CP (2011) Latest detector developments with HgCdTe grown by MOVPE on GaAs substrates. Proc SPIE 8012:801236
2. Akhavan ND, Jolley G, Umana-Membreno GA, Antoszewski J, Faraone L (2014) Performance modeling of bandgap engineered HgCdTe-based nBn infrared detectors. IEEE Trans Electron Dev 61:3691–3698
3. Antoszewski J, Akhavan ND, Umana-Membreno G, Gu R, Lei W, Faraone L (2015) Recent developments in Mercury Cadmium Telluride IR detector technology. ECS Trans 69(14):61–75
4. Antoszewski J, Musca CA, Dell JM, Faraone L (2003) Small two-dimensional arrays of mid-wavelength infrared HgCdTe diodes fabricated by reactive ion-induced p-to-n type conversion. J Electron Mater 32:627–632
5. Armstrong JM, Skokan MR, Kinch MA, Luttmer JD (2014) HDVIP five-micron pitch HgCdTe focal plane arrays. Proc SPIE 9070:907033

6. Baker I, Maxey C, Hipwood L, Weller H, Thorne P (2012) Developments in MOVPE HgCdTe arrays for passive and active infrared imaging. Proc SPIE 8542:85421A
7. Baker IM, Ballinga RA (1984) Photovoltaic CdHgTe-silicon hybrid focal planes. Proc SPIE 510:121–129
8. Bangs J, Langell M, Reddy M, Melkonian L, Johnson S, Elizondo L et al (2011) Large format high operability SWIR and MWIR focal plane array performance and capabilities. Proc SPIE 8012:801234
9. Bhan RK, Dhar V (2019) Recent infrared detector technologies, applications, trends and development of HgCdTe based cooled infrared focal plane arrays and their characterization. Opto-Electron Rev 27(2):174–193
10. Boulade O, Moreau V, Mulet P, Gravrand O, Cervera C, Zanatta J-P et al (2016) Development activities on NIR large format MCT detectors for astrophysics and space science at CEA and SOFRADIR. Proc SPIE 9915:99150C
11. Brellier D, Gout E, Gaude G, Pelenc D, Ballet P, Miguet T, Manzato MC (2014) Bulk growth of CdZnTe: quality improvement and size increase. J Electron Mater 43(8):2901–2907
12. Brill G, Chen Y, Wijewarnasuriya P, Dhar NK (2009) Infrared focal plane array technology utilizing HgCdTe/Si: successes, roadblocks and material improvements. Proc SPIE 7419:74190L
13. Buurma C, Ciani AJ, Pimpinella RE, Feldman JS, Grein CH, Guyot-Sionnes P (2017) Advances in HgTe colloidal quantum dots for infrared detectors. J Electron Mater 46:6685–6688
14. Caulfield JT, Wilson JA, Dhar NK (2014) Benefits of oversampled small pixel focal plane arrays. Proc SPIE 9070:907035
15. Cervera C, Boulade O, Gravrand O, Lobre C, Guellec F, Sanson E, Castelein P (2017) Ultra-low dark current HgCdTe detector in SWIR for space applications. J Electron Mater 46(10):6142–6149
16. Chatterjee A, Babu PN, Jagtap A, Koteswara Rao KSR (2019) Uncooled mid-wave infrared focal plane array using band gap engineered mercury cadmium telluride quantum dot coated silicon ROIC. e-J Surf Sci Nanotechnol 17:95–100
17. Chen Y, Farrell S, Brill G, Wijewarnasuriya P, Dhar NK (2008) Dislocation reduction in CdTe/Si by molecular beam epitaxy through in-situ annealing. J Crystal Growth 310(24):5303–5307
18. Costard E, Bois P, De Rossi A, Nedelcu A, Cocle O, Gauthier F-H, Audier F (2003) QWIP detectors and thermal imagers. C R Physique 4:1089–1102
19. Delaunay PY, Nosho BZ, Gurga AR, Terterian S, Rajavel RD (2017) Advances in III-V based dual-band MWIR/LWIR FPAs at HRL. Proc SPIE 10177:101770T
20. Dhar NK, Dat R, Sood AK (2013) Advances in infrared detector array technology. In: Pyshkin S (ed) Optoelectronics – advanced materials and devices. INTECH, pp 149–190
21. Dorn ML, Piphera JL, McMurtry C, Hartman S, Mainzer A, McKelvey M et al (2016) Proton irradiation results for long-wave HgCdTe infrared detector arrays for near-earth object camera. J Astronom Telesc Instrum Syst 2(3):036002
22. D'Souza AI, Robinson E, Ionescu AC, Okerlund D, de Lyon TJ, Rajavel RD et al (2012) MWIR InAs$_{1-x}$Sb$_x$ nCBn detectors data and analysis. Proc SPIE 8353:835333
23. Gravrand O, Rothman J, Cervera C, Baier N, Lobre C, Zanatta JP et al (2016b) HgCdTe detectors for space and science imaging: general issues and latest achievements. J Electron Mater 45(9):4532–4541
24. Gravrand O, Rothman J, Castelein P, Cervera C, Baier N, Lobre C et al (2016a) Latest achievements on MCT IR detectors for space and science imaging. Proc SPIE 9819:98191W
25. Gu R, Kala H, Antoszewski J, Umana-Membreno G, Dehdashtiakhavan N, Madni I, Faraone L (2018) Recent advances in IR imaging focal plane arrays technology at UWA. In: Proc. of the Conference on Optoelectronic and Microelectronic Materials and Devices (COMMAD), 9–13 Dec. 2018. Perth, WA, Australia, 18673702, pp 11–12. https://doi.org/10.1109/COMMAD.2018.8715245
26. Guellec F, Boulade O, Cervera C, Moreau V, Gravrand O, Rothman J, Zanatta J (2014) ROIC development at CEA for SWIR detectors: pixel circuit architectures for space applications and trade-offs, ICSO conference. Proc SPIE 10563:105630K

27. Gunapala SD, Bandara SV, Hill CJ, Ting DZ, Liu JK, Rafol SB et al (2007) Demonstration of 640 x 512 pixels long-wavelength infrared (LWIR) quantum dot infrared photodetector (QDIP) imaging focal plane array. Infrared Phys Technol 50:149–155

28. Guyot-Sionnest P, Roberts JA (2015) Background limited mid-infrared photodetection with photovoltaic HgTe colloidal quantum dots. Appl Phys Lett 107:91115

29. Fièque B, Martineau L, Sanson E, Chorier P, Boulade O, Moreau V, Geoffray H (2011) Infrared ROIC for very low flux and very low noise applications. Proc SPIE 8176:81761I

30. Fièque B, Lamoure A, Salvetti F, Aufranc S, Gravrand O, Badano G et al (2018) Development of astronomy large focal plane array "ALFA" at Sofradir and CEA. Proc SPIE 10709:1070905

31. Figgemeier H, Hanna S, Eich D, Mahlein K-M, Fick W, Schirmacher W, Thöt R (2016) State of the art of AIM LWIR and VLWIR MCT 2D focal plane detector arrays for higher operating temperatures. Proc SPIE 9819:98191C

32. Filachov AM, Taubkin IL, Trishenkov MA (2015) A review on advances in the solid-state photoelectronics. Uspehi Prikladnoi Phiziki 3(2):162–168. (in Russian)

33. Finger G, Dorn R, Meyer M, Mehrgan L, Moorwood AFM, Stegmeier J (2006) Interpixel capacitance in large format CMOS hybrid arrays. Proc SPIE 6276:62760F

34. Hafz SB, Scimeca M, Sahu A, Ko D-K (2019) Colloidal quantum dots for thermal infrared sensing and imaging. Nano Convergence 6:7

35. Ishimwe R, Abutaleb K, Ahmed F (2014) Applications of thermal imaging in agriculture—a review. Adv Remote Sens 3:128–140

36. Jerram P, Beletic J (2019) Teledyne's high performance infrared detectors for space missions. Proc SPIE 11180:111803D

37. Jiang J, Tsao S, Mi K, Razeghi M, Brown GJ, Jelen C, Tidrow MZ (2005) Advanced monolithic quantum well infrared photodetector focal plane array integrated with silicon readout integrated circuit. Infr Phys Technol 46:199–207

38. Joshi AM (1998) The next generation of monolithic infrared detector arrays. AIP Conf Proc 420:67

39. Kinch MA (2014) State-of-the-art infrared detector technology. SPIE Press, Bellingham

40. Kinch MA (2010) HgCdTe: recent trends in the ultimate IR semiconductor. J Electron Mater 39(7):1043–1052

41. Klipstein PC, Avnon E, Azulai D, Benny Y, Fraenkel R, Glozman A et al (2016) Type II superlattice technology for LWIR detectors. Proc SPIE 9819:98190T

42. Knowles P, Hipwood L, Pillans L, Ash R, Abbott P (2011) MCT FPAs at high operating temperatures. Proc SPIE 8185:818505

43. Kopytko M, Kębłowski A, Gawron W, Kowalewski A, Rogalski A (2014) MOCVD grown HgCdTe barrier structures for HOT conditions. IEEE Trans Electron Dev 61(11):3803–3807

44. Kozlov AI (2010) Design features and some implementations of silicon multiplexers for IR photodetectors. J Opt Technol 77(7):421–428

45. Kulchitsky NA, Naumov AV, Startsev VV (2020a) Infrared focal plane array detectors: "post pandemic" development trends. Part I. Photonics Russia 14(3):234–244

46. Kulchitsky NA, Naumov AV, Startsev VV (2020b) Infrared focal plane array detectors: "post pandemic" development trends. Part II. Photonics Russia 14:320–330

47. Lahiri BB, Bagavathiappan S, Jayakumar T, Philip J (2012) Medical applications of infrared thermography: a review. Infr Phys Technol 55:221–235

48. Liu M, Wang C, Zhou L-Q (2019) Development of small pixel HgCdTe infrared detectors. Chin Phys B 28(3):037804

49. McEwen RK, Jeckells D, Bains S, Weller H (2015) Developments in reduced pixel geometries with MOVPE grown MCT arrays. Proc SPIE 9451:94512D

50. Patten EA, Goetz PM, Viela FA, Olsson K, Lofgrren DF, Vodicka JG, Johnson SM (2010) High-performance MWIR/LWIR dual-band 640 × 480 HgCdTe/Si FPA's. J Electron Mater 39(10):2215–2219

51. Péré-Laperne N, Rubaldo L, Kerlain A, Carrère E, Dargent L, Taalat R, Berthoz J (2015) 10 μm pitch design of HgCdTe diode array in Sofradir. Proc SPIE 9370:937022

52. Peric D, Livada B, Peric M, Vujic S (2019) Thermal imager range: predictions, expectations, and reality. Sensors 19:3313
53. Perrais G, Gravrand O, Baylet J, Destefanis G, Rothman J (2007) Gain and dark current characteristics of planar HgCdTe avalanche photo diodes. J Electron Mater 36(8):963–970
54. Popov V (2020) Modern cooled IR photodetectors. Systemi Bezopasnosti 3:68–70. (in Russian)
55. Pusino V, Xie C, Khalid A, Steer MJ, Sorel M, Thayne IG, Cumming DRS (2016) InSb photodiodes for monolithic active focal plane arrays on GaAs substrates. IEEE Trans Electron Dev 63(8):3135–3142
56. Radford WA, Patten EA, King DF, Pierce GK, Vodicka J, Goetz P, Venzor G et al (2005) Third generation FPA development status at Raytheon vision systems. Proc SPIE 5783:331–339
57. Robinson J, Kinch M, Marquis M, Littlejohn D, Jeppson K (2014) Case for small pixels: system perspectives and FPA challenges. Proc SPIE 9100:91000I
58. Rogalski A, Martyniuk P, Kopytko M, Hu W (2021) Trends in performance limits of the HOT infrared photodetectors. Appl Sci 11(2):501
59. Rogalski A, Martyniuk P, Kopytko M (2017) InAs/GaSb type-II superlattice infrared detectors: future prospect. Appl Phys Rev 4:031304
60. Rogalski A (2009) Infrared detectors for the future. Acta Phys Polonica A 116(3):389–405
61. Rogalski A (2008) New material systems for third generation infrared photodetectors. Opto-Electron Rev 16(4):458–482
62. Rogalski A (2005) HgCdTe infrared detector material: history, status and outlook. Rep Prog Phys 68:2267–2336
63. Rogalski A (2004) Optical detectors for focal plane arrays. Opto-Electron Rev 12(2):221–245
64. Rogalski A (2000) Infrared detectors at the beginning of the next millennium. Sens Mater 12(5):233–288
65. Szajewska A (2017) Development of the thermal imaging camera (TIC) technology. Procedia Eng 172:1067–1072
66. Sánchez FJ, Rodrigo MT, Vergara G, Lozano M, Santander J, Torquemada MC et al (2005) Progress on monolithic integration of cheap IR FPAs of polycrystalline PbSe. Proc SPIE 5783:441–447
67. Shkedy L, Brumer M, Klipstein P, Nitzani M, Avnon E, Kodriano Y et al (2016) Development of 10μm pitch XBn detector for Low SWaP MWIR applications. Proc SPIE 9819:98191D
68. Singh A, Shukla AK, Pal R (2015) HgCdTe e-avalanche photodiode detector arrays. AIP Adv 5:087172
69. Sizov F (2015) IR-photoelectronics: photon or thermal detectors? Outlooks. Sens Electron Microelectron Technol 12(1):26–53. (in Russian)
70. Smith KD, Wehner JGA, Graham RW, Randolph JE, Ramirez AM, Venzor GM et al (2012) High operating temperature mid-wavelength infrared HgCdTe photon trapping focal plane arrays. Proc SPIE 8353:83532R
71. Smith EPG, Venzor GM, Gallagher AM, Reddy M, Petterson JM, Lofgreen DD, Randolph JE (2011) Large-format HgCdTe dual-band long-wavelength infrared focal-plane arrays. J Electron Mater 40(8):1630–1636
72. Starr B, Mears L, Fulk C, Getty J, Beuville E, Boe R et al (2016) RVS large format arrays for astronomy. Proc SPIE 9915:99152X
73. Startsev V, Naumov A (2018) Modern photodetectors of the infrared spectrum and development trends. Tehnologii Zashiti 5:66–70. (in Russian)
74. Strong RL, Kinch MA, Armstrong J (2013) Performance of 12-μm- to 15-μm-pitch MWIR and LWIR HgCdTe FPAs at elevated temperatures. J Electron Mater 42:3103–3107
75. Sun X, Abshire JB, Krainak MA, Lu W, Beck JD, Sullivan WW III et al (2019) HgCdTe avalanche photodiode array detectors with single photon sensitivity and integrated detector cooler assemblies for space lidar applications. Opt Eng 58(6):067103
76. Sun X, Abshire JB, Beck JD (2014) HgCdTe e-APD detector arrays with single photon sensitivity for space lidar applications. Proc SPIE 9114:91140K

77. Tennant WE, Gulbransen DJ, Roll A, Carmody M, Edwall D, Julius A et al (2014) Small-pitch HgCdTe photodetectors. J Electron Mater 43:3041–3046
78. Tennant WE (2012) Interpreting mid-wave infrared MWIR HgCdTe photodetectors. J Prog Quantum Electron 36:273–292
79. Thorne P, Gordon J, Hipwood LG, Bradford A (2013) 16 Megapixel 12 μm array developments at Selex ES. Proc SPIE 8704:87042M
80. Xie C, Aziz M, Pusino V, Khalid A, Steer M, Thayne IG, M. Sorel M., Cumming D.R.S. (2017) Single-chip, mid-infrared array for room temperature video rate imaging. Optica 4(12):1498–1502
81. Xie C, Pusino V, Khalid A, Aziz M, Steer MJ, Cumming DRS (2016) A new monolithic approach for mid-IR focal plane arrays. Proc SPIE 9987:99870T
82. Zanio K (1990) HgCdTe on Si for hybrid and monolithic FPAs. Proc SPIE 1308:180–193
83. Zverev AV, Makarov Yu S, Mikhantiev EA, Sabinina IV, Sidorov GY, Dvoretskiy SA (2015) 384 × 288 readout integrated circuit for MWIR and LWIR HgCdTe based FPA. J Phys Conf Series 643:012055

Chapter 5
New Trends and Approaches in the Development of Photonic IR Detector Technology

Ghenadii Korotcenkov and Igor Pronin

5.1 Introduction

As will be shown in subsequent chapters (Chaps. 4 and 15, Vol. 1), HgCdTe is an expensive material. Therefore, in an attempt to replace this material with cheaper ones in certain applications, new approaches to the development of new materials sensitive to infrared radiation have been proposed. At the same time, the task of increasing operating temperatures and increasing the efficiency of devices was solved [67, 72]. As a result of these developments, new technologies have appeared that make it possible to form structures with unique photoelectric properties on the basis of III-V and II-VI compounds. These include superlattices, barrier photoconductors (so-called barrier structures), and structures with multiple quantum wells (MQWs) [35, 49]. The technology of using quantum dots (QDs) in the manufacture of infrared detectors has also received significant development [72, 87].

5.2 High Operating Temperature (HOT) Detectors

One of the main disadvantages of photonic IR photodetectors is the need to cool them down to low temperatures to achieve the required sensitivity parameters. As a rule, photodetectors operate at temperatures from 8 to 80 K. Such low temperatures are necessary to reduce the dark current density, which is high in all detectors with a narrow band gap. However, this greatly complicates their use and increases the

G. Korotcenkov (✉)
Department of Physics and Engineering, Moldova State University, Chisinau, Moldova
e-mail: ghkoro@yahoo.com

I. Pronin
Department of Nano- and Microelectronics, Penza State University, Penza, Russia

© The Author(s), under exclusive license to Springer Nature Switzerland AG 2023
G. Korotcenkov (ed.), *Handbook of II-VI Semiconductor-Based Sensors and Radiation Detectors*, https://doi.org/10.1007/978-3-031-20510-1_5

cost of the devices being developed. Therefore, the search for solutions that allow increasing the operating temperatures up to 150–300 K is an important task, the solution of which gives a greater economic effect. For example, when using a Stirling cooling motor for cooling, simply raising the operating temperature from 80 K to 150 K results in approximately a half of energy consumption. The cooling time of the device to the set temperature is also reduced. Higher temperatures also significantly increase the mean time between failures of the Stirling cooling engine.

A number of concepts to increase the operating temperature of photodetectors have been proposed [35, 60, 67]. However, all of them, as a rule, are aimed at reducing the dark current, which can be classified into two groups [39]:

1. inherent mechanisms, which depend only on the intrinsic material properties:

   (a) diffusion current due to Auger or radiative recombination in the n-region or p-region, and
   (b) band-to-band tunneling current;

2. defect-related mechanisms, which require surface or bulk defects, located within the depletion region or within a diffusion length of either side of the depletion region:

   (a) diffusion current due to SRH recombination in the n-region or p-region,
   (b) generation–recombination within the depletion region,
   (c) trap-assisted tunneling, and
   (d) surface generation current from surface states.

Therefore, any approaches, which can suppress or eliminate one or more of these dark current mechanisms, will be helpful in enhancing the operating temperature of photodetectors and FPAs. In particular, significant improvements have been obtained by suppression of Auger thermal generation in excluded photoconductors and extracted photodiodes. As is known, thermal generation and recombination in narrow gap semiconductors at near room temperature is determined by the Auger mechanism [14]. It was established that this problem can be solved using reverse biased N^+-p-P^+ photodiodes with lightly doped absorber and non-equilibrium mode of operation. Under strong depletion, the majority carrier concentration saturates at the extrinsic level while the concentration of minority carriers is reduced below the extrinsic level. A device operating in non-equilibrium mode was first proposed by Ashley and Elliott [4] in the mid-1980s. However, these non-equilibrium devices require significant bias currents and exhibit excessive low frequency 1/f noise that extends up to MHz rang [15, 37].

At low temperatures, the diffusion-limited component of the dark current dominates in the dark current, while at high temperatures; the dark current components associated with defects dominate in the dark current. As the operating temperature rises, more and more defects become electrically active, which leads to a significant increase in the dark current [39]. Therefore, if there is a task to increase the operating temperatures of the detectors, there is a strong incentive to reduce and eliminate the effect of defects on the dark current. It would seem that the most reliable way to solve this problem is to improve growth and post-growth processing techniques. However, it turned out that this approach is not optimal. Experiments and simulations have shown that significantly better results can be achieved with new approaches to band-gap engineering of various compound semiconductors, which allows the development of new detector architectures that are less sensitive to the presence of defects. New emerging strategies include barrier structures such as nBn detectors, low-dimensional structures such as T2SLs, photon trapping detectors, and multistage/cascade infrared devices [35, 49].

5.3 Quantum Well Infrared Photodetectors

In their principle of operation, quantum well infrared photodetectors (QWIPs) are fundamentally different from the previously considered bulk detectors [13]. If the operation of QWIPs is based on quantum-scale physical effects, then bulk detectors operate on larger scale effects. In 1987, Levine et al. [40] demonstrated the convincing merit of this approach on the example of structures based on GaAlAs superlattices. It was shown that efficient IR photodetectors for 10 μm spectral range can be developed on the basis of wide-gap semiconductors. QWIPs in their most basic form are a periodic repetition of layers of two materials with dissimilar band gaps (see Fig. 5.1a). A material with a lower band gap is usually called a well layer, and a material with a higher band gap is called a barrier layer. The well layers are doped such that without illumination there will be carrier electrons present in the ground energy state. These carriers are then excited by the incident photons into an energy state near the edge of the conduction band of the barrier material, where an applied voltage moves the carrier from the well to the contacts (Fig. 5.1b). In principle, QWIPs are man-made extrinsic photoconductors, such as Ge:Au or Si:As, in which quantum wells replace impurity atoms. The ability to vary the binding energy of

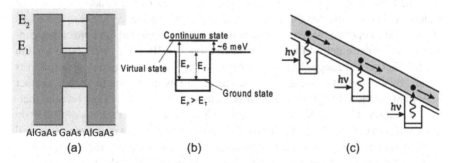

Fig. 5.1 (**a, b**) A band diagram of a quantum well in a QWIP. Carriers are excited by incident photons from the ground energy state of the well (E_1) to the excited energy state of the well (E_2). (**c**) Photonic mechanisms of excitation of the electron subsystem in a multiple quantum wells. (**a**) Reprinted from Ref. [13]. Published 2013 by MDPI as open access. (**b, c**) (Adapted with permission from [66]. Copyright 2003: Pergamon)

electrons in QWIPs to match the desired IR response by changing quantum well depth and width is an important advantage of such structures.

In the early stages of the development of QWIPs, AlGaAs/InGaAs, InGaAs/ GaInP, InGaAs/InP, GaSb/AlGaSb, and AlGaAs/GaAs were considered as the main materials suitable for these applications [13]. However, after numerous studies, they came to the conclusion that the most promising system is AlGaAs/GaAs. It was found that GaAs/AlGaAs QWIR detectors are the most advances due to the almost perfect natural lattice match between GaAs and AlGaAs. Besides that, GaA/AlGaAs quantum well devices were able to utilize standard manufacturing and processing technologies developed for GaAs. They had high uniformity and well-controlled growth during molecular beam epitaxy on six-inch GaAs wafers, high yield and hence low cost, and better thermal and radiation resistance. As a result, the uniformity of GaA/AlGaAs-based QWIP devices is very high, and their performance approaches theoretical limits. Currently, based on such GaAs/AlGaAs structures, QWIPs have been developed for the spectral ranges of 3–5 μm and 8–14 μm [85].

When comparing QWIPs with other IR detectors, QWIPs based on III-V compounds have a number of positive and negative performance characteristics [13]. Compared to MCT devices, QWIPs can have lower dark currents, higher detectivity, and higher NETDs. They use widely applied III-V material processing techniques, making them easier to fabricate than MCTs-based devices. QWIPs also generally have higher radiation resistance than narrow band gap materials such as MCT and InSb. However, there are also significant disadvantages. The most obvious disadvantage is associated with the use of QWIP quantum constraint. Due to the limitations imposed by quantum mechanics, QWIP can only absorb light falling on it only when there is quantum limitation along one of the perpendicular axes. This leads to a big problem, as due to conventional growing methods, most QWIPs are not able to absorb normally perpendicular incident light. To mitigate this problem, most QWIP devices use some form of frontal diffusion filter (optical coupler) that redirects normally incident light so that it can be absorbed.

It should also be noted that the maximum photon energy that QWIP can absorb is limited by the energy difference between the edges of the conduction band of materials. This means that there must be a very large difference between the edges of the band to operate at shorter wavelengths. Finding materials that meet this condition can be extremely difficult. For this reason, QWIPs have rarely been used in the SWIR spectral range [81]. In addition, the optical cross-section absorption is limited by the dopant concentrations, leading to a low quantum efficiency. Another important problem of QWIP-based IR detectors is the relatively high level of thermal excitation at T > 40 K, and especially at T > 70 K. These features of QWIP do not allow it to compete with MCT-based detectors in this temperature range, as well as in applications where a short accumulation time is required. However, despite such a comparison, in the ultra-long-wavelength range (VLWIR) and at low temperatures, the QWIP array demonstrates excellent performance, including in conditions with a low background level.

5.4   Type-II Strained-Layer Superlattice

Strained-layer superlattice (SLS), while initially appearing to have a similar structure to QWIPs, actually operates using dramatically different physical principles [13]. The misleading similarities arise from a superlattice with extremely thin layer thicknesses (approximately single nanometers) as an active absorbing layer. However, while QWIPs generally utilize relatively thick barrier layers (approximately ten of nanometers), all active layers in SLS have a thickness of the same order of magnitude [64]. Like QWIP, the most common SLS structure is interleaving layers of two different materials, but some later devices utilize more complex heterostructures [8].

Type-II strained-layer superlattice is usually formed from monolayers of InAs and GaSb, repeated with a certain period [19, 64], although other combinations are possible such as InAs/GaInSb, InAsSb/GaSb or InAs/InAsSb [8]. However, the InAs/GaSb combination turned out to be more promising [13]. The layers can be interleaved more than 300 times to form an IR absorbing region. These structures are mainly grown on GaSb (100) substrates. Although the lattice mismatch between InAs and GaSb is less than 1%, InAs deforms when stretched. These quantum structures are characterized by a broken band gap with type II band alignment, leading to spatially indirect transitions between hole states localized in GaSb layers and delocalized electronic states in InAs layers (see Fig. 5.2a). An important advantage of these structures is the ability to vary the band gap, as is done in HgCdTe. The effective band gap of these structures can be changed from 0.3 eV to values below 0.1 eV. It is important that, in contrast to HgCdTe, this is done not by changing the composition of the compound, but by changing the thickness and composition of the layers that form the superlattice. However, the band structure in SLS causes low overlap integral between the electron and hole wave-functions, and hence a low IR absorption of such structures. In recent years, it has been proposed to improve the

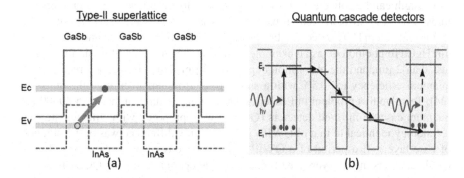

Fig. 5.2 (a) Energy band diagram of type-II superlattice, showing electronic transition. (b) A band diagram of a quantum cascade detector. (a) Reprinted from Ref. [30]. Published 2013 by IOP as open access. (b) Reproduced with permission from (Buffaz et al., State of the art of quantum cascade photodetectors. *Proc. SPIE* 7660 (2010), 76603Q). Copyright 2010: SPIE

basic structure of SLS by introducing a very thin (a few angstroms) layer of wider bandgap material (AlSb) as a barrier to electrons of the main charge carriers. This opens up great opportunities for varying the band gap and improving performance. Detectors, using type-II SLS, are photovoltaic devices.

It should be noted that the type-II strain layer superlattice is the most advanced among the new IR technologies. Successful demonstrations of MWIR imaging arrays, operating at 120 K with efficiencies close to 40% have been reported. There are also reports of single pixel LWIR detectors based on type-II strain layer super-lattice. For example, Rehm et al. [64] reported about InAs/GaSb-based SLS designed for 8–12 μm (LWIR) spectral range. The main difficulties in this technology are the formation of ideal boundaries in the superlattice and passivation of device. Research has shown that dark current, consisting of bulk leakage and surface leakage, is an important problem in SLS-based photodetectors. The bulk component depends on the quality of the formed SLS, while the surface component depends on the manufacturing conditions of the device itself, including etching and passivation. In the shorter wavelength region of the spectrum, it appears that due to the overlap of the band structure of different materials in devices, the effective band gap between minibands tends to be quite narrow. This tends to make it difficult to use these devices for SWIR and shorter wavelength applications [13].

5.5   Multi-Stage or Cascade IR Detectors

In an optimally designed photodiode, the sensitivity and diffusion length are closely related. In particular, increasing the thickness of the absorber does not always lead to the desired improvement in the signal-to-noise (S/N) ratio. With a thickness of absorber much greater than the diffusion length, only a limited fraction of the photogenerated charge carriers contributes to the quantum efficiency, since only charge carriers that are photogenerated at a distance from the junction less than the diffusion length can be collected. This effect is especially pronounced at high temperatures, when the diffusion length usually decreases. This is precisely the situation that is realized in HOT detectors. For example, calculations using an uncooled 10.6-μm HgCdTe photodiode as an example show that the ambipolar diffusion length is less than 2 μm, and the absorption depth is ~13 μm. This discrepancy reduces the quantum efficiency to ~15% with a single pass of radiation through the detector [24]. To solve this problem in PDs operating at temperatures higher than 40–80 K, quantum cascade infrared detectors (QCIDs) were proposed with a manufacturing technology identical to that of QWIPs [49]. The only difference was in the number of base layers. If a period of a QWIP contains only two layers, each period of a QCID contains many layers (see Fig. 5.2b). The operation of such QCDs is based on the principles of multi-stage detection. Gendron et al. [20] first demonstrated the QCID in 2004. Quantum cascade detectors were originally developed as photoelectric detectors, that is, they operate without an applied bias. QCIDs contain several discrete absorbers that form a series of cascade stages. In such structures with carefully selected layer thicknesses incident light excites a carrier from the ground energy state of the absorbing well to an excited energy state. The carrier then tunnels through the barrier layers into the adjacent wells. This continues until the carrier reaches the ground state of the next period of the superlattice. Thus, the photoexcited electrons are transported from one active well to the next one by phonon emission through cascaded levels. It is important to note that the thickness of each step can be less than the diffusion length, while the total thickness of all absorbers can be comparable or even greater than the diffusion length. As a result, we were able to drastically reduce the generated dark current and increase the operating temperatures of the IR detectors [23]. Thus, QCIDs are a good photovoltaic alternative to QWIPs.

Currently, several options for the implementation of cascade IR detectors have been developed. They are usually grouped into two main classes: (i) so called intersubband (IS) unipolar quantum cascade IR detectors, and (ii) interband (IB) ambipolar QCIDs. To describe the performance of IS QCIDs, it is convenient to use the formalism originally developed for QWIPs detectors [75]. Well-established semiconductor material systems are currently available to implement IS QCIDs, such as InGaAs/AlAsSb (near IR), InGaAs/InAlAs (mid IR), and GaAs/AlGaAs (long IR up to THz) [9, 25]. These detectors are cryogenically cooled. As for IB QCIDs, one of the best candidates for realizing such devices is the InAs/GaSb T2SL material system [45]. These IB QCIDs were able to operate at temperatures up to room temperature [61]. For example, the T2SL cascade detectors have demonstrated high operating temperatures, up to 400 K, which cannot be achieved with photodetectors based on the HgCdTe material [49].

Due to the stringent requirements for layer thickness in the structure used in QWIP and QCID, such devices are usually grown using molecular beam epitaxy (MBE) [9]. QCIDs are currently being developed for all spectral bands from SWIR to VLWIR [9].

5.6 Unipolar/Monovalent Barrier IR Detectors

The term "unipolar barrier" was proposed to describe a barrier that can block the flow of one type of charge carriers (electrons or holes), but does not impede the flow of carriers of another type (see Fig. 5.3a, b). Currently, many versions of IR detectors have been proposed that use such barriers to optimize their parameters [33, 34, 49]. Barrier layers were investigated for both the conduction band (nBn structures) and the valence band (pBp structures), but the most popular are nBn type detectors [34, 69]. For the first time, Maimon and Wicks [47] proposed detectors based on nBn structures. In such nBn structures, the n-type semiconductor on one side of the barrier forms a contact layer through which voltage is applied to the device, while the narrow-gap n-type semiconductor on the other side of the barrier forms a photon absorbing layer. For efficient collection of photogenerated carriers, the thickness of this layer must be comparable to the absorption length of light. Usually it is a few

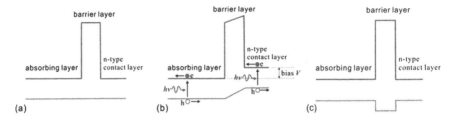

Fig. 5.3 Schematic energy band diagram of an ideal nBn detector under (**a**) zero bias and (**b**) illumination and low reverse bias V. (**c**) Schematic energy band diagram of an nBn HgCdTe detector with high x value $Hg_{1-x}Cd_xTe$ alloy as the barrier layer. (Reproduced with permission from [39]. Copyright 2015: AIP Publishing LLC)

microns. The same type of doping in the barrier and active layers is the key to maintaining a low diffusion limited dark current. Essentially, the nBn detector allows photogenerated holes to flow to a contact (cathode). Thus, the nBn detector operates as a minority carrier device, selectively blocking electrons in the contact layer by channeling holes out of the absorber. The optically generated carriers in the absorber collect at opposite contacts, where collection efficiency can be improved if the device is operated with a slight bias. At the same time, the dark current of the main charge carriers associated with the Shockley-Read-Hall (SRH) processes and the surface leakage current are blocked. As a result, the noise level is sharply reduced. This is because in the nB_nn the depletion region in the active layer is almost completely absent, where generating processes are usually activated by the SRH mechanism.

For maximum effect, the barrier must be carefully designed. This means that the material forming the barrier must have good lattice matching with the surrounding material and have zero offset in one band, and large offset in the other band. It is believed that the height of the potential barrier should be sufficient (>nkT) to block thermally induced electrons and thick enough to prevent tunneling. It was found that barriers with a thickness of more than 100 nm are sufficient for this. In addition, this barrier should be located near the minority carrier collector and away from the optical absorption region. This structure can be optimized by adjusting the bias voltage and doping profiles. Unfortunately, little or no valence band offset was difficult to realize using standard infrared detector materials such as InSb and HgCdTe. At the same time, the absence of a depletion region in absorber offers a way for materials with relatively poor SRH lifetimes, such as all III-V compounds, to overcome such disadvantage as large depletion dark currents.

The principles of operation of nBn-based detectors are described in detail in the literature [34, 47, 74]. Although the idea for nBn-based detectors originated in the development of detectors based on bulk materials (InAs [47]), the greatest success in its implementation was achieved with the use of materials based on T2SL [65]. Simulation and experiment have shown that the nBn detector offers two important advantages: (1) nBn-based detector should exhibit a higher signal-to-noise ratio than a conventional diode operating at the same temperature, and (2) such detectors will operate at a higher temperature than a conventional diode with the same dark current.

A conventional p-n photodiode architecture can also be optimized through the formation of unipolar barriers in them [73, 74]. For example, placing a barrier in a p-type layer blocks the surface leakage current. At the same time, currents associated with diffusion, generation-recombination processes, trap-assisted tunneling and interband tunneling cannot be blocked [74]. If the barrier is located in the n-type region, then the currents generated by the p-n junction, as well as surface leakage currents, are effectively filtered out (see Fig. 5.4a, b).

Experiment has shown that unipolar barriers can significantly improve the performance of infrared p-n photodiodes, especially in low temperature range [74]. For example, as shown in Fig. 5.4c, an n-side unipolar barrier InAs-based photodiode (nBp structure) has an R_0A value (detector resistance multiplied by active area) in the low temperature range six orders of magnitude higher than that of a conventional p-n junction.

Despite the fact that the barrier detector can be implemented in different semiconductor materials, including InAs [74], InAsSb [33], InAs/GaSb (T2SL) [63] and HgCdTe [26, 36], the most promising materials for barrier detector structures are InAs(InAsSb)/B-AlAsSb and InAs/GaSb due to nearly zero valence band offset (VBO) with respect to AlAsSb barriers. The research results presented in [31] show how important it is for nBn detectors to have zero VBO. It was shown that in InAsSb nBn devices the signal-to-noise ratio increases due to a decrease in the valence band offset.

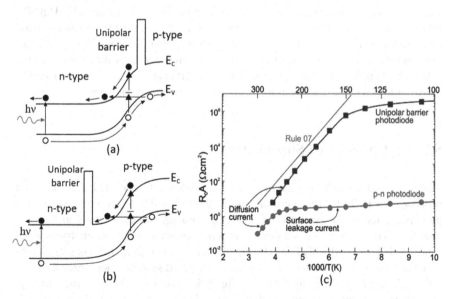

Fig. 5.4 (**a, b**) Band diagrams of a p-side (**a**) and an n-side (**b**) unipolar photodiode under bias, (**c**) R_0A product of a conventional InAs photodiode and a comparable n-side barrier photodiode. (Reprinted with permission from Ref. [74]. Copyright 2013: Elsevier)

For materials in which large conduction band offset cannot be achieved, the pBn architecture may be preferable. The same architecture can also be used in devices that require operation with zero bias [32]. For information, the traditional nBn structure requires biased device operation. Because of the inevitable changes in the growth process, most unipolar/monovalent nBn detectors operate at slightly higher bias voltage than other devices to overcome the unintended barrier that can arise between the barrier layer and the rest of the device [56].

As for the barrier structures based on HgCdTe, in contrast to III-V compounds, in the HgCdTe structures with uniform n-type doping, it is impossible to create conditions under which there is no valence band offset (VBO $\approx$ 0 eV) between the absorber and the barrier. Usually V_{BO} < 200 meV (T = 200 K), depending on both the composition and doping of HgCdTe layers, (see Fig. 5.3c). This feature of is a key limiting factor in HgCdTe-based barrier detector performance [27]. Such structures require a relatively high bias (so-called "turn on" voltage) to be applied to the device to collect the photogenerated carriers. This leads to strong band-to-band tunneling (BTB) and trap-assisted tunneling (TAT) effects due to a high electric field at the barrier-absorber heterojunction. Proper p-type doping of the heterojunctions should reduce the VBO in HgCdTe-based barriers structures. However, p-type doping is a technological challenge, associated with the activation of the dopant after molecular beam epitaxy (MBE) growth. Metalorganic chemical vapor deposition (MOCVD) is considered more favorable technology because it allows both in situ donor and acceptor doping. This seems to be more attractive in terms of the growth of the pB_pn and pB_pp HgCdTe barrier structures. In these structures, B_p is a p-type barrier. Barrier structures with p-type-doped constituent layers, grown by MOCVD, were presented by Kopytko et al. [37]. A further development strategy for HgCdTe nBn detectors should be focused on reducing or even eliminating the valence band offset in the barrier layer, which will lead to a lower operating bias, lower dark current, and the ability to operate at higher temperatures. Methods for eliminating the valence band offset have been proposed by Schubert et al. [76] and have been undertaken for HgCdTe barrier detectors by appropriate bandgap engineering [1].

5.7 HgCdTe-Based Superlattice

Schulman and McGill [78] as an alternative material for infrared detection first proposed the HgTe-CdTe superlattice in 1979. HgTe-CdTe superlattice has unique properties, due to the inversion of the conduction and valence bands in the well materials. This system unites a direct-gap semiconductor with a symmetry-induced semimetal. That is why HgTe-CdTe superlattice places it in a different class, compared to other II-VI and III-V superlattices – a type-III system. These three different types of superlattice are illustrated in Fig. 5.5. Type 1 and II are most common superlattices. They were reviewed in Sects. 5.3 and 5.4 as QWIP and type-II strained-layer superlattices.

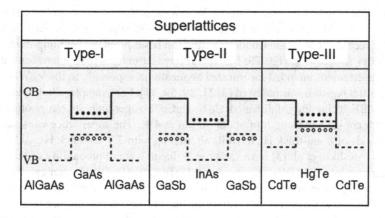

Fig. 5.5 The conduction and valance band profiles of the three types of superlattices. Holes are shown as hollow dots, and electrons as black dots

Predictions by Schulman and McGill [50, 77, 78] suggested that this new material would have a number of advantages over the HgCdTe compounds, such as: (1) The cut-off wavelength would be easier to control than the alloy at longer wavelengths, (2) Tunneling current prevalent in HgCdTe would be vastly reduced in the superlattice, and (3) VLWIR operation would be possible without the need for cooling to extremely low temperatures. It is also important that HgTe and CdTe are nearly lattice-matched to HgCdTe, guaranteeing the epitaxial growth of high quality HgCdTe heterostructures.

It was shown that in these structures the cut-off wavelength depends on the HgTe width. An increase in thickness within 1–6 nm leads to an increase in the cut-off wavelength from 2 to 100 µm. In addition, the HgTe-CdTe superlattice is characterised by a large absorption coefficient due to intrinsically high density of states present in the semimetal HgTe. The increased absorption coefficient is important because it allows the use of thinner absorber layers while maintaining high quantum efficiency. Thinner layers result in less thermal generation of charge carriers in the detector and therefore lower diffusion dark currents. Another effect of reducing the thickness of the absorber layer is the reduction of epilayer growth times and simplified fabrication technology.

Faurie et al. [18] have grew the first HgTe-CdTe superlattice by MBE in 1982. However, extensive growth, characterization, and theoretical modeling of the HgTe-CdTe superlattice only continued from the 1980s to the mid-nineties. As a result of the studies carried out, it was found that in the manufacture of high-quality HgTe-CdTe superlattices, significant difficulties arose associated with surface passivation and layer interdiffusion at typical processing temperatures. In particular, Zhou et al. [91] have found that annealing at 225 °C even for 30 min generates large numbers of defects, presumably mercury vacancies, which leads to a large number of defects, observed mainly in HgTe layers and, therefore, reduces the 77 K charge carries mobility in superlattice by two orders of magnitude. This is why, despite the first growth of the HgTe-CdTe superlattices 30 years ago, the number of published studies of IR photodetectors made from the HgTe-CdTe superlattice is very limited. Only in recent decades, after understanding the basic problems and improving the technology for growing HgCdTe-based multilayer structures [88], interest in CdTe-HgTe superlattices intended for infrared applications, especially in the VLWIR and FIR spectral regions, was renewed [3, 21, 29, 39, 91]. For example, Zhou et al. [91] using MBE technology, fabricated HgTe-HgCdTe superlattice-based photodetectors with cut-off wavelength of about 30 μm at 4 K. The superlattice consisted of 100 periods of 8 nm-thick HgTe wells alternating with 7.7 nm-thick $Hg_{0.05}Cd_{0.95}Te$ barriers. Aleshkin et al. [3] have shown that due to a small probability of the electron capture into the QWs, the interband HgTe-CdHgTe QWIPs can exhibit very high gain in photoconductivity. Akhavan et al. [2], based on their calculations also found that the electron effective mass in the SL absorber is higher than in the HgCdTe absorber, which results in a lower tunneling dark current. Based on all of the above, we can agree with Aleshkin et al. [3], that HgTe-HgCdTe superlattice-based IR detectors really have significant potential advantages compared to the conventional HgCdTe photodetectors and the III-V heterostructures. However, it is not clear when it will be possible to fully realize the advantages of HgTe-CdTe superlattices, since due to intense diffusion at the interfaces, only very low-temperature processes can be used when growing these structures and fabricating devices.

5.8   Quantum Dot Infrared Photodetectors (QDIPs)

Interest in quantum dots (QDs) arose in the 1970s. These are zero dimensional semiconductor structures with unique optical and electronic properties as a result of quantum confinement. The development of quantum dots-based infrared photodetectors (QDIPs) has actually been stimulated by the advances in quantum wells IR photodetectors (QWIP). QDIPs are similar to QWIPs, but they have additional benefits due to the 3D constraint in QDs. The detection mechanism in QDIP is also based on intersubband transitions between quantized QDs energy levels and continuous states. IR absorption, especially in LWIR spectral range, can also be done via plasmonic transitions [44]. One of the manufacturing options for single pixel QDIP is shown in Fig. 5.6. IR technology based on QDs has become a promising technology for the development of third generation thermal imaging cameras, and this technology has developed rapidly over the past decade. QDIPs based on II-VI compounds are considered in details in the Chap. 17, Vol. 2. Prospects for further development are considered in [44, 72].

QDIPs have three main advantages over QWIPs [87], which are the following:

1. QDIPs are inherently sensitive to infrared radiation at normal incidence due to violation of the polarization selection rule. This eliminates the need to manufacture a grating coupler on the surface of imaging arrays;

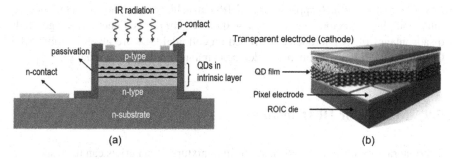

Fig. 5.6 (a) Schematic of a QD based pin diode device. (b) IR monolithic array structure based on CQDs. (a) Reprinted from Ref. [30]. Published 2013 by IOP as open access. (b) Reprinted from Ref. [72]. Published 2021 by MDPI as open access

Table 5.1 Colloidal QD (CQDs) photodetectors advantages and disadvantages in comparison with single crystal photodetectors

Advantages	Disadvantages
Control of dot synthesis and absorption spectrum by ability of QD size-filtering, leading to highly-uniform ensembles; Much stronger absorption than in Stranski-Krastanov grown QDs due to close-packed of CDs; Considerable elimination of strains influencing the growth of epitaxial QDs by better selection of absorber materials; Reduction of cost fabrication (using e.g., such solution as spin coating, inject printing, doctor blade or roll-to-roll printing) compared to epitaxial growth; Deposition methods are compatible with a variety of flexible substrates and sensing technologies such as CMOS (e.g., direct coating on silicon electronics for imaging)	Inferior chemical stability and electronic passivation of the nanomaterials in comparison with epitaxial materials; Bipolar, interband (or excitonic) transitions across the CQD bandgap (e.g., electrons hopping among QDs and holes transport through the polymer) contrary to the intraband transitions in the epitaxial QDs; Insulating behaviour due to slow electron transfer through many barrier interfaces in a nanomaterial; Problems with long term stability due to the large density of interfaces with atoms presenting different or weaker binding; High level of 1/f noise due to disordered granular systems

Source: Reprinted from Ref. [72]. Published 2021 by MDPI as open access

2. QDIPs have a lower dark current than QWIPs due to weaker thermionic emission from QDs with three-dimensional quantum confinement of carriers, and
3. discrete energy levels in QDs have no dispersion, which reduces phonon scattering and can lead to an increase in the carrier lifetime (> 100 ps).

The advantages of QDIPs in comparison with QWIPs include also the ability to work at elevated temperatures, multicolour detection and low cost [30].

QDIPs have already demonstrated MWIR and LWIR imaging. However, QDIPs currently suffer from lower quantum absorption efficiency compared to interband type photodetectors [48]. QDIPs have lower absorption quantum efficiency due to the small fill factor area of QDs and large inhomogeneous broadening of the self-assembled QDs [10]. However, if there are applications in which the incident photon flux is large, then in such applications QDIPs can achieve the same characteristics as interband photodetectors due to very low dark current levels. Other advantages and disadvantages of IR photodetectors based on colloidal QDs are given in Table 5.1. Colloidal QD PDs are compared to detectors manufactured by epitaxial techniques.

5.9   Multicolor IR Detectors

Thermal radiation from objects made from a mixture of materials can have spectral variations similar to color in the visible range. Therefore, for reliable interpretation of the observed image, it is always preferable to have a detector sensitive in several spectral ranges. As we indicated earlier in Sect. 5.2, detection in various atmospheric windows has its advantages and disadvantages. Multi-range detection allows to combine these advantages to achieve the best image quality. Multicolour detector technology is now required in applications such as remote sensing and imaging, military, and medical imaging [58]. Of course, it is possible to develop FPAs for these purposes, which contain several types of pixels, with sensitivity in different spectral ranges. But this significantly complicates the design and manufacturing technology. Therefore, an approach where the same pixel has these capabilities is more efficient.

Currently, technologies have already been developed that implement this possibility. These are two-color and three-color array technologies [28, 58]. Their description in relation to HgCdTe can be found in [5] and Chap. 19, Vol. 2. It is important to note that QWIP technology is particularly well suited to complex pixel designs for multi-color photodetectors array (Fig. 5.7b, c). One approach used to make two-color pixels is shown in Fig. 5.7a. Selective patterning and etching techniques are used to separately contact the two detectors as indicated in Fig. 5.7a.

Various FPAs have already been developed based on two-color detectors [58]. However, such "simultaneous" two colour photovoltaic detector structures suffer

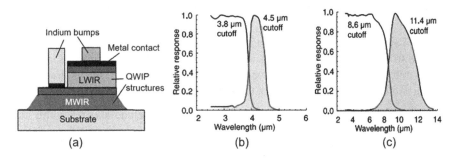

Fig. 5.7 (**a**) Schematic diagram of the structure of two-color pixel, and (**b, c**) Examples of HgCdTe two-color detectors. Note that the high absorption coefficient of HgCdTe in Band 1 limits the spectral crosstalk from Band 2 to low values. (Reprinted from Ref. [57]. Published 2002 by PAS as open access)

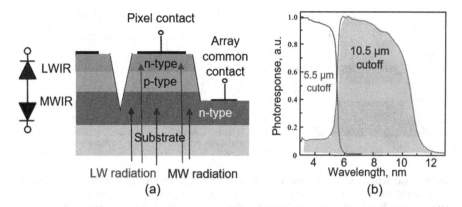

Fig. 5.8 (a) Schematic device structure and (b) top-view SEM image of sequential two-colour detectors; (b) Photoresponse spectra of a sequential LWIR and MWIR two-colour HgCdTe infrared detector with cutoff wavelengths of $\lambda_{1,\text{cutof}} = 5.5$ µm and $\lambda_{2,\text{cutoff}} = 10.5$ µm. (Adapted with permission from [82]. Copyright 2006: Springer)

from a major disadvantage: a large pixel size due to the requirement of two contacts per pixel, which limits the pixel density and thus the array format size. Therefore, the developers are looking for other approaches to solving this problem [39]. One such alternative approach is the "sequential" approach. Detectors made in accordance with this approach, are also called a bias-selectable detector [82, 83]. Such detectors have only one contact per pixel, and their configuration does not prevent pixel size reduction. Figure 5.8 shows the schematic structure of such a detector. Typical devices include p-n-n-p or n-p-p-n sandwich structures optimized for a specific wavelength. The wavelength band can be selected by the bias polarity applied between the two contacts. Such sequential bi-color HgCdTe-based detectors have been successfully implemented [82, 83]. Smith et al. [82] demonstrated sequential two-colour (LWIR/MWIR) HgCdTe FPA (256 × 256 pixels) with an MWIR cutoff wavelength of 5.5 µm and an LWIR cutoff wavelength of 10.5 µm. The spectral characteristics of these detectors are shown in Fig. 5.8b. This is undoubtedly an elegant method, but it suffers from the operational disadvantage of non-simultaneous integration.

Another approach that multicolor detection can provide is the integration of a photodetector with a tunable MEMS filter [17, 39, 54], in the development of which significant progress has been made in recent years [22, 53]. The MEMS optical filters are electrostatically actuated Fabry–Perot tunable filters. Figure 5.9 shows the general concept and an example of optical spectral transmission of tunable MEMS-based filter [17, 54]. A Fabry-Perot filter requires that its two mirrors be placed parallel to each other with a spacing (d), which determines the specific wavelengths transmitted through the filter. This means that by changing the spacing, or optical cavity length (d), you can control the wavelength of light transmitted through the filter. In the tunable optical filter design, this change is carried out by applying a voltage between two electrodes, which generates an electrostatic attraction that

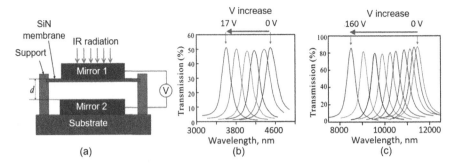

Fig. 5.9 (a) Schematic concept and (**b**, **c**) simulated transmission spectra with different spacing d for an (**b**) MWIR and (**c**) LWIR tunable MEMS optical filters developed for multi-spectral imaging applications. (Reprinted with permission from Ref. [39]. Copyright 2015: AIP publishing LLC)

moves the upper mirror relatively to the fixed lower mirror. Thus, by controlling the voltage applied to the MEMS filter, one can control the spectral sensitivity of the IR detector, which can be conventional broadband multi-color detector.

As for the problem of integrating MEMS filter technology with HgCdTe FPA, there are two main approaches: hybridization and monolithic integration. Despite significant progress, the design and manufacture of monolithic integrated filter/detector structures are very complex and require a lot of effort. In the case of a hybrid integrated filter/detector, this integration is much easier, since a large-area tunable MEMS optical filter can be fabricated separately and then hybridized onto the FPA or a section of the FPA. Currently, work is underway to transfer the existing technology [22, 53] to a large area tunable filter design to provide wavelength tuning capabilities for large-area FPAs [17, 54]. The realization of adaptive concepts offers the potential approach to achieving real-time tunable multiband detection. Despite the great progress made, fabrication of such HgCdTe-based MPAs with tunable spectral characteristics is still a very difficult task [39].

5.10 Photon Trapping Detectors

Another approach that makes it possible to reduce dark currents and, therefore, to increase the operating temperatures of IR detectors is to reduce the volume of the active region of the detectors [49]. Experiments and theoretical simulations have shown that this goal can be achieved using the concept of photon trapping (PT). In this case, a decrease in the dark current is achieved without degrading the quantum efficiency. However, a decrease in the volume of detectors' active region is possible up to a certain limit, after which photon collection begins to decrease faster than the noise, and, therefore, the overall performance degrades.

Photon trapping detectors have been demonstrated independently in II-VI [84, 90] and III-V [12, 80] based IR detectors. To improve the photon collection, it was

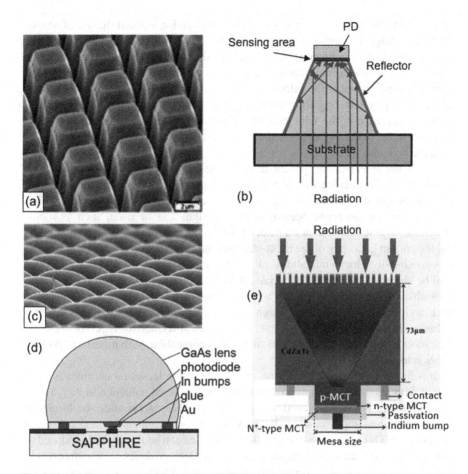

Fig. 5.10 (a) Examples of photon trapping HgCdTe microstructures; (b) Schematic cross-section of optically immersed mesa HgCdTe-based photodiode for operation at 200–300 K; (c) SEM picture of 2D array of 2D immersion lenses photodetectors manufactured by the Vigo Systems. The size of the individual lens is 50 × 50 μm; (d) schematic diagram of HgCdTe photodiode with GaAs lens; (e) Schematic diagram of the meta-lens integrated HgCdTe infrared detector. The meta-lens forms at the top by etching into the CdZnTe substrate. (a) Reprinted with permission from Ref. [90]. Copyright 2011: Springer. (c, d) Reprinted from Ref. [59]. Published 2004 by PAS as open access. (e) Reprinted from Ref. [42]. Published 2020 by Nature as open access

proposed to incorporate 3D photonic structures into the IR photodetector. These structures can be of various shapes. They can have shapes such as pyramidal, sinusoidal, or rectangular. An example of such a structure is shown in Fig. 5.10a. Theoretical estimates have shown that PT photodetector arrays should have significantly better device performance than non-PT PD arrays, especially for small pixel pitches [79]. In addition, PT structures should have superior resolving capability compared to non-PT structures. Taking into account the general trend towards decreasing pixel size in FPAs, it is obvious that this technology is an effective means

of increasing the quantum efficiency of photoconversion without the use of antireflection coatings. It is important to note that the results of theoretical modeling have
been experimentally confirmed. As a result of a decrease in the volume of detector's
active region, both an improvement in device performance and an increase in the
operating temperature of the detector arrays were observed [84]. For example, Dhar
and Dat [12] by reducing the volume of the detector's active region, observed a
threefold decrease in the dark current without degrading the quantum efficiency
when using the pyramidal structured diodes in comparison with conventional diodes
with the bulk absorber.

As we see in Fig. 5.10a, photon-trapping structures are usually several microns
in size and require a rather complex process to manufacture them. At the same time,
the experiment showed that to improve the absorption of light in detectors with a
reduced area of the active region, simpler designs can be used, such as mesa-
structures. Figure 5.10b shows a detector with a mesa geometry design. By choosing the correct geometry of the mesa-structure and using reflectors on the sidewalls,
it is possible to achieve that almost all the radiation incident on the bottom surface
will be absorbed by sensing materials. Baker et al. [6], reported that they reached
90% absorption for HgCdTe FPAs with mesa-structure. They have also shown that
such a simple mesa geometry design provides a reliable way to reduce material
volume while maintaining a high level of quantum efficiency for HgCdTe detectors
and, most importantly, the proposed structure is compatible with modern FPAs fabrication technology.

It is important to note that the use of the immersion lenses monolithically integrated with photodiode can also contribute to a significant reduction in the volume
of the detector's active region (see Fig. 5.10c–e). By concentrating the luminous
flux, a significant improvement in the performance enhancement of the detector can
be achieved. This means that the photosensitive area can be significantly reduced to
achieve the required output signal. This significantly reduces dark currents and
noise levels, and, as we indicated earlier, allows the detector to operate at a higher
temperature [42, 59]. For example, Li et al. [42] reported that compared to the pristine device, the integration of the meta-lens together with the reduction in photosensitive area enhances the detectivity (D^*) by 3.2–5.5 times.

5.11 Nano Wire-Based Photodetectors

Many believe that the development of nanowires-based IR photodetectors can also
lead to significant performance improvements [87]. Semiconductor nanowires
(NWs) possess unique photonic and electrical properties due to their unique anisotropic geometry and high surface-to-volume ratio. To date, the technology for synthesizing NWs has been developed for almost all semiconductor compounds,
including complex compounds, and therefore NWs with a band gap varying over a
wide range are available at this time. Semiconductor NWs III-V with a narrow band
gap, due to their excellent transport properties and ease of manufacture, are considered as promising candidates for creating IR photodetectors (PD). In addition, their

low capacity results in high operating speeds. Recently, there have been reports of nanowire-based IR PD, such as InAs, GaAsSb, InGaAs, InPAs and InGaSb [41, 46]. For example, it was reported that $In_{0.65}Ga_{0.35}As$ nanowires have demonstrated sensitivity 6.5×10^3 A/W in the range of 1.1–2.0 μm [51], and for InAs NW photodetectors with a Schottky-Ohmic contact, a sensitivity of 5.3×10^3 A/W was recorded at a wavelength of about 1.5 μm [16]. However, most research related to nanowire-based photodetectors is limited to the visible and ultraviolet regions of the spectrum (read Chap. 16, Vol. 2). In addition, the use of NWs in the manufacture of devices is accompanied by significant technological difficulties [38] that significantly limit the possibility of their use in devices intended for the market.

5.12 New Emerging Nanomaterials for Detection

In recent years, interest in so-called 2D materials has grown significantly [68, 71, 72, 87]. Graphene was the first 2D material to attract the attention of many scientists due to its suitable properties for nanophotonic applications. Due to linear dispersion near the Dirac point and various forms of interaction of light with substance, graphene has a high optical sensitivity in a wide spectral range. However, graphene has low absorption as an IR PD material. This is due to the short carrier lifetime in graphene and the zero band gap [71]. In other words, the use of graphene-based photodetectors is limited by the lower external quantum efficiency and photoresponsivity in comparison with traditional photodetectors [52]. At the same time, the use of graphene as an element of photonic photodetectors based on other IR materials, or as a sensitive element in a thermal detector can improve the parameters of IR photodetectors [11, 86]. For example, encouraging results have been obtained in the development of a MEMS thermopile with a graphene sensitive layer [11], and when using hybrid photodetectors based on graphene and HgCdTe [7, 72] and graphene-QDs [70].

Another very interesting family of 2D materials is the single-layer transition metal dichalcogenides (TMDC) [71, 89], such as molybdenum disulfide (MoS_2) and tungsten diselenide (WSe_2). Unlike graphene, the TMDC family has transitions from an indirect bandgap to a straight bandgap, which occur as the material thickness decreases from multilayer to monolayer. Moreover, TMDCs can be easily included into a wide variety of atomic-level controlled heterostructures to achieve higher IR PD performance. However, despite the large amount of funding and research invested in 2D materials, there is a very limited set of 2D materials, covering the infrared region [43]. The most studied TMDCs have a band gap in the range of 1.0–2.5 eV. In addition, these new materials suffer from high environmental sensitivity and large manufacturing area.

Rogalski et al. [72] formulated other limitations that exist in the way of widespread use of 2D nanomaterials in the development of IR detectors. They concluded that:

- In general, the performance of 2D materials-based infrared detectors is lower compared to commercially available detectors, especially detectors based on HgCdTe and emerging III-V compounds, including detectors using T2SL;

- The improvement in responsivity due to the use of a combination of 2D materials with bulk materials (hybrid photodetectors) owing to the photogating effect causes the limited linear dynamic range due to the charge relaxation time, which leads to a decrease in sensitivity with decreasing optical power;
- Responsivity of hybrid and chemically functionalized 2D material photodetectors is comparable to that of detectors on the world market; however, a significant decrease in operating speed (bandwidth) is observed; in general, their response time (millisecond range and longer) is three orders of magnitude longer compared to commercially available photodetectors (microsecond range and shorter) [70]; and
- The potential for commercialization will depend not only on detector performance, but also on whether high quality 2D materials can be produced on a large scale at a low cost.

5.13 Summary

As a conclusion, Tables 5.2, 5.3 and 5.4, are given, in which various authors [62, 66, 87] summarize the advantages and disadvantages of different approached and materials used for the development of IR photodetectors, as well as the current state of their applicability for fabricating large area FPAs.

Table 5.2 Comparison of infrared detectors

Detector type		Advantages	Disadvantages
Thermal	Thermopile, bolometers, pyroelectric	Light, rugged, reliable, and low cost; Room temperature operation	Low detectivity at high frequency; Slow response (ms order)
Photon			
Intrinsic	IV-VI (PbS, PbSe, PbSnTe)	Available low-gap materials; Well studied; Easier to prepare; More stable materials	Very high thermal Expansion coefficient; Large permittivity
	II-VI (HgCdTe)	Easy band-gap tailoring; Well-developed theory and exp.; Multicolour detectors	Non-uniformity over large area; High cost in growth and processing; Surface instability
	III-V (InGaAs, InAs, InSb, InAsSb)	Good material and dopants; Advanced technology; Possible monolithic integration	Heteroepitaxy with large lattice mismatch; Long wavelength cutoff limited to 7 μm (at 77 K)
Photon Extrinsic	Si:Ga, Si:As, Ge:Cu, Ge: Hg, Ge:Au	Very long wavelength operation; Relatively simple technology	High thermal generation; Extremely low temperature operation

(continued)

Table 5.2 (continued)

Detector type		Advantages	Disadvantages
Free carriers	PtSi-Si, IrSi-Si	Low-cost, high yields; Large and close packed 2D arrays	Low quantum efficiency; Low temperature operation
Quantum wells			
Type I	GaAs/AlGaAs, InGaAs/ AlGaAs	Matured material growth; Good uniformity over large area; Multicolour detectors	High thermal generation; Low quantum efficiency; Complicated design and growth
Type II	InAs/InGaSb, InAs/ InAsSb	Low auger recombination rate; Easy wavelength control	Complicated design and growth; Sensitive to the interfaces
Quantum dots	InAs/GaAs, InGaAs/ InGaP, Ge/Si	Normal incidence of light normal incidence of light; Low thermal generation	Complicated design and growth

Source: Reprinted with permission from Ref. [66]. Copyright 2003: Pergamon Press

Table 5.3 Comparison of IR detectors based on different materials designed for LWIR spectral range

	Bolometer	HgCdTe	QWIP	Type – II SLs	QDIP
Status	TRL 9	TRL 9	TRL 8	TRL 2–3	TRL 1–2
	Designed for applications requiring medium to low performance	Material for applications requiring high performance	Commercial	Research and development stage	Research and development stage
Advantages	Low cost; requires no active cooling; uses standard Si manufacturing equipment	Close to theoretical performance; basic material for IR detectors for the next 10–15 years.	Very uniform material; uses commercial manufacturing processes; low cost applications	Theoretically better than HgCdTe at >14 µm cutoff; uses commercial III-V manufacturing techniques	Not sufficient data to characterize material advantages
Military system examples	Weapon sight; night vision goggles; missile seekers; small UAV sensors; unattended ground detectors, etc.	Missile intercept; tactical ground and air born imaging; hyper spectral detection; missile seeker; missile tracking; space based sensing, etc.	Being evaluated for some military applications	Is at the stage of development and assessment of the potential for military use	Very early stages of development

(continued)

Table 5.3 (continued)

	Bolometer	HgCdTe	QWIP	Type − II SLs	QDIP
Limitations	Low sensitivity; long time constants	Performances are sensitive to changes in manufacturing process; difficult to expand to >14 μm cutoff	Narrow bandwith; low sensitivity	Requires a significant investment and additional fundamental study	Narrow bandwith; low sensitivity

Source: Data extracted from [55]

TRL technology readiness level. The highest level of TRL (ideal maturity) achieves value of 10

Table 5.4 Summary of the advantages and disadvantages of the current nanostructure-enhanced IR photodetector

Nanostructure IR PD	Advantages	Disadvantage	Possibility for FPA fabrication
Quantum well structure	Mature GaAs growth and fabrication process; High uniformity; Low cost; Covers MWIR to WLWIR and THz	Low quantum efficiency; Cannot absorb normal incident angles; Poor performance at elevated temperatures	Mature technology available for IR FPA production
Type II superlattice structure	High operating temperature; Covers SWIR to WLWIR; High absorption coefficient	High fabrication and processing cost; Low yield in large array fabrication	Mature technology available for IR FPA production
Quantum dot	Sensitive to infrared irradiation at normal incidence; Lower dark current than QWIPs; Higher operating temperature; Longer carrier lifetime	Low absorption quantum efficiency; Large inhomogeneous broadening of the self-assembled QDs	Research on going for IR FPA production
Nanowires and nanopillar	Higher light sensitivity; Antireflection and light trapping properties; Can be integrated with CMOS technology	Non-uniformity over large area fabrication; Difficult to form large detector arrays; Difficult to fabricate; Repeatability issues	Possible for IR FPA production
Graphene-based photodetector	High internal quantum efficiency; Low cost; Ease of processing; Ultrafast process	Weak light absorption in a single layer; Zero bandgap	Possible for IR FPA production

(continued)

Table 5.4 (continued)

Nanostructure IR PD	Advantages	Disadvantage	Possibility for FPA fabrication
Transition metal dichalcogenides	Low cost; High optical absorption coefficient; Ease of processing	Low speed; Do not easily achieve LWIR and MWIR; No proven technology for large area fabrication	Not possible for IR FPA production
Colloidal quantum dot	Low cost; Scalable for focal plane arrays; Photoconductive gain	Relatively low quantum efficiency; Relatively high dark current	Research on going for IR FPA production

Source: Reprinted from Ref. [87]. Published 2018 by De Gruyter as open access

Acknowledgments This research was funded by the State Program of the Republic of Moldova, project 20.80009.5007.02.

References

1. Akhavan ND, Jolley G, Umana-Membreno G, Antoszewski J, Faraone L (2014) Performance modelling of bandgap engineered HgCdTe-based nBn infrared detectors. IEEE Trans Electron Dev 61(11):3691–3698
2. Akhavan ND, Umana-Membreno GA, Gu R, Antoszewski J, Faraone L (2019) Interdiffusion effects on bandstructure in HgTe-CdTe superlattices for VLWIR imaging application. J Electron Mater 48:6159–6168
3. Aleshkin VY, Dubinov AA, Morozov SV, Ryzhii M, Otsuji T, Mitin V et al (2018) Interband infrared photodetectors based on HgTe–CdHgTe quantum-well heterostructures. Opt Mater Exprss 8(5):1349–1358
4. Ashley T, Elliot CT (1985) Non-equilibrium devices for infra-red detection. Electron Lett 21(2):451–452
5. Baker IM (2017) II-VI narrow bandgap semiconductors: optoelectronics. In: Kasap S, Capper P (eds) Springer handbook of electronic and photonic materials. Springer, pp 867–896
6. Baker I, Maxey C, Hipwood L, Weller H, Thorne P (2012) Developments in MOVPE HgCdTe arrays for passive and active infrared imaging. Proc SPIE 8542:85421A
7. Bansal S, Das A, Jain P, Prakash K, Sharma K, Kumar N et al (2019) Enhanced optoelectronic properties of bilayer graphene/HgCdTe-based single- and dual-junction photodetectors in long infrared regime. IEEE Trans Nanotechnol 18:781–789
8. Brown GJ (2005) Type-II InAs/GaInSb superlattices for infrared detection: an overview. Proc SPIE 5783:65–77
9. Buffaz A, Carras M, Doyennette L, Nedelcu A, Bois P, Berger V (2010) State of the art of quantum cascade photodetectors. Proc SPIE 7660:76603Q
10. Chakrabarti S, Stiff-Roberts A, Su X, Bhattacharya P, Ariyawansa G, Perera A (2005) High-performance mid-infrared quantum dot infrared photodetectors. J Phys D Appl Phys 38:2135–2141
11. Chen S-J, Chen B (2020) Research on a CMOS-MEMS infrared sensor with reduced graphene oxide. Sensors 20:4007
12. Dhar NK, Dat R (2012) Advanced imaging research and development at DARPA. Proc SPIE 8353:835302

13. Downs C, Vandervelde TE (2013) Progress in infrared photodetectors since 2000. Sensors 13:5054–5098
14. Elliott CT (1990) Non-equilibrium modes of operation of narrow-gap semiconductor devices. Semicond Sci Technol 5(1990):S30–S37
15. Elliott CT (2001) Photoconductive and non-equilibrium devices in HgCdTe and related alloys. In: Capper P, Elliott CT (eds) Infrared detectors and emitters: materials and devices. Kluwer Academic Publishers, Boston, pp 279–312
16. Fang H, Hu W, Wang P, Guo N, Luo W, Zheng D et al. (2016) Visible light-assisted high-performance mid-infrared photodetectors based on single InAs nanowire. Nano Lett 16(10):6416–6424
17. Faraone L (2005) MEMS for tunable multi-spectral infrared sensor arrays. Proc SPIE 5957:59570F
18. Faurie JP, Million A, Piaguet J (1982) CdTe-HgTe multilayer grown by molecular beam epitaxy. Appl Phys Lett 41:713–715
19. Gautam N, Myers S, Barve AV, Klein B, Smith EP, Rhiger DR et al (2013) Barrier engineered infrared photodetectors based on type-II InAs/GaSb strained layer superlattices. IEEE J Quantum Electron 49:211–217
20. Gendron L, Carras M, Huynh A, Ortiz V, Koeniguer C, Berger V (2004) Quantum cascade photodetector. Appl Phys Lett 85(14):2824–2826
21. Grein CH, Jung H, Singh R, Flatte ME (2005) Comparison of normal and inverted band structure HgTe/CdTe superlattice for very long wavelength infrared detectors. J Electron Mater 34(6):905–908
22. Gunning W, Lauxtermann S, Durmas H, Xu M, Stupar P, Borwick R et al (2009) MEMS-based tunable filters for compact IR spectral imaging. Proc SPIE 7298:72982I
23. Hinds S, Buchanan M, Dubek R, Haffouz S, Laframboise S, Wasilewski Z et al (2011) Near-room-temperature mid-infrared quantum well photodetector. Adv Mater 23(46):5536–5539
24. Hinkey RT, Yang RQ (2013) Theory of multiple-stage interband photovoltaic devices and ultimate performance limit comparison of multiple-stage and single-stage interband infrared detectors. J Appl Phys 114:104506-1–104506-18
25. Hofstetter D, Giorgetta FR, Baumann E, Yang Q, Manz C, Kohler K (2010) Mid-infrared quantum cascade detectors for applications in spectroscopy and pyrometry. Appl Phys B Lasers Opt 100:313–320
26. Itsuno AM, Philips JD, Velicu S (2011) Design and modeling of HgCdTe nBn detectors. J Electron Mater 40:1624–1629
27. Itsuno AM, Phillips JD, Velicu S (2012) Mid-wave infrared HgCdTe nBn photodetector. Appl Phys Lett 100:161102
28. Jozwikowski K, Rogalski A (2007) Numerical analysis of three-colour HgCdTe detectors. Opto-Electron Rev 15(4):215–222
29. Jung HS, Boieriu P, Greain CH (2006) P-type HgTe/CdTe superlattices for very-long wavelength infrared detectors. J Electron Mater 35:1341–1345
30. Karim A, Andersson JY (2013) Infrared detectors: advances, challenges and new technologies. IOP Conf Ser Mater Sci Eng 51:012001
31. Khoshakhlagh A, Myers S, Plis E, Kutty MN, Klein B, Gautam N et al (2010) Mid-wavelength InAsSb detectors based on nBn design. Proc SPIE 7660:76602Z
32. Klem JF, Kim JK, Cich MJ, Hawkins SD, Fortune TR, Rienstra JL (2010) Comparison of nBn and nBp mid-wave barrier infrared photodetectors. Proc SPIE 7608:76081P
33. Klipstein P (2008) "XBn" barrier photodetectors for high sensitivity and high operating temperature infrared sensors. Proc SPIE 6940:69402U
34. Kopytko M (2014) Design and modeling of high-operating temperature MWIR HgCdTe nBn detector with n- and p-type barriers. Infrared Phys Technol 64:47–55
35. Kopytko M, Martyniuk P (2016) HgCdTe mid- and long-wave barrier infrared detectors for higher operating temperature condition. In: Beg A, Akbar NS (eds) Modeling and simulation in engineering sciences. Intech, London, pp 72–90

36. Kopytko M, Kebłowski A, Gawron W, Madejczyk P, Kowalewski A, Jozwikowski K (2013) High-operating temperature MWIR nBn HgCdTe detector grown by MOCVD. Opto-Electron Rev 21(42):402–405
37. Kopytko M, Jóźwikowski K, Rogalski A (2014) Fundamental limits of MWIR HgCdTe barrier detectors operating under non-equilibrium mode. Solid State Electron 100:20–26
38. Korotcenkov G (2020) Current trends in nanomaterials for metal oxide-based conductometric gas sensors: advantages and limitations. Part 1: 1D and 2D nanostructures. Nanomaterials 10:1392
39. Lei W, Antoszewski J, Faraone L (2015) Progress, challenges, and opportunities for HgCdTe infrared materials and detectors. Appl Phys Rev 2:041303
40. Levine BF, Choi KK, Bethea CG, Walker J, Malik R (1987) New 10 um infrared detector using intersubband absorption in resonant tunneling GaAlAs superlattices. Appl Phys Lett 50:1092–1094
41. Li Z, Yuan X, Fu L, Peng K, Wang F, Fu X et al (2015) Room temperature GaAsSb single nanowire infrared photodetectors. Nanotechnology 26:445202
42. Li F, Deng J, Zhou J, Chu Z, Yu Y, Dai X et al (2020) HgCdTe mid-infrared photo response enhanced by monolithically integrated meta-lenses. Sci Rep 10:6472
43. Liu F, Shimotani H, Shang H, Kanagasekara T, Zólyomi V, Drummond N et al (2013) High-sensitivity photodetectors based on multilayer GaTe flakes. ACS Nano 8:752–760
44. Livache C, Martinez B, Goubet N, Ramade J, Lhuiller E (2018) Road map for nanocrystal based infrared photodetectors. Front Chem 6:575
45. Lotfi H, Hinkey RT, Li L, Yang RQ, Klem JF, Johnson MB (2013) Narrow-bandgap photo-voltaic devices operating at room temperature and above with high open-circuit voltage. Appl Phys Lett 102:211103
46. Ma L, Hu W, Zhang Q, Ren P, Zhuang X (2014) Room-temperature near-infrared photodetectors based on single heterojunction nanowires. Nano Lett 14:694–698
47. Maimon S, Wicks G (2006) nBn detector, an infrared detector with reduced dark current and higher operating temperature. Appl Phys Lett 89:151109
48. Martyniuk P, Rogalski A (2008) Quantum–dot infrared photodetectors: status and outlook. Prog Quantum Electron 32:89–120
49. Martyniuk P, Antoszewski J, Martyniuk M, Faraone L, Rogalski A (2014) New concepts in infrared photodetector designs. Appl Phys Rev 1:041102
50. McGill TC, Wu GY, Hetzler SR (1986) Superlattice: progress and prospects. J Vac Sci Technol A 4(4):2091–2095
51. Miao J, Hu W, Gu N, Lu Z, Zhou X, Liao L et al (2014) Single InAs nanowire room-temperature near-infrared photodetectors. ACS Nano 8:3628–3635
52. Mueller T, Xia F, Avouris P (2010) Graphene photodetectors for high speed optical communications. Nat Photon 4:297–301
53. Musca CA, Antoszewski J, Winchester KJ, Keating AJ, Nguyen T, Silva KKMBD et al (2005) Monolithic integration of an infrared photon detector with a MEMS-based tunable filter. IEEE Electron Dev Lett 26:888–890
54. Musca CA, Antoszewski J, Keating AJ, Winchester KJ, Silva KKMBD, Nguyen T et al (2007) MEMS-based micro-spectrometers for infrared sensing. In: Proceedings of 2007 IEEE/LEOS international conference on optical MEMS and nanophotonics, Hualien, 12–16 August 2007, pp 137–138
55. National Research Council (NRC) (2010) Seeing photons: Progress and limits of visible and infrared sensor arrays. The National Academies Press, Washington, DC. http://www.nap.edu/catalog/12896.html
56. Nguyen B-M, Chen G, Hoang AM, Abdollahi Pour S, Bogdanov S, Razeghi M (2011) Effect of contact doping in superlattice-based minority carrier unipolar detectors. Appl Phys Lett 99:033501
57. Norton P (2002) HgCdTe infrared detectors. Opto-Electron Rev 10(3):159–174
58. Norton PR (2006) Third-generation sensors for night vision. Opto-Electron Rev 14:283–296

59. Piotrowski J (2004) Uncooled operation of IR photodetectors. Opto-Electron Rev 12(1):111–122
60. Piotrowski J, Rogalski A (2007) High-operating-temperature infrared photodetectors. SPIE, Bellingham
61. Pusz W, Kowalewski A, Gawron W, Plis E, Krishna S, Rogalski A (2013) MWIR type-II InAs/GaSb superllatice interband cascade photodetectors. Proc SPIE 8868:88680M
62. Razeghi M (1998) Current status and future trends of infrared detectors. Opto-Electron Rev 6:155–194
63. Razeghi M, Pour SA, Huang EK-W, Chen G, Haddadi B-MN et al (2011) High-operating temperature MWIR photon detectors based on Type II InAs/GaSb superlattice. Proc SPIE 8012:80122Q
64. Rehm R, Masur M, Schmitz J, Daumer V, Niemasz J, Vandervelde T et al (2013) InAs/GaSb superlattice infrared detectors. Infrared Phys Technol 59:6–11
65. Rodriguez JB, Plis E, Bishop G, Sharma YD, Kim H, Dawson LR, Krishna S (2007) nBn structure based on InAs/GaSb type-II strained layer superlattices. Appl Phys Lett 91:043514
66. Rogalski A (2003) Infrared detectors: status and trends. Prog Quant Electron 27:59–210
67. Rogalski A (2009) Infrared detectors for the future. Acta Phys Pol A 116(3):389–405
68. Rogalski A (2019) Graphene-based materials in the infrared and terahertz detector families: a tutorial. Adv Opt Photon 11(2):314–379
69. Rogalski A, Martyniuk P (2014) Mid-wavelength infrared nBn for HOT detectors. J Electron Mater 43(8):2963–2969
70. Rogalski A, Kopytko M, Martyniuk P (2019) Two dimensional infrared and terahertz detectors: outlook and status. Appl Phys Rev 6:021316
71. Rogalski A, Kopytko M, Martyniuk P (2020) 2D material infrared and terahertz detectors: status and outlook. Opto-Electron Rev 28:107–154
72. Rogalski A, Martyniuk P, Kopytko M, Hu W (2021) Trends in performance limits of the HOT infrared photodetectors. Appl Sci 11:501
73. Savich GR, Pedrazzani JR, Sidor DE, Maimon S, Wicks GW (2011) Dark current filtering in unipolar barrier infrared detectors. Appl Phys Lett 99:121112
74. Savich GR, Pedrazzani JR, Sidor DE, Wicks GW (2013) Benefits and limitations of unipolar barriers in infrared photodetectors. Infrared Phys Technol 59:152–155
75. Schneider H, Liu HC (2007) Quantum well infrared photodetectors. Springer, Berlin
76. Schubert EF, Tu LW, Zydzik GJ, Kopf RF, Benvenuti A, Pinto MR (1992) Elimination of heterojunction band discontinuities by modulation doping. Appl Phys Lett 60:466–468
77. Schulman JN, Chang Y-C (1985) Hg-CdTe superlattice bang-gap enhancement due to interdiffusion. Appl Phys Lett 46(6):571–573
78. Schulman JN, McGill TC (1979) The CdTe/HgTe superlattice: proposal for a new infrared material. Appl Phys Lett 34(10):663–665
79. Schuster J, Bellotti E (2013) Numerical simulation of crosstalk in reduced pitch HgCdTe photon-trapping structure pixel arrays. Opt Express 21(12):14712
80. Sharifi H, Roebuck M, De Lyon T, Nguyen H, Cline M, Chang D et al (2013) Fabrication of high operating temperature (HOT), visible to MWIR, nCBn photon-trap detector arrays. Proc SPIE 8704:87041U
81. Sherliker B, Halsall M, Kasalynas I, Seliuta D, Valusis G, Vengris M et al (2007) Room temperature operation of AlGaN/GaN quantum well infrared photodetectors at a 3–4 μm wavelength range. Semicond Sci Technol 22:1240–1244
82. Smith EPG, Patten EA, Goetz PM, Venzor GM, Roth JA, Nosho BZ et al (2006) Fabrication and characterization of two-color midwavelength/long wavelength HgCdTe infrared detectors. J Electron Mater 35:1145–1152
83. Smith EPG, Venzor GM, Gallagher AM, Reddy M, Peterson JM, Lofgreen DD, Randolph JE (2011) Large-format HgCdTe dual-band long-wavelength infrared focal-plane arrays. J Electron Mater 40:1630–1636

84. Smith KD, Wehner JGA, Graham RW, Randolph JE, Ramirez AM, Venzor GM et al (2012) High operating temperature mid-wavelength infrared HgCdTe photon trapping focal plane arrays. Proc SPIE 8353:83532R
85. Smuk S, Kochanov Y, Petroshenko MP, Solomitskii D (2014) IRnova long-wavelength infrared sensors based on quantum wells. Komponenti Tehnologia 1:20–25. (in Russian)
86. Sood AK, Zeller JW, Ghuman P, Babu S, Dhar NK, Ganguly S et al (2019) Development of high-performance detector technology for UV and IR applications. Proc SPIE 11151:1115113
87. Tan CL, Mohseni H (2018) Emerging technologies for high performance infrared detectors. Nano 7(1):169–197
88. Wang C, Wang X, Zhao J, Chang Y, Grein CH, Sivananthan S, Smith DJ (2007) Microstructure of interfacial HgTe/CdTe superlattice layers for growth of HgCdTe on CdZnTe (211)B substrates. J Crystal Growth 309:153–157
89. Wang QH, Kalantar-Zadeh K, Kis A, Coleman JN, Strano MS (2012) Electronics and optoelectronics of two-dimensional transition metal dichalcogenides. Nat Nanotechnol 7:699–712
90. Wehner JGA, Smith EPG, Venzor GM, Smith KD, Ramirez AM, Kolasa BP et al (2011) HgCdTe photon trapping structure for broadband mid-wavelength infrared absorption. J Electron Mater 40:1840–1846
91. Zhou YD, Becker CR, Selamet Y, Chang Y, Ashokan R, Boreiko RT et al (2003) Far-infrared detector based on HgTe/HgCdTe superlattices. J Electron Mater 32:608–614

Chapter 6
II-VI Semiconductor-Based Unipolar Barrier Structures for Infrared Photodetector Arrays

A. V. Voitsekhovskii, S. N. Nesmelov, S. M. Dzyadukh, D. I. Gorn, S. A. Dvoretsky, N. N. Mikhailov, G. Y. Sidorov, and M. V. Yakushev

6.1 Introduction

The rapid development of thermal imaging technology requires a radical improvement in the technology of infrared photodetectors in the mid wave (MWIR, 3–5 μm) and long wave (LWIR, 8–14 μm) regions of the infrared (IR) range. Today, there is an urgent need to develop MWIR and LWIR array photodetectors of the third generation, which are subject to increased requirements for photosensitive elements, in particular, for operating temperatures, weight, dimensions, and power consumption.

One of the main ways to improve the performance of such photosensitive device structures is to increase the operating temperature of cooled photosensitive layer in the photodetectors without losing temperature sensitivity and infrared image quality. An increase in the cooling temperature makes it possible to use microcryogenic systems with significantly reduced weight, dimensions and power consumption, making them cheaper. This trend is directly related to the development and implementation of new photosensitive semiconductor structures that provide low dark currents and, as a result, low intrinsic noise. This is achieved through the creation of semiconductor heterostructures by epitaxial methods, which make it possible to grow photosensitive structures with a complex architecture of layers – with shut-off, buffer, barrier and other functional layers. The use of zone engineering methods in theory makes it possible to eliminate individual components of dark currents and, as a result, to achieve a significant increase in the photosensitive characteristics of

A. V. Voitsekhovskii · S. N. Nesmelov · S. M. Dzyadukh · D. I. Gorn (✉)
National Research Tomsk State University, Tomsk, Russia
e-mail: vav43@mail.tsu.ru; gorn.di@gmail.com

S. A. Dvoretsky · N. N. Mikhailov · G. Y. Sidorov · M. V. Yakushev
Rzhanov Institute of Semiconductor Physics of the Siberian Branch of the RAS, Novosibirsk, Russia

© The Author(s), under exclusive license to Springer Nature Switzerland AG 2023
G. Korotcenkov (ed.), *Handbook of II-VI Semiconductor-Based Sensors and Radiation Detectors*, https://doi.org/10.1007/978-3-031-20510-1_6

photodetectors [1, 2]. Currently, work on the creation of high operating temperature focal plane array (HOT FPA) is being actively carried out by leading manufacturers of optoelectronic equipment from France, USA, Germany, Great Britain, Israel, China and other countries.

In the field of creating HOT FPA, two main directions can be distinguished. The first is the development of p-n photodiode arrays based on high-quality HgCdTe (MCT) heteroepitaxial structures with p on n architecture (local regions of p-type conductivity are formed in the base active layer of the electronic type of conductivity by external doping with an acceptor impurity, for example, arsenic) or n^+ on p (the p-type base active layer is vacancy-doped, and n^+ is obtained by radiation doping, for example, with boron ions) [1, 3]. The second direction is the use of so-called unipolar xBn barrier structures, where x is a contact semiconductor layer of n- or p-type conductivity, B is a barrier layer, and n is an absorbing layer of n-type of conductivity. To date, the most studied are unipolar barrier structures in the nBn configuration.

At present, research and development of nBn structures for FPAs are carried out both on the basis of III-V and II-VI materials. In the case of II-VI materials, a semiconducting HgCdTe solid solution is used. From a fundamental point of view, MCT is an ideal material for creating IR detectors. This is due, firstly, to the dependence of the band gap on the CdTe content, secondly, to large optical absorption coefficients, which leads to high quantum efficiencies, thirdly, to recombination mechanisms that provide long lifetimes of charge carriers and a relatively high operating cooling temperature, and, fourthly, an extremely weak dependence of the lattice constant on the composition, which makes it possible to grow high-quality multilayer heterostructures. Due to the listed properties, MCT is widely used in the development of highly sensitive IR detectors for various spectral regions [2].

This chapter will review the latest advances in the development and fabrication of unipolar barrier structures based on HgCdTe.

6.2 Basics of Barrier Detectors Based on II-VI Semiconductors

In 2006, Maimon and Wicks [4] proposed the concept of the so-called unipolar barrier nBn photosensitive structure, which is often structurally compared with the classical p-n-photodiode, in which the space charge region of the p-n junction is replaced by a wide-gap barrier B, and the p-region is replaced by n-type contact. The second n-layer plays the role of an active absorbing region. Due to the introduction of a wide-gap barrier at a negative bias of the structure (when a negative potential is applied to the contact layer), a potential barrier is created for the main charge carriers (electrons in the case of the nBn structure) and the dark currents caused by the main carriers are suppressed. But this does not create a potential barrier for the photogenerated minority charge carriers [5]. Thus, due to the exclusion of charge carriers of the same sign from the current transfer process and the suppression of the generation-recombination process, a decrease in the dark currents of the photodetector is achieved. The introduction of a wide gap barrier instead of the space charge region of the p-n photodiode also makes it possible to reduce the contribution of the Shockley-Read-Hall generation-recombination mechanism, as well as the surface leakage mechanism, to the dark current.

The characteristic of the dark current of a photodiode with a p-n junction has two segments with different slopes, which represent the diffusion and generation-recombination components of the dark current. Figure 6.1 shows the dependence $\log(I) = f(1/T)$ of a photodiode with p-n junction and nBn structure. The slope change point corresponds to the temperature T_c at which the diffusion and generation-recombination components of the dark current are equal. In the nBn-type structure, there is no generation-recombination current; therefore, the dark current consists only of a diffusion component. Consequently, at the same temperatures, the dark currents of the nBn structure will be significantly less than that of a standard photodiode with p-n junction, and the photoelectric parameters will be much higher. By eliminating the generation-recombination component of the dark current in barrier nBn structures, it is possible to increase the threshold photoelectric parameters at the cryogenic temperature of 77 K or significantly increase the cooling temperature of photosensitive elements to 100 K and higher, while ensuring dark currents less than the background current. The noise equivalent temperature difference (NETD) characterizing the temperature sensitivity will be the same as at 77 K.

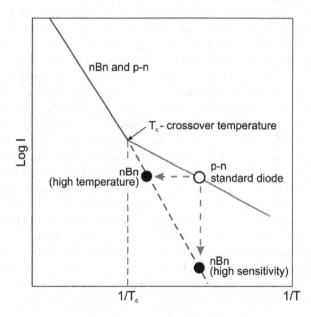

Fig. 6.1 Temperature dependence of the dark current $\log(I) = f(1/T)$ for a photodiode with p-n junction and for nBn structure. (Reprinted from [6]. Published 2014 by Springer Nature as open access)

FPAs for MWIR and LWIR spectral ranges usually have a hybrid architecture: a matrix of sensitive elements based on p-n or n-p photodiodes made of HgCdTe is coupled with a silicon readout circuit. The operating temperatures of the FPAs, at which the mode of limiting the threshold characteristics by background radiation noise is realized, are dictated by the generation-recombination mechanisms in HgCdTe, which determine the magnitude of dark currents (noise). Dark currents are suppressed by cooling the sensitive elements of photodetectors based on HgCdTe to sufficiently low temperatures (for example, up to 77 K when detecting in LWIR). The key condition for creating a HOT detector is to minimize thermal generation in the active region without reducing the quantum efficiency. Among all the mechanisms of thermal generation-recombination (GR) in narrow-gap semiconductors, the main role is played by interband Auger processes (Auger 1 and Auger 7, which have the lowest threshold energies), as well as GR by the Shockley-Reed-Hall (SRH) mechanism through trap levels. Radiative recombination usually does not limit the performance of properly designed photodetectors based on HgCdTe [7].

Progress in the creation of IR detectors based on HgCdTe nBn structures is based on the development of such technologies as molecular beam epitaxy (MBE) and metal-organic chemical vapor deposition (MOCVD), which make it possible to grow layers with a precisely controlled distribution of the component composition and dopant concentration over thickness. This makes it possible to develop new architectures of device structures that provide an increase in the operating temperature of the detectors and, in some cases, a simplification of the technological cycle of their manufacture [8].

At present, it is common to create traditional n on p HgCdTe photodiodes by ion implantation into an epitaxial film, for example, boron ions (without annealing), as well as p on n HgCdTe photodiodes by implanting arsenic ions and subsequent activation annealing. The technology of MCT photodiodes with dual-layer planar heterojunctions (DLPH) is being actively developed. The use of the nBn architecture for HgCdTe can provide a significant advantage over traditional photodiodes due to an increase in the quality of the material due to the absence of post-implantation defects [9, 10]. Eliminating the need for ion implantation (sometimes followed by annealing) will simplify the technology for detectors creating.

The concept of a barrier photosensitive structure was first implemented in practice for detectors based on InAs compounds (group III-V) with type II heterojunctions. The structure consisted only of n-type layers with an undoped InAsSb (or AlAsSb) barrier [11]. A feature of such heterostructures is the close to zero discontinuity (band offset) of the valence band at the heterojunctions boundaries of the barrier layer with the contact and absorbing layers at high energy barrier height (>1 eV) in the conduction band. Among the III-V compounds, the greatest progress in the creation of nBn structures has been achieved based on the InAsSb solid solution [12], which combines the advantages of the InSb compound with the possibility of achieving new properties upon transition to InAsSb epitaxial layers with a barrier based on the $InAs_{1-x}Sb_x/InAs_{1-y}Sb_y$ heterojunction. It has been theoretically and experimentally shown that nBn detectors based on III-V compounds with full realization of their potential advantages are capable to compete with traditional HgCdTe detectors [13], especially in MWIR.

In contrast to the III-V compounds, in the HgCdTe solid solution, which is the II-VI compound, type I heterojunctions are realized, which are characterized by the presence of an energy barrier ΔE_v for holes in the valence band. The presence of this barrier adversely affects the sensitivity of the detector structure and the threshold characteristics of devices and is considered one of the main fundamental obstacles to the practical implementation of barrier detectors based on HgCdTe. The search for ways to eliminate the influence of this barrier on the hole photocurrent in such structures is the subject of most of the works in this direction.

Figure 6.2 shows the layout of layers and a schematic band diagram of a unipolar barrier n^+Bn photosensitive structure, which demonstrates the mechanism of selective blocking of the electron current of the main charge carriers.

To date, many different architectures of barrier detectors based on HgCdTe have been proposed, among which unipolar configurations seem to be the most promising [14, 15].

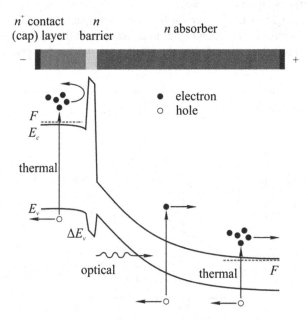

Fig. 6.2 Schematic structure and band diagram of n^+Bn HgCdTe detector under reverse bias and optical radiation

6.3 HgCdTe Based nBn Unipolar Barrier Structures

As mentioned above, the practical realization of the potential advantages of barrier structures based on HgCdTe is significantly limited by the fundamental presence of a potential barrier for holes formed by the B layer.

The most obvious solution to the problem of the non-zero valence band offset is to increase the external bias on the photosensitive structure. In [16], a numerical simulation of the energy diagrams of photosensitive $nB_n n$ structure with $Cd_{0.275}Hg_{0.725}Te$ absorbing layer designed for the MWIR and with complex three-layer barrier including central layer with $Cd_{0.6}Hg_{0.4}Te$ and two surrounding layers with variable composition was carried out. The authors also analyzed the influence of the composition of the central part of the barrier layer on the values of the detectivity. Figure 6.3 shows the calculated dependences of the detectivity on the magnitude of the external bias at the temperature of 200 K for various values of the barrier layer composition. Here, N_d is the donor impurity concentration in the corresponding layer of the structure.

It is clearly seen from the presented curves that the presence of a barrier for holes in the valence band determines the low values of the detectivity at low values of the external bias. As the negative external bias increases, the geometry of the potential barriers undergoes significant changes for both electrons and holes. As the bias increases, the height of the barrier for holes becomes smaller, which reduces the barrier for the minority current and leads to an increase in the detectivity. However, at too high bias values (~ 0.4 V), the change in the barrier geometry in the

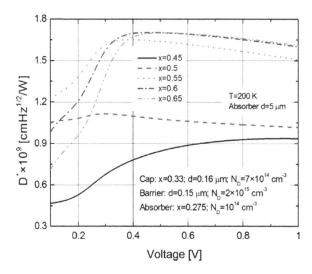

Fig. 6.3 Calculated dependences of the detectivity on the value of the external bias for the barrier photosensitive $nB_n n$ structure at the temperature of 200 K for various values of the composition of the central layer of the barrier. (Reprinted from [16]. Published 2013 by Polish Academy of Sciences as open access

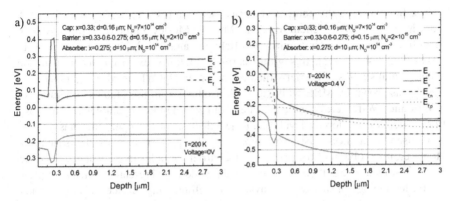

Fig. 6.4 Energy band diagrams of the structure in (**a**) the absence and (**b**) the presence of the bias for the barrier photosensitive nB_nn structure at the temperature of 200 K. (Reprinted from [16]. Published 2013 by Polish Academy of Sciences as open access)

conduction band becomes significant: its shape tends to triangular, the tunnel transparency of the barrier increases, and the shielding efficiency of the majority carrier current decreases, which leads to subsequent decrease in the values of detectivity (see Fig. 6.4). Here E_c, E_v are the energies of the edges of the conduction band and the valence band, E_f is the energy of the Fermi level, while $E_{f,n}$ and $E_{f,p}$ are the energies of non-equilibrium quasi-Fermi levels.

Another important conclusion can be drawn from Fig. 6.3. It can be seen that the dependence of the detectivity on the composition in the barrier layer is ambiguous and also depends on the bias voltage. At external biases >0.45 V, barrier layers with a larger composition provide better device detectivity values, which is explained by more efficient shielding of the current of majority charge carriers in a structure with a wider gap barrier. However, at bias voltage <0.45 V another situation is observed: an increase in the values of detectivity with an increase in the composition of the barrier at a certain value is replaced by a decrease. This is explained by the fact that in this range of biases, the values of the detectivity will be significantly affected by the value of the barrier for holes. An increase in the composition of the barrier increases the potential barrier both for the majority carriers in the structure and for the minor ones. It is obvious that there is some optimal ratio between the composition of the barrier layer and the value of the external bias, which provides the best values of the detectivity.

In the example described above, the optimization of the parameters of the structure and operating conditions was carried out only in relation to the external bias and the composition of the barrier layer. When designing real photosensitive structures, optimization must be carried out with respect to many structural parameters, such as composition profiles, thicknesses, as well as profiles and doping levels of the contact, barrier, and absorbing layers. Since 2011, various scientific groups have been working on finding the optimal ratios of the parameters of the layers of photosensitive structures.

The first work on this topic was carried out by a research group from the USA Anne M. Itsuno, Silviu Velicu, Jun Zhao, Michael Morley, Jamie D. Phillips, Angelo S. Gilmore from the University of Michigan and EPIR Technologies Inc. In 2011–2012 [14, 17–22], this group carried out fairly extensive theoretical and experimental studies of barrier structures based on HgCdTe in nBn and other configurations grown by MBE with Riber 32 installation. In this studies structures for MWIR and LWIR were considered. The optimization of layer parameters (compositions, thicknesses, doping levels) and the selection of the operating bias voltage were used as a mechanism for improving the performance of the structures. However, the problem of the presence of a barrier for holes in the valence band has not been solved. The authors noted the need for additional research aimed at improving the technology of structure passivation and eliminating surface leakage currents. The necessity of improving the theoretical model (taking into account tunneling effects, surface properties, Shockley-Read-Hall generation-recombination mechanisms, inhomogeneities of the stoichiometric composition and dopant concentration) is noted in order to achieve simulation results that would adequately describe real structures grown by the MBE.

A number of works [23–37] were carried out by scientific groups from Australia (The University of Western Australia), China (Shanghai Institute of Technical Physics, University of Chinese Academy of Sciences and National University of Defense Technology), another group from the USA (Consultant on Infrared Detectors and Boston University), as well as groups from France (CEA-Leti-Minatec), Turkey (Middle East Technical University), and Iran (Aerospace Research Institute). All of them carried out mainly theoretical and occasionally experimental studies aimed at improving the characteristics of the barrier structure based on HgCdTe by developing efficient architectures and optimizing the parameters of the structure layers. However, as far as we know, to date, none of these works has yielded breakthrough and practically realizable results.

Despite a significant number of publications devoted to the theoretical substantiation of the potential advantages of nBn structures, there are few attempts to implement nBn HgCdTe detectors in practice. For the first time, MWIR nBn detectors based on HgCdTe with the cut-off wavelength of 5.7 µm at 77 K were grown by MBE (Riber 32 MBE system) on a bulk CdZnTe substrate [20, 22]. It was found that the current-voltage characteristic (CVC) is determined by the shape of the barrier and depends on the applied bias. In the temperature range of 180–250 K, the diffusion limitation of the dark current was observed (the current density was 1–3 A/cm^2); at lower temperatures, the current was limited by generation through surface traps. The photoresponse depends on the applied bias. According to estimates, the maximum internal quantum efficiency was 66%. The authors do not give absolute values of the spectral sensitivity, which makes it difficult to assess the quality of the device. During the second development of the HgCdTe nBn structure using the MBE method, significantly lower (by about 5 orders of magnitude) values of the dark current density (3.74×10^{-6} A/cm^2 at 77 K and the bias of 0.5 V) were obtained

[19]. Such a big scatter in the obtained results indicates the need for systematic research to optimize the parameters of structures and technological processes.

A scientific group from France also attempted to create MWIR nBn structure based on MBE HgCdTe with an asymmetric barrier layer and uniform doping [28]. Despite the fact that the quantum efficiency, according to estimates, reached 60%, large current densities were observed at 77 K (of the order of 10^{-3} A/cm^2). The CVC had a form that differed significantly from that predicted by the simulation, and studies using secondary ion mass spectroscopy showed that the actual composition distribution over the film thickness differed significantly from the planned one.

Attempt was made by Itsuno et al. [14] to create a planar LWIR nBn structure from HgCdTe, for which high values of the dark current density (about 50 A/cm^2 at 77 K) were observed. Illumination resulted in a slight increase in current with reverse bias. The authors explained the results by the presence of surface leakage currents. The current density values for fabricated MWIR and LWIR nBn structures were several orders of magnitude higher than the values predicted by the expression Rule 07 for the dark current of IR detectors [38], which may indicate high leakage currents along the perimeter of the structure due to insufficient passivation quality.

As far as it is known from open sources, at present, systematic studies of unipolar photosensitive MBE nBn structures based on HgCdTe, including experimental ones, are being carried out by Voitsekhovskii et al. [39–51] at Tomsk State University (Tomsk, Russia) and Institute of Semiconductor Physics of Siberian Branch of Russian Academy of Sciences (ISP SB RAS, Novosibirsk, Russia).

For example, in [42], the authors fabricated and studied the MWIR structure, in which the diffusion limitation of dark current was found. MWIR nBn structures based on MBE HgCdTe/GaAs were fabricated using the plasma-enhanced atomic layer deposition (PE-ALD) Al$_2$O$_3$ dielectric to passivate the mesa structure. The diameter of the mesa structures was in the range from 20 to 500 μm. In a wide temperature range, the dark CVC of fabricated nBn structures based on MBE HgCdTe were studied. It was found that the reverse current of the nBn structure based on MBE HgCdTe increases significantly when illuminated by infrared radiation. According to preliminary estimates, the responsivity of fabricated non-optimized nBn structures at the temperature of 220 K and the voltage of −1 V is about 0.15 A W^{-1}. It is shown that the dark current is determined by the volume component, which is much larger than the surface leakage component. From the temperature dependences of the dark current, the activation energy is determined, which is close to the energy of the full band gap of the absorbing layer. In the temperature range of 180–300 K, the dark current values slightly exceed the dark current values for high-quality p-n photodiodes based on HgCdTe according to the empirical model Rule 07. It is shown that a diffusion-limited dark current is observed for the studied samples in the temperature range of 180–300 K at the reverse voltages around −1 V.

Figure 6.5 shows the CVC of this nBn structure at different temperatures in the dark mode and under illumination.

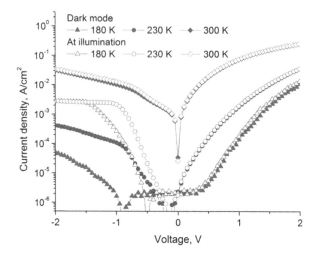

Fig. 6.5 CVC of the nBn structure based on MBE HgCdTe in dark mode and under illumination

6.4 HgCdTe Based Unipolar Barrier Structures with P-Type Layers

Another approach to minimizing the valence band offset is to create a p-type barrier. This mechanism was described, for example, in [5, 52]. According to theoretical and experimental studies described in the literature, structures with p-type barriers demonstrate better performance than structures with n-type barriers.

In [52], it was shown that the barrier for holes can be almost completely eliminated at a certain bias by precise acceptor doping of the barrier region. This approach is unpromising when using MBE and is possible only when heterostructures are grown by the MOCVD method, which makes it possible to obtain in situ MCT material of both donor and acceptor types of conductivity. In addition, the formation of p-type barrier layer will lead to the appearance of the space charge regions near the heterointerfaces of the barrier layer, which will cause intense generation according to the SRH mechanism. Therefore, the advantage of using a barrier doped with an acceptor impurity is not obvious.

Figure 6.6 shows the band diagrams calculated by the authors of [52] for photosensitive structures with n- and p-type barriers. It can be seen from the above figures that in the case of acceptor doping of the barrier layer at the external bias of −0.6 V, it is possible to almost completely eliminate the barrier for holes.

Since 2013, a group of scientists from Poland (M. Kopytko, A. Jóźwikowska, K. Jóźwikowski, A. Martyniuk, J. Antoszewski, A. Rogalski and others from Military University of Technology, Vigo System S.A. and University of Life Science) has been actively engaged in the problem of creating barrier detectors based on HgCdTe, including layers of p-type conductivity. In 2014–2018 this group have published a large number of theoretical, review and analytical works [52–76].

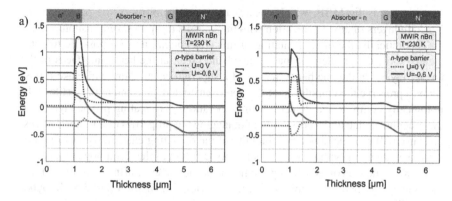

Fig. 6.6 Calculated band diagrams of photosensitive structures with (**a**) n- and (**b**) p-type barriers at the temperature of 230 K in the presence and absence of an external bias. (Reprinted with permission from [52]. Copyright 2014: Elsevier)

The main ways to optimize barrier structures considered by this group were the selection of the parameters of the layers of the barrier structure, as well as the use of complex multilayer bipolar configurations with p-type barriers. The experimental samples studied in these works were grown mainly by the MOCVD method on an Aixtron AIX-200 setup.

6.5   HgCdTe Based nBn Unipolar structures with Superlattice Barriers

To date, none of the approaches aimed at eliminating the energy barrier for minority charge carriers in nBn structures based on HgCdTe has made it possible to reduce it to values that provide acceptable values of the sensitivity of the structure while maintaining the efficiency of blocking majority charge carriers in the conduction band. The number of works devoted to this problem in the literature has been systematically decreasing for several years. This, on the one hand, is due to the fact that the parametric optimization of HgCdTe nBn structures has exhausted its possibilities in theoretical studies, and, on the other hand, is due to the small number and non-systematic nature of experimental work related to the fabrication of real nBn HgCdTe structures by the MBE method and studying their properties.

At present, one of the most promising methods for increasing the photosensitivity and quantum efficiency of nBn structures based on HgCdTe is the use of a superlattice with specified parameters as the barrier layer that effectively blocks the dark electron current in the conduction band and unimpeded hole current flow in the valence band. The level of technological development of methods for the epitaxial growth of HgCdTe nanoheterostructures makes it possible to fabricate such structures only by MBE, which ensures the reproducibility of the parameters of HgCdTe layers of any composition with thicknesses of up to several nanometers.

Theoretically, the implementation of this concept can make it possible to completely solve the problem of the valence band break at the heterointerfaces of the barrier layer. On the one hand, the use of a superlattice structure will make it possible to preserve the unipolarity of the photosensitive device and not resort to technologically complex and defect-forming operations for creating p-type layers. On the other hand, due to the possibility of managing the properties of the superlattices, their use will provide high controllability of the operating characteristics of photosensitive structures without fundamental modification of the technological mode of its growth. The idea of using a superlattice as barrier layer of nBn structure to eliminate the energy barrier for minority charge carriers was first proposed by Kopytko et al. [77].

Figure 6.7a shows the calculated band diagram of a similar structure, in which uniform layers of n- and p-type conductivity act as a barrier [77]. Calculation of the energy band diagram of the structure, in which the third type superlattice (T3SL) HgTe/Cd$_{0.95}$Hg$_{0.05}$Te plays the role of a barrier, is shown in Fig. 6.7b. The thickness of the Cd$_{0.95}$Hg$_{0.05}$Te layers in all cases considered by the authors was 28 nm, while the thicknesses of the HgTe layers varied during the calculations. Modeling showed that in such system with the thickness of HgTe layers of 5 nm, the potential barrier in the valence band is completely eliminated without the formation of space charge region (Fig. 6.7b, curves 5/28). Unfortunately, the authors do not provide information on how many periods of the superlattice they included in the calculations.

This topic was further developed in the works [78–83] of a group of scientists from The University of Western Australia (N.D. Akhavan, G. Jolley, G. A. Umana-Membreno, J. Antoszewski, L. Faraone et al.). Calculations carried out in [82],

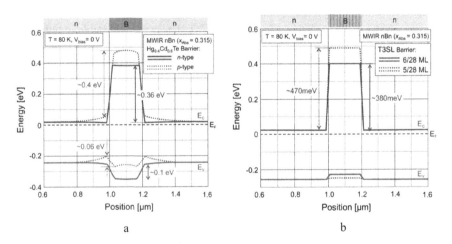

a   b

Fig. 6.7 Calculated band diagrams of structures with barriers of n- and p-types (**a**) and with the barrier in the form of the superlattice (**b**) at the temperature of 80 K at zero external bias. (Reprinted from [77]. Published 2015 by Springer as open access)

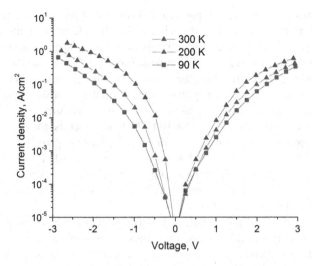

Fig. 6.8 CVCs of nBn structure including superlattice of 18 periods $Hg_{0.20}Cd_{0.80}Te$ (9 nm) – HgTe (2 nm) as selective barrier, measured at different temperatures

taking into account the quantum mechanical nature of the superlattice in the barrier layer, showed that at the thickness of the CdTe barriers and HgTe wells equal to 1.3 and 3.7 nm, respectively, the maximum ratio of the hole current to the electron current is reached, if the number of layers in the superlattice exceeds 12. It was also theoretically shown in the works of this group that a reduction in the barrier in the valence band is possible with a non-uniform distribution of the composition and dopant in the barrier layer [31]. However, these studies were predominantly theoretical. Only one of the works known to us [79] presents the results of measurements of the dark current of a structure with HgTe/CdTe superlattice as the barrier layer at the temperature of 155 K, as well as the corresponding calculation. The experimental structure was grown by MBE at The University of Western Australia.

In [84] Voitsekhovskii et al. have studied the electrophysical properties of nBn structure with a superlattice grown by the MBE method at the ISP SB RAS by admittance spectroscopy. The structure included superlattice of 18 periods $Hg_{0.20}Cd_{0.80}Te$ (9 nm) – HgTe (2 nm) as a selective barrier for majority charge carriers. Figure 6.8 shows the CVC of this structure measured at various temperatures.

6.6 HgCdTe Based NBνN HOT Unipolar Structures

One way to suppress Auger processes that does not require semiconductor cooling is the creation of non-equilibrium depletion [85]. Dark currents, which determine the operating temperature of highly sensitive IR detectors based on HgCdTe, are limited by Auger generation processes. It is not a trivial task to reduce the requirements for cooling detectors without compromising performance. The concept of

HOT structures is based on the suppression of Auger GR processes by reducing the thermal concentration of charge carriers in the absorbing layer to values that are less than equilibrium. In HOT detectors, a lightly doped narrow-gap absorbing layer is used, on one side of which a heterojunction is formed, which ensures the exclusion of charge carriers, and on the other, the extraction of charge carriers. When a reverse bias is applied, the concentration of charge carriers in the absorbing layer decreases, which manifests itself in the suppression of Auger processes. Estimates of the threshold characteristics show that for HOT MWIR detectors, the mode of limitation by background radiation noise is realized at 203 K (compared to $T = 155$ K for an ideal DLPH detector), and at 145 K for LWIR detector (102 K for DLPH detector). The dark current in such HOT structures, according to calculations, should be less than in photodiodes with p-n junctions [86], and the operating characteristics at a given temperature should be better.

In the case of n-HgCdTe MBE, a promising architecture of unipolar barrier detectors is the NBνN configuration, which makes it possible to increase the operating temperature of detectors by suppressing Auger GR processes. Theoretically, NBνN-architecture HgCdTe detectors retain the technological advantages of the nBn configuration, but can show lower dark currents than p-n photodiodes and nBn detectors.

The NBνN device contains four n-type layers: a heavily doped cover layer, a lightly doped barrier layer, a lightly doped absorber layer, and a heavily doped bottom layer. The offset of the conduction band at the boundary of the absorbing and barrier layers prevents the current of electrons from the covering to the absorbing layer. Under equilibrium conditions, there is a barrier in the valence band at the boundary between the barrier and absorbing layers, which blocks the transfer of holes in the direction of the coating layer. When a reverse bias is applied, the barrier in the valence band decreases. As a result, holes that appear in the absorbing layer due to thermal and optical generation are collected in the upper layer. The boundary between the absorbing and barrier (or covering) layer provides hole extraction, and the boundary between the absorbing and lower layers provides exclusion. As a result, thermally generated holes are effectively removed from the absorbing layer, and the hole concentration becomes much lower than the equilibrium values. To maintain electrical neutrality in the absorbing layer, the electron concentration also decreases below the equilibrium value. GR processes by the Auger mechanism are suppressed due to a decrease in the total carrier concentration in the active layer. According to the simulation results, the use of the NBνN configuration makes it possible to reduce the dark current values, increase the detectivity and operating temperature (compared to nBn and DLPH), and also eliminate the technological problem of forming p-type regions.

The first work on this topic was carried out by a research group from the USA Anne M. Itsuno, Silviu Velicu at al [19]. from the University of Michigan and EPIR Technologies Inc. In [19], the characteristics of IR detectors based on HgCdTe with NBνN architecture were simulated. The calculated values of the dark current density of the detectors for the MWIR and LWIR ranges are an order of magnitude (or more) lower than for the nBn or DLPH detectors in the temperature range from 50 to 225 K. The calculated values of the peak detectivity ($D*$) at the maximum sensitivity for the MWIR detector were 6.0×10^{14} and 2.4×10^{10} cm $\times$ W^{-1} $\times$ Hz$^{1/2}$ at the temperatures of 95 and 225 K, respectively. The $D*$ values for the LWIR detector were 2.4×10^{14} and 2.3×10^{11} cm $\times$ W^{-1} $\times$ Hz$^{1/2}$ at 50 and 95 K, respectively. Detectivity values were estimated from the maximum sensitivity calculated in earlier works [17].

6.7 Summary

Based on the review, we can conclude that unipolar structures based on II-VI compounds are promising and can lead to the creation of IR FPAs with improved characteristics, operating at higher cooling temperatures. A large theoretical background in the study of nBn structures based on HgCdTe for both MWIR and LWIR shows some advantages over photodiodes based on p-n junction in HgCdTe. If for the medium-wavelength IR range the creation of FPAs based on barrier structures of III-V compounds has reached the practical level and mass production, then the practical implementation of devices based on II-VI compounds is hindered by a large number of unsolved fundamental, design and technological problems. The presence of a barrier for holes in the valence band in nBn structures based on MCT material requires a number of technological solutions, which are the use of large external biases, control of the barrier layer parameters, including acceptor doping of the barrier, and the use of multilayer structures with a complex barrier layer design, including barriers in the form of superlattices. Therefore, despite a certain theoretical background, the developed technological implementations of the devices have not yet made it possible to obtain practically significant results realizing the theoretically predicted advantages of the nBn MCT structure in the development of the FPAs.

The authors of one of the latest reviews of the achievements of nBn infrared detectors [87] conclude that for the case of HgCdTe, the technology of barrier detectors is the most promising for achieving operation at close to room temperature.

Acknowledgments Study was partially supported by the Tomsk State University Development Program («Priority-2030»).

References

1. Rogalski A (2019) Infrared and terahertz detectors, 3rd edn. Taylor & Francis Group, LLC
2. Kinch MA (2015) The future of infrared; III–Vs or HgCdTe? J Electron Mater 44(9):2969–2976
3. Gu R, Antoszewski J, Lei W, Madni I, Umana-Membrenao G, Faraone L (2017) MBE growth of HgCdTe on GaSb substrates for application in next generation infrared detectors. J Cryst Growth 468:216–219

4. Maimon S, Wicks GW (2006) nBn detector, an infrared detector with reduced dark current and higher operating temperature. Appl Phys Lett 89:151109
5. Kopytko M, Keblowski A, Gawron W, Madejczyk P (2015) Different cap-barrier design for MOCVD grown HOT HgCdTe barrier detectors. Opto-Electron Rev 23(2):143–148
6. Rogalski A, Martyniuk P (2014) Mid-wavelength infrared nBn for HOT detectors. J Electron Mater 43(8):2963–2969
7. Kopytko M, Kębłowski A, Gawron W, Pusz W (2016) LWIR HgCdTe barrier photodiode with Auger-suppression. Semicond Sci Technol 31(3):035025
8. Kopytko M, Rogalski A (2016) HgCdTe barrier infrared detectors. Prog Quantum Electron 47:1–18
9. Bubulac LO (1988) Defects, diffusion and activation in ion implanted HgCdTe. J Crystal Growth 86(1–4):723–734
10. Talipov N, Voitsekhovskii A (2018) Annealing kinetics of radiation defects in boron-implanted p-Hg1-xCdxTe. Semicond Sci Technol 33(6):065009
11. Pedrazzani JR, Maimon S, Wicks GW (2008) Use of nBn structures to suppress surface leakage currents in unpassivated InAs infrared photodetectors. Electron Lett 44(25):1487–1488
12. Soibel A, Keo SA, Fisher A, Hill CJ, Luong E, Ting DZ, Gunapala SD, Lubyshev D, Qiu Y, Fastenau JM, Liu AWK (2018) High operating temperature nBn detector with monolithically integrated microlens. Appl Phys Lett 112(4):041105
13. Martyniuk P, Kopytko M, Rogalski A (2014) Barrier infrared detectors. Opto-Electron Rev 22(2):127–146
14. Itsuno AM (2012) Bandgap-engineered mercury cadmium telluride infrared detector structures for reduced cooling requirements. PhD thesis, University of Michigan
15. Iakovleva NI (2019) Unipolar MCT-based nBn-structure for a MWIR FPA. Appl Phys 3:53–60
16. Martyniuk P, Rogalski A (2013) Theoretical modelling of MWIR thermoelectrically cooled nBn HgCdTe detector. Bull Polish Acad Sci, Techn Sci 61(1):211–220
17. Itsuno AM, Phillips JD, Velicu S (2011) Design and modeling of HgCdTe nBn detectors. J Electron Mater 40(8):1624–1629
18. Itsuno AM, Phillips JD, Gilmore AS, Velicu S (2011) Calculated performance of an Auger suppressed unipolar HgCdTe photodetector for high temperature operation. Proc SPIE 8155:81550J
19. Itsuno AM, Phillips JD, Velicu S (2012) Design of an Auger-Suppressed Unipolar HgCdTe NBvN photodetector. J Electron Mater 41(10):2886–2892
20. Velicu S, Zhao J, Morley M, Itsuno AM, Phillips JD (2012) Theoretical and experimental investigation of MWIR HgCdTe nBn detectors. Proc SPIE 8268:82682X
21. Itsuno AM, Phillips JD, Velicu S (2012) Unipolar barrier-integrated HgCdTe infrared detectors. In: Proceeding of 70th device research conference, 18–20 June 2012. University Park, PA, USA, 12908883
22. Itsuno AM, Phillips JD, Velicu S (2012) Mid-wave infrared HgCdTe nBn photodetector. Appl Phys Lett 100:161102
23. Akhavan ND, Jolley G, Umana-Membreno GA, Antoszewski J, Faraone L (2015) Theoretical study of Midwave infrared HgCdTe nBn detectors operating at elevated temperatures. J Electron Mater 44:3044–3055
24. Ye ZH, Chen YY, Zhang P, Lin C, Hu XN, Ding RJ, He L (2014) Modeling of LWIR nBn HgCdTe photodetector. Proc SPIE 9070:90701L
25. Akhavan ND, Umana-Membreno GA, Jolley G, Antoszewski J, Faraone L (2014) A method of removing the valence band discontinuity in HgCdTe-based nBn detectors. Appl Phys Lett 105:121110
26. Akhavan ND, Jolley G, Membreno GU, Antoszewski J, Faraone L (2014) Band-to-band tunnelling (BTBT) in HgCdTe-based nBn detectors for LWIR applications. In: Proceedings of the conference on optoelectronic and microelectronic materials & devices, 14–17 December 2014. Perth, WA, Australia, 14920679

27. Akhavan ND, Jolley G, Umana-Membreno GA, Antoszewski J, Faraone L (2014) Performance modeling of bandgap engineered HgCdTe-based nBn infrared detectors. IEEE Trans Electron Devices 61(11):3691–3698
28. Gravrand O, Boulard F, Ferron A, Ballet P, Hassis W (2015) A new nBn IR detection concept using HgCdTe material. J Electron Mater 44(9):3069–3075
29. Akhavan ND, Jolley G, Umana-Membreno GA, Antoszewski J, Faraone L (2015) Design of Band Engineered HgCdTe nBn detectors for MWIR and LWIR applications. IEEE Trans Electron Devices 62(3):722–728
30. Ting DZ, Soibel A, Khoshakhlagh A, Gunapala SD (2017) Theoretical analysis of nBn infrared photodetectors. Opt Eng 56(9):091606
31. Akhavan ND, Umana-Membreno GA, Gu R, Antoszewski J, Faraone L (2018) Delta doping in HgCdTe based unipolar barrier photodetectors. IEEE Trans Electron Devices 65(1):4340–4345
32. He J, Hu W (2018) Numerical simulation of HgCdTe nBn longwavelength infrared detector. In: Proceedings of international conference on numerical simulation of optoelectronic devices (NUSOD), 05–09 November 2018. Hong Kong, China, 18310399
33. Uzgur F, Kocaman S (2019) A dual-band HgCdTe nBn infrared detector design. Proc SPIE 11129:1112903
34. Uzgur F, Kocaman S (2019) Barrier engineering for HgCdTe unipolar detectors on alternative substrates. Infrared Phys Technol 97:123–128
35. Sharma I, Srivastava T, Kaushik R, Goyal A (2019) Design and analysis of HgCdTe infrared photodetector. In: Proceedings of the international conference on signal processing and communication (ICSC), 07–09 March 2019. Noida, India, 19256588
36. He J, Wang P, Li Q, Wang F, Gu Y, Shen C, Lu Chen P, Martyniuk A, Rogalski XC, Wei L, Hu W (2010) Enhanced performance of HgCdTe long-wavelength infrared photodetectors with nBn design. IEEE Trans Electron Devices 67(5):2001–2007
37. Shaveisi M, Aliparast P (2021) Dark current evaluation in HgCdTe-based nBn infrared detectors. In: 2021 Iranian international conference on microelectronics (IICM2021), 22–24 December 2021. IEEE. https://doi.org/10.1109/IICM55040.2021.9730154
38. Tennant WE, Lee D, Zandian M, Piquette E, Carmody M (2008) MBE HgCdTe technology: a very general solution to IR detection, described by «Rule 07», a very convenient heuristic. J Electron Mater 37(9):1406–1410
39. Voitsekhovskii AV, Nesmelov SN, Dzyadukh SM, Dvoretsky SA, Mikhailov NN, Sidorov GY (2019) Admittance characteristics of nBn structures based on HgCdTe grown by molecular beam epitaxy. Russ Phys J 62(5):818–826
40. Voitsekhovskii AV, Nesmelov SN, Dzyadukh SM, Dvoretsky SA, Mikhailov NN, Sidorov GY, Yakushev MV (2019) Admittance dependences of the mid-wave infrared barrier structure based on HgCdTe grown by molecular beam epitaxy. Mater Res Express 6:116411
41. Voitsekhovskii AV, Nesmelov SN, Dzyadukh SM, Dvoretsky SA, Mikhailov NN, Sidorov GY (2019) Current-voltage characteristics of nBn structures based on mercury cadmium telluride epitaxial films. Russ Phys J 62(6):1054–1061
42. Voitsekhovskii AV, Nesmelov SN, Dzyadukh SM, Dvoretsky SA, Mikhailov NN, Sidorov GY, Yakushev MV (2020) Diffusion-limited dark currents in mid-wave infrared HgCdTd-based nBn structures with Al_2O_3 passivation. J Phys D Appl Phys 53:53 055107
43. Voitsekhovskii AV, Nesmelov SN, Dzyadukh SM, Dvoretsky SA, Mikhailov NN, Sidorov GY (2019) Electrical properties of nBn structures based on HgCdTe grown by molecular beam epitaxy on GaAs substrates. Infrared Phys Technol 102:103035
44. Izhnin II, Voitsekhovskii AV, Nesmelov SN, Dzyadukh SM, Dvoretsky SA, Mikhailov NN, Sidorov GY, Yakushev MV (2022) Admittance of barrier nanostructures based on MBE HgCdTe. Appl Nanosci 12:403–409
45. Voitsekhovskii AV, Nesmelov SN, Dzyadukh SM, Dvoretsky SA, Mikhailov NN, Sidorov GY, Yakushev MV (2020) Admittance of barrier structures based on mercury cadmium telluride. Russ Phys J 63(3):432–445

46. Voitsekhovskii AV, Nesmelov SN, Dzyadukh SM, Dvoretsky SA, Mikhailov NN, Sidorov GY, Yakushev MV (2020) Admittance of metal-insulator-semiconductor devices based on HgCdTe nBn structures. Semicond Sci Technol 35:055026

47. Voitsekhovskii AV, Nesmelov SN, Dzyadukh SM, Dvoretsky SA, Mikhailov NN, Sidorov GY, Yakushev MV (2020) Impedance of mis devices based on nBn structures from mercury cadmium telluride. Russ Phys J 63(6):907–916

48. Voitsekhovskii AV, Nesmelov SN, Dzyadukh SM, Dvoretsky SA, Mikhailov NN, Sidorov GY, Yakushev MV (2021) Admittance of MIS structures based on nBn systems of epitaxial HgCdTe for detection in the 3-5 μm spectral range. Tech Phys Lett 47(6):616–619

49. Voitsekhovskii AV, Nesmelov SN, Dzyadukh SM, Dvoretsky SA, Mikhailov NN, Sidorov GY, Yakushev MV (2021) An experimental study of the dynamic resistance in surface leakage limited nBn structures based on HgCdTe grown by molecular beam epitaxy. J Electron Mater 50:4599–4605

50. Voitsekhovskii AV, Nesmelov SN, Dzyadukh SM, Dvoretsky SA, Mikhailov NN, Sidorov GY, Yakushev MV (2021) Dark currents of unipolar barrier structures based on mercury cadmium telluride for long-wave IR detectors. Russ Phys J 64(5):763–769

51. Burlakov ID, Kulchitsky NA, Voitsekhovskii AV, Nesmelov SN, Dzyadukh SM, Gorn DI (2021) Unipolar semiconductor barrier structures for infrared photodetector arrays (review). J Commun Technol Electron 66(9):1084–1091

52. Kopytko M (2014) Design and modelling of high-operating temperature MWIR HgCdTe nBn detector with n- and p-type barriers. Infrared Phys Technol 64:47–55

53. Kopytko M, Keblowski A, Gawron W, Madejczyk P, Kowalewski A, Jozwikowski K (2013) High-operating temperature MWIR nBn HgCdTe detector grown by MOCVD. Opto-Electron Rev 21(4):402–405

54. Kopytko M, Jozwikowski K (2013) Numerical Analysis of Current–Voltage Characteristics of LWIR nBn and p-on-n HgCdTe Photodetectors. J Electron Mater 42(11):3211–3216

55. Kopytko M, Jozwikowski K, Rogalski A (2014) Fundamental limits of MWIR HgCdTe barrier detectors operating under non-equilibrium mode. Solid State Electron 100:20–26

56. Martyniuk P, Gawron W (2014) Barrier detectors versus homojunction photodiode. Metrol Meas Syst XXI (4):675–684

57. Martyniuk P (2014) HOT HgCdTe infrared detectors. In: IEEE Xplore. Numerical simulation of optoelectronic devices. https://doi.org/10.1109/NUSOD.2014.6935412

58. Martyniuk P (2015) HOT mid-wave HgCdTe nBn and pBp infrared detectors. Opt Quant Electron 47:1311–1318

59. Martyniuk P, Gawron W, Pusz W, Stanaszek D, Rogalski A (2014) Modeling of HOT (111) HgCdTe MWIR detector for fast response operation. Opt Quant Electron 46:1303–1312

60. Kopytko M, Jozwikowska A, Jozwikowska K, Martyniuk A, Antoszewski J, Faraone L (2014) Numerical analysis of fluctuation phenomena in HOT HgCdTe barrier detectors. In: IEEE Xplore. 2014 conference on optoelectronic and microelectronic materials & devices. https://doi.org/10.1109/COMMAD.2014.7038687

61. Martyniuk P, Gawron W, Stanaszek D, Pusz W, Rogalski A (2014) Theoretical modelling of mercury cadmium telluride mid-wave detector for high temperature operation. IET Optoelectron 8(6):239–244

62. Qiu WC, Jiang T, Cheng XA (2015) A bandgap-engineered HgCdTe PBπn long-wavelength infrared detector. J Appl Phys 118:124504

63. Kopytko M, Jozwikowski K (2015) Generation-recombination effect in MWIR HgCdTe barrier detectors for high-temperature operation. IEEE Trans Electron Devices 62(7):2278–2284

64. Kopytko M, Keblowski A, Gawron W, Martyniuk P, Madejczyk P, Jozwikowski K, Kowalewski A, Markowska O, Rogalski A (2015) MOCVD grown HgCdTe barrier detectors for MWIR high-operating temperature operation. Opt Eng 54(10):105105

65. Chen Y, Ye Z, Zhang P, Hu X, Ding R, He L (2016) A barrier structure optimization for widening processing window in dual-band HgCdTe IRFPAs detectors. Opt Quant Electron 48:294

66. Martyniuk P, Gawron W, Madejczyk P, Kopytko M, Grodecki K, Gomulka E (2017) Theoretical utmost performance of (100) mid-wave HgCdTe photodetectors. Opt Quant Electron 49:20
67. Madejczyk P, Gawron W, Martyniuk P, Keblowski A, Pusz W, Pawluczyk J, Kopytko M, Rutkowski J, Rogalski A, Piotrowski J (2017) Engineering steps for optimizing high temperature LWIR HgCdTe photodiodes. Infrared Phys Technol 81:276–281
68. Kopytko M, Keblowski A, Madejczyk P, Martyniuk P, Piotrowski J, Gawron W, Grodecki K, Jozwikowski K, Rutkowski J (2017) Optimization of a HOT LWIR HgCdTe photodiode for fast response and high detectivity in zero-bias operation mode. J Electron Mater 46(10):6045–6055
69. Martyniuk P, Gawron W, Mikolajczyk J (2017) The development of the room temperature LWIR HgCdTe detectors for free space optics communication systems. In: Proceedings of the SPIE 10437, advanced free-space optical communication techniques and applications III, 104370G (6 October 2017). https://doi.org/10.1117/12.2278628
70. Jozwikowski K, Piotrowski J, Jozwikowska A, Kopytko M, Martyniuk P, Gawron W, Madejczyk P, Kowalewski A, Markowska O, Martyniuk A, Rogalski A (2017) The numerical-experimental enhanced analysis of HOT MCT barrier infrared detectors. J Electron Mater 46:5471–5478
71. Kopytko M (2017) Theoretical performance of mid wavelength HgCdTe(100) heterostructure infrared detector. Solid State Electron 137:102–108
72. Kopytko M, Gomolka E, Michalczewski K, Martyniuk P, Rutkowski J, Rogalski A (2018) Investigation of surface leakage current in MWIR HgCdTe and InAsSb barrier detectors. Semicond Sci Technol 33:125010
73. Kopytko M, Gawron W, Keblowski A, Stępien D, Martyniuk P, Jozwikowska K (2018) Numerical analysis of HgCdTe dual-band infrared detector. Opt Quant Electron 51:62
74. Martyniuk P, Madejczyk P, Kopytko M, Henig AM, Grodecki K, Gawron W, Rutkowski J (2018) Theoretical simulation of the thermoelectrically cooled HgCdTe LWIR detector for fast response operating under unbiased conditions. IET Optoelectron 12(4):161–167
75. Kopytko M, Gawron W, Keblowski A, Stepien D, Martyniuk P, Jozwikowska K (2019) Numerical analysis of HgCdTe dual-band infrared detector. Opt Quant Electron 51:62
76. He J, Li Q, Wang P, Wang F, Yue G, Shen C, Luo M, Yu C, Lu Chen X, Chen W, Lu WH (2020) Design of a bandgap-engineered barrier-blocking HOT HgCdTe long-wavelength infrared avalanche photodiode. Opt Express 28(22):33556–33563
77. Kopytko M, Wrobel J, Jozwikowska K, Rogalski A, Antoszewski J, Akhavan ND, Umana-Membreno GA, Faraone L, Becker CR (2015) Engineering the bandgap of unipolar HgCdTe-based nBn infrared photodetectors. J Electron Mater 44(1):158–166
78. Benyaya J, Martyniuk P, Kopytko M, Antoszewski J, Gawron W, Madejczyk P (2015) nBn HgCdTe infrared detector with HgTe/CdTe SLs barrier. In: IEEE Xplore, 2015 international conference on numerical simulation of optoelectronic devices (NUSOD). https://doi.org/10.1109/NUSOD.2015.7292881
79. Benyahia D, Martyniuk P, Kopytko M, Antoszewski J, Gawron W, Madejczyk P, Rutkowski J, Gu R, Faraone L (2016) nBn HgCdTe infrared detector with HgTe(HgCdTe)/CdTe SLs barrier. Opt Quant Electron 48:215
80. Gu R, Lei W, Antoszewski J, Madni I, Umana-Menbreno G, Faraone L (2016) Recent progress in MBE grown HgCdTe materials and devices at UWA. Proc SPIE 9819:98191Z
81. Akhavan ND, Umana-Membreno GA, Antoszweski J, Faraone L (2016) Self consistent carrier transport in band engineered HgCdTe nBn detector. In: IEEE Xplore. 2016 International conference on numerical simulation of optoelectronic devices (NUSOD). https://doi.org/10.1109/NUSOD.2016.7547060
82. Akhavan ND, Umana-Membreno GA, Gu R, Asadnia M, Antoszewski J, Faraone L (2016) Superlattice barrier HgCdTe nBn infrared photodetectors: validation of the effective mass approximation. IEEE Trans Electron Devices 63(12):4811–4818
83. Akhavan ND, Umana-Membreno GA, Gu R (2018) Optimization of superlattice barrier HgCdTe nBn infrared photodetectors based on an NEGF approach. IEEE Trans Electron Devices 65(2):591–598

84. Izhnin II, Kurbanov KR, Voitsekhovskii AV, Nesmelov SN, Dzyadukh SM, Dvoretsky SA, Mikhailov NN, Sidorov GY, Yakushev MV (2020) Unipolar superlattice structures based on MBE HgCdTe for infrared detection. Appl Nanosci 10:4571–4576
85. Ashley T, Elliott CT (1985) Nonequilibrium devices for infra-red detection. Electron Lett 21(10):451–452
86. Schaake HF, Kinch MA, Chandra D, Aqariden F, Liao PK, Weirauch DF, Wan C-F, Scritchfield RE, Sullivan WW, Teherani JT, Shih HD (2008) High-operating-temperature MWIR detector diodes. J Electron Mater 37(9):1401–1405
87. Shi Q, Zhang S-K, Wang J-L, Chu J-H (2022) Progress on nBn infrared detectors. J Infrared Millim Waves 41(1):139–150

Chapter 7
Infrared Sensing Using Mercury Chalcogenide Nanocrystals

Emmanuel Lhuillier, Tung Huu Dang, Mariarosa Cavallo, Claire Abadie, Adrien Khalili, John C. Peterson, and Charlie Gréboval

7.1 Introduction

II-VI NCs are bright, narrow-band emitters, that achieve many interesting properties through quantum confinement. Commercially, they have been used as a light down converter for displays, but NCs have also been applied in ways that make central use of the materials, such as in single photon emitters [1], in light emitting diodes [2] (LED), or as biolabels [3].

Solar cells have been the main application of NC materials as light sensors. Thanks to their size-tunable band gap, lead sulfide (PbS) NCs can have the optimal band gap for single junction solar cells, around 1.3 eV. In addition, they may be tuned to the near-infrared part of the solar spectrum, a region that remains unaddressed by organic solar cells, the other low-cost alternative to silicon-based solar cells. The interest in NC-based solar cells has also been motivated by multi-exciton generation [4–6]. In a semiconductor, high-energy photons may result in the generation of more than one electron-hole pair if carrier multiplication occurs before their thermalization. In NCs, the threshold energy for this process to occur is drastically reduced compared to the bulk material. This motivate infrared sensing based on narrower band gap lead chalcogenides NCs.

Thanks to their small or absent bulk band gap, mercury chalcogenides also offer a nice playground as infrared optoelectronic building blocks. Mercury chalcogenide NCs [7] offer a broadly tunable absorption from the visible range up to the THz (Fig. 7.1) while being synthetically similar to cadmium chalcogenides, the preparation of which has been extensively studied. Initially, the interest in HgTe

E. Lhuillier (✉) · T. H. Dang · M. Cavallo · C. Abadie · A. Khalili · C. Gréboval
Sorbonne Université, CNRS, Institut des NanoSciences de Paris, INSP, Paris, France
e-mail: el@insp.jussieu.fr

J. C. Peterson
The James Franck Institute, The University of Chicago, Chicago, IL, USA

© The Author(s), under exclusive license to Springer Nature Switzerland AG 2023
G. Korotcenkov (ed.), *Handbook of II-VI Semiconductor-Based Sensors and Radiation Detectors*, https://doi.org/10.1007/978-3-031-20510-1_7

Fig. 7.1 Size tunability of HgTe NCs absorption edge. Wavelength of the lowest energy absorption feature (interband edge for small size and intraband peak for largest size) as a function of particle size for HgTe NCs

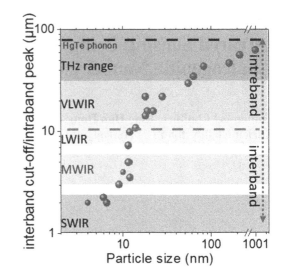

nanocrystals had been driven by their use as broadband optical amplifiers [8], however after the collapse of the telecom bubble in the early 2000s, the use of this material has shifted toward detection [9]. Bulk mercury chalcogenides are already used for infrared sensing, as in the alloy HgCdTe, where the Cd content is used to tune the band gap. In NCs, the quantum confinement plays the same role, see Fig. 7.1. Note that this chapter is mainly focused on HgTe since, among mercury chalcogenides, it has generated the most interest by far. Integration of HgS and HgSe was extremely limited until the observation of intraband absorption in the doped form of these NCs. A section is dedicated to intraband devices at the end of the chapter.

Though the first report on the preparation of monodisperse NCs was published in 1993 [10], it took nearly 10 years before NCs progressed from being optically active materials to building blocks for optoelectronics [11]. One major challenge to this effect is adressing the initial poor charge transport properties of NC films, as discussed in Sect. 7.2 of this chapter. Surface ligands are required to obtain monodisperse NC solution with well-passivated surfaces, but these often act as a barrier to charge transport, leading to poor charge carrier mobility. Thus, a pristine film of NCs typically presents a low mobility ($<10^{-6}$ cm^2.V^{-1}.s^{-1}), seemingly limiting the application of these materials. It is only as the field has matured that such challenges have been overcome, and NC device performances have increased.

7.2 Enabling Transport and Photoconduction in Nanocrystal Films

NCs are typically colloidally synthesized: grown in solution in the presence of ligands. These ligands are necessary. During the synthesis, they reduce the availability of the NC surface, slowing particle growth. Once the NCs are synthesized, these ligands also aid colloidal stability and prevent aggregation. They also help to electronically passivate the surface through hybridization with dangling bonds. However, from a device perspective, ligands act as energetic barriers to charge transport. The barrier width roughly matches the ligand length, while the height is typically a few eV. As a result, the NCs are only weakly coupled with one another. Macroscopically, this leads to a low carrier mobility. Typical values for films of pristine NCs (*i.e.*, without further processing) fall in the 10^{-6} cm^2.V^{-1}.s^{-1} range limiting their use in optoelectronic devices. Due to this weak local coupling, charge transport in a nanocrystal film may be modeled as a hopping process, where charge wave functions are localized to individual NCs, which tunnel through the barrier formed by the ligands with some hopping time. Since single particle devices remain uncommon [12], and the electrode spacing ranges from 100 nm to a few 10 s of μm while the particle size is typically 5 to 20 nm, charges must tunnel through anywhere from 10 to a few thousand barriers. To improve the charge transport, the barrier must be modified to minimize its height and width. This is the purpose of the ligand exchange procedure.

7.2.1 Solid-State Ligand Exchange

A film of deposited NCs may be treated by a solution of short ligands in the solid state to improve the film conductivity. NCs are initially spread onto a surface using different methods such as spin-coating, dip-coating or even drop-casting. Complementary methods, such as spray-coating [13, 14], inkjet printing [9], or electrophoretic deposition [15] have also been developed, but these will not be described here. Synthetized NCs are often capped with long, organic ligands (acid, amine, thiol and phosphine are the most typical ligands at this stage) that bond to the NC surface, see Fig. 7.2a. These ligands typically include an alkyl chain, commonly 12–18 carbons long. This results in a typical length of the ligand of 1.5–2 nm, depending on the exact number of carbons and possible ligand interdigitation.

This preformed film of NCs is dipped into a solution of shorter ligands in order to maximize nanoparticle coupling. The shorter the ligand is, the stronger the NC coupling will be, leading to higher mobility. The Law's group (Fig. 7.2e) has been able to measure a scaling law that relates the ligand length and the mobility, and the relationship is approximately exponential. Also, note that since the nature of the ligand is not changed (acid in this case), one may extrapolate the mobility to zero ligand length estimated to be around a few cm^2 V^{-1} s^{-1}. With EDT, the most common ligand, the mobility is rather in the 10^{-4}–10^{-1} cm^2 V^{-1} s^{-1} range, see Fig. 7.2e, f. To further increase the carrier mobility, it is necessary to also play on the height of the tunnel barrier and consequently on the nature of the ligand. This can be done by introducing inorganic ligands [18] or very short ions (Cl$^-$ [19], S^{2-} [20], As$_2$S$_3$ [17, 21]…), see Fig. 7.2f.

This solid-state ligand exchange procedure relies on the diffusion of the new ligand from the solution to the inner part of the NC film and on the reverse

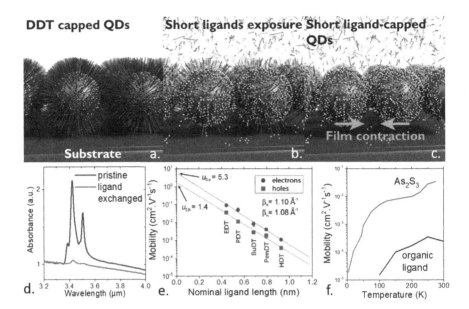

Fig. 7.2 Principle and consequence of the solid-state ligand exchange. (**a**) Schematic of a NC thin film deposited onto a substrate and capped with their native ligands. (**b**) Schematic of the previous film once the film is exposed to short ligands. (**c**) Schematic of the previous film after the ligand exchange process. (**d**) Infrared absorption spectrum in the region of the C-H bond resonance before and after ligand exchange. (**e**) Mobility as a function of the ligand length for a thin film of PbSe NCs capped with various thiol ligands. Part (e) is adapted with permission from Ref. [16]. Copyright 2010: American Chemical Society. (**f**) Mobility as a function of temperature for a HgTe NC thin film capped with organic and inorganic ligands. Part (f) is adapted with permission from Ref. [17]. (Copyright 2013: WILEY-VCH Verlag GmbH & Co. KGaA)

procedure for the native ligand. As a result, this process is only efficient for thin films (20–50 nm).

The polarity and general solvent affinity of the native ligand drive the efficiency of the ligand exchange. Solvents in which the native ligands are more soluble will lead to more efficient extraction. Care must be taken to select a solvent in which the NCs themselves are not soluble. The completion of the procedure can easily be followed by infrared spectroscopy (see Fig. 7.2d) by tracing the height of the resonance associated with the C-H bond appearing around 2900–3000 cm^{-1} (the doublet at 3.4 μm in Fig. 7.2d). The magnitude of this peak is drastically reduced at the end of the procedure as the free ligands are removed, and the ligand length is reduced.

Another consequence of the ligand exchange is a contraction of the film. A 5 nm NC, capped with 1.5 nm long ligands, presents an initial equivalent size of 8 nm, which shrinks to 6 nm once capped with EDT. This 20% diameter reduction potentially leads to a 60% volume reduction, and consequently, cracks may be formed, see Fig. 7.2c. The presence of these cracks reduces the film quality and makes the percolation path for the carrier to reach the electrodes even longer. This, in addition to the diffusive character of the solid-state ligand exchange, requires that the film deposition should be conducted through a multilayer procedure. New layers may cover cracks from the previous layers, improving the overall film quality. Each layer will become poorly soluble in the non-polar solvents used to deposit the NCs once the ligands are exchanged, which allows this multilayer approach. The procedure is typically repeated around 10 times to form a 200 nm thick film. Beyond this, the film quality may suffer.

7.2.2 Ink Preparation

In addition to the solid-state ligand exchange, the procedure can also be conducted in the liquid phase. Ions (from $HgCl_2$ and mercaptoethanol in the case of HgTe NCs [22, 23]) are dissolved in a polar solvent such as dimethyl formamide. This ion solution is mixed with the NCs dissolved in their immiscible non-polar solvent. As the two solutions are stirred, the native ligands are removed, and the ions cap the NC surface. The NCs are now charged and thus soluble in the polar solvent. As short ligands are often polar, this method can result in an even thinner barrier than the solid-state method, further increasing carrier transport. Mobilities of 1 cm^2 V^{-1} s^{-1} are commonly obtained [20] (Fig. 7.3b) with record values of above 100 cm^2 V^{-1} s^{-1} [25], albeit with an additional sintering step. This improved mobility has led to an increased specific detectivity (D*) in photoconductive devices [24], see Fig. 7.3c. The constraint of this approach is that the boiling point of a polar solvent is generally high, requiring the preparation of highly concentrated solutions of NCs (>200 g. L^{-1} typically) for practical removal. By doing so, however, it is possible to deposit a film of NCs with a tunable thickness in the 100–600 nm range (see Fig. 7.3a) through a single step of deposition, which considerably eases the fabrication process when compared to the solid-state ligand exchange.

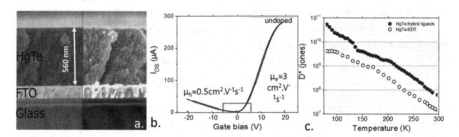

Fig. 7.3 Benefit of a thin film prepared from an ink. (**a**) SEM image with false colors of a diode including a HgTe NC thin film obtained from a single deposition step. Part (**a**) is adapted with permission from Ref. [23]. Copyright 2019 WILEY-VCH Verlag GmbH & Co. KGaA, Weinheim. (**b**) Transfer curve (i.e., drain current as a function of the applied gate bias under constant drain source bias) for a HgTe NC thin film obtained from an ink solution. (**c**) Specific detectivity for a HgTe NC thin film obtained from an ink solution and from the same nanocrystal film processed from a solid-state ligand exchange strategy. Parts (**b**) and (**c**) are adapted with permission from Ref. [24]. (Copyright 2010 American Chemical Society)

7.3 From Proof of Concept to High Performances Sensors

Now that the strategies to transform a NC solution into a conductive film have been established, the next step is to design a device with a geometry that allows for photoconduction.

7.3.1 Photoconductive Devices

While in the case of CdSe, the first demonstration of charge conduction in a NC array started around the year 2000 [11], the investigation in the case of mercury chalcogenides started in 2007 with the pioneering work from the Heiss [9] and Kim [26, 27] groups. Planar photoconductive geometry is certainly the easiest strategy to probe transport. This geometry appears more tolerant to defects (to NC aggregation in particular) than the vertical geometry discussed later. A simple implementation is the use of interdigitated electrodes, which can be seen in Fig. 7.4a. At room temperature, the measured IV curve for a HgTe NC film is often linear, as in the case of a gold electrode. This is due to the fact that the Schottky barrier remains weak for both electrons and holes, because of the narrow band-gap nature of HgTe NCs. Upon illumination, an increase in the current at a given bias is observed (*i.e.,* a rise of the IV curve slope, see Fig. 7.4b). For a band gap at 6000 cm^{-1} (720 meV), the current modulation induced by the light can be large (factor of 10) but decreases as the band gap is reduced (just a few percent for a 2000 cm^{-1} (250 meV)) band gap). These numbers have to be considered as order of magnitude only, since the exact current modulation depends on the incident light power, the device geometry and the thin film preparation. In general, the magnitude of the photocurrent does not scale linearly with the incident light irradiance in NC films. Generally, a higher power is associated with a weaker response (relative to the input), even in a regime where the illumination leads to less than 1 carrier per particle, see Fig. 7.4c.

Note that in photoconductive mode, the noise [30] has always been reported to be $1/f$ noise [31–33] limited, see Fig. 7.4d. In this case, the Hooge's empirical law [32] provides a scaling law to connect the noise magnitude (S_1) to the current flow, $SI = \dfrac{\alpha I^2}{Nf}$. Here, α is a proportionality constant, f is the signal frequency, I is the current and N is the number of free charge carriers. It is worth pointing out that the α value is typically significantly larger than the value measured in the bulk [30]. The value of the responsivity (R_λ) in this device configuration ranges from 10 μA·W^{-1} to

---

Fig. 7.4 (continued) Responsivity of a HgTe NC thin film under illumination ($\lambda = 1.55$ μm) as a function of the laser power used for illumination for various operating temperature. (**d**) Noise current spectral density for a HgTe NC thin film as a function of the signal frequency and for various operating temperature. (**e**) Specific detectivity (D*) as a function of the electrode spacing for a HgTe NC thin film (measured at 200 K for 1.55 μm illumination and for a signal at 1 kHz). (**f**) Schematic of a nanotrench device used to probe photoconduction at the few tens of nm scale (i.e., few NC size). (Parts (**b**) and (**f**) are adapted from Ref. [29])

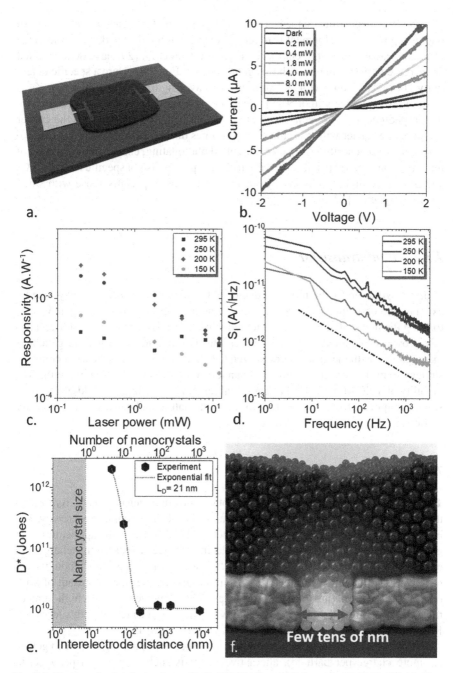

Fig. 7.4 Photoconductive device from a HgTe NC thin film. (**a**) Schematic of a HgTe NC thin film deposited onto interdigitated electrodes. Parts (**a**) is adapted with permission from Ref. [28]. Copyright 2019: American Chemical Society. (**b**) I-V curves for a HgTe NC ($\lambda_{\text{cut-off}}$ = 2 μm) thin film under dark condition and under illumination (λ = 1.55 μm) at room temperature. (**c**)Fig. 7.4

$100\ mA\cdot W^{-1}$, depending on the exact geometry, film thickness, and doping level. An interesting recent development is to reduce the size of the device down to the range of the the carrier diffusion length, see Fig. 7.4f. Chu et al. [29] have demonstrated an increase in performance with this geometry, see Fig. 7.4e. Here a specific detectivity above 10^{12} Jones (for 2.5 μm cut-off wavelength, at 1 kHz and 200 K) is achieved for a device with a 50 nm electrode spacing. Such high performances are due to photoconductive gain. As the transit time becomes shorter than the carrier lifetime, one carrier will recirculate several times during the lifetime of the other to ensure the film neutrality. This process, called photogating, enables the collection of more than one carrier per absorbed photon. This process is not specific to HgTe NCs and is very commonly observed in nanocrystal films, especially those with small electrode spacing [12, 34].

7.3.2 Phototransistor

A direct evolution of the photoconductor is the phototransistor. One may take an existing photoconductor and add a gate to tune the carrier density. The gate is used to bring the material close to intrinsic, reducing the dark current. While for epitaxially grown semiconductor-based devices, the introduction of a gate is generally incompatible with the growth process, the field-effect transistor is the most common strategy to probe carrier mobility in a nanocrystal array [35, 36]. Owing to the low mobility of a NC film, the Hall effect signal remains weak, and the field-effect measurement appears to be a better tool to probe the nature and density of the majority carriers.

7.3.2.1 Gating Technology

Traditionally, the gating of a NC film is obtained from a dielectric gate. By far, the use of commercially available Si/SiO_2 wafers is the most common strategy, see Fig. 7.5a. The SiO_2 layer is used as a dielectric [17, 37]. Upon the application of a gate bias, charges are generated on the dielectric surface. To screen this charge, the electrode will inject some charge with the NC array as in a capacitor. Though easy to implement, the silica gating suffers from the limited dielectric constant of SiO_2 ($\varepsilon_{SiO2} = 3.9$) necessitating the a large gate bias. An alternative is to use a high-κ material such as alumina. In the case of HgTe NCs, this has been tested with success by Kim et al. [38, 39] and, more recently, by Chee et al. [40, 41].

When even stronger gating is required (to reduce the operating bias or to generate more charge per particle), alternative methods such as quantum paraelectric [42], ionic glass [28, 43, 44] and electrolyte [45–47] have also been used to generate a gate effect on a HgTe NC film. While most of gating methods are surface methods,

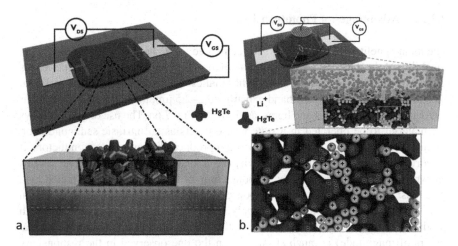

Fig. 7.5 Schematic of a NC-based field-effect transistor. (**a**) Schematic of a back-gate transistor. (**b**) Schematic of a top-gated electrolytic transistor. Parts (**a, b**) are adapted with permission from Ref. [28]. (Copyright 2019: American Chemical Society)

Table 7.1 Gate technology for a HgTe NC film

Gate technology	Dielectric	Quantum paraelectric	Ionic glass	Electrolyte
Material associated with this technology	SiO_2, Al_2O_3	$SrTiO_3$	LaF_3	Li based electrolyte in liquid or ion gel matrix
Basic concept	Capacitance formation	Divergence of the dielectric constant in the vicinity of the Curie temperature	Ionic gating	Ionic gating
Temperature range	4–300 K	Below 100 K for $SrTiO_3$	180–260 K	300 K
Sweep-rate range	Fast (several V·s^{-1})	Fast (several V·s^{-1})	Intermediate (0.1 V·s^{-1})	Slow (1 mV·s^{-1})
Subthreshold slope	3400 mV/decade		1200 mV/decade	152 mV/decade
Gate voltage range	< 60 V (dielectric breakdown)	Up to 200 V without breakdown	Up to 10 V at 200 K	< 3 V (electrochemical stability of the electrolyte)

the use of an electrolyte where the ions can diffuse in the NC array enables the gating of thick films, see Fig. 7.5b. Table 7.1 summarizes the different gate technologies reported for HgTe NC film, along with their benefits and limitations.

7.3.2.2 Advantages of Phototransistors

The main benefit of the phototransistor compared to the bare photoconductor is the tunability of the dark current. Note that with the size, HgTe nanocrystal experiences a crossover from a p-type behavior for the smallest NC to an n-type behavior for the largest particle [48, 49]. For a band gap around 2–2.5 μm, the material is ambipolar, conducting both electrons and holes, as shown in Fig. 7.6a. The use of gating allows the material to be operated in a state where it acts as an intrinsic semiconductor, minimizing the dark current and improving reproducibility over photoconductors. It is worth pointing out that the responsivity (R_λ) is also modulated by the gate bias, as shown in Fig. 7.6b. However, the response time of the device is also affected by a gate bias; a bias yielding a larger response will also result in a slower response. In other words, the gate application has only a moderate effect on the gain bandwidth product. On the other hand, it is clear that the dark current modulation (by several orders of magnitude) is much stronger than the one observed in the responsivity (roughly 1 decade). This means that the detectivity (D^*) can be improved by gating. Chen et al. [37] used this concept to design a 2 μm cut-off phototransistor with a 2×10^{10} jones detectivity at room temperature. Using a $SrTiO_3$ quantum paraelectric gate operated at low temperature (30 K), Gréboval et al. [42] have demonstrated a 2.5 μm cut-off sensor with a detectivity reaching 10^{12} jones. These high performances come at the price of high driving voltage (>100 V) and very low operating temperature.

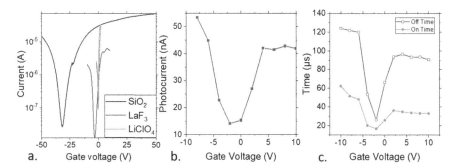

Fig. 7.6 Performances of a HgTe NC-based phototransistor. (**a**) Transfer curve (dark current as a function of the applied gate bias) for a HgTe NC thin film gated by various gate materials. (**b**) Responsivity as a function of the applied gate bias for a HgTe NC thin film on a LaF_3 substrate used as a back-gate. (**c**) Turn-on and -off time in response to a 1.5 μm pulse of light as a function of the applied gate bias for a HgTe NC thin film on a LaF_3 substrate used as back-gate. Parts (**a–c**) are adapted with permission from Ref. [28]. (Copyright 2019: American Chemical Society)

7.4 Beyond the Control of the Dark Current

As the operation of a phototransistor involves the modulation of the charge carrier density, this concept may readily be applied to control the doping profile in devices. For example, it is possible to form a p-n junction through gating without material modification. The simplest way to design a *p-n* junction from a field-effect transistor is to use a single gate device and operate in a regime where the drain-source bias is greater compared to the gate bias. This generates a homogeneous doping where the gate-induced charge might be opposite for the drain and the source electrodes. In the case of HgTe NC by Noumbé et al. [43], they proposed to use a HgTe NC film coupled to graphene electrodes and a LaF$_3$ ionic gate to achieve this. In particular, they demonstrate that the introduction of a graphene electrode, thanks to the material's work function tunability and to the partial transparency to the gate-induced electric field, enables the generation of a planar *p-n* junction, which is then used to enhance charge dissociation.

Later on, to gain greater control over the carrier density landscape, Chee et al. [40] proposed a dual gate geometry, see Fig. 7.7a, b. Using two independent gates,

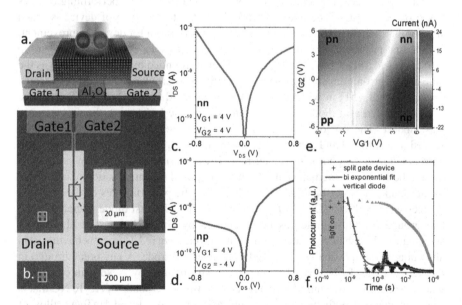

Fig. 7.7 Gate induced planar p-n junction. (**a**) Schematic of a dual gated HgTe NC thin film. (**b**) False color microscopy image of the device depicted in part (**a**). (**c**) (resp (**d**)) IV curve while the device is operated in the transistor (resp diode) mode $V_{G1} = V_{G2}$ (resp. $V_{G1} = -V_{G2}$). Parts (**a**–**d**) are adapted with permission from Ref. [40]. Copyright 2021: American Chemical Society. (**e**) Photocurrent under 0 V drain source bias, for a dual gated device which drain and source electrodes are made of graphene as a function of the two gates biases. (**f**) Photocurrent in response to a 1 ns long pulse of light at 1573 nm as a function of time for a planar p-n junction and for a vertical geometry diode, also made of HgTe NC with the same cut-off wavelength. Parts (**e**–**f**) are adapted with permission from Ref. [41]. (Copyright 2021: American Chemical Society)

it is possible to generate a *p-n* junction between the drain source electrodes. By carefully choosing the bias over the two gates, the device can behave as a traditional field-effect transistor ($V_{G1} = V_{G2}$, see Fig. 7.7c) or as a diode ($V_{G1} = -V_{G2}$, see Fig. 7.7d). The latter mode can then operate the device under a zero-drain source bias to minimize the dark current, see Fig. 7.7e. A key advantage of this approach compared to the vertical diode, described in the next paragraph, is its lower capacitance, enabling faster response times. Currently, the time response of a vertical geometry diode remains above 100 ns, while for the planar configuration time responses as short as 3 ns have been reported [41], see Fig. 7.7f. On the other hand, a key limitation of the phototransistor remains its low absorption. Gate effect, for all gate types except for electrolyte, requires thin films (t < 200 nm and even thinner), which lead to weak light absorption and thus reduced responsivity (R_λ is in the 1–10 mA·W^{-1} range usually).

7.4.1 Photodiode

To date, the vertical photodiode geometry has led to the best-performing devices based on HgTe NCs. The body of work dedicated to this type of device is much smaller than for PbS NCs, and there is certainly still room for device improvement.

The first proposed photodiodes were very close to solar cells. The Heiss' group [50] proposed a stack based on HgTe NC layer surrounded by P3HT and TiO$_2$ as the hole and electron transporting layers, respectively. A similar device has been tested by Jagtap et al. [51] in which HgTe NCs are coupled to a TiO$_2$ layer. Though the device operated as a diode (*i.e.* it had an asymmetric I-V curve), the performance was modest. The use of TiO$_2$ is certainly related to the overall weak response. Indeed, the narrow band gap of HgTe means the TiO$_2$ filters not only the hole dark current but also the photoelectron current. Thus, alternative charge transport layers have to be explored. Jagtpap et al. [48] have proposed the introduction of a unipolar barrier made of larger band gap HgTe NCs, but the performances of the resulting devices were still low (D* ≈ 10^8 Jones at room T). The first diode operated in the background limited regime has been reported by Guyot-Sionnest et al. [52] by stacking a CaF$_2$ substrate coated with a NiCr layer, a HgTe NC layer behaving as an absorbing layer and finally connected to a silver electrode. A key breakthrough in the design of HgTe NC-based diodes has been obtained by Ackerman et al. [53] where the authors have introduced a hole extraction layer made of Ag$_2$Te NC that is cation exchanged with mercury, forming a layer of Ag-doped HgTe/an alloy of AgHgTe with an unclear stoichiometry. A schematic of such a diode is shown in Fig. 7.8a, while the band diagram is shown in Fig. 7.8b. The associated IV curve of such a device is clearly rectifying, as shown in Fig. 7.8c. This layer seems to be well-suited to design a diode using both SWIR (short-wave infrared) [54, 55] and MWIR (mid-wave infrared) [53] HgTe NCs. Though for the mid-wave, the

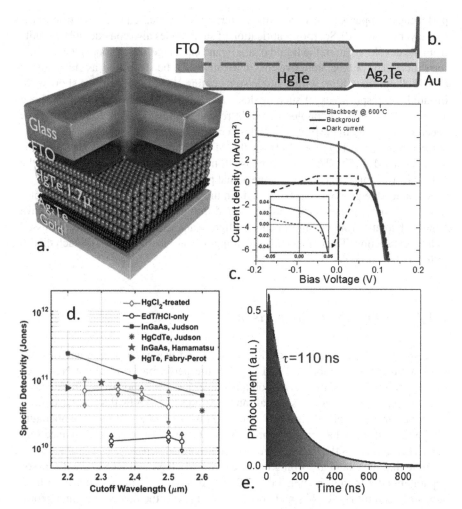

Fig. 7.8 HgTe NC-based vertical geometry photodiode. (**a**) Schematic of a HgTe NC-based vertical geometry photodiode which hole transport layer is based on Ag₂Te NCs. (**b**) Energy profile determined by photoemission for the device depicted in part (**a**). The dashed line represents the Fermi level. (**c**) IV curve in the dark and under illumination by a blackbody for an extended short-wave operating diode based on the structure depicted in part (**a**). Part (**c**) is adapted with permission from Ref. [53]. Copyright 2018: American Chemical Society. (**d**) Specific detectivity as a function of the cut-off wavelength for the device depicted in part (**a**) and its comparison with commercial device performances. Part (**d**) is adapted with permission from Ref. [54]. Copyright 2020: AIP Publishing. (**e**) Time-response to a 1 ns pulse of light from the device depicted in part (**a**). Parts (**a**) and € adapted with permission from Ref. [55]. (Copyright 2021: WILEY-VCH Verlag GmbH & Co. KGaA)

performance appears higher when the material is also coupled to an electron transport layer made of Bi_2Se_3 [56]. Fabrication of such diodes also considerably benefits from the development of the liquid phase ligand exchange. This process leads to the formation of a photoconductive ink that can easily be deposited as thick layers (200–500 nm thick film) with a single spin coating step [23, 55]. In the short-wave infrared, the responsivity of such diodes reaches 0.5 $A \cdot W^{-1}$ and can even be above 1 $A.W^{-1}$ if the diode is coupled to a light resonator [55, 57]. This corresponds to a specific detectivity in the high 10^{10} Jones at room temperature in the extended short-wave infrared, see Fig. 7.8d [54]. In the MWIR, D* reaching 4×10^{11} Jones is achieved at 80 K [57]. The time-response of this diode is generally measured in the 100 ns to 1 µs range depending on the actual pixel size, see Fig. 7.8e.

Considerable progress has been made in the performance of HgTe-based photodiodes. As vertical geometries have required thin devices, these photodiodes have benefited greatly from the introduction of light management strategies, as discussed in the next section. The main challenge going forward is the transfer of such diodes in focal plane arrays.

7.4.2 Performance Comparison

To compare the performances of devices obtained using various geometries, we have summarized in Table 7.2 and Fig. 7.9 the performance of HgTe NC-based devices. Since most of the devices obtained operate in the short-wave infrared, we have chosen to focus on this spectral range. At present, MWIR devices have much lower performance than their SWIR counterparts, particularly at room-temperature where these devices have the potential to lead to low-cost, high-performance sensors. Therefore, MWIR detectors will not be discussed here.

Note that intraband absorbing materials (mostly HgS and HgSe) are discussed separately later in this chapter. Table 7.2 provides typical figures of merit for light sensors based on HgTe NCs and correlates them with the device type and geometry. Figure 7.9 shows that the responsivities of these devices are now commonly in the 0.2–2 $A.W^{-1}$ range, which corresponds to external quantum efficiencies between 10 and 100%. Devices with gains larger than unity have also been reported in the photoconductive geometry and with high detectivity ($>10^{12}$ jones for 200 K operation and 2.5 µm cut-off wavelength), meaning that high-performing devices are not limited to photodiodes. However, it remains that photodiodes constitute most of the best-performing devices. The number of results in the near-IR is weak and device performances are generally poor compared to those reported for PbS NC-based [66] devices or InGaAs technology. The toxicity of the material and the current level of performance for HgTe NC based devices limit their scope of application to a spectral range where it has no highly developed competitor (*i.e.* above 1.7 µm).

Table 7.2 Figures of merit for SWIR HgTe NC-based light sensors

Cut-off λ (μm)	Operating mode	R_λ (A.W^{-1})	Response time	D* (Jones)	T_{oper} (K)	Specific feature	References
2.5	PC	0.1	10 μs	3.5×10^{10}	230	As$_2$S$_3$ surface chemistry	[17]
2.5	PC	1000	20 μs	2×10^{12}	200	Nanotrench	[29]
2.5	PC	150	1.5 ms	6×10^8	80	HgTe decorated graphene channel	[58]
2.4	PC	0.9	264 μs decay time	8×10^9	300	Spray coating with patterning	[59]
2.4	PC	0.22	2.2 ms	3.5×10^8	300	Multicolor pixel	[60]
2.5	PT	6.5×10^{-3}	10 μs	10^9	220	Graphene electrode	[43]
2.5	PT	2.0×10^{-3}	14 μs	10^{12}	30	STO gate+resonator	[42]
2	PT	<0.5	≈10 μs	3×10^{10}	300	SiO$_2$ back gate	[37]
2.4	PT	1	1.5 μs	10^{10}	300	Hybrid polymer: HgTe	[61]
2.5	PT	–	15 ms	–	300	Doped-graphene/ HgTe	[62]
2.5	PT	0.08	10 μs	>10^8 <10^{10}	250	Planar *pn* junction based on dual gate	[40]
2.5	PD	2.5×10^{-3}	370 ns	3×10^9	300	HgTe ink	[23]
2.5	PD	0.25	260 ns decay time	3×10^{10} (without cavity) 7.5×10^{10} (with cavity)	300	Flexible substrate	[56]
2.2	PD	1	1.4 μs decay time	6×10^{10}	300	HgCl$_2$ treatment	[54]
1.8	PD	0.13	110 ns	2×10^{10}	300	With resonator	[55]
2.5	PD	0.6	–	4×10^{11}	85	Resonator grating + fabry perot	[57]
2.4	PD	0.45	13 ns	10^{10}	300	Si/graphene/HgTe	[63]
2.5	PD	0.28	2.5 μs	6×10^{10}	300	Bi$_2$Se$_3$/HgTe/ Ag$_2$Te	[64]
2	PD	0.8	170 ns	9×10^{11} 9×10^{10}	200 300	CdSe/HgTe/Ag$_2$Te	[65]

7.5 Light Management in HgX Nanocrystal Films

The design of NC-based light sensors faces a trade-off. On the one hand, thick films are required to absorb most of the incident light. The absorption depth of HgTe NC with 2.5 μm cut-off wavelength is several μm, and a 200 nm thick film only absorbs 10% of the incident light. On the other hand, these films exhibit low carrier mobility

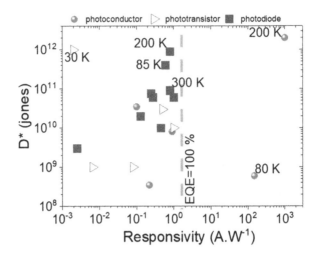

Fig. 7.9 HgTe NC based performances for short wave infrared sensing. Specific detectivity as a function of device responsivity for HgTe NC-based device operated in the SWIR

and so have a short diffusion length [29] (<50 nm). As a result, charge collection is not efficient over several μm. In addition, the fabrication of such thick films using NCs remains extremely challenging. So, in order to help the low absorption of the thinner films, light management strategies have been developed in order to "focus" the light on a thin semiconductor slab Here, we would like to review some of the recent developments in this direction [67].

7.5.1 Enhancement of Absorption

Many of the concepts that are being explored for NCs were first demonstrated in epitaxially grown semiconductors. The use of a grating to tune the light-matter coupling is well-established for quantum well-based infrared detectors (QWIP). Here, the grating is etched on the top of the pixel to break the selection rule forbidding absorption under normal incidence. The first attempt to tune light-matter coupling in a HgTe NC array was made by Chen et al. [68], where they introduced gold nanorods in a diode geometry, see Fig. 7.10a. The gold nanorods have one resonance in the visible (around 530 nm) and a second one in the near-infrared, its energy is driven by the rod length. The latter resonance was spectrally matched with the NC absorption. In the vicinity of a rod's tip, the electromagnetic field is locally enhanced, boosting light absorption. When coupling excitons and plasmons in this way, the absorption enhancement must occur within the nanocrystal rather than in the metal used to induce the plasmonic behavior; otherwise, substantial thermal losses result. Therefore, in their design, the authors have also been careful not to directly locate the HgTe NCs on the top of the gold in order to avoid any exciton

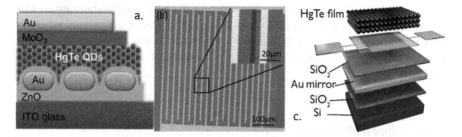

Fig. 7.10 Strategy to enhance the light absorption in HgTe NC-based devices. (**a**) Schematic of HgTe NC based photodiode coupled to gold nanorods. Part (**a**) is adapted with permission from Ref. [68]. Copyright 2014: American Chemical Society. (**b**) Microscopy image of interdigitated electrode in which the substrate between the electrode is functionalized by a plasmonic array. Part (**b**) is adapted with permission from Ref. [69]. Copyright 2017: AIP publishing. (**c**) Schematic of a HgTe NC film coupled to a guided more resonator. Part (c) is adapted with permission from Ref. [70]. (Copyright 2019: American Chemical Society)

quenching. While this first strategy is based on an all-colloidal-material approach, most of the following efforts have focused on incorporating conventional materials through microfabrication. Note that in order to avoid any damage to the NCs, much of the fabrication (lithography, annealing step) is conducted before the NC deposition. This prevents interparticle sintering and possible oxidation/ligand removal. Hopefully, in the future, such constraints will be removed with the design of more stable NCs.

Yifat et al. [69] proposed to functionalize the area between the interdigitated electrodes by an array of metallic dots, the period is chosen to generate a resonance matching the NC absorption band edge, see Fig. 7.10b. By doing so, they observe a change in the photocurrent spectrum and a three times enhancement of the absorption around the plasmonic resonance. Later on, Chu et al. [70] proposed an evolution of this strategy where an optical grating is directly used as a set of interdigitated electrodes, see Fig. 7.10c. The benefit of this strategy is the reduction of electrode spacing which generates photoconductive gain. While the impact on the absorption is likewise a factor of 3, the enhancement of the photocurrent is considerably greater: between 100 and 1000, depending on the material and the cut-off wavelength. The same approach was later extended to other device geometries, such as phototransistors [42] and vertical diodes [55]. Devices based on exciton-plasmon coupling generally present a low-quality factor, which is appealing for broadband applications such as imaging, but not for specific spectroscopic applications such as gas sensing, where narrow band enhancement is desirable. Tang et al. [71] have proposed to couple a HgTe NC film to a Bragg mirror and the obtained resonance is much narrower (down to 30 cm^{-1}). The same group has also demonstrated that vertical Fabry-Perot interference can be used to enhance light absorption [57], a technique which is compatible with flexible substrates [56].

While much effort has been focused on the enhancement of the absorption, noise reduction similarly increases the device performance. Zhu et al. [72] proposed an approach based on size reduction. Noise in sensors scales like the device electrical volume. Reducing the device size while maintaining the absorption unchanged will thus result in greater signal-to-noise ratio. To reach this goal, they have placed HgTe NCs in a small waveguide where the gold edges act as the device electrodes. As devices move to pixel arrays with small pixel pitch, such approaches will become necessary.

7.5.2 Spectral Shaping

As a first approximation, one may assume that the response spectrum of a detector is the same as the absorption spectrum of the active layer. There are, of course, always optical effects arising from the material interfaces as well as scattering from non-uniformity and charge transport effects. When parts of a device are smaller than the wavelength of the light absorbed, the geometry of the device becomes particularly important in determining the shape of the response spectrum. For example, Chu et al. [70] demonstrated a 500 cm^{-1} shift of the spectral response of a HgTe NC film at 2.5 μm by tuning the period of the grating in a structure such as the one depicted in Fig. 7.10c.

We have already mentioned that the absorption cross-section can be focused on a very narrow band using a Bragg mirror. For imaging applications, it is also interesting to achieve broadband absorption enhancement. The latter can be obtained by combining several broad individual resonances, see Fig. 7.11a. Gréboval et al. [42], for instance designed a grating that shows two shifted guided-mode resonances. Tang et al. [57] chose to combine a Fabry-Perot mode with a metallic grating to reach the same goal.

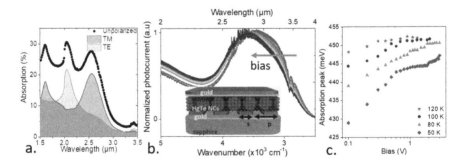

Fig. 7.11 Spectral shaping of the photoresponse through a coupling to a resonator. (**a**) Absorption spectrum of a HgTe NC film coupled to three resonances to achieve broadband absorption in the near- and short- wave infrared. Part a is adapted with permission from Ref. [42]. Copyright 2021: American Chemical Society. (**b**) Photocurrent spectra of HgTe thin film integrated in the device depicted as inset for various applied biases. (**c**) Energy of the absorption peak as a function of the applied bias for various temperature for the device of part (**b**). Parts (**b**) and (**c**) are adapted with permission from Ref. [73]. (Copyright 2021: American Chemical Society)

All the devices discussed so far have had a single, unchanging response spectrum. Active reconfigurability would be even more interesting. This can be achieved at the device level [74], for example, by stacking diodes. Tang et al. [64] proposed to stack a SWIR and a MWIR diode where the spectral response of the full device is determined by the bias polarity. Dang et al. [73] demonstrated another form of active device, showing a shift of the spectral response with bias. Interestingly, the spectral response shows a blueshift with bias (Fig. 7.11b, c), allowing the exclusion of the Stark effect as a possible mechanism to explain the observed bias-tunable photoresponse.

To observe their bias-tunable photoresponse, they included the NC layer into a cavity, as shown in the inset of Fig. 7.11b. The role of the cavity is to generate a spatially inhomogeneous electromagnetic map. Then the charge collection length will be tuned by the bias. Under a low electric field, the charges are collected only in the very vicinity of the electrode, while under a larger field, longer distance charge collection can occur. A tunable photoresponse can be obtained if the spectral response within the cavity differs close and far from the electrode.

7.6   From Single Pixel to Focal Plane Array

Now that single-element detectors have reached a certain maturity, a new challenge will be to transfer the technology to focal plane arrays. NCs present several potential advantages compared to epitaxially grown semiconductors. They are inexpensive to synthesize, and there is no need to epitaxially match them to an expensive wafer. In the case of InGaAs, the III-V layer is responsible for 1/3 of the final cost.

Moreover, NCs are not restricted to backside illumination through an absorbing substrate, so they can be used for broadband detection (*i.e.,* from band edge to ultraviolet). Thus, an infrared NC-based camera is also responsive in the visible range, contrary to InGaAs grown on InP. The solution-processed nature of NCs may also ease the hybridization to read-out circuit. Epitaxially grown semiconductors are generally coupled to a CMOS circuit with the use of an array of Indium bumps, see Fig. 7.12a. NCs can be deposited directly on the top of the read-out-circuit, see Fig. 7.12b. Aside from removing a complex step of fabrication, NCs are also highly promising to design smaller pixel pitch closer to the diffraction limit and improve the image quality. Using near-infrared or short-wave infrared absorbing PbS NCs, a pixel pitch as small as 1.8 µm have been reported [77]. In the case of HgTe, both SWIR [78] (2 µm cut-off, see Fig. 7.12c) and MWIR [79] (5 µm cut-off - see Fig. 7.12d) focal plane arrays have been reported. Bossavit et al. [76] recently reported a basic proof of concept of an all-nanocrystal-based communication setup where a HgTe NC light emitting diode (LED) is imaged using a HgTe NC-based focal plane array, see Fig. 7.12e–g. To date, only few performances have been reported. Buurma et al. [79] reported a sub-100 mK NETD (noise equivalent temperature difference) for their MWIR sensor, but 8 years later, there has been little progress, and updated data will be of utmost interest.

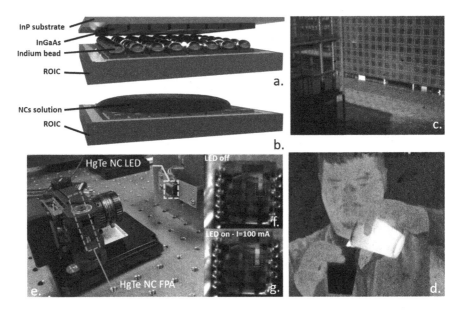

Fig. 7.12 Imaging using a HgTe NC thin film as the active layer. (**a**) Schematic of an epitaxially grown semiconductor coupled to a read-out circuit. (**b**) Schematic of a HgTe NC thin film coupled to a read-out circuit. Parts (**a**) and (**b**) are adapted with permission from Ref. [7]. Copyright 2021: American Chemical Society. (**c**) extended short-wave infrared image obtained from a HgTe NC thin film. (**d**) Mid-wave infrared image obtained from a HgTe NC thin film. Part (**d**) is adapted from Ref. [75]. (**e**) Image of a camera which active layer is based on HgTe NCs imaging an HgTe NC based LED ($\lambda = 1.3$ μm). (**f**) (resp (**g**)) Obtained image from the setup depicted in part (**e**) while the two top left pixels are off (resp on). Parts (**e**–**g**) are adapted with permission from Ref. [76]. (Copyright 2021: WILEY-VCH Verlag GmbH & Co. KGaA)

7.7 Intraband Device

The first sections of this chapter have only been focused on HgTe NCs. The synthesis and device fabrication using HgTe NCs are far more advanced compared to other mercury chalcogenides (HgS and HgSe). Nevertheless, as stated in the volume 1 Chap. 6, HgS and HgSe NCs have demonstrated intraband absorption resulting from their doped character [80, 81]. This absorption is the 0D counterpart of the intersubband transition in quantum well structures. A clear signature of this intraband absorption is an infrared peak appearing in the mid- or long- wave infrared, see Fig. 7.13a, which is clearly offset compared to the interband transition appearing at higher energy. For a small size (5 nm) of HgS and HgSe NCs, this peak overlaps well with the 3–5 μm band. Though intraband has been mostly observed in heavy metal-containing materials (HgS [82], HgSe [74, 83–85], HgTe, PbS [86]), the fact that it has also been observed in Ag_2Se [87–91] may raise clear expectation to achieve mid-infrared absorption in wider band gap materials, even greener materials.

Interestingly the intraband transition not only leads to absorption but also to some photoconduction, first observed by Deng et al. [92]. However, an array of

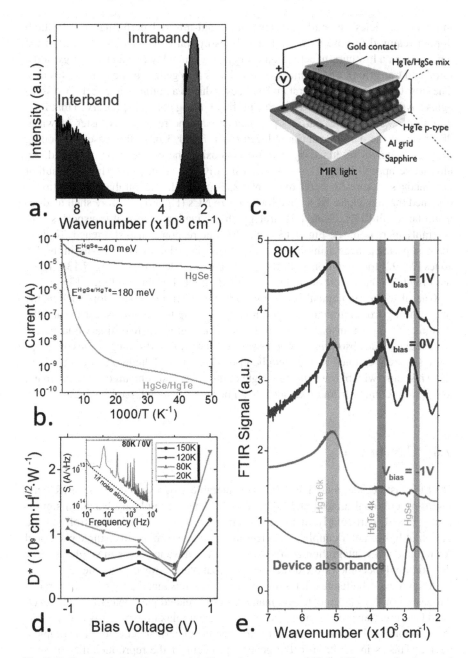

Fig. 7.13 Intraband photodetection. (**a**) Absorption spectrum of HgSe NCs presenting inter and intraband absorption. (**b**) Current as a function of temperature for HgSe NC film and for a HgSe/HgTe NC mixture. (**c**) Schematic of an intraband photodiode based on a HgSe/HgTe NC mixture as an active layer. (**d**) Specific detectivity as a function of the applied voltage under various operating conditions. The inset shows the 1/f noise limited noise current spectral density. (**e**) Absorption and photocurrent spectrum under different biases at 80 K from the device depicted in part (**c**). (Parts (**b–e**) are adapted from Ref. [74])

intraband particles generally presents poor mid-infrared performance. The high doping required to observe the intraband absorption generates an increase of the dark current, a low current activation energy (*i.e.,* much lower than the energy of the intraband transition), and a time response that is also generally slow (ms or even s). The slow time response of intraband devices is due to a bolometric effect. Two strategies have been proposed to solve this problem. Deng et al. [92] observed that for a doping of exactly n = 2 carriers per nanocrystal, the response is faster (down to 200 ns) and the dark current is reduced (by a factor 30). In this case, the doping magnitude has to be tuned by carefully choosing the proper capping ligand. An alternative approach has been proposed by Livache et al. [74], where the author designed a structure close to a dye-sensitized solar cell. The intraband absorption is obtained through HgSe NCs mixed with HgTe NCs (HgSe/HgTe core shell leads to a similar result [93]). Using this strategy, the activation energy of transport is considerably increased, see Fig. 7.13b, and the time response also shortened. To date, the best performances using intraband materials have led to a detectivity in the 10^9 Jones [74, 92] for operation at 80 K (5 µm cut off wavelength), see Fig. 7.13d. There has only been one report of an intraband photodiode thus far, see Fig. 7.13c.

One of the technological issues raised by mid-infrared absorption, such as the one resulting from intraband transition is the lighting through the substrate and the design of partially transparent electrodes at wavelengths where ordinary transparent conductive oxide absorption becomes strong. One proposed strategy relies on a sapphire substrate coupled with a metallic grid. The obtained diode offers a reconfigurable response with bias where the weight of the intraband compared to the interband absorption (coming from other layers) can be strongly tuned, see Fig. 7.13e.

7.8   Conclusion

Over the last decade, there have been considerable progress in the field of IR sensing using colloidal nanocrystals. The level of device complexity has been significantly raised, particularly in the achievement of carrier density control and the design of light-matter coupling. Devices are no longer limited to a single pixel, and focal plane array integration is now a reality. Any advancements in single-element detection will become future directions for focal plane array development.

A next-level challenge relates to modeling, which is mandatory to conduct rational design of device. To date, too many material-related parameters (doping level, band bending, trap cross section and density, carrier relaxation pathway and dynamics…) remain unknown to accurately simulate the electrical behavior of a complete device. This is in addition to the general problem of the reproducibility of such parameters. Thus, progress at the device level cannot be decorrelated from a deeper material investigation.

References

1. Pietryga JM, Park Y-S, Lim J, Fidler AF, Bae WK, Brovelli S et al (2016) Spectroscopic and device aspects of nanocrystal quantum dots. Chem Rev 116:10513–10622. https://doi.org/10.1021/acs.chemrev.6b00169
2. Wood V, Bulović V (2010) Colloidal quantum dot light-emitting devices. Nano Rev 1:5202. https://doi.org/10.3402/nano.v1i0.5202
3. Dubertret B, Skourides P, Norris DJ, Noireaux V, Brivanlou AH, Libchaber A (2002) In vivo imaging of quantum dots encapsulated in phospholipid micelles. Science 298:1759–1762. https://doi.org/10.1126/science.1077194
4. Schaller RD, Klimov VI (2004) High efficiency carrier multiplication in PbSe nanocrystals: implications for solar energy conversion. Phys Rev Lett 92:186601. https://doi.org/10.1103/PhysRevLett.92.186601
5. Semonin OE, Luther JM, Choi S, Chen H-Y, Gao J, Nozik AJ et al (2011) Peak external photocurrent quantum efficiency exceeding 100% via MEG in a quantum dot solar cell. Science 334:1530–1533. https://doi.org/10.1126/science.1209845
6. Pijpers JJH, Ulbricht R, Tielrooij KJ, Osherov A, Golan Y, Delerue C et al (2009) Assessment of carrier-multiplication efficiency in bulk PbSe and PbS. Nat Phys 5:811–814. https://doi.org/10.1038/nphys1393
7. Gréboval C, Chu A, Goubet N, Livache C, Ithurria S, Lhuillier E (2021) Mercury chalcogenide quantum dots: material perspective for device integration. Chem Rev 121:3627–3700. https://doi.org/10.1021/acs.chemrev.0c01120
8. Rogach A, Kershaw SV, Burt M, Harrison MT, Kornowski A, Eychmüller A et al (1999) Colloidally prepared HgTe nanocrystals with strong room-temperature infrared luminescence. Adv Mater 11:552–555. https://doi.org/10.1002/(SICI)1521-4095(199905)11:7<552::AID-ADMA552>3.0.CO;2-Q
9. Boeberl M, Kovalenko M, Gamerith S, List-Kratochvil E, Heiss W (2007) Inkjet-printed nanocrystal photodetectors operating up to 3 μm wavelengths. Adv Mater 19:3574–3578. https://doi.org/10.1002/adma.200700111
10. Murray CB, Norris DJ, Bawendi MG (1993) Synthesis and characterization of nearly monodisperse CdE (E = sulfur, selenium, tellurium) semiconductor nanocrystallites. J Am Chem Soc 115:8706–8715. https://doi.org/10.1021/ja00072a025
11. Leatherdale CA, Kagan CR, Morgan NY, Empedocles SA, Kastner MA, Bawendi MG (2000) Photoconductivity in CdSe quantum dot solids. Phys Rev B 62:2669–2680. https://doi.org/10.1103/PhysRevB.62.2669
12. Wang H, Lhuillier E, Yu Q, Zimmers A, Dubertret B, Ulysse C et al (2017) Transport in a single self-doped nanocrystal. ACS Nano 11. https://doi.org/10.1021/acsnano.6b07898
13. Kramer IJ, Minor JC, Moreno-Bautista G, Rollny L, Kanjanaboos P, Kopilovic D et al (2015) Efficient spray-coated colloidal quantum dot solar cells. Adv Mater 27:116–121. https://doi.org/10.1002/adma.201403281
14. Zhang S, Chen M, Mu G, Li J, Hao Q, Tang X (2021) Spray-stencil lithography enabled large-scale fabrication of multispectral colloidal quantum-dot infrared detectors. Adv Mater Technol 7:2101132. https://doi.org/10.1002/admt.202101132
15. Lhuillier E, Hease P, Ithurria S, Dubertret B (2014) Selective electrophoretic deposition of CdSe Nanoplatelets. Chem Mater 26:4514–4520. https://doi.org/10.1021/cm501713s
16. Liu Y, Gibbs M, Puthussery J, Gaik S, Ihly R, Hillhouse HW et al (2010) Dependence of carrier mobility on nanocrystal size and ligand length in PbSe nanocrystal solids. Nano Lett 10:1960–1969. https://doi.org/10.1021/nl101284k
17. Lhuillier E, Keuleyan S, Zolotavin P, Guyot-Sionnest P (2013) Mid-infrared HgTe/As2S3 field effect transistors and photodetectors. Adv Mater 25:137–141. https://doi.org/10.1002/adma.201203012

18. Kovalenko MV, Scheele M, Talapin DV (2009) Colloidal nanocrystals with molecular metal chalcogenide surface ligands. Science 324:1417–1420. https://doi.org/10.1126/science.1170524

19. Tang J, Kemp KW, Hoogland S, Jeong KS, Liu H, Levina L et al (2011) Colloidal-quantum-dot photovoltaics using atomic-ligand passivation. Nat Mater 10:765–771. https://doi.org/10.1038/nmat3118

20. Nag A, Kovalenko MV, Lee J-S, Liu W, Spokoyny B, Talapin DV (2011) Metal-free inorganic ligands for colloidal nanocrystals: S^{2-}, HS^-, Se^{2-}, HSe^-, Te^{2-}, HTe^-, TeS_3^{2-}, OH^-, and NH_2^- as surface ligands. J Am Chem Soc 133:10612–10620. https://doi.org/10.1021/ja2029415

21. Yakunin S, Dirin DN, Protesescu L, Sytnyk M, Tollabimazraehno S, Humer M et al (2014) High infrared photoconductivity in films of arsenic-sulfide-encapsulated Lead-sulfide nanocrystals. ACS Nano 8:12883–12894. https://doi.org/10.1021/nn5067478

22. Lan X, Chen M, Hudson MH, Kamysbayev V, Wang Y, Guyot-Sionnest P et al (2020) Quantum dot solids showing state-resolved band-like transport. Nat Mater 19:323–329. https://doi.org/10.1038/s41563-019-0582-2

23. Martinez B, Ramade J, Livache C, Goubet N, Chu A, Gréboval C et al (2019) HgTe nanocrystal inks for extended short-wave infrared detection. Adv Opt Mater 7:1900348. https://doi.org/10.1002/adom.201900348

24. Chen M, Lan X, Tang X, Wang Y, Hudson MH, Talapin DV et al (2019) High carrier mobility in HgTe quantum dot solids improves mid-IR photodetectors. ACS Photonics 6:2358–2365. https://doi.org/10.1021/acsphotonics.9b01050

25. Dolzhnikov DS, Zhang H, Jang J, Son JS, Panthani MG, Shibata T et al (2015) Composition-matched molecular "solders" for semiconductors. Science 347:425–428. https://doi.org/10.1126/science.1260501

26. Kim S, Kim T, Im SH, Seok SI, Kim KW, Kim S et al (2011) Bandgap engineered monodisperse and stable mercury telluride quantum dots and their application for near-infrared photodetection. J Mater Chem 21:15232–15236. https://doi.org/10.1039/C1JM12436F

27. Seong H, Cho K, Kim S (2008) Photocurrent characteristics of solution-processed HgTe nanoparticle thin films under the illumination of 1.3 μm wavelength light. Semicond Sci Technol 23:5

28. Gréboval C, Noumbe U, Goubet N, Livache C, Ramade J, Qu J et al (2019) Field-effect transistor and photo-transistor of narrow-band-gap nanocrystal arrays using ionic glasses. Nano Lett 19:3981–3986. https://doi.org/10.1021/acs.nanolett.9b01305

29. Chu A, Gréboval C, Prado Y, Majjad H, Delerue C, Dayen J-F et al (2021) Infrared photoconduction at the diffusion length limit in HgTe nanocrystal arrays. Nat Commun 12:1794. https://doi.org/10.1038/s41467-021-21959-x

30. Liu H, Lhuillier E, Guyot-Sionnest P (2014) 1/f noise in semiconductor and metal nanocrystal solids. J Appl Phys 115:154309. https://doi.org/10.1063/1.4871682

31. Balandin AA (2013) Low-frequency 1/f noise in graphene devices. Nat Nanotechnol 8:549–555. https://doi.org/10.1038/nnano.2013.144

32. Hooge FN (1994) 1/f noise sources. IEEE Trans Electron Devices 41:1926–1935. https://doi.org/10.1109/16.333808

33. Lai Y, Li H, Kim DK, Diroll BT, Murray CB, Kagan CR (2014) Low-frequency (1/f) noise in nanocrystal field-effect transistors. ACS Nano 8:9664–9672. https://doi.org/10.1021/nn504303b

34. Konstantatos G, Sargent EH (2007) PbS colloidal quantum dot photoconductive photodetectors: transport, traps, and gain. Appl Phys Lett 91:173505. https://doi.org/10.1063/1.2800805

35. Talapin DV, Murray CB (2005) PbSe nanocrystal solids for n- and p-channel thin film field-effect transistors. Science 310:86–89. https://doi.org/10.1126/science.1116703

36. Hetsch F, Zhao N, Kershaw SV, Rogach AL (2013) Quantum dot field effect transistors. Mater Today 16:312–325. https://doi.org/10.1016/j.mattod.2013.08.011

37. Chen M, Lu H, Abdelazim NM, Zhu Y, Wang Z, Ren W et al (2017) Mercury telluride quantum dot based phototransistor enabling high-sensitivity room-temperature Photodetection at 2000 nm. ACS Nano 11:5614–5622. https://doi.org/10.1021/acsnano.7b00972

38. Kim H, Cho K, Kim D-W, Lee H-R, Kim S (2006) Bottom- and top-gate field-effect thin-film transistors with p channels of sintered HgTe nanocrystals. Appl Phys Lett 89:173107. https://doi.org/10.1063/1.2364153

39. Kim D-W, Jang J, Kim H, Cho K, Kim S (2008) Electrical characteristics of HgTe nanocrystal-based thin film transistors fabricated on flexible plastic substrates. Thin Solid Films 516:7715–7719. https://doi.org/10.1016/j.tsf.2008.04.044

40. Chee S-S, Gréboval C, Magalhaes DV, Ramade J, Chu A, Qu J et al (2021) Correlating structure and detection properties in HgTe nanocrystal films. Nano Lett 21:4145–4151. https://doi.org/10.1021/acs.nanolett.0c04346

41. Gréboval C, Dabard C, Konstantinov N, Cavallo M, Chee S-S, Chu A et al (2021) Split-gate photodiode based on graphene/HgTe Heterostructures with a few nanosecond Photoresponse. ACS Appl Electron Mater. https://doi.org/10.1021/acsaelm.1c00442

42. Gréboval C, Chu A, Magalhaes DV, Ramade J, Qu J, Rastogi P et al (2021) Ferroelectric gating of narrow band-gap nanocrystal arrays with enhanced light–matter coupling. ACS Photonics 8:259–268. https://doi.org/10.1021/acsphotonics.0c01464

43. Noumbé UN, Gréboval C, Livache C, Chu A, Majjad H, Parra López LE et al (2020) Reconfigurable 2D/0D p–n graphene/HgTe nanocrystal Heterostructure for infrared detection. ACS Nano 14:4567–4576. https://doi.org/10.1021/acsnano.0c00103

44. Gréboval C, Noumbé UN, Chu A, Prado Y, Khalili A, Dabard C et al (2020) Gate tunable vertical geometry phototransistor based on infrared HgTe nanocrystals. Appl Phys Lett 117:251104. https://doi.org/10.1063/5.0032622

45. Liu H, Keuleyan S, Guyot-Sionnest P (2012) n- and p-type HgTe quantum dot films. J Phys Chem C 116:1344–1349. https://doi.org/10.1021/jp2109169

46. Livache C, Izquierdo E, Martinez B, Dufour M, Pierucci D, Keuleyan S et al (2017) Charge dynamics and Optolectronic properties in HgTe colloidal quantum Wells. Nano Lett 17:4067–4074. https://doi.org/10.1021/acs.nanolett.7b00683

47. Martinez B, Livache C, Goubet N, Jagtap A, Cruguel H, Ouerghi A et al (2018) Probing charge carrier dynamics to unveil the role of surface ligands in HgTe narrow band gap nanocrystals. J Phys Chem C 122:859–865. https://doi.org/10.1021/acs.jpcc.7b09972

48. Jagtap A, Martinez B, Goubet N, Chu A, Livache C, Gréboval C et al (2018) Design of a Unipolar Barrier for a nanocrystal-based short-wave infrared photodiode. ACS Photonics 5:4569–4576. https://doi.org/10.1021/acsphotonics.8b01032

49. Goubet N, Jagtap A, Livache C, Martinez B, Portalès H, Xu XZ et al (2018) Terahertz HgTe nanocrystals: beyond confinement. J Am Chem Soc 140:5033–5036. https://doi.org/10.1021/jacs.8b02039

50. Günes S, Neugebauer H, Sariciftci NS, Roither J, Kovalenko M, Pillwein G et al (2006) Hybrid solar cells using HgTe nanocrystals and Nanoporous TiO$_2$ electrodes. Adv Funct Mater 16:1095–1099. https://doi.org/10.1002/adfm.200500638

51. Jagtap A, Goubet N, Livache C, Chu A, Martinez B, Gréboval C et al (2018) Short wave infrared devices based on HgTe nanocrystals with air stable performances. J Phys Chem C 122:14979–14985. https://doi.org/10.1021/acs.jpcc.8b03276

52. Guyot-Sionnest P, Roberts JA (2015) Background limited mid-infrared photodetection with photovoltaic HgTe colloidal quantum dots. Appl Phys Lett 107:253104. https://doi.org/10.1063/1.4938135

53. Ackerman MM, Tang X, Guyot-Sionnest P (2018) Fast and sensitive colloidal quantum dot mid-wave infrared photodetectors. ACS Nano 12:7264–7271. https://doi.org/10.1021/acsnano.8b03425

54. Ackerman MM, Chen M, Guyot-Sionnest P (2020) HgTe colloidal quantum dot photodiodes for extended short-wave infrared detection. Appl Phys Lett 116:083502. https://doi.org/10.1063/1.5143252

55. Rastogi P, Chu A, Dang TH, Prado Y, Gréboval C, Qu J et al (2021) Complex optical index of HgTe nanocrystal infrared thin films and its use for short wave infrared photodiode design. Adv Opt Mater 9:2002066. https://doi.org/10.1002/adom.202002066

56. Tang X, Ackerman MM, Shen G, Guyot-Sionnest P (2019) Towards infrared electronic eyes: flexible colloidal quantum dot photovoltaic detectors enhanced by resonant cavity. Small 15:1804920. https://doi.org/10.1002/smll.201804920

57. Tang X, Ackerman MM, Guyot-Sionnest P (2018) Thermal imaging with Plasmon resonance enhanced HgTe colloidal quantum dot photovoltaic devices. ACS Nano 12:7362–7370. https://doi.org/10.1021/acsnano.8b03871

58. Grotevent MJ, Hail CU, Yakunin S, Bachmann D, Calame M, Poulikakos D et al (2021) Colloidal HgTe quantum dot/graphene phototransistor with a spectral sensitivity beyond 3 μm. Adv Sci 8:2003360. https://doi.org/10.1002/advs.202003360

59. Cryer ME, Halpert JE (2018) 300 nm spectral resolution in the mid-infrared with robust, high responsivity flexible colloidal quantum dot devices at room temperature. ACS Photonics 5:3009–3015. https://doi.org/10.1021/acsphotonics.8b00738

60. Cryer ME, Browning LA, Plank NOV, Halpert JE (2020) Large Photogain in multicolor nanocrystal photodetector arrays enabling room-temperature detection of targets above 100 °C. ACS Photonics 7:3078–3085. https://doi.org/10.1021/acsphotonics.0c01156

61. Dong Y, Chen M, Yiu WK, Zhu Q, Zhou G, Kershaw SV et al (2020) Solution processed hybrid polymer: HgTe quantum dot phototransistor with high sensitivity and fast infrared response up to 2400 nm at room temperature. Adv Sci 7:2000068. https://doi.org/10.1002/advs.202000068

62. Tang X, Lai KWC (2019) Graphene/HgTe quantum-dot photodetectors with gate-tunable infrared response. ACS Appl Nano Mater 2:6701–6706. https://doi.org/10.1021/acsanm.9b01587

63. Tang X, Chen M, Kamath A, Ackerman MM, Guyot-Sionnest P (2020) Colloidal quantum-dots/graphene/silicon Dual-Channel detection of visible light and short-wave infrared. ACS Photonics 7:1117–1121. https://doi.org/10.1021/acsphotonics.0c00247

64. Tang X, Ackerman MM, Chen M, Guyot-Sionnest P (2019) Dual-band infrared imaging using stacked colloidal quantum dot photodiodes. Nat Photonics 13:277–282. https://doi.org/10.1038/s41566-019-0362-1

65. Rastogi P, Izquierdo E, Gréboval C, Cavallo M, Chu A, Dang TH et al (2022) Extended short-wave photodiode based on CdSe/HgTe/Ag$_2$Te stack with Unity internal efficiency. submitted

66. Vafaie M, Fan JZ, Morteza Najarian A, Ouellette O, Sagar LK, Bertens K et al (2021) Colloidal quantum dot photodetectors with 10-ns response time and 80% quantum efficiency at 1,550 nm. Matter 4:1042–1053. https://doi.org/10.1016/j.matt.2020.12.017

67. Chen M, Lu L, Yu H, Li C, Zhao N (2021) Integration of colloidal quantum dots with photonic structures for optoelectronic and optical devices. Adv Sci 8:2101560. https://doi.org/10.1002/advs.202101560

68. Chen M, Shao L, Kershaw SV, Yu H, Wang J, Rogach AL et al (2014) Photocurrent enhancement of HgTe quantum dot photodiodes by Plasmonic gold Nanorod structures. ACS Nano 8:8208–8216. https://doi.org/10.1021/nn502510u

69. Yifat Y, Ackerman M, Guyot-Sionnest P (2017) Mid-IR colloidal quantum dot detectors enhanced by optical nano-antennas. Appl Phys Lett 110:041106. https://doi.org/10.1063/1.4975058

70. Chu A, Gréboval C, Goubet N, Martinez B, Livache C, Qu J et al (2019) Near Unity absorption in nanocrystal based short wave infrared photodetectors using guided mode resonators. ACS Photonics 6:2553–2561. https://doi.org/10.1021/acsphotonics.9b01015

71. Tang X, Ackerman MM, Guyot-Sionnest P (2019) Acquisition of Hyperspectral Data with colloidal quantum dots. Laser Photonics Rev 13:1900165. https://doi.org/10.1002/lpor.201900165

72. Zhu B, Chen M, Zhu Q, Zhou G, Abdelazim NM, Zhou W et al (2019) Integrated Plasmonic infrared photodetector based on colloidal HgTe quantum dots. Adv Mater Technol 4:1900354. https://doi.org/10.1002/admt.201900354

73. Dang TH, Vasanelli A, Todorov Y, Sirtori C, Prado Y, Chu A et al (2021) Bias tunable spectral response of nanocrystal array in a plasmonic cavity. Nano Lett 21:6671–6677. https://doi.org/10.1021/acs.nanolett.1c02193

74. Livache C, Martinez B, Goubet N, Gréboval C, Qu J, Chu A et al (2019) A colloidal quantum dot infrared photodetector and its use for intraband detection. Nat Commun 10:2125. https://doi.org/10.1038/s41467-019-10170-8

75. Lhuillier E, Guyot-Sionnest P (2017) Recent progresses in mid infrared nanocrystal opto-electronics. IEEE J Sel Top Quantum Electron 23:6000208. https://doi.org/10.1109/JSTQE.2017.2690838
76. Bossavit E, Qu J, Abadie C, Dabard C, Dang TH, Izquierdo E et al (2021) Optimized infrared LED and its use in an all-HgTe nanocrystal-based active imaging setup. Adv Opt Mater 2101755. https://doi.org/10.1002/adom.202101755
77. SWIR cost cut: Imec achieves 1.82 µm pixels | Imaging and Machine Vision Europe. https://www.imveurope.com/feature/swir-cost-cut-imec-achieves-182-m-pixels. Accessed 15 Nov 2021
78. Chu A, Martinez B, Ferré S, Noguier V, Gréboval C, Livache C et al (2019) HgTe nanocrystals for SWIR detection and their integration up to the focal plane Array. ACS Appl Mater Interfaces 11:33116–33123. https://doi.org/10.1021/acsami.9b09954
79. Buurma C, Pimpinella RE, Ciani AJ, Feldman JS, Grein CH, Guyot-Sionnest P (2016) MWIR imaging with low cost colloidal quantum dot films. In: Optical sensing, imaging, and photon counting: nanostructured devices and applications 2016. SPIE, p 993303
80. Kim J, Choi D, Jeong KS (2018) Self-doped colloidal semiconductor nanocrystals with intraband transitions in steady state. Chem Commun 54:8435–8445. https://doi.org/10.1039/C8CC02488J
81. Jagtap A, Livache C, Martinez B, Qu J, Chu A, Gréboval C et al (2018) Emergence of intraband transitions in colloidal nanocrystals. Opt Mater Express 8:1174–1183. https://doi.org/10.1364/OME.8.001174
82. Jeong KS, Deng Z, Keuleyan S, Liu H, Guyot-Sionnest P (2014) Air-stable n-doped colloidal HgS quantum dots. J Phys Chem Lett 5:1139–1143. https://doi.org/10.1021/jz500436x
83. Robin A, Livache C, Ithurria S, Lacaze E, Dubertret B, Lhuillier E (2016) Surface control of doping in self-doped nanocrystals. ACS Appl Mater Interfaces 8:27122–27128. https://doi.org/10.1021/acsami.6b09530
84. Lhuillier E, Scarafagio M, Hease P, Nadal B, Aubin H, Xu XZ et al (2016) Infrared Photodetection based on colloidal quantum-dot films with high mobility and optical absorption up to THz. Nano Lett 16:1282–1286. https://doi.org/10.1021/acs.nanolett.5b04616
85. Jeong J, Yoon B, Kwon Y-W, Choi D, Jeong KS (2017) Singly and doubly occupied higher quantum states in nanocrystals. Nano Lett 17:1187–1193. https://doi.org/10.1021/acs.nanolett.6b04915
86. Ramiro I, Özdemir O, Christodoulou S, Gupta S, Dalmases M, Torre I et al (2020) Mid- and long-wave infrared optoelectronics via Intraband transitions in PbS colloidal quantum dots. Nano Lett 20:1003–1008. https://doi.org/10.1021/acs.nanolett.9b04130
87. Qu J, Goubet N, Livache C, Martinez B, Amelot D, Gréboval C et al (2018) Intraband mid-infrared transitions in Ag₂Se nanocrystals: potential and limitations for hg-free low-cost Photodetection. J Phys Chem C 122:18161–18167. https://doi.org/10.1021/acs.jpcc.8b05699
88. Hafiz SB, Scimeca MR, Zhao P, Paredes IJ, Sahu A, Ko D-K (2019) Silver selenide colloidal quantum dots for mid-wavelength infrared Photodetection. ACS Appl Nano Mater 2:1631–1636. https://doi.org/10.1021/acsanm.9b00069
89. Hafiz SB, Al Mahfuz MM, Ko D-K (2021) Vertically stacked Intraband quantum dot devices for mid-wavelength infrared Photodetection. ACS Appl Mater Interfaces 13:937–943. https://doi.org/10.1021/acsami.0c19450
90. Hafiz SB, Al Mahfuz MM, Lee S, Ko D-K (2021) Midwavelength infrared p–n heterojunction diodes based on Intraband colloidal quantum dots. ACS Appl Mater Interfaces 13:49043–49049. https://doi.org/10.1021/acsami.1c14749
91. Sahu A, Khare A, Deng DD, Norris DJ (2012) Quantum confinement in silver selenide semiconductor nanocrystals. Chem Commun 48:5458–5460. https://doi.org/10.1039/C2CC30539A
92. Deng Z, Jeong KS, Guyot-Sionnest P (2014) Colloidal quantum dots Intraband photodetectors. ACS Nano 8:11707–11714. https://doi.org/10.1021/nn505092a
93. Goubet N, Livache C, Martinez B, Xu XZ, Ithurria S, Royer S et al (2018) Wave-function engineering in HgSe/HgTe colloidal Heterostructures to enhance mid-infrared photoconductive properties. Nano Lett 18:4590–4597. https://doi.org/10.1021/acs.nanolett.8b01861

Chapter 8
Graphene/HgCdTe Heterojunction-Based IR Detectors

Shonak Bansal, M. Muthukumar, and Sandeep Kumar

8.1 Introduction

Infrared (IR) detection technology discovery dates back to the early eighteenth century (February 11, 1800), yet the first IR detectors (IRDs) were developed late twentieth century [1]. Detection and sensing of IR radiation is tremendously important in various domains such as military, medical, imaging, and geological sciences including NASA earth science. Nowadays, IRDs are utilized in diverse products and have profound applications in optical communications, remote control, night vision, motion detection, satellite remote sensing, gas detection, fire alarming, biomedical and thermal imaging, navigational aids, missile guidance, telecommunications, spectroscopy, security, chemical analysis, and agriculture, etc. [2–8]. The imaging and night vision of thermal sources require IRDs with a spectral sensitivity of about 10 μm, whereas IR spectroscopy of molecules and gases requires a spectral sensitivity of about 2-15 μm [9]. To achieve this epitaxially grown narrow bandgap mature photosensitive semiconductor materials are required which includes; Si, Ge [10–12], II–VI (HgCdTe, CdZnTe, CdSeTe) [2, 13–16], III–V (InAsSb, InSb, GaAs, InAs) [2, 17–19], and IV–VI (PbSnTe) [2]. These photosensitive materials absorb the incident light energy corresponding to their energy bandgaps which result in the net photocurrent. The fundamental properties of narrow bandgap materials (such as high electron mobility, high optical absorption coefficient, and low thermal generation rate), along with the bandgap engineering, make these alloys almost ideal for a wide range of IRDs. Thus, these materials provide an exceptional degree of freedom in the development of IRDs.

S. Bansal (✉)
Electronics and Communication Engineering Department, Chandigarh University, Gharuan, India
e-mail: shonakk@gmail.com

M. Muthukumar · S. Kumar
ICAR-Central Institute for Subtropical Horticulture, Lucknow, Uttar Pradesh, India

© The Author(s), under exclusive license to Springer Nature Switzerland AG 2023 183
G. Korotcenkov (ed.), *Handbook of II-VI Semiconductor-Based Sensors and Radiation Detectors*, https://doi.org/10.1007/978-3-031-20510-1_8

The bandgap energy tunability results in IR detector applications that cover all IR spectral regimes. Generally, IR spectral regime can be differentiated into the near-IR (NIR: 0.75–1.1 μm), short-wave IR (SWIR: 1–3 μm), mid-wave IR (MWIR: 3–5 μm), long-wave IR (LWIR: 8–12 μm), very long-wave IR (VLWIR: 12–30 μm), and far IR (FIR: 30–1000 μm) ranges [17, 20]. These spectral regimes have different conventions, for example, 1.55 μm is considered as SWIR, in the Department of Defense community, but is considered NIR for the astronomy community. The spectral regimes ranging from MWIR to FIR are widely used for astronomy and free-space communications because the high transparency of the atmosphere at these spectral regimes permits lossless transmission. The same spectral regimes are also used for military applications including natural resources management and environmental monitoring, cloud properties, sea surface temperatures, forest fires, and volcanic activities [17, 21, 22]. Nowadays, numerous military airplanes are furnished with high performance IR cameras for scanning the battlefield during poor visibility conditions.

Rapid developments are being made in evolving cost-effective narrow bandgap semiconductor detectors with enhanced sensitivity and longer wavelengths. Till date, different IR detector's architectures, namely, p-n [7, 23–27], p-i-n [4, 27–29], dual-band detector [5], metal-semiconductor-metal [28], avalanche photodetector (APD) [30], carbon nanotube detector [31], and nanowire photodetector [31] have been reported. Similarly, novel design architectures, namely, quantum dots [31, 32], quantum well [33], and Schottky-barrier photoemissive detector [34–41] based IRDs are developed with improved performances. However, such device fabrication processes are more expensive and include sophisticated equipment that are not commercially feasible.

The excellent optoelectronic properties of graphene and HgCdTe makes these materials a favorable choice for the development of high performance IRD to support and further advance a variety of applications. Initially, graphene has been proposed as a replacement for the indium tin oxide (ITO) as transparent electrodes for optical devices such as LCD and LED devices [28]. The high carrier mobility, wide spectral regime from ultraviolet (UV) to IR due to linear dispersion nearby the Dirac point of graphene, and the outstanding light absorption properties of other semiconductor materials enables graphene an excellent candidate choice for the development of next-generation optoelectronic devices such as photodetectors. This chapter will present firstly the key properties of HgCdTe and graphene materials, challenges, developments, and the modeling approach on graphene/HgCdTe based IRDs in the subsequent sections.

8.2 Graphene/HgCdTe Heterojunction Based IRDs

8.2.1 Early Generation IRD Technologies

The detector technologies that were initially developed uses the compound semiconductor materials. PbS was the first material used for the IRD in SWIR spectral regime for military applications [17, 42]. Later, the detection range was extended up to MWIR spectral regime by employing compound semiconductor materials such as InSb, PbSe, and PbTe, etc. The development of semiconductor alloy materials on groups II-VI, III-V, and IV-IV was started in the 1950's. The better stability, low leakage current, low thermal generation rate, relatively high absorption coefficient, and tunable bandgap make the HgCdTe most well-regarded semiconductor material after Si and GaAs for the development of high-performance broadband IR and THz detectors in military applications to cover a spectral regime from the NIR to the FIR [4, 17, 43–46]. The reason for this is that HgCdTe can be tuned to the desired IR spectral regime by varying the Cd concentration.

8.2.2   Existing and Next Generation Technologies, Challenges, and Prospects of Effective IR Detection

HgCdTe can be epitaxially grown as a chemical ternary alloy of HgTe and CdTe. HgTe is a semi-metal chemical compound with high conductivity and zero bandgap, whereas CdTe is a semiconductor with a bandgap of about 1.6 eV. These alloys have almost the same lattice constant, which allows the growth of HgCdTe in any composition without defects introduced by lattice mismatch. The extremely small change of lattice constant with Cd composition makes it possible to grow high-quality layers and heterostructures. Mixing these two materials allows obtaining any energy bandgap between 0 and 1.6 eV [17]. Furthermore, the growth of HgCdTe material layer can be controlled by metal-organic chemical vapor deposition (MOCVD), molecular beam epitaxy (MBE), and liquid phase epitaxy (LPE) [47–50]. LPE is the most widely used for industrial applications for many years, while MBE, a vapor phase epitaxy (VPE) process leads to high accuracy in the deposition of detector material structures resulting in good lateral homogeneity, high quantum efficiency, and abrupt doping profiles in HgCdTe-based IR detectors [51, 52]. The bandgap tunability of HgCdTe resulted into the numerous HgCdTe-based high performance IRDs with different configurations such as p-n [25, 53, 54], p-i-n [4, 29], APD [30], and dual-band IRD [5, 55] at cryogenic temperatures. The need for cryogenic cooling facilities to reduce thermally generated dark current adds a significant amount of energy/power consumption, cost, and weight to such IRDs, thereby requiring a bulky imaging instrument. HgCdTe technology development continues to be primarily for military-related security or safety applications for the detection of a condition or an object. The main motivations for replacing HgCdTe are the technological disadvantages of this material. Apart from this high Auger recombination process results in high dark currents [4, 41, 56]. Therefore, it is essential to design and develop highly efficient HgCdTe based IRD above cryogenic temperature operations for low weight, size, power, and cost space applications. Nevertheless, it remains the leading IRD technology for applications with multiple spectral regimes, where cost is not a major issue. HgCdTe based IRD performance at higher operating temperatures is significantly affected by the dislocations and defects arising from lattice mismatches, ensuring residual stress that further reduces lifetimes and minimizes the mobility of the carriers. Conversely, methods such as thermal cycle annealing (TCA) reduces the dislocation density down to the saturation limit, resulting in the improved high temperature operations of HgCdTe based IRDs. The utilization of epitaxially growth thin substrates can further reduce the dislocations towards the surface. Moreover, for HgCdTe epitaxial growth the Si (112) orientation is preferable choice due to Hg defect control, consumption, and doping issues [22].

The synthesis of novel 2D nanomaterials are emerging as enabling technologies of the future as they provide extended spectral regime in the photonic devices [57]. Graphene is the choicest and most favorable material for the development of high-operating detectors in the UV to IR, and THz regimes due to its excellent electrical and optical properties [58, 59]. It is utilized to form a suitable heterojunction with other semiconductor materials due to enormous mobility ($\sim 10^5$ cm^2/Vs), high sensitivity, high current-density carrying capacity ($\sim 10^9$ A/cm^2), high Fermi velocity (10^8 cm/s), better strength, flexibility, high thermal conductivity, low resistivity, tunable Fermi-level, broadband light absorption, and so on [60–62]. The high carrier mobility in graphene permits it to be used in developing an efficient high-speed charge carrier collector. Ultrafast carrier multiplication (CM) has also been observed in bilayer graphene (BLG) recently [7, 27]. The graphene has been successfully composited with different materials, namely, Si [63], ZnO [64–66], CdS [67] (in UV spectral regime), CdSe [68], GaN [69] (in visible spectral regime), and Si [10, 66, 70], Ge [12], GaAs [71], PbS [72, 73], and HgCdTe [7, 26, 27, 29, 34] (in IR spectral regime). The heterostructure of graphene with other materials demonstrates a low dark current, small parasitics, low power dissipation, high response speed, and higher breakdown voltage than that of conventional homostructures. Such heterojunctions offer a high electric field beneficial for fast separation of the photogenerated carriers without application of any external bias, enabling self-powered and higher performance detectors [74]. Till date, numerous graphene-based heterostructures are reported with improved performances [63, 64, 67, 69, 71]. Additionally, such heterostructures are utilized as a fundamental component of high-performance nanoelectronic devices to estimate the device performance such as response time, open-circuit voltages in solar cells, the ON/OFF ratios, and ON-state current [74]. The graphene layer helps in efficient carrier separation due to the tunable Fermi-level [61] and also acts as an antireflection coating to reduce the light reflection by ~80% in the IR spectral regime [75].

The graphene can be synthesized by the processes such as chemical vapor deposition (CVD), oxidation reduction growth process, epitaxial growth on silicon carbide, and liquid phase stripping. Out of these, the CVD is a widely used process for fabricating graphene layers with better structural integrity and crystalline quality on copper substrates without any contamination issues [76]. But copper and other metallic substrates are not suitable for many applications as they need the employment of graphene sheets directly on semiconductors or metal oxide [51]. Therefore, there is a need for the improvement in effective processes over which graphene can be effectively and directly transported onto any desirable substrate, while also avoiding wrinkles, cracks, and any kind of contamination.

However, the photodetectors utilizing single layer of graphene suffer from the poor photodetection response due to the zero bandgap and small optical absorption (~2.3%) of graphene [19, 58, 63, 77–80]. Several efforts including the use of graphene quantum dots, a few graphene layers, inducing bandgap in graphene are made to improve the detection ability [3, 72, 73, 81–84]. The chemical doping in graphene shifts the Fermi-level, creating a bandgap [85]. Furthermore, the integration of high carrier mobility graphene layer with HgCdTe, the detector's operational capabilities and performance can be further improved.

8.2.3 Graphene/HgCdTe Detector Fabrication and Operating Principle

MBE growth method allows the high crystalline quality HgCdTe material on CdZnTe or Si substrates with precise control of the detector material structure parameters. On the other hand, HgCdTe is an expensive technology primarily due to the fact that it can be grown on a close lattice-matched CdZnTe substrate. The IR detection performance of HgCdTe epitaxially grown on Si substrate is also analogous to that of HgCdTe deposited on CdZnTe substrate and is thermally compatible with the current readout integrated circuit (ROICs) [22, 86, 87]. CdZnTe substrate is preferred as HgCdTe/CdZnTe interface offers less interface trap charge as compared to CdTe, Si, and Ge substrates [54] and does not require a buffer layer between the absorber and substrate. But, CdZnTe substrates present cost-related challenges that require the search for an alternate viable substrate for HgCdTe growth. GaAs is high-quality, inexpensive, and easily available substrate material. The lattice mismatch of HgCdTe with Si and GaAs layer involves the utilization of an extra CdTe buffer layer [52, 86]. Using the MBE method, n-type doping of $5 \times 10^{14} - 2 \times 10^{15}$ and up to 10^{19} cm^{-3} were obtained by using CdZnTe [88] and Si [86] substrates, respectively. On the other hand, by using the MOCVD method, the donor and acceptor doping levels of $5 \times 10^{14} - 5 \times 10^{17}$ cm^{-3} were achieved. The HgCdTe absorbing layer is epitaxially grown on the buffer layer with the desired Cd composition and thickness. Then the graphene layer is grown on the absorbing layer followed by deposition of the top metal contacts.

In terms of material structure, the graphene/HgCdTe detector is mostly composted of the three principal layers:

(i) Gate: Substrate layer is used as gate terminal which provides electrical field aiding carrier transport,

(ii) Absorber/active layer: The absorber layer (composed of HgCdTe) is the active optical layer where the process of carrier photogeneration takes place, and

(iii) Channel: High mobility, low noise graphene channel transfers the photogenerated carriers to the electrical readouts.

IR detector operation process can be divided into three main phases;

   (i) Carrier generation and separation: In the first step the incident IR photons are transmitted to HgCdTe absorber layer resulting in the production of electron-hole pairs, or excitons,

  (ii) Carrier transport and injection: The carriers are transported through the absorber and get injected into graphene channel, and

 (iii) Carrier transport in graphene channel: Injected carriers transported to and collected by ROIC.

8.2.4 Graphene/HgCdTe Heterojunction Based IRD Structures and Operation

The graphene/HgCdTe based detector technology involves the incorporation of HgCdTe materials along with graphene, which allows for higher IR detection performance as compared with detectors using HgCdTe material solely. The graphene layer conformably covers a narrow bandgap, lightly doped HgCdTe active/absorbing layer wherein the photogeneration of carrier takes place to form a heterojunction IRD. The lightly doped HgCdTe generates total dark-current and photo-current densities. The bandgap and doping concentration of the active region is tuned to allow better absorption of IR radiations suppressing thermally generated carriers. Furthermore, the active layer estimates the carrier lifetime, quantum efficiency, photogeneration rate, and which mutually govern the overall performance of the detector. The intrinsic interfacial barrier between the graphene and HgCdTe active layer is designed to efficiently minimize the recombination of photogenerated carriers in the detector. The graphene functions as low noise, high mobility channel that drifts away the photogenerated carriers from the active layer to the electrical contacts, and consequently to the ROIC for electrical readout before their recombination. This will further contribute to the IR performance of graphene/HgCdTe detectors compared to that of conventional HgCdTe detectors [7, 22, 27, 86]. HgCdTe is a material with a relatively high dielectric constant (~14) which makes it an amenable substrate for providing relatively high electric conductance by transferring or coupling with graphene film at room temperature. It was experimentally demonstrated that the integration of 5-10 graphene layers onto HgCdTe substrate resulted in 25 times higher electrical conductance than that of HgCdTe and 80% optical transmittance in the IR regime at 77 and 300 K which was slightly lower than that on the SiO_2/Si substrate in the visible regime [89]. This experimental study proposed that graphene is considered a good candidate for transparent electrodes as a replacement of metal electrodes while developing better and cost-effective HgCdTe based IRDs with smooth surfaces without any crack or folding.

Figure 8.1a, b show the schematic structures of the BLG composite $Hg_{1-x}Cd_xTe$ based p^+- n^- single heterojunction and p^+-n^--n^+ dual junctions IRDs, respectively [7, 26, 27]. For the single-heterojunction IRD formation, highly doped thin p^+-BLG

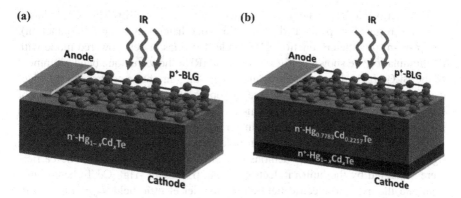

Fig. 8.1 The schematic of BLG/Hg$_{1-x}$Cd$_x$Te based (a) p$^+$-n$^-$ single-heterojunction and (b) p$^+$-n$^-$-n$^+$ dual-junction IR detectors

was employed on the lightly doped wide n$^-$-Hg$_{1-x}$Cd$_x$Te active layer as shown in Fig. 8.1a. For the p$^+$-BLG/n$^-$-Hg$_{0.7783}$Cd$_{0.2217}$Te/n$^+$-Hg$_{1-x}$Cd$_x$Te dual-junction LWIR detectors formation, the device dimensions were kept constant same as that of the single-heterojunction IRD (Fig. 8.1a), except the inclusion of a thin heavily doped n$^+$-Hg$_{1-x}$Cd$_x$Te below active layer as illustrated in Fig. 8.1b. The p$^+$-doping of the BLG can be achieved through the chemical agents via the CVD technique [7, 90]. The BLG allows the opening of bandgap in addition to other excellent optoelectronic properties [7, 91, 92]. In order to analyze the effect of bandgap tuning on the detector performance, the Cd composition in n$^+$-Hg$_{1-x}$Cd$_x$Te was optimized to 0.2217 and 0.32 in the p$^+$-BLG/n$^-$-Hg$_{0.7783}$Cd$_{0.2217}$Te/n$^+$-Hg$_{1-x}$Cd$_x$Te dual-junction LWIR detectors. This resulted in the formation of two different dual-junction devices, named, p$^+$-BLG/n$^-$-Hg$_{0.7783}$Cd$_{0.2217}$Te/n$^+$-Hg$_{0.7783}$Cd$_{0.2217}$Te (consisting of one hetero- and one homojunction) and p$^+$-BLG/n$^-$-Hg$_{0.7783}$Cd$_{0.2217}$Te/n$^+$-Hg$_{0.68}$Cd$_{0.32}$Te (active layer was sandwiched between two different bandgap materials to form the dual heterojunctions). The dual heterojunctions device helps to reduce the thermal generation and parasitic impedances. The optimized Cd composition in n$^+$-Hg$_{1-x}$Cd$_x$Te facilitated high built-in electric field $\left(\overrightarrow{E_{field}} \right)$ at n$^-$-n$^+$ heterojunction. Such a large built-in $\overrightarrow{E_{field}}$ drifts the photogenerated carriers to the electrodes, resulting in higher effective photocurrent through the heterojunction which further reduces the tunneling current in reverse bias condition. The bandgap and doping concentration of active layer was selected for optimum absorption of IR radiations suppressing thermally generated carriers.

In real fabricated devices, the ohmic contacts can be made by using gold/molybdenum (Au/Mo) and palladium (Pd) with lower contact resistance [93, 94]. The IR radiations were incident from the BLG cladding window over the narrow bandgap active layer. The BLG was used as a light absorber. During the IR illuminations on the p$^+$-BLG/n$^-$-Hg$_{1-x}$Cd$_x$Te heterojunction, the hot photocarriers transport from the active layer to the BLG layer due to the built-in $\overrightarrow{E_{field}}$ [95], resulting in the change in graphene conduction.

The energy bandgap diagram and electric field profile for BLG/$Hg_{1-x}Cd_xTe$ based p$^+$-n$^-$ heterojunction, p$^+$-n$^-$-n$^+$ dual-junction (one hetero- and one homojunction), and p$^+$-n$^-$-n$^+$ dual-heterojunction IRDs under 0 V bias or self-powered mode with IR illumination are shown in Fig. 8.2a–c. The IRDs show a unique band-alignment of the BLG/$Hg_{1-x}Cd_xTe$ heterostructure, suggesting an ohmic contact for the electrons. The small bandgap in BLG due to doping will change the work function of BLG resulting in the shifting of Fermi-level towards the valence band. This makes graphene different than the other 2D materials [90].

Under IR illumination, the electron-hole pairs are produced in the bandgap of a lightly doped active layer as demonstrated in Fig. 8.2a–c for IRDs. The carriers were separated by the built-in electric field at p$^+$-BLG/n$^-$-$Hg_{1-x}Cd_xTe$ heterojunction, and the external electric field aligns the net electric field E_{p-n} towards its direction. Consequently, photogenerated carriers move in the opposite direction providing net photocurrent to the external circuit.

Figure 8.2b, c show the energy bandgap diagram wherein Cd composition of the lightly doped active layer ($x = 0.2217$) is tuned for the LWIR operation. In p$^+$-n$^-$-n$^+$ dual-junction IRDs under IR illumination, both the p$^+$-BLG and n$^-$-$Hg_{1-x}Cd_xTe$ layers absorb the incident photons and generate electron-hole pairs as shown in

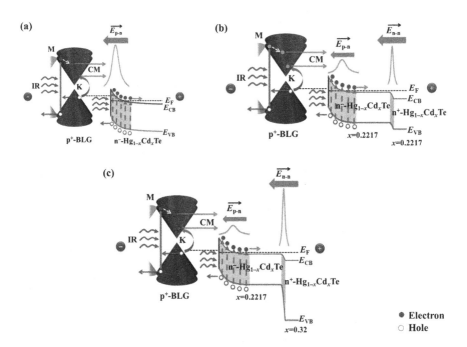

Fig. 8.2 The energy bandgap diagram and electric field profile of BLG/HgCdTe based (**a**) p$^+$-n$^-$ heterojunction; (**b**) p$^+$-n$^-$-n$^+$ dual-junction (one hetero- and one homojunction), and (**c**) p$^+$- n$^-$-n$^+$ dual-heterojunction IR detectors under 0 V bias or self-powered mode demonstrating light absorption and carrier multiplication process. Here, E_{CB}, E_F, and E_{VB} represent conduction band, Fermi-level, and valence band energies, respectively; M– and K– are the points of Brillouin zone in BLG resulting in a higher CM effect. (Idea from [63])

Fig. 8.2b, c. The built-in electric field $\overrightarrow{E_{p-n}}$ and $\overrightarrow{E_{n-n}}$ at the p$^+$-BLG/n$^-$-Hg$_{1-x}$Cd$_{x=0.2217}$Te and n$^-$-Hg$_{1-x}$Cd$_{x=0.2217}$Te/n$^+$-Hg$_{1-x}$Cd$_x$Te junction separates the photogenerated carriers. The external field dominates over the junction and results in the drift of photogenerated carriers towards n$^+$-Hg$_{1-x}$Cd$_x$Te and p$^+$-BLG, respectively, thereby resulting in a net photocurrent. Due to the existence of dual heterojunctions a huge built-in electric field exists at the n$^-$-Hg$_{1-x}$Cd$_{x=0.2217}$Te/n$^+$-Hg$_{1-x}$Cd$_{x=0.32}$Te hetero-interface which effectively separates the photogenerated carriers so as to produce a better photoresponse. It is evident that under IR illumination, the photogenerated carriers accumulate in the potential well, raising the Fermi-level and increasing the conductivity of the detector [7, 27]. The various electrical and optical characteristic parameters were computed and analyzed using computer simulations. The results were further validated by an analytical model based on drift-diffusion, tunneling, and Chu's methods [7, 96–98], suggesting potential applications in next-generation high-performance, ultra-low-power, and cost-effective IRDs for opto-electronics devices. The dark current density was reduced due to the lower thermo-generation rate of BLG and Hg$_{1-x}$Cd$_x$Te heterostructure, offering improved performance. The heterostructure of BLG with Hg$_{1-x}$Cd$_x$Te offered excellent photo-detection in the IR regime along with the near room temperature external quantum efficiency (QE_{ext}) of more than 100% attributed to the hot carrier multiplication mechanism in graphene [99–105].

In the VLWIR spectral regime, the p$^+$-BLG/n$^-$-Hg$_{1-x}$Cd$_{x=0.1867}$Te heterojunction photodetector [7] offers the photosensitivity of 4.8×10^{14}, response time of 9.4 ps, 3-dB cut-off frequency of 36.16 GHz, and QE_{ext} of 89% at -0.5 V and 77 K. Contrarily in the LWIR spectral regime, p$^+$-BLG/n$^-$-Hg$_{1-x}$Cd$_{x=0.2217}$Te heterojunction photode-tector [26, 27] demonstrated the photosensitivity of 6.9×10^5, rise (fall) time of 7.4 (5.7) ps, 3-dB cut-off frequency of 73 GHz, and QE_{ext} of 54.45% with -0.5 V bias at 77 K. Moreover, the carrier multiplication and external biasing effects in p$^+$-BLG/n$^-$-Hg$_{1-x}$Cd$_x$Te heterojunction resulted in the near room temperature QE_{ext} of ~3337.70 and ~ 892% for the VLWIR and LWIR detector, respectively.

The dual-heterojunction detector exhibited low dark current density and higher temperature operation due to the lower thermal generation rate than the single-heterojunction based detector [27]. The p$^+$-BLG/n$^-$-Hg$_{1-x}$Cd$_{x=0.2217}$Te/n$^+$-Hg$_{1-x}$Cd$_{x=0.32}$Te dual-heterojunction LWIR detector showed rapid photocurrent switching with rise (fall) time of ~0.05 (0.013) ps. Under self-powered mode, the photocapaci-tance of 7.21×10^{-13} F/μm was obtained for the dual-heterojunction detector sug-gesting the effective photogeneration of carriers and partial screening of the electric field. The highest QE_{ext} of 85.8% at 10.6 μm cut-off wavelength at 77 K, was achieved for the dual-heterojunction detector [27]. Such improved performance was due to the existence of a huge built-in $\overrightarrow{E_{field}}$ at n$^-$-n$^+$ heterojunction. Moreover, the temperature-dependent QE_{ext} was observed to be ~101%, for dual-heterojunction LWIR detector at near room temperature due to the carrier multiplication effect in BLG.

Inspired by the physical properties of BLG such as hot carrier injection, tunable bandgap, and carrier multiplication, a new device comprising a BLG over the top of a thin CdTe passivation layer on a HgCdTe absorbing layer (Fig. 8.3a) is proposed

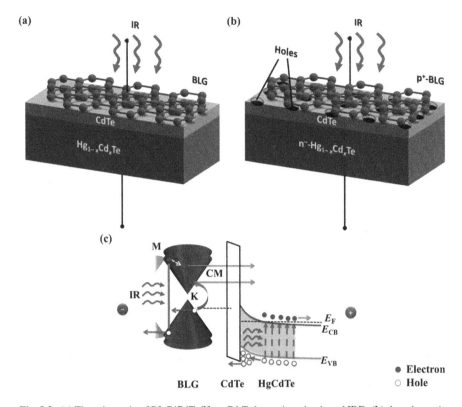

Fig. 8.3 (**a**) The schematic of BLG/CdTe/Hg$_{1-x}$Cd$_x$Te heterojunction based IRD, (**b**) the schematic of p$^+$-BLG/CdTe/n$^-$-Hg$_{1-x}$Cd$_x$Te heterojunction based IRD with micro-holes, and (**c**) the band diagram of BLG/CdTe/HgCdTe heterojunction based IRD demonstrating light absorption and carrier multiplication process

by using a quantum mechanical approach [106]. Using the same design architecture as shown in Fig. 8.3a, a p$^+$-BLG/CdTe/n$^-$-Hg$_{1-x}$Cd$_x$Te heterojunction detector design that uses photon-trapping micro-holes in addition to the carrier multiplication effect in graphene is proposed and is shown in Fig. 8.3b [107]. As shown in Fig. 8.3c, the CdTe layer allows the passage of electrons in the conduction bands but shows a significant barrier for the holes due to the existence of a huge Schottky barrier. This confirms the transport of carriers from ambipolar to unipolar. Using the non-equilibrium Green's function (NEGF) it was found that the electrons are the dominant charge carriers over holes which reduces the carrier recombination, increases carrier lifetimes, and hence results in improved performance in comparison to conventional HgCdTe based IRDs.

Figure 8.4 shows the graphene/HgCdTe heterojunction based MWIR on a Si substrate having a CdTe buffer layer deposited on it [22, 86]. The Si substrate act as the electrode terminal that provides an electric field $\overrightarrow{E_{\text{field}}}$ in the vertical direction, resulting in transport of the photogenerated electron-hole pairs in the vertical direction as shown in Fig. 8.4a. The IR radiations were incident through the substrate and

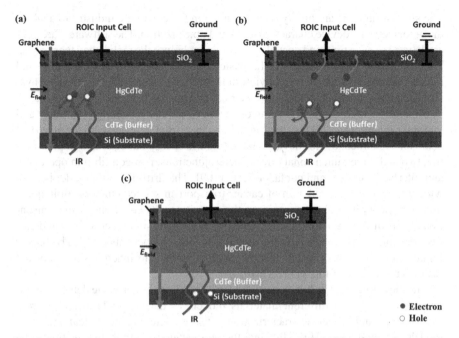

Fig. 8.4 The schematic of graphene/HgCdTe heterostructure based IRD on Si substrate: (**a**) Generation of carriers due to incident IR radiations and their separation due to the built-in electric field, (**b**) Transportation and injection of photogenerated carrier into graphene, and (**c**) Transportation of photogenerated carriers in graphene horizontally

buffer layers into the active region where they are absorbed to yield photogenerated charge carriers. The $\overrightarrow{E}_{\text{field}}$ separates the photogenerated charge carriers and further suppresses Auger recombination process in the active layer so as to reduce the loss of photogenerated charge carriers. The photogenerated electrons drifts from the active layer to the graphene layer and then injected into it (Fig. 8.4b). The injected electrons into the graphene layer are transported in the lateral direction and are collected by the ROIC terminal as illustrated in Fig. 8.4c. This indicates that the graphene/HgCdTe heterojunction provides the improved performance than that of conventional HgCdTe based IRDs.

8.2.5 Graphene/HgCdTe Heterojunction Based IRD Modelling Approach

The general modeling approach is based on numerous components that together constitute a comprehensive model for the optoelectronic technology. The state-of-art optoelectronic technology research focusses on improving the detection capabilities by utilizing the novel structures. Therefore, analytical modeling and simulation tools became a crucial part of the design and development procedure.

The simulation tools are highly recommended in the electronic industry to analyze the optoelectronic device characterization and optimization. The software's are generally used to save time and cost of an experiment before the real device fabrication. The simulation tools depend on the basic semiconductor physics equations and mechanism that provide an insight view of the device to understand the device physics. The goal is to develop accurate electrical and optical behavior of the device.

A solution of fundamental semiconductor equations including continuity and Poisson's equations [108] for electrons and holes is used to understand the electrical behavior of semiconductor devices, including the energy bandgap and doping profile, two and three dimensional effects, heterojunctions, non-equilibrium operation, and interface, surface, and contact effects [109]. The drift-diffusion model is most widely used for the simulation of carrier transport in semiconductors. Boltzmann transport theory is used to approximate the current density equations by the conventional drift-diffusion model, expressed them as a function of electric field and carrier concentrations, consisting of drift and diffusion components [108]. The reason for choosing the drift-diffusion charge transport model is its inherent simplicity and the better accuracy of this model [110].

The modeling approach includes modularly building the complete detector simulation platform from different materials models as data is made available from experiments and device characterizations. These including individual materials models for graphene, HgCdTe, and the graphene/HgCdTe heterounction. The Silvaco TCAD software was utilized to design and evaluate the optoelectronic performances of BLG/HgCdTe based IR detectors [7, 27]. The computer simulations for optoelectronic characteristics of the p^+-BLG/n^--Hg$_{1-x}$Cd$_x$Te IR detector were carried out by solving the continuity, carrier transport, and Poisson's equations with optimized boundary conditions based on the Boltzmann's transport model [5, 108]. To reproduce the carrier transport, a drift-diffusion approach was implemented to the degenerate semiconductor and parabolic shape of the conduction band [111]. The Newton-Richardson iteration method and concentration-dependent Analytic model were used to estimate the carrier mobility in the detector [5, 54]. Further to solve these equations, in order to characterize the carrier lifetime, and the dark current density in the p^+-BLG/n^--Hg$_{1-x}$Cd$_x$Te IR detector, Shockley-Read-Hall, Auger, optical, surface recombination rates, and standard tunneling mechanism models were considered. The doping and carrier densities were evaluated using Fermi-Dirac statistics [5, 54, 108, 112]. Accordingly, the total dark current density as a function of applied voltage and ambient temperature arise due to the existence of diffusion, generation-recombination, and tunneling components [7].

The optical characterizations were performed by coupling the optical and basic semiconductor equations. The optical absorption coefficient of Hg$_{1-x}$Cd$_x$Te material within the Kane region was calculated by Chu's empirical relation [96, 97]. To compute the optical characteristics, wavelength-dependent complex refractive indices for both the BLG [79, 113] and Hg$_{1-x}$Cd$_x$Te [54, 98] materials were described in IR regime. However, the total QE_{ext} of the detector comprises of neutral $p^+ - (QE_{ext})_{p^+}$, neutral $n^- - (QE_{ext})_{n^-}$, and the depletion $(QE_{ext})_{dep}$ regions as given by [7]:

$$QE_{ext} = CM(V,T) \times \left[(QE_{ext})_{p^+} + (QE_{ext})_{n^-} + (QE_{ext})_{dep} \right] \tag{8.1}$$

where, CM(V,T) represents the voltage (V) and temperature (T) dependent hot carrier multiplication factor [7, 99, 101], that can be estimated from the simulated current density-voltage characteristics under dark and illumination conditions at different temperatures as [30]:

$$CM(V,T) = \frac{J_{light}(V,T) - J_{dark}(V,T)}{J_{light}(V=0,T) - J_{dark}(V=0,T)} \tag{8.2}$$

J_{dark} and J_{light} represents the dark current and photocurrent density, respectively.

8.3 Conclusion and Future Prospects

Graphene is considered as the most favorable material choice for the development of high-operating detectors in the UV to IR, and THz regimes. A high-performance graphene/HgCdTe detector utilizes the properties of both materials for the development of IR detectors. This chapter presented the overview of the graphene/HgCdTe-based IR detectors, their fabrication, principles, optimization procedures, and the modeling approach. The graphene functions as high carrier mobility channel to drift the photogenerated charge carriers away before they can recombine. Therefore, the hetero-interface between the graphene layer and HgCdTe active layer acts as a tunable rectifier that minimizes the recombination of photogenerated carriers in the detector. The presented study in this chapter suggest that the successful integration of graphene film on to HgCdTe layer can develop the high-performance IR detectors as compared to HgCdTe detectors. The room temperature operation capability of graphene/HgCdTe IR detectors can be beneficial in various NASA earth Science applications. Furthermore, with recent developments and advancements made in nanoscience and nanotechnology, novel fabrication designs and development of heterostructures of desired interests are done with the use of novel alloy combinations. Utilization of latest technological advancements in materials sciences, it is possible to refine and fine tune the existing detector technology with cost-effective alternatives and this could be scaled up towards use in different applications *viz.*, development of sensors, lasers, optoelectronic devices like IR detectors and IR light sources.

References

1. Rogalski A, Kopytko M, Martyniuk P, Hu W (2020) Comparison of performance limits of HOT HgCdTe photodiodes with 2D material infrared photodetectors. Opto-Electron Rev 28(2):82–92. https://doi.org/10.24425/opelre.2020.132504
2. Rogalski A (2003) Infrared detectors: status and trends. Prog Quantum Electron 27(2–3):59–210. https://doi.org/10.1016/S0079-6727(02)00024-1
3. Ryzhii V, Ryzhii M (2009) Graphene bilayer field-effect phototransistor for terahertz and infrared detection. Phys Rev B Condens Matter Mater Phys 79(24):245311-1-245311–8. https://doi.org/10.1103/PhysRevB.79.245311
4. Saxena PK (2011) Modeling and simulation of HgCdTe based p^+-n-n^+ LWIR photodetector. Infrared Phys Technol 54(1):25–33. https://doi.org/10.1016/j.infrared.2010.10.005
5. Saxena PK (2017) Numerical study of dual band (MW/LW) IR detector for performance improvement. Def Sci J 67(2):141–148. https://doi.org/10.14429/dsj.67.11177
6. Zhuge F, Zheng Z, Luo P, Lv L, Huang Y, Li H, Zhai T (2017) Nanostructured materials and architectures for advanced infrared photodetection. Adv Mater Technol 2(8):1700005-1-1700005–26. https://doi.org/10.1002/admt.201700005
7. Bansal S, Sharma K, Jain P, Sardana N, Kumar S, Gupta N, Singh AK (2018) Bilayer graphene/HgCdTe based very long infrared photodetector with superior external quantum efficiency, responsivity, and detectivity. RSC Adv 8(69):39579–39592. https://doi.org/10.1039/c8ra07683a
8. Yao J, Yang G (2020) 2D material broadband photodetectors. Nanoscale 12(2):454–476. https://doi.org/10.1039/c9nr09070c
9. Grotevent MJ, Hail CU, Yakunin S, Bachmann D, Calame M, Poulikakos D, Kovalenko MV, Shorubalko I (2021) Colloidal HgTe quantum dot/graphene phototransistor with a spectral sensitivity beyond 3 μm. Adv Sci 8:1–7. https://doi.org/10.1002/advs.202003360
10. Amirmazlaghani M, Raissi F, Habibpour O, Vukusic J, Stake J (2013) Graphene-Si Schottky IR detector. IEEE J Quantum Electron 49(7):589–594. https://doi.org/10.1109/JQE.2013.2261472
11. Assefa S, Xia F, Vlasov YA (2010) Reinventing germanium avalanche photodetector for nanophotonic on-chip optical interconnects. Nature 464(7285):80–84. https://doi.org/10.1038/nature08813
12. Zeng LH, Wang MZ, Hu H, Nie B, Yu YQ, Wu CY, Wang L, Hu JG, Xie C, Liang FX, Luo LB (2013) Monolayer graphene/germanium Schottky junction as high-performance self-driven infrared light photodetector. ACS Appl Mater Interfaces 5(19):9362–9366. https://doi.org/10.1021/am4026505
13. Norton P (2002) HgCdTe infrared detectors. Opto-Electron Rev 10(3):159–174
14. Rogalski A (2004) Toward third generation HgCdTe infrared detectors. J Alloys Compd 371(1–2):53–57. https://doi.org/10.1016/j.jallcom.2003.06.005
15. Wijewarnasuriya PS, Chen Y, Brill G, Zandi B, Dhar NK (2010) High-performance long-wavelength infrared HgCdTe focal plane arrays fabricated on CdSeTe compliant Si substrates. IEEE Trans Electron Devices 57(4):782–787. https://doi.org/10.1109/TED.2010.2041511
16. Wang J, Chen X, Hu W, Wang L, Lu W, Xu F, Zhao J, Shi Y, Ji R (2011) Amorphous HgCdTe infrared photoconductive detector with high detectivity above 200 K. Appl Phys Lett 99(11):113508–1–113508–3. https://doi.org/10.1063/1.3638459
17. Rogalski A (2005) HgCdTe infrared detector material: history, status and outlook. Rep Prog Phys 68(10):2267–2336. https://doi.org/10.1088/0034-4885/68/10/R01
18. Yoon J, Jo S, Chun IS, Jung I, Kim HS, Meitl M, Menard E, Li X, Coleman JJ, Paik U, Rogers JA (2010) GaAs photovoltaics and optoelectronics using releasable multilayer epitaxial assemblies. Nature 465(7296):329–333. https://doi.org/10.1038/nature09054
19. Miao J, Hu W, Guo N, Lu Z, Liu X, Liao L, Chen P, Jiang T, Wu S, Ho JC, Wang L, Chen X, Lu W (2015) High-responsivity graphene/InAs nanowire heterojunction near-infrared

photodetectors with distinct photocurrent on/off ratios. Small 11(8):936–942. https://doi. org/10.1002/smll.201402312

20. Long M, Wang P, Fang H, Hu W (2018) Progress, challenges, and opportunities for 2D material based photodetectors. Adv Funct Mater 29(19):1803807-1-1803807–28. https://doi. org/10.1002/adfm.201803807

21. Tan CL, Mohseni H (2018) Emerging technologies for high performance infrared detectors. Nano 7(1):169–197

22. Sood AK, Zeller JW, Ghuman P, Babu S, Dhar NK, Ganguly S, Ghosh AW, Dupuis RD (2019) Development of high-performance detector technology for UV and IR applications. In: Proceedings of SPIE 11151, sensors, systems, and next-generation satellites XXIII, vol 11151, p 1115113-1-1115113–11. https://doi.org/10.1109/IGARSS.2019.8897813

23. Saxena PK, Chakrabarti P (2008) Analytical simulation of HgCdTe photovoltaic detector for long wavelength infrared (LWIR) applications. Optoelectron Adv Mater Rapid Commun 2(3):140–147

24. Dwivedi ADD (2011) Analytical modeling and atlas simulation of p^+-$Hg_{0.78}Cd_{0.22}Te$/ $nHg_{0.78}Cd_{0.22}Te$/CdZnTe homojunction photodetector for lwir free space optical communication system. J Electron Devices 9:396–404

25. Bansal S, Sharma K, Jain P, Gupta N, Singh AK (2018) Atlas simulation of a long-infrared P^+-N homojunction photodiode. In: 2018 6th edition of international conference on Wireless Networks & Embedded Systems (WECON), Rajpura (near Chandigarh), India, pp 19–22. https://doi.org/10.1109/WECON.2018.8782077

26. Bansal S, Jain P, Kumar N, Kumar S, Sardana N, Gupta N, Singh AK (2018) A highly efficient bilayer graphene HgCdTe heterojunction based p^+-n photodetector for long wavelength infrared (LWIR). In: 2018 IEEE 13th Nanotechnology Materials and Devices Conference (NMDC), Portland, OR, USA, pp 1–4. https://doi.org/10.1109/NMDC.2018.8605848

27. Bansal S, Das A, Jain P, Prakash K, Sharma K, Kumar N, Sardana N, Gupta N, Kumar S, Singh AK (2019) Enhanced optoelectronic properties of bilayer graphene/HgCdTe based single- and dual-junction photodetectors in long infrared regime. IEEE Trans Nanotechnol 18:781–789. https://doi.org/10.1109/TNANO.2019.2931814

28. Song S, Wen L, Chen Q (2015) Graphene composites based photodetectors. In: Sadasivuni K, Ponnamma D, Kim J, Thomas S (eds) Graphene-based polymer nanocomposites in electronics. Springer International Publishing, pp 193–222

29. Bansal S, Sharma K, Soni K, Gupta N, Ghosh K, Singh AK (2017) $Hg_{1-x}Cd_xTe$ based p-i-n IR photodetector for free space optical communication. In: 2017 Progress In Electromagnetics Research Symposium-Spring (PIERS), St Petersburg, Russia, pp 544–547. https://doi. org/10.1109/PIERS.2017.8261800

30. Singh A, Shukla AK, Pal R (2017) Performance of graded bandgap HgCdTe avalanche photodiode. IEEE Trans Electron Devices 64(3):1146–1152. https://doi.org/10.1109/ TED.2017.2650412

31. Shin D, Choi S-H (2018) Graphene-based semiconductor heterostructures for photodetectors. Micromachines 9(7):1–29. https://doi.org/10.3390/mi9070350

32. Asgari A, Razi S (2010) High performances III-nitride quantum dot infrared photodetector operating at room temperature. Opt Express 18(14):14604–14615. https://doi.org/10.1364/ OE.18.014604

33. Hao MR, Yang Y, Zhang S, Shen WZ, Schneider H, Liu HC (2014) Near-room-temperature photon-noise-limited quantum well infrared photodetector. Laser Photonics Rev 8(2):297–302. https://doi.org/10.1002/lpor.201300147

34. Bansal S, Prakash K, Sardana N, Kumar S, Sharma K, Jain P, Gupta N, Singh AK (2019) Bilayer graphene/HgCdTe based self-powered mid-wave IR nBn photodetector. In: 2019 IEEE 14th Nanotechnology Materials and Devices Conference (NMDC), Stockholm, Sweden, pp 1–4. https://doi.org/10.1109/NMDC47361.2019.9083985

35. Martyniuk P (2015) HOT mid-wave HgCdTe nBn and pBp infrared detectors. Opt Quant Electron 47(6):1311–1318. https://doi.org/10.1007/s11082-014-0044-7

36. Craig AP, Thompson MD, Tian Z-B, Krishna S, Krier A, Marshall ARJ (2015) InAsSb-based nBn photodetectors : lattice mismatched growth on GaAs and low- frequency noise performance. Semicond Sci Technol 30(10):105011-1-105011–7. https://doi.org/10.1088/0268-1242/30/10/105011

37. Haddadi A, Dehzangi A, Chevallier R, Adhikary S, Razeghi M (2017) Bias-selectable nBn dual-band long-/very long-wavelength infrared photodetectors based on InAs/InAs$_{1-x}$Sb$_x$/AlAs$_{1-x}$Sb$_x$ type-II superlattices. Sci Rep 7(3339):1–7. https://doi.org/10.1038/s41598-017-03238-2

38. Nguyen TD, Kim JO, Kim YH, Kim ET, Nguyen QL, Lee SJ (2018) Dual-color short-wavelength infrared photodetector based on InGaAsSb/GaSb heterostructure. AIP Adv 8(2):025015-1-025015–7. https://doi.org/10.1063/1.5020532

39. Madejczyk P, Gawron W, Keblowski A, Mlynarczyk K, Stepien D, Martyniuk P, Rogalski A, Rutkowski J, Piotrowski J (2020) Higher operating temperature IR detectors of the MOCVD grown HgCdTe Heterostructures. J Electron Mater:1–10. https://doi.org/10.1007/s11664-020-08369-3

40. Casalino M, Sirleto L, Iodice M, Saffioti N, Gioffr M, Rendina I, Coppola G (2010) Cu/p-Si Schottky barrier-based near infrared photodetector integrated with a silicon-on-insulator waveguide. Appl Phys Lett 96(24):241112-1-241112-1–3. https://doi.org/10.1063/1.3455339

41. Mohammadian M, Saghai HR (2015) Room temperature performance analysis of bilayer graphene terahertz photodetector. Optik-Int J Light Electron Optics 126(11–12):1156–1160. https://doi.org/10.1016/j.ijleo.2015.03.021

42. Corsi C (2010) History highlights and future trends of infrared sensors. J Mod Opt 57(18):1663–1686. https://doi.org/10.1080/09500341003693011

43. Devarakonda V, Dwivedi ADD, Pandey A, Chakrabarti P (2020) Performance analysis of N$^+$-CdTe/n^0-Hg$_{0.824675}$Cd$_{0.175325}$Te/p$^+$-Hg$_{0.824675}$Cd$_{0.175325}$Te n–i–p photodetector operating at 30 μm wavelength for terahertz applications. Opt Quant Electron 52(340):1–19. https://doi.org/10.1007/s11082-020-02450-1

44. Akhavan ND, Umana-Membreno GA, Gu R, Asadnia M, Antoszewski J, Faraone L (2016) Superlattice barrier HgCdTe nBn infrared photodetectors: validation of the effective mass approximation. IEEE Trans Electron Devices 63(12):4811–4818. https://doi.org/10.1109/TED.2016.2614677

45. Kopytko M, Keblowski A, Gawron W, Kowalewski A, Rogalski A (2014) MOCVD grown HgCdTe barrier structures for hot conditions. IEEE Trans Electron Devices 61(11):3803–3807. https://doi.org/10.1109/TED.2014.2359224

46. Akhavan ND, Umana-membreno GA, Gu R, Antoszewski J, Faraone L (2018) Optimization of superlattice barrier HgCdTe nBn infrared photodetectors based on an NEGF approach. IEEE Trans Electron Devices 65(2):591–598. https://doi.org/10.1109/TED.2017.2785827

47. Reine MB (2001) HgCdTe photodiodes for IR detection : a review. Proc SPIE Int Soc Opt Eng 4288:266–277

48. Vasilyev VV, Ovsyuk VN, Sidorov YG (2003) IR photodetectors based on MBE-grown MCT layers. Proc SPIE Int Soc Opt Eng 5065:39–46

49. Chorter P, Tribolet P, Pelletan C (2001) High performance HgCdTe SWIR detectors development at Sofradir. Proc SPIE Int Soc Opt Eng 4369:698–712

50. Piotrowski J, Orman Z, Nowak Z, Pawluczyk J, Pietrzak J, Piotrowski A, Szabra D (2005) Uncooled long wave infrared photodetectors with optimized spectral response at selected spectral ranges. Proc SPIE Int Soc Opt Eng 5783:616–624. https://doi.org/10.1117/12.606244

51. Sood AK, Zeller JW, Ghuman P, Babu S, Dhar NK, Jacobs RN, Chaudhary LS, Efstathiadis H, Ganguly S, Ghosh AW, Ahmed SZ, Tonni FF (2022) Doping and transfer of high mobility graphene bilayers for room temperature mid-wave infrared photodetectors. In: 21st century nanostructured materials – physics, chemistry, classification, and applications in industry and biomedical [Working title]. IntechOpen, London

52. Gawron W, Sobieski J, Manyk T, Kopytko M, Madejczyk P, Rutkowski J (2021) MOCVD grown HgCdTe Heterostructures for medium wave infrared detectors. Coatings 11(5):1–13. https://doi.org/10.3390/coatings11050611

53. Saxena PK, Chakrabarti P (2009) Computer modeling of MWIR single heterojunction photodetector based on mercury cadmium telluride. Infrared Phys Technol 52(5):196–203. https://doi.org/10.1016/j.infrared.2009.07.009

54. Dwivedi ADD (2011) Analytical modeling and numerical simulation of P^+-$Hg_{0.69}Cd_{0.31}Te$/ n-$Hg_{0.78}Cd_{0.22}Te$/CdZnTe heterojunction photodetector for a long-wavelength infrared free space optical communication system. J Appl Phys 110(4):043101-1-043101–10. https://doi.org/10.1063/1.3615967

55. Bellotti E, D'Orsogna D (2006) Numerical analysis of HgCdTe simultaneous two-color photovoltaic infrared detectors. IEEE J Quantum Electron 42(4):418–426. https://doi.org/10.1109/JQE.2006.871555

56. Piotrowski A, Madejczyk P, Gawron W, Kłos K, Pawluczyk J, Rutkowski J, Piotrowski J, Rogalski A (2007) Progress in MOCVD growth of HgCdTe heterostructures for uncooled infrared photodetectors. Infrared Phys Technol 49(3):173–182. https://doi.org/10.1016/j.infrared.2006.06.026

57. Bablich A, Kataria S, Lemme MC (2016) Graphene and two-dimensional materials for optoelectronic applications. Electronics 5(1):1–16. https://doi.org/10.3390/electronics5010013

58. Xia F, Mueller T, Lin YM, Valdes-Garcia A, Avouris P (2009) Ultrafast graphene photodetector. Nat Nanotechnol 4(12):839–843. https://doi.org/10.1038/nnano.2009.292

59. Rogalski A, Kopytko M, Martyniuk P (2019) Two-dimensional infrared and terahertz detectors: outlook and status. Appl Phys Rev 6(2):021316-1-021316–23. https://doi.org/10.1063/1.5088578

60. Boruah BD, Ferry DB, Mukherjee A, Misra A (2015) Few-layer graphene/ZnO nanowires based high performance UV photodetector. Nanotechnology 26(23):235703-1-235703–7. https://doi.org/10.1088/0957-4484/26/23/235703

61. Cheng CC, Zhan JY, Liao YM, Lin TY, Hsieh YP, Chen YF (2016) Self-powered and broadband photodetectors based on graphene/ZnO/silicon triple junctions. Appl Phys Lett 109(5):053501-1-053501–5. https://doi.org/10.1063/1.4960357

62. Awasthi S, Gopinathan PS, Rajanikanth A, Bansal C (2018) Current–voltage characteristics of electrochemically synthesized multi-layer graphene with polyaniline. J Sci Adv Mater Devices 3(1):37–43. https://doi.org/10.1016/j.jsamd.2018.01.003

63. Wan X, Xu Y, Guo H, Shehzad K, Ali A, Liu Y, et al. (2017) A self-powered high-performance graphene/silicon ultraviolet photodetector with ultra-shallow junction: breaking the limit of silicon? npj 2D Mater Appl 1(4):1–8. https://doi.org/10.1038/s41699-017-0008-4

64. Dhar S, Majumder T, Mondal SP (2016) Graphene quantum dot-sensitized ZnO nanorod/ polymer Schottky junction UV detector with superior external quantum efficiency, detectivity, and responsivity. ACS Appl Mater Interfaces 8(46):31822–31831. https://doi.org/10.1021/acsami.6b09766

65. Nie B, Hu JG, Luo LB, Xie C, Zeng LH, Lv P, et al. (2013) Monolayer graphene film on ZnO nanorod array for high-performance schottky junction ultraviolet photodetectors. Small 9(17):2872–2879. https://doi.org/10.1002/smll.201203188

66. Bansal S, Prakash K, Sharma K, Sardana N, Kumar S, Gupta N, Singh AK (2020) A highly efficient bilayer graphene/ZnO/silicon nanowire based heterojunction photodetector with broadband spectral response. Nanotechnology 31(40):405205–1–405205–10. https://doi.org/10.1088/1361-6528/ab9da8

67. Spirito D, Kudera S, Miseikis V, Giansante C, Coletti C, Krahne R (2015) UV light detection from CdS nanocrystal sensitized graphene photodetectors at kHz frequencies. J Phys Chem C 119(42):23859–23864. https://doi.org/10.1021/acs.jpcc.5b07895

68. Gao Z, Jin W, Zhou Y, Dai Y, Yu B, Liu C, Xu W, Li Y, Peng H, Liu Z, Dai L (2013) Self-powered flexible and transparent photovoltaic detectors based on CdSe nanobelt/graphene Schottky junctions. Nanoscale 5(12):5576–5581. https://doi.org/10.1039/c3nr34335a

69. Lin F, Chen SW, Meng J, Tse G, Fu XW, Xu FJ, Shen B, Liao ZM, Yu DP (2014) Graphene/GaN diodes for ultraviolet and visible photodetectors. Appl Phys Lett 105(7):073103-1-073103–5. https://doi.org/10.1063/1.4893609

70. Yu X, Dong Z, Liu Y, Liu T, Tao J, Zeng Y, Yang JKW, Wang QJ (2016) A high performance, visible to mid-infrared photodetector based on graphene nanoribbons passivated with HfO_2. Nanoscale 8(1):327–332. https://doi.org/10.1039/C5NR06869J

71. Luo L-B, Hu H, Wang X-H, Lu R, Zou Y-F, Yu Y-Q, Liang F-X (2015) A graphene/GaAs near-infrared photodetector enabled by interfacial passivation with fast response and high sensitivity. J Mater Chem C 3(18):4723–4728. https://doi.org/10.1039/C5TC00449G

72. Sun Z, Liu Z, Li J, Tai GA, Lau SP, Yan F (2012) Infrared photodetectors based on CVD-grown graphene and PbS quantum dots with ultrahigh responsivity. Adv Mater 24(43):5878–5883. https://doi.org/10.1002/adma.201202220

73. Konstantatos G, Badioli M, Gaudreau L, Osmond J, Bernechea M, De Arquer FPG, Gatti F, Koppens FHL (2012) Hybrid graphene-quantum dot phototransistors with ultrahigh gain. Nat Nanotechnol 7(6):363–368. https://doi.org/10.1038/nnano.2012.60

74. Periyanagounder D, Gnanasekar P, Varadhan P, He JH, Kulandaivel J (2018) High performance, self-powered photodetectors based on a graphene/silicon Schottky junction diode. J Mater Chem C 6(35):9545–9551. https://doi.org/10.1039/c8tc02786b

75. Fan G, Zhu H, Wang K, Wei J, Li X, Shu Q, Guo N, Wu D (2011) Graphene/silicon nanowire Schottky junction for enhanced light harvesting. ACS Appl Mater Interfaces 3(3):721–725. https://doi.org/10.1021/am1010354

76. Tai L, Zhu D, Liu X, Yang T, Wang L, Wang R, Jiang S, Chen Z, Xu Z, Li X (2018) Direct growth of graphene on silicon by metal-free chemical vapor deposition. Nano-Micro Lett 10(20):1–9. https://doi.org/10.1007/s40820-017-0173-1

77. Mueller T, Xia F, Avouris P (2010) Graphene photodetectors for high-speed optical communications. Nat Photonics 4(5):297–301. https://doi.org/10.1038/nphoton.2010.40

78. Gan X, Shiue RJ, Gao Y, Meric I, Heinz TF, Shepard K, Hone J, Assefa S, Englund D (2013) Chip-integrated ultrafast graphene photodetector with high responsivity. Nat Photonics 7(11):883–887. https://doi.org/10.1038/nphoton.2013.253

79. Pospischil A, Humer M, Furchi MM, Bachmann D, Guider R, Fromherz T, Mueller T (2013) CMOS-compatible graphene photodetector covering all optical communication bands. Nat Photonics 7(11):892–896. https://doi.org/10.1038/nphoton.2013.240

80. Rogalski A (2019) Graphene-based materials in the infrared and terahertz detector families: a tutorial. Adv Opt Photon 11(2):314–379. https://doi.org/10.1364/aop.11.000314

81. Liu N, Tian H, Schwartz G, Tok JBH, Ren TL, Bao Z (2014) Large-area, transparent, and flexible infrared photodetector fabricated using P-N junctions formed by N-doping chemical vapor deposition grown graphene. Nano Lett 14(7):3702–3708. https://doi.org/10.1021/nl500443j

82. Ryzhii V, Ryzhii M, Mitin V, Otsuji T (2010) Terahertz and infrared photodetection using p-i-n multiple-graphene-layer structures. J Appl Phys 107(5):054512-1-054512–7. https://doi.org/10.1063/1.3327441

83. Ryzhii M, Otsuji T, Mitin V, Ryzhii V (2011) Characteristics of p-i-n terahertz and infrared photodiodes based on multiple graphene layer structures. Jpn J Appl Phys 50(7):070117-1-070117–6. https://doi.org/10.1143/JJAP.50.070117

84. Pykal M, Jurečka P, Karlický F, Otyepka M (2016) Modelling of graphene functionalization. Phys Chem Chem Phys 18(9):6351–6372. https://doi.org/10.1039/C5CP03599F

85. Zhao S, Xue J (2012) Tuning the band gap of bilayer graphene by ion implantation: insight from computational studies. Phys Rev B Condens Matter Mater Phys 86(16):165428-1-165428–10. https://doi.org/10.1103/PhysRevB.86.165428

86. Sood AK, Zeller JW, Ghuman P, Babu S, Dhar NK, Ganguly S, Ghosh A (2021) Development of high-performance graphene-HgCdTe detector Technology for mid-wave Infrared Applications. In: Proceedings of SPIE 11530, infrared sensors, devices, and applications XI, vol 11530, p 115300I-1-115300I–11. https://doi.org/10.1117/12.2572904

87. Sood AK, Zeller JW, Welser RE, Puri YR, Lewis J, Mto D, Street NR (2015) Development of GaN/AlGaN UVAPDs for ultraviolet sensor applications. Int J Phys Appl 7(1):49–58
88. Vilela MF, Olsson KR, Rybnicek K, Bangs JW, Jones KA, Harris SF, Smith KD, Lofgreen DD (2014) Higher dislocation density of arsenic-doped HgCdTe material. J Electron Mater 43(8):3018–3024. https://doi.org/10.1007/s11664-014-3180-8
89. Xu W, Gong Y, Liu L, Qin H, Shi Y (2011) Can graphene make better HgCdTe infrared detectors? Nanoscale Res Lett 6(1):250. https://doi.org/10.1186/1556-276X-6-250
90. Patel K, Tyagi PK (2017) P-type multilayer graphene as a highly efficient transparent conducting electrode in silicon heterojunction solar cells. Carbon 116:744–752. https://doi.org/10.1016/j.carbon.2017.02.042
91. McCann E, Koshino M (2013) The electronic properties of bilayer graphene. Rep Prog Phys 76(5):056503-1-056503–28. https://doi.org/10.1088/0034-4885/76/5/056503
92. Schmitz M, Engels S, Banszerus L, Watanabe K, Taniguchi T, Stampfer C, Beschoten B (2017) High mobility dry-transferred CVD bilayer graphene. Appl Phys Lett 110(26):263110–1–263110–5. https://doi.org/10.1063/1.4990390
93. Liu D, Lin C, Zhou S, Hu X (2016) Ohmic contact of Au/Mo on $Hg_{1-x}Cd_xTe$. J Electron Mater 45(6):2802–2807. https://doi.org/10.1007/s11664-016-4375-y
94. Song SM, Park JK, Sul OJ, Cho BJ (2012) Determination of work function of graphene under a metal electrode and its role in contact resistance. Nano Lett 12(8):3887–3892. https://doi.org/10.1021/nl300266p
95. Suhail A, Pan G, Jenkins D, Islam K (2018) Improved efficiency of graphene/Si Schottky junction solar cell based on back contact structure and DUV treatment. Carbon 129:520–526. https://doi.org/10.1016/J.CARBON.2017.12.053
96. Chu J, Mi Z, Tang D (1992) Band-to-band optical absorption in narrow-gap $Hg_{1-x}Cd_xTe$ semiconductors. J Appl Phys 71(8):3955–3961. https://doi.org/10.1063/1.350867
97. Chu J, Li B, Liu K, Tang D (1994) Empirical rule of intrinsic absorption spectroscopy in $Hg_{1-x}Cd_xTe$. J Appl Phys 75(2):1234–1235. https://doi.org/10.1063/1.356464
98. Liu K, Chu JH, Tang DY (1994) Composition and temperature dependence of the refractive index in $Hg_{1-x}Cd_xTe$. J Appl Phys 75(8):4176–4179. https://doi.org/10.1063/1.356001
99. Zhang BY, Liu T, Meng B, Li X, Liang G, Hu X, Wang QJ (2013) Broadband high photoresponse from pure monolayer graphene photodetector. Nat Commun 4(1811):1–11. https://doi.org/10.1038/ncomms2830
100. Tielrooij KJ, Song JCW, Jensen SA, Centeno A, Pesquera A, Zurutuza Elorza A, Bonn M, Levitov LS, Koppens FHL (2013) Photoexcitation cascade and multiple hot-carrier generation in graphene. Nat Phys 9(4):248–252. https://doi.org/10.1038/nphys2564
101. Lee YK, Choi H, Lee H, Lee C, Choi JS, Choi CG, Hwang E, Park JY (2016) Hot carrier multiplication on graphene/TiO_2 Schottky nanodiodes. Sci Rep 6(27549):1–9. https://doi.org/10.1038/srep27549
102. Ploetzing T, Winzer T, Malic E, Neumaier D, Knorr A, Kurz H (2014) Experimental verification of carrier multiplication in graphene. Nano Lett 14(9):5371–5375. https://doi.org/10.1021/nl502114w
103. Johannsen JC, Ulstrup S, Crepaldi A, Cilento F, Zacchigna M, Miwa JA, et al. (2015) Tunable carrier multiplication and cooling in graphene. Nano Lett 15(1):326–331. https://doi.org/10.1021/nl503614v
104. Kadi F, Winzer T, Knorr A, Malic E (2015) Impact of doping on the carrier dynamics in graphene. Sci Rep 5(16841):1–7. https://doi.org/10.1038/srep16841
105. Winzer T, Knorr A, Malic E (2010) Carrier multiplication in graphene. Nano Lett 10(12):4839–4843. https://doi.org/10.1021/nl1024485
106. Ganguly S, Tonni FF, Ahmed SZ, Ghuman P, Babu S, Dhar NK, Sood AK (2021) Dissipative quantum transport study of a bi-layer graphene-CdTe-HgCdTe Heterostructure for MWIR photodetector. In: IEEE research and applications of photonics in defense conference (RAPID), pp 1–2. https://doi.org/10.1109/RAPID51799.2021.9521427

107. Ahmed SZ, Tonni FF, Ganguly S, Ghosh AW, Ghuman P, Babu S, Dhar NK, Sood AK (2020) Using novel properties of graphene for designing efficient infrared photodetectors. In: Graphene & 2D materials international conference and exhibition, p 34
108. ATLAS user's manual version 5.20.2.R, SILVACO International, Santa Clara, CA, USA. 2016
109. Rogalski A (2010) Infrared detectors, 2nd edn. CRC Press
110. Dwivedi ADD, Pranav A, Gupta G, Chakrabarti P (2015) Numerical simulation of HgCdTe based simultaneous MWIR/LWIR photodetector for free space optical communication. Int J Adv Appl Phys Res 2(1):37–45. https://doi.org/10.15379/2408-977X.2015.02.01.5
111. Ancona MG (2010) Electron transport in graphene from a diffusion-drift perspective. IEEE Trans Electron Devices 57(3):681–689. https://doi.org/10.1109/TED.2009.2038644
112. Dwivedi ADD, Chakrabarti P (2007) Modeling and analysis of photoconductive detectors based on $Hg_{1-x}Cd_xTe$ for free space optical communication. Opt Quant Electron 39(8):627–641. https://doi.org/10.1007/s11082-007-9122-4
113. Bruna M, Borini S (2009) Optical constants of graphene layers in the visible range. Appl Phys Lett 94(3):031901-1-031901–3. https://doi.org/10.1063/1.3073717

Part II
II–VI Semiconductors–Based Detectors for Visible and UV Spectral Regions

Chapter 9
CdTe-Based Photodetectors and Solar Cells

Alessio Bosio

9.1 Introduction

Cadmium telluride has been a known compound since the mid-1800s. In 1879 it was prepared by the French chemist Margottet by making Te react with metals at red-heat. In 1888 the enthalpy of formation of the compound was obtained and this favored the production of the material in crystalline form [1]. The ease of preparation has been a known feature of CdTe from the very beginning and the basic method of preparation has changed little to date.

For a long time, the interest for CdTe was purely academic and the only reported use was as a dye, but in 1946 a publication appeared in which the great photosensitivity of CdTe was highlighted, especially towards the β and γ radiation [2]. The material was readily proposed as a γ-ray detector if coupled with suitable amplifiers (scintillators). In the 1950s, photoelectric cells, based on CdTe were already investigated [3]. These studies highlighted that the Cd and Te stoichiometric excess had great influence in the operation of these devices [4]. In particular, the heat treatments carried out after the material growth assumed great importance and several detailed investigations into the behavior of CdTe single crystal were made. During the same years, the role of dopants was clarified and explained [5] in an organic model, opening to the opportunity to produce new devices such for example as photodiodes, photoconductive sensors, infrared windows, image intensifiers, camera

---

This chapter is dedicated to my wife Eugenia and our sons Marco and Silvia for all their forbearance during its writing.

A. Bosio (✉)
Department of Mathematical, Physical and Computer Sciences, University of Parma v.le delle Scienze, Parma, Italy
e-mail: alessio.bosio@unipr.it

© The Author(s), under exclusive license to Springer Nature Switzerland AG 2023   205
G. Korotcenkov (ed.), *Handbook of II-VI Semiconductor-Based Sensors and Radiation Detectors*, https://doi.org/10.1007/978-3-031-20510-1_9

sensors, photovoltaic cells, X-ray dosimeters, γ-ray (with scintillators) and ultraviolet-visible (Uv-Vis) detectors, etc...

Since only some applications found great interest, in this chapter we will describe only some of these, leaving out the detectors sensitive outside the visible light range, which will be broadly described in another chapter (IR, X- and γ-ray), focusing on some special uses/devices and especially, on photodetectors exploiting the photo-voltaic effect (solar cells).

Effectively, considering its direct forbidden band, this material is considered very suitable for the manufacture of solar cells. Because of the direct energy gap, the absorption edge is very sharp allowing more than 90% of the incident light to be absorbed in a thickness of a few micrometers. Moreover, its direct bandgap of 1.5 eV, at room temperature, perfectly matches the requirement for highly efficient sunlight energy conversion [6]. The maximum photocurrent that can be produced by a CdTe-based solar cell under a standard global spectrum light with a power density of 100 mW/cm^2 is 30.5 mA/cm^2, allowing a theoretically predicted maximum pho-tovoltaic conversion efficiency (PCE) of around 32% in the Shockley–Queisser limit and 30.5% if reflectance is considered [7].

Despite these remarkable properties, it was quickly realized that high conversion efficiencies were limited by intrinsic defects that formed in II-VI heterojunctions. Not surprisingly, in the early 1980s efficiencies of around 10% were achieved in homo- and hetero-junction-based devices using CdTe single crystal [8–11]. In fact, the higher reported efficiency of 13.4% concerns a n-ITO/p-CdTe single crystal buried homojunction [12]. On the contrary, all the II-VI heterojunctions show appreciable lattice mismatch (except for n-CdSe/p-CdTe); in particular, for the n-CdS/p-CdTe system the lattice mismatch is around 10%. The defects introduced by the lattice mismatch were responsible for the recombination losses associated with the junction interface and the conversion efficiency was limited to a value not exceeding 10%. However, it was not clear why the homojunctions behaved like the heterojunctions, not having the problem of the lattice mismatch and therefore the associated defects.

An initial explication to this remark was given considering the severe difficulties to obtain a low-resistance ohmic contacts to p-CdTe and this problem was common to every device based on this material. In the same years, the scientific community observed, not without surprise, that solar cells based on n-CdS/p-CdTe heterojunc-tions were made by means of the thin-film technology [13]. Soon, this technology took over, achieving excellent results in terms of conversion efficiency. In fact, the 10% efficiency value was overcome in 1982 [14] followed by an efficiency of 15.8% in 1993 [15]. In 2001, the National Renewable Energy Laboratory (NREL) reported an efficiency of 16.5% [16]. Since then, other notable advances have been made and in 2016 a solar cell based on a CdTe$_{(1-x)}$(S, Se)$_x$ thin film exhibited a world effi-ciency record of (22.1 ± 0.5) % [17]. Nowadays, commercial modules with an effi-ciency of (19 ± 0.9) % are available on the market [18].

Since after several years, a convincing explanation of many of the outstanding points has been given, in the following, we will report the most important clarifica-tions to involve the reader into the so varied and apparently contradictory world of the CdTe-based devices. So, we will discover together how the problem of ohmic contact on the *p*-type CdTe was worked out, how the grain boundaries in polycrystalline thin films were passivated and how the defects formed in the heterojunctions were made harmless, allowing to reach very high efficiencies comparable to the higher ones obtained with single- and/or multi-crystalline silicon [18].

9.2 Noteworthy Applications

9.2.1 Infrared Window

Considering a prohibited energy gap of 1.5 eV, this material was immediately considered particularly suitable as a window material for infrared radiation. In the 1960s there was a great demand for transparent materials in the IR, which allowed the development of CO_2 power lasers. The first system made available on the market was based on vacuum pressed CdTe powder (Irtran 6 by Kodak), but it was only with the advent of the monocrystalline material that a great boost took place. CdTe single crystals, grown by Bridgman method, exhibited a very flat transmittance spectrum up to 40 μm (see Fig. 9.1), which is mandatory for use as window/modulator in high power CO_2 lasers, considering the CdTe low thermal conductivity [19] as well as low mechanical strength. However, it is quite clear, that the extremely flat transmission (up to ≈ 40 μm) is not easily matched with any other material with similar energy gap (GaAs), chemical bond or lattice structure.

Thanks to the great demand of IR windows for military use, in the 80–90s the CdTe production was gradually replaced by two other materials of the II-VI family:

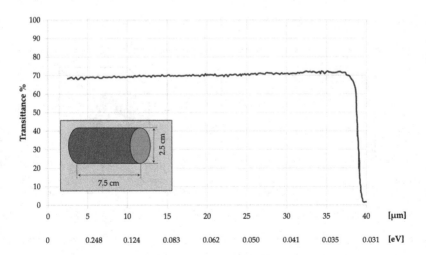

Fig. 9.1 Optical transmission of a CdTe polished single crystal grown by Bridgman method between 2.5 and 40 μm. In the inset a sketch of the crystal ingot without any anti-reflecting coating (ARC). Parallelism: 3 arcmin; Flatness: 2 waves at 633 nm

ZnSe and ZnS. These two semiconductors, although not having the same characteristics as CdTe, have been widely used, since it is possible to grow them in form of single crystals with the chemical vapor deposition (CVD) technique [20]. This technique allows to obtain crystals of excellent quality with growth rates much higher than those typical of the Bridgman method. Today, polycrystalline CdTe IR windows are still available when it is necessary to have good transmittance in the far infrared and, at the same time, to be completely opaque in the visible region of the solar spectrum (solar blind feature). This is a characteristic of CdTe which, together with the transparency width in the IR spectrum, is not reached by any other semiconductor currently available on the market. On the contrary, the high cost places the use of this material only in niche applications [21].

9.2.2 Electro-Optical Modulator

In order to exploit the excellent electro-optical characteristics of CdTe, both in form of bulk crystalline material or as an epitaxial thin film, the material must be free of any segregation, with a limited concentration of impurities and with very little crystalline stress. Under these conditions there aren't any phenomena of birefringence, scattering or absorption. Furthermore, CdTe crystallizes in the zincblende phase described by the space group $\bar{4}3m$; this group highlights that no center of symmetry is present, allowing the CdTe system to exhibit the linear electro-optic effect (Pockels effect). By applying an electric field through the CdTe crystal, it is observed a variation in the refractive index with a linear behavior with respect to the applied voltage (see Fig. 9.2).

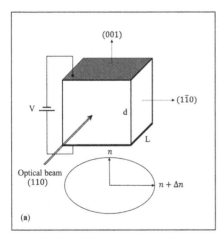 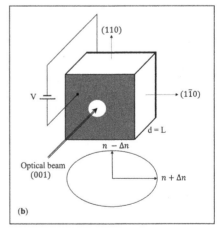

Fig. 9.2 Schematic of the Pockels effect in cubic non-centrosymmetric media (CdTe). (**a**) transverse geometry; (**b**) longitudinal geometry; in both cases the electric field $\vec{E}$ is parallel to the (001) direction of the cubic crystal

This anisotropic variation is typically described by a tensor, but in case of the $\overline{4}3m$ symmetry, the description of the electro-optical properties can be reduced to the r_{41} tensorial element only. As a consequence of the change of the refractive index, a phase modification of the electromagnetic wave, propagating inside the crystal, is induced by the applied electrical potential.

The induced phase change Δ (λ), as a function of the wavelength, in terms of the applied voltage V is (Eq. 9.1):

$$\Delta(\lambda) = \frac{2\pi n^3(\lambda) r_{41}(\lambda) VL}{\lambda d},$$

(9.1)

where λ, $n(\lambda)$, $r_{41}(\lambda)$, V, L and d, are the wavelength, the refraction index, the electro-optical coefficient, the applied bias, the crystal length along the direction of light propagation, the distance between the electrical contacts respectively. Therefore, a CdTe-based modulator is well-described in terms of $n^3(\lambda) \cdot r_{41}(\lambda)$ and, considering $\lambda = 10.6$ µm, we have: $n^3(10.6) \cdot r_{41}(10.6) = 1.2 \cdot 10^{-11}$ m/V. Frequently, modulators are specified in terms of the voltage acquired to achieve a phase change $\Delta(\lambda) = \pi$ corresponding to a maximum retardance of a half-wave. The voltage producing a π phase shift is given by Eq. (9.2):

$$V_\pi = \frac{\lambda d}{2n^3(\lambda) r_{41}(\lambda) L},$$

(9.2)

At a fixed $\lambda = 10.6$ µm and d/L = 1 (i.e., the distance between contacts is equal to the light path length) $V_\pi \approx 53$ kV, while for low power modulators d = $2.5 \cdot 10^{-2}$ cm, L = 6 cm (d/L = $4 \cdot 10^{-3}$) and $V_\pi \approx 200$ V, which is a very intriguing result from the device point of view. For this reason, CdTe can be considered a good candidate for the assembly of electro-optical modulators (Pockels cells). Nowadays, very efficient Pockels cells are made with a lot of different non-linear materials (LiNbO$_3$, LiTaO$_3$, NH$_4$H$_2$PO$_4$ etc....), but in mid-infrared applications, CdTe is the best suited material, thanks to its very low absorption coefficient from the band edge at $\approx$ 0.83 µm up to $\approx$ 35 µm (see Table 9.1). In this wavelength range the optical absorption coefficient of CdTe is lower than that of many other materials, since the fundamental lattice vibration located at $\approx$ 70 µm ($\approx$ 141 cm^{-1}) is sufficiently far away from the wavelength of interest ($\approx$ 40 µm). The transmittance in the mid-infrared region is also due to the low reflectivity since the reststrahlen band is sufficiently far from the upper limit of the wavelength range. Moreover, anti-reflecting coatings that give 99% transmission are possible in this wavelength range.

At high light power CdTe crystals exhibit a non-linear behavior, which occurs in generation of harmonics and frequency intermixing; the driving parameter is the non-linear susceptibility coefficient. In this case, the polarization density $\vec{P}$ becomes a function of the square of the electric field $\vec{E}$:

$$\vec{P}(\omega) = \varepsilon_0(\chi^{(1)}\vec{E}(\omega) + \chi^{(2)}\vec{E}^2(\omega) + ...,$$

(9.3)

Table 9.1 Electrooptical properties of some different materials; the values of the linear electrooptical coefficient and of the refractive index are referred to the specified wavelength

Material	Symmetry (system)	λ [μm]	Δλ [μm]	$r_{i,j}$ [pm/V]	n	References
SiO₂ Silicon dioxide	32 (trigonal)	0.633	0.19–2.30	r_{11}=0.20 r_{41}=0.93	1.542	[22]
GaAs Gallium arsenide	$\bar{4}$3 m (cubic)	1.150	1.00–11.0	r_{41}=1.51	3.43	[23]
CdTe Cadmium telluride	$\bar{4}$3 m (cubic)	10.60	0.83–35.0	r_{41}=6.80	2.6	[24]
KH₂PO₄ (KDP) Potassium dihydrogen phosphate	$\bar{4}$2 m (tetragonal)	0.633	0.20–1.50	r_{41}=8.60 r_{63}=10.6	1.507	[25]
KD₂PO₄ (KD*P) Potassium dideuterium phosphate	$\bar{4}$2 m (tetragonal)	0.633	0.20–2.10	r_{41}=8.80 r_{63}=26.4	1.493	[26]
NH₄H₂PO₄ (ADP) Ammonium dihydrogen phosphate	$\bar{4}$2 m (tetragonal)	0.633	0.18–1.53	r_{41}=8.50 r_{63}=24.1	1.521	[27]
LiNbO₃ Lithium niobate	3 m (trigonal)	0.633	0.33–4.50	r_{13}=8.60 r_{22}=3.40 r_{33}=30.8 r_{51}=28.0	2.286	[28]
LiTaO₃ Lithium tantalate	3 m (trigonal)	0.633	0.28–4.00	r_{13}=8.20 r_{22}=0.50 r_{33}=35.0 r_{15}=20.0	2.176	[29]
CdS Cadmium sulphide	6 mm (hexagonal)	10.60	0.60–14	r_{13}=2.45 r_{33}=2.75 r_{42}=1.70	2.226	[23]
LiIO₃ Lithium iodate	6 (hexagonal)	0.633	0.30–5.5	r_{13}=4.10 r_{23}=6.40 r_{41}=1.40 r_{51}=3.30	1.883	[30]

Abbreviations: n refractive index, $r_{i,j}$ linear electrooptical coefficient, λ wavelength, Δλ transparency range

stopping the expression for $\vec{P}(\omega)$ to the second order. The quantity $\chi^{(2)}$ is the so-called second-order non-linear optical susceptibility and ε_0 is the permittivity of free space. In particular, $\chi^{(2)}$ is a second order tensor, which could be greatly simplified by means of the crystal symmetry. CdTe crystals, at a wavelength of 10.6 μm (0.117 eV), exhibit a non-linear susceptibility coefficient $\chi^{(2)} = 1.7 \times 10^{-10} m/V$ [31] that is one of the greater values among the harmonic generator materials, comparable with $\chi^{(2)} = 1.88 \times 10^{-10} m/V$ typical value for GaAs. From the symmetry of the CdTe crystal, the most efficient second harmonic generation occurs for the input electromagnetic wave along (110) with the electric field polarized in (110).

It is worth remembering here, that the power of the second harmonic (at frequency 2ω) is proportional to the square of the non-linear susceptibility coefficient

$\chi^{(2)}$ as well as to the square of the power density of the incident light beam (at a frequency ω). For example, KDP, KD*P and ADP exhibit a non-linear susceptibility coefficient of about $(0.5 \div 0.6) \times 10^{-10} m/V$ [32]. With the same incident light power, by using CdTe or GaAs crystals as second harmonic generators, a power density 10 times greater, could obtained.

9.2.3   UV-Vis-Photodetector

Many CdTe-based devices are realized using bulk crystals, but where it is important to exploit large areas, such as in solar cells, the preferred technology is thin film. Furthermore, low-dimensional semiconductor nanostructures, such as quantum dots (QDs), nanowires (NWs), nanorods (NRs), nanotubes (NTs) and nanobelts (NBs) [33–37] are considered the most sensitive and fast responsive materials for photon sensors due to their large surface-to-volume ratio and direct pathway for charge transport. Nanostructures are very attractive due to their unique optical properties, which are generated by the quantum confinement effect. In fact, by modifying the size of the nanoparticles (NPs) it is possible to change the charge carrier confinement while band gap engineering can be achieved by changing the distance between the NPs. CdTe nanostructures have been synthetized via many techniques, including physical vapor deposition (PVD), close-spaced deposition (CSS), chemical vapor deposition (CVD), molecular beam epitaxy (MBE) and RF magnetron sputtering (RFMS) [38–42].

CdTe-based photodetectors have been made with different structures such as Schottky barriers, p-n junctions, pure photoconductive metal-semiconductor-metal devices and field-effect transistors (FETs). Among these deposition techniques and device configurations it is hard to find the best combination, since each of them has its advantages, but also some drawbacks. In fact, for a high-performance photodetector a proper spectral range, high-signal-noise ratio, high responsivity and fast response are important requirements, but the most characterizing parameter is undoubtedly the photo-gain, defined as the number of photogenerated carriers per incident photon collected at the external contacts. This is the case of CdTe-based photoconductive detectors which are highly sensitive thanks to the presence of deep traps in the semiconductor layer, which causes a long recombination lifetime τ for one type of charge carriers. As it is well explained by Dan et al. in [43] the photo-gain depends on the ratio between the charge trapping lifetime τ and the transit time τ_t and a long recombination lifetime leads to a long response time, which restrains the photodetector application.

It's a matter of fact that a high photo-gain is coupled with a low response speed and a good photodetector, depending on its use, is the result of the best compromise between these two parameters. From this point of view, nanostructures have an advantage over bulk materials or thin films since, at the nanoscale, it is virtually easier to extract the photogenerated carriers, to obtain large photogain and contemporarily fast response time, defined as the rise time from dark current to 90% of the maximum current. Effectively, CdTe-based nanowires and nanoribbons devices show response time on the order of 1 and 3.3 s [44, 45], respectively and a photogain in the range of 10^3–10^4, while CdTe thin film photodetector exhibits average response times, on the order of tenth of seconds. So long response times are generally due to the detrapping lifetime of the deep traps. It has been proposed to remove the deep traps keeping the shallow traps could maintain the high gain and, at the same time, a fast response speed. In fact, the shallow traps can be thermally activated, promoting the release of the charge carriers more quickly. As an example of this technique, a CdTe QDs-based photodetector capped with poly(3-hexylthiophene) (P3HT), with enhanced behavior through trap engineering is reported in [46]. In that case the Cd_i^{2+} deep traps (trap depth of 0.64 eV), on the surface of CdTe QDs, were passivated by P3HT coordination, while shallow traps can still activate a high photogain of 50 and a short response time of 2 μs. For the same device, without any surface passivation of the QDs, the exhibited response time was about 1–2 s. CdTe microwire (μW)-based ultraviolet photodetectors reported in [47], not having any passivation of the deep-traps, showed a pohotogain of ≈ 50 and an average response time of about 7.7 s.

In [45] a CdTe-based Field Effect Transistor, by using single-crystalline NW, was realized. Exploiting the good properties of the CdTe single crystal an average photogain of 250 and a response time of 0,7 s were obtained. Despite an excellent crystalline quality, the photoresponse was just sufficient, indicating that the performance of this device could be enhanced if a deep traps engineering will be adopted.

Other photodetector structures are realized for obtaining high response speed, as in the case of CdS-CdS$_x$Te$_{1-x}$-CdTe core-shell NB detector [48]. This promising device exploits the built-in potential which is formed at the interface between the three materials. In CdTe- and CdS-based NB detectors, photoresponse is principally ruled by defects heavily influenced by absorption and desorption of oxygen atoms on the surface of the nanoparticles, in the above-mentioned core-shell NB-based devices, the photogenerated carriers are quickly separated by the built-in potential, naturally formed at the CdS-CdS$_x$Te$_{1-x}$ and CdS$_x$Te$_{1-x}$-CdTe interfaces. The absorption/desorption of atoms, on the surface of the nanoparticles is very slow, if compared with the separation of charge carriers due to the interface potential. As a result, a response time of the order of 11 μs for the rise time and 23 μs for the fall time is obtained. For comparison, CdS NB photodetector exhibits typical response time of 16 and 367 μs respectively. By the way, this core-shell NB detector shows an excellent photogain of about 4.7×10^3 mainly due to the simultaneous absorption of photons, with energy larger than the energy gap of CdTe, by all the three constituent materials. Table 9.2 summarizes important results obtained by some photodetector described above.

Table 9.2 Typical parameters of different CdTe-based photodetectors. All the reported values are optimized by varying wavelengths and voltages as indicated in the reported references

Photodetector	Responsivity [A/W]	Photogain	Photoresponse rise/ decay time [s]	Spectral range [nm]	References
CdTe – NR	780	2400	1.1/3.3	400–800	[44]
CdTe – NW FET	80.1	250	0.7/1	400–	[45]
CdTe – QD P3HT-capped	13.6	50	$0.2 \cdot 10^{-6}$/–	390–	[46]
CdTe – µW	–	50	7.7/0.06	365–	[47]
CdTe core-shell NB	1520	4700	$11 \cdot 10^{-6}$/$23 \cdot 10^{-6}$	355–785	[48]

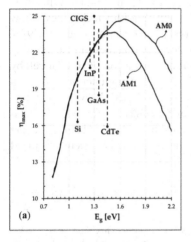
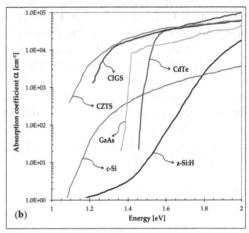

Fig. 9.3 (a) maximum efficiency (η_{max}) vs energy gap (E_g) for two different atmosphere absorption; AM 1 = sea level, sun at zenith; AM 0 = outside the earth atmosphere; AM = air mass. The vertical dashed lines indicate different semiconductors industrially used as absorbers in solar cells. (Adapted from Ref. [49]. With the permission of AIP Publishing). (b) The absorption coefficient α of different semiconductors as a function of energy

9.3 Solar Cells

CdTe, the most commercially successful TF technology, finds its fortune in some physical and chemical peculiarities:

1. direct energy band gap of 1.45 eV near the maximum of the solar spectrum (Fig. 9.3a) [49];
2. absorption coefficient in the visible part of the solar spectrum in the range ($10^4 \div 10^5$) cm^{-1}, which means that 1 µm thick layer is sufficient to capture all visible light (Fig. 9.3b);
3. it shows a high enthalpy of formation (100 kJ mol^{-1}), which means great thermodynamic stability;

4. it sublimates and evaporates congruently by means of the equilibrium reaction $CdTe \leftrightarrow Cd + \frac{1}{2}Te_2$. The high stability and congruent evaporation allow growth by very different preparation methods;
5. it naturally grows with inherent stoichiometry defects making it moderately p-type;
6. post-growth treatment reduces defects, increases crystalline quality by making grain boundaries electrically inactive.

Historically, the best performance has been achieved with heterojunctions where the n-type partner was cadmium sulfide (CdS). A few different attempts have been made when p-type CdTe single crystals were paired with In_2O_3 [12], ZnO [9] or a very thin n-type CdTe layer [11], obtaining a 13.8% maximum efficiency. Surprisingly, CdTe-based solar cells, produced using thin-film technology, show higher efficiencies than those made using monocrystalline materials (see Table 9.3).

It is a matter of fact that the success of this material was achieved by exploiting one of its best features, namely the possibility of producing a complete solar cell by using the thin film technology. Total thin-film CdTe/CdS heterogiunction resulted in 6% conversion efficiency as it has been known since 1972 [13]. However, the psychological limit of 10% efficiency was exceeded in the 1980s only after the application of a heat treatment in a chlorine atmosphere to the stacked CdTe/CdS layers [14]. In the following 10 years, devices with efficiency close to 17% were optimized by developing new front and back contacts [16, 50].

Table 9.3 Representative data for single crystal and all thin film CdTe-based solar cells

Type of cell	Open-circuit voltage Voc [V]	Short-circuit current density [mA/cm^2]	Energy conversion efficiency, h [%]	References
CdTe single crystal				
Buried homojunction: n-ITO/p-CdTe	890	20.0[a]	13.4	[12]
Heterojunction: n-ZnO/p-CdTe	540	19.5[a]	8.8	[9]
CdTe homojunction: p-CdTe(CSVT)/n-CdTe single crystal	820	21.0[a]	10.7	[11]
Thin films				
All thin film CdTe-based cells			6.0	[13]
CdS and CdTe (low T-CSS)	750	17.0[b]	10.5	[14]
CdS (CBD) – CdTe (high T-CSS)	843	25.1[a]	15.8	[15]
CdS (CBD) – CdTe (low T-CSS)	845	25.9[a]	16.5	[16]

CSVT Close-Spaced Vapor Transport, *CSS* Close-Spaced Sublimation
[a]Under simulated AM 1.5 solar illumination at 100 mW/cm^2
[b]Under simulated AM 2 solar illumination at 75 mW/cm^2

These noceably succcess was followed by a period in which many researchers have dedicated their activities to the technology transfer from laboratory to industrial scale of the production processes, slowing down the achievement of ever higher device efficiencies. It was increasingly realized that some difficulties lay mainly in the stability over time of the back contact and in the real impossibility of extrinsically doping the polycrystalline thin films of CdTe. Around 2010, PCE started to increase again, quickly reaching values close to 20%. The continuous increase in conversion efficiency of that devices, was principally due to the optimization of the anti-reflective coating (ARC), the transparent electrical contact and the window layer, in addition to a careful choice of the glass substrate. This resulted in a considerable increase in photocurrent, going from 26.1 mA/cm² for a cell with 17.6% efficiency [16] to a photocurrent of 28.59 mA/cm² for a cell with an efficiency of 19.6% corresponding only to a slightly increase in photovoltage [51]. Unfortunately, there are no details in the literature regarding these impressive results. Nevertheless, it immediately appeared evident that the increase in photocurrent is not only due to an excellent optimization of light collection, but also to an accurate management of the energy gap of the CdTe to extend the absorption of light to longer wavelengths.

CdTe$_{(1-x)}$(S,Se)$_x$ compounds show energy bandgap values lower than CdTe when X ≤ 0.05, corresponding to a maximum increase in the cut-off wavelength of approximately 15 nm. The corresponding increase in the photocurrent can be estimated at 1 mA/cm². Furthermore, the use of CdTe$_{(1-x)}$(S,Se)$_x$ alloy, by reducing the lattice mismatch between the window and the absorber materials in the region of the metallurgical junction, results in improved charge transport properties since fewer recombination centers and killer levels due to interface states are present.

In 2015, a solar cell with a thin CdTe$_{(1-x)}$(S,Se)$_x$ intermixed layer in the junction region showed a world record efficiency of (22.1 ± 0.5)%, exhibiting the following parameters measured with the AM1.5G spectrum (1000 W/m²) at 25 °C: V_{oc} = 0.8872 V, I_{sc} = 31.69 mA/cm², fill factor = 0.785 over a designated exposed area of 0.4798 cm² [51, 52]. Nowadays, the most used and accepted architecture of the CdTe thin film solar cell is sketched in Fig. 9.4.

The electrical in-series connections of adjacent cells, monolithically integrated inside the production process, by means of a robotic laser scribing, made possible a fully automated large-scale in-line production process. In 2002, by exploiting the electrodeposition technique for CdTe-deposition and laser scribing for the elecrical in-serie connections a 11% efficient module was realized [53]. In 2010–2011, the number of factories able to produce tens of megawatts/years of CdTe-based modules were about 10 units. The CdTe films deposition was mainly based on the close-spaced vapor transport (CSVT) or close-spaced sublimation (CSS) techniques and large-area modules, exhibiting efficiency in the 10% to 12% range was obtained [54]. In the following years, a 14.4% efficient device was reported, which became 16.1% at the beginning of 2013. Immediately after (2014), a 17.5% efficient module was obtained, followed by a world efficiency record of (18.6 ± 0.5)% in 2015. The photovoltaic parameters of such a module, taken under the global AM 1.5 spectrum (1000 W/m²) at a cell temperature of 25 °C, are: V_{oc} = 110.6 V, I_{sc} = 1.533 A and fill factor = 0.742 over a designated illumination area of 7038 cm² [51]. Nowadays,

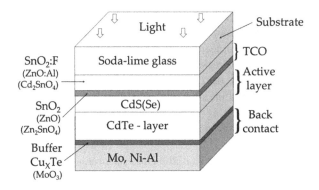

Fig. 9.4 The CdTe/CdS solar cell in superstrate configuration (light enters through the substrate). (Reprinted from "A. Bosio, S. Pasini, N. Romeo, The History of Photovoltaics with Emphasis on CdTe Solar Cells and Modules, Coatings 2020, 10, 344". Published 2020 by MDPI as open access)

Table 9.4 Photovoltaic Conversion Efficiency (PCE) of the best solar cells and commercial modules collected for different technologies

Parameter	Solar cells			Modules		
	PCE, %	Area, cm^2	Producer	PCE, %	Area, cm^2	Producer
CdTe	22.1	0.4798*	First solar	19.0	23,573*	First solar
CIGS	23.35	1043*	Solar Frontier	18.6	10,858**	Miasolé
CIS	15.4	100.0***	IPE	–	–	–
CZTS	12.6	0.4209**	IBM	–	–	–
Cu$_2$S	10.0	≈1.0	IEC	–	–	–
GaAs 6-J	47.1§	0.099	NREL	–	–	–
GaAs 3-J	37.9	1047**	Sharp	31.2	968*	Sharp
GaAs SJ	32.8	1000**	LG electronics	–	–	–
Si single- crystal	26.7	79*	Kaneka	24.4	13,177*	Kaneka
Si multi-crystal	23.2	247.79***	Trina solar	20.4	14,818**	Hanwa Q CellsH

The confirmed PCE data are measured under the global AM 1.5 spectrum (1000 W/m^2) at 25 °C (IEC 60904–3: 2008, ASTM G-173-03 global)
*(da) = designated illumination area; ** (ap) = aperture area; *** (t) = total area; § 6-J = six-junction with concentrator (143X)

commercial modules with an efficiency close to 19.0% are commonly available on the market [55].

In today's photovoltaic panorama, the technology based on CdTe shows conversion efficiencies comparable to those of CIGS and of both single- and poly-crystalline Si, the most used material in photovoltaic modules production. An overview of the best performing solar cells and commercial modules is shown in Table 9.4.

A CdTe-based solar cell is typically made with very thin overlapping stacked layers, arranged to form a high-quality heterojunction. When high-temperature deposition techniques are used, generally *p*-type CdTe thin films are naturally obtained, forcing to select an *n*-type partner to make the *p-n* junction. The high-efficiency CdTe-based solar cells are manufactured in a superstrate configuration, which means that light passes through the substrate and the front contact is as transparent and conductive as possible. On the other hand, the back-contact, which is generally opaque, must ensure the ohmicity with *p*-type CdTe to efficiently collect all the photogenerated carriers.

To date, the most widely used system is the CdS/CdTe heterojunction, where CdS constitutes the "window", and CdTe is the "absorber". A pressing demand for this device is that the window layer must be as transparent as possible, while the absorber layer must be thick enough to fully absorb visible light. A thickness of a few microns is enough for CdTe, while thin films up to a few tens of nm are used for CdS. At the operating temperature, the concentration of free carriers of both CdS and CdTe films ensures that the electric field falls mainly into the absorber material, so that all the photogenerated electron-hole pairs can be separated and pushed through the material, enhancing the collection of the charge carriers. The electrodes complete the device, ensuring the passage of the photocurrent. Now, we will review the main requirements the individual materials should have.

9.3.1 The Substrate

When realized in superstrate configuration, high efficiency CdTe-based solar cells and modules normally make use of the so-called "soda-lime glass" (SLG), which is the common window glass. Since sunlight passes through the substrate, the glass must be as transparent as possible to minimize the parasitic absorption of the visible light. Inexpensive minerals such as trona, sand, and feldspar are typically used in place of pure chemicals, making soda lime glass cost-effective. Unfortunately, starting from minerals, some impurities are inadvertently introduced, such as Fe_2O_3 and MnO_2, which are responsible for a decrease in transparency in the visible part of the solar spectrum. For this reason, iron-free glasses are normally used in CdTe technology, obtaining 8% more transparency at short wavelengths (λ <300 nm) [56].

Attempts to employ flexible substrates have been made using polymers, such as Dupont's polyimide, which enables high-speed roll-to-roll technology to produce large-area and very light cost-effective photovoltaic devices. Figure 9.5 shows the optical transparency of standard SLG and polyimide, which are comparable at long wavelengths, but the transmittance of polyimide below 530 nm is not sufficient for photovoltaic application, despite using thin foils (typical thickness 7 μm). This results in a photocurrent loss of at least 3 mA/cm². For this reason, together with the limitation of the process temperature, solar cells made with this polymer have not, until now, presented an efficiency greater than 14% [57].

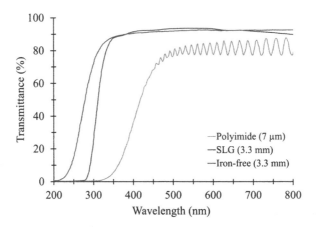

Fig. 9.5 Transmittance spectra of commonly used substrates such as soda-lime glass (SLG) and polyimide in CdTe-based production process of solar cells and modules

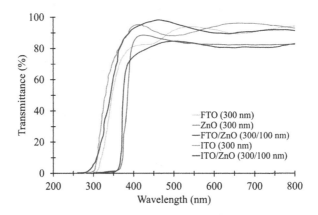

Fig. 9.6 Transmittance spectra of SLG (3.3 mm thick) covered with only ITO or FTO and SLG covered with ITO or FTO coupled with ZnO as a high-resistivity transparent (HRT) layer. (Reprinted from "A. Bosio, S. Pasini, N. Romeo, The History of Photovoltaics with Emphasis on CdTe Solar Cells and Modules, Coatings 2020, 10, 344". Published 2020 by MDPI as open access)

9.3.2 The Front Contact

In thin-film photovoltaic technology, one of the most stringent needs is the use of transparent and conductive electrical contacts, capable of passing light and to be electrically conductive to effectively collect the photogenerated charge carriers without introducing unnecessary sheet resistance. For this purpose, high-performance transparent and conductive oxides (TCOs), exhibiting transparencies to visible light close to 90% (see Fig. 9.6) and electrical conductivity up to $10^4 \, \Omega^{-1} \cdot cm^{-1}$ are normally used. These two opposing requirements are achieved by considering sufficiently high energy gaps (larger than 3 eV) in heavily doped degenerate semiconductor characterized by the Fermi level inside the conduction band. This particularly condition promotes the Burstein–Moss effect widening the energy gap and increasing, consequently, the transparency [58].

Near-degenerate semiconductors exhibit the typical free-carrier absorption in the near-infrared (NIR) making some of these TCOs not widely used in large-scale PV production. To overcome this drawback, semiconductors exhibiting high mobility of the charge carriers are commonly used, thus obviating the need to be degenerate to obtain high electrical conductivities.

The most common TCOs used in CdTe technology, are: Sn-doped In_2O_3 (ITO), F-doped SnO_2 (FTO), Al-doped ZnO (AZO) and Cd_2SnO_4 (CTO). All these TCOs are generally coupled with high-resistivity buffer layers (HRT) for reducing shunt effects caused by pinholes in active layers and to hinder the diffusion of impurities from the TCO layers or from the substrate. Un-doped SnO_2, In_2O_3 and ZnO are HRT layers commonly coupled with ITO, while FTO is generally paired with pure SnO_2 and CTO is combined with Zn_2SnO_4 (ZTO) [59]. The typical double layer structure of these TCOs modifies the chemical and physical interaction between the front contact and the window layer, as seen when ZnO and CdS, heat-treated at high temperature, form a mixing layer, which changes the optical properties of both films.

9.3.3   The Window Layer

The most widely used n-type partner with CdTe is cadmium sulfide (CdS), which exhibits n-type conduction due to stoichiometric defects, such as sulfur vacancies, which form during film growth. With an energy gap of 2.42 eV, it allows sunlight to pass up to a wavelength of 512 nm, cutting the wavelengths of the near ultraviolet. With a typical dark resistivity of the order of $(10^6–10^7)$ $\Omega \cdot cm$, CdS is not particularly suited as window material in solar cells because, in an efficient p/n junction, the electric field must principally fall into the p-type region (CdTe). This important requirement is satisfied if the spatial density of the p-type carrier in CdTe is significantly less than the density of the n-type carriers in CdS. Under sunlight this condition is satisfied thanks to the photoconductivity of CdS which helps to distribute the electric field into the p-type film. Considering the transparency, since nothing can be done about the absorption coefficient, very thin films, with a thickness of only 100 nm, give optical density suitable to be used as window layer in CdTe-based solar cells.

RF sputtering, chemical bath deposition (CBD) and high vacuum thermal evaporation (HVTE) belong to low-temperature (L-T) processes, while close-spaced sublimation (CSS) and close-spaced vapor transport (CSVT) are high-temperature (H-T) deposition procedures. As a consequence, CdS is normally deposited at a substrate temperature lower or higher than 200 °C. The choice of the deposition technique is crucial, since the quality of the solar cell relies on the interaction between the active layers, which in turn strongly depends on the deposition temperature of all the layers. The CBD process produces high-quality pin-hole free CdS films characterized by a very high density and compactness. A heat-treatment at 400 °C is needed to remove the Cd and S excesses naturally present in the CBD deposited films. While this deposition method offers excellent results, in industrial manufacturing, sputtering and H-T processes are generally preferred as CBD is a low-speed process producing a large amount of waste, which must be costly recycled.

Among the L-T technique, sputtering is the best suited for large scale in-line production. This technique is considered an L-T deposition technique even though the growing film surface is continuously bombarded with impinging electrons and atoms, which exchange their kinetic energy and promote the typical effects of high temperatures. Indeed, RF sputter deposition is not suitable for producing CdS films with the quality and chemical stability suitable for use in CdTe-based solar cells (see Fig. 9.7). Only reactive RF sputtering, which introduces oxidizing atoms into the process chamber, such as gases containing fluorine or oxygen, produces high quality CdS films that can form a good CdS/CdTe heterojunction. What happens in the sputter discharge when a hydrofluorocarbon or oxygen gas is introduced into the process gas (Ar) is described in [60].

Between the H-T deposition techniques, CSS is the most used. CdS films obtained with this process show superior quality, even if depositions carried out in pure Ar provide low density films with many pinholes. If oxygen is added to the process chamber, the deposition equilibrium is substantially altered, growth slows down, since grain boundaries of the growing film are decorated with oxides such as $CdSO_4$ and $CdSO_3$. As a result, a denser film without pinholes is obtained, but the surface of the CdS film is covered with an oxide layer. For this reason, high efficiency solar cells are obtained when CSS-deposited CdS films in Ar + O_2 atmosphere, are heat-treated at high temperature (400 °C) in presence of hydrogen.

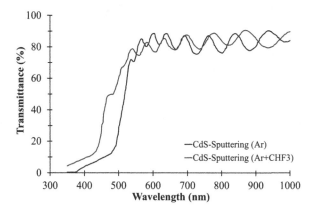

Fig. 9.7 Transmittance spectra of sputtered CdS film: deposited in pure argon or deposited in argon + CHF_3

9.3.4   The Absorber Layer

The Close-Spaced Sublimation of CdTe is made possible since at high temperature, CdTe dissociates into its elements Cd and Te, which can recombine on the substrate to form the CdTe film. Generally, with this technique the CdTe films are deposited in the temperature range of (500–600) °C in an Ar atmosphere at a pressure in the (1–100) mbar range. If soda lime glass is used a temperature of 520 °C cannot be exceeded.

CdTe film thickness of 2 μm is more than enough to absorb all the visible light, since CdTe exhibits an absorption coefficient in the range of $(10^4/10^5)$ cm^{-1}. This thickness is optimal both for optical and electrical requirements, even though it's very difficult to obtain CSS-deposited pinhole-free films with proper qualities and compactness. For this reason, oxygen is generally added to the inert gas in the CSS or CSVT deposition chamber. Due to the presence of CdO and TeO$_2$ species, a greater superficial diffusion of the incoming Cd and Te atoms is expected. The presence of oxygen increases the number of nucleation sites promoting a denser growth of the CdTe film (see Fig. 9.8). In these conditions a small quantity of CdTeO$_3$ is formed, which has an appreciated effect of passivation of the grain boundaries [61].

As already mentioned, CdS/CdTe junction can never work because a 9.7% lattice mismatch between the two materials generates too many interface defects, which can capture the photogenerated carriers crossing the junction. It is now established that a good way to overcome this drawback is to create in the junction region a mixed compound, namely CdS$_X$Te$_{(1-X)}$, between the two active materials. In this way the lattice mismatch is progressively adapted, and the number of defects at the interface can be significantly reduced. Since the CSS-deposition of CdTe on top of CdS occurs at a high temperature, the beneficial intermixing between these materials begins to take place. Moreover, under AM 1.5 G solar light a solar cell reaches its maximum efficiency when the energy gap of the absorber is 1.34 eV. The energy gap of CdTe is a little bit wider, being 1.5 eV, but could be adjusted exploiting the

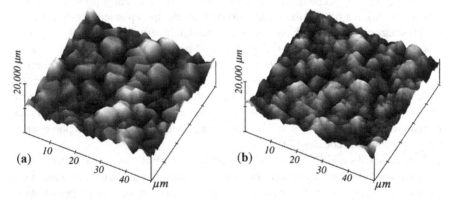

Fig. 9.8 Atomic Force Microscope (AFM) image of the surface of a CdTe film deposited by CSS; (**a**) deposition performed in pure Ar, (**b**) in Ar + 10%O$_2$

favorable Cd-Te-Se phase diagram. In fact, according to [62] a mixed $CdTe_{(1-x)}Se_x$ compound with $X \approx 0.4$, exhibits an energy gap $E_g = 1.4$ eV. With the $CdTe_{0.6}Se_{0.4}$ alloy a better collection of the solar light spectrum, extended to a longer wavelength, is obtained. Nowadays, an engineered profile of the energy gap of the CdS/$CdTe_{(1-x)}Se_x$ is beneficial to improve the photocurrent maintaining the high photovoltage typical of the CdTe absorber.

Similarly, to what is done in CIGS-based cells, the small gap material is used only in a thin surface layer of the absorber in such a way to not compromise the typical photovoltage of the CdTe bulk. Besides, when a $CdS/CdTe_{(1-x)}Se_x$ heterojunction is made, it was found that the thickness of the CdS layer can be drastically reduced or the layer can be even avoided, decreasing the parasitic absorption in the high-energy region of the solar spectrum and obtaining an increasing photocurrent [63]. Additional improvements can be acquired if the CdS is replaced with MgZnO, a material with which CdTe and $CdS/CdTe_{(1-x)}Se_x$ have a better alignment to the energy band. This material is characterized by a better transparency in the high-energy region with respect to CdS [64, 65]. Furthermore, a greater passivation effect of the interface defects due to the presence of Se is effective for a longer lifetime of the charge carrier in $CdS/CdTe_{(1-x)}Se_x$ compared to CdTe [66]. By adopting these methods, solar cells with an efficiency of over 22% were realized, including the world record [51].

High-efficiency solar cells have also been made on flexible polymeric substrates, where a L-T process is essential. CdTe and CdS films, deposited by HVTE at a temperature of about 200 °C with a typical thickness of (2/3) μm, show excellent smoothness and compactness with a characteristic grain size of (100/500) nm [67]. The size of the crystallites suggests that the mobilities and lifetime of the charge carriers are very low, unsuitable for forming an efficient *p-n* junction capable of collecting the photogenerated carriers. Never, as in this case, is a post-deposition treatment necessary, the so-called "chlorine treatment". This annealing in a chlorine atmosphere can reduce stacking defects, mismatch dislocations and grain boundaries, thereby increasing the crystal grain size [68].

Among the L-T techniques, electrodeposition was originally presented in the 1970s [69, 70], but only in the early 1980s was an electrodeposited thin-film CdTe-based solar cell developed, showing a remarkable efficiency of 8% [71–73]. Normally, the electrodeposited CdTe layer exhibits *n*-type conductivity. As a result, an innovative SLG/FTO/n-CdS/n-CdTe/p-CdTe/Au layer sequence was implemented. The main improvement is represented by a buried homojunction near the back-contact. The presence of a thin *p*-type CdTe film close to the back contact allows a better control of the junction position within the absorber layer, with an increase of the barrier height and a consequent good collection of the photogenerated electron-hole couples [74].

When the CdS and CdTe layers are deposited by sputtering, the electro-optical properties of the films strongly depend on the sputtering parameters, as well as on the sputtering power density, on the Ar and reactive gas pressure, on the polarization voltage, on the substrate temperature and on the target-substrate distance. These numerous varieties of process parameters allow a very fine tuning of the physical characteristics of the growing film. Taking advantage of the excellent coverage and high density of the sputter-deposited films, a 2 μm thick layer of CdTe is used to realize very thin solar cells [75]. By appropriately regulating the chlorine treatment, it would be possible to produce CdTe thin films with crystalline grains large enough to obtain high efficiency solar cells. Following this philosophy, a 14% efficient solar cell was made in the early 2000s [76].

9.3.5 The Heat Treatment in Chlorine Atmosphere

It is commonly accepted by the scientific community that chlorine treatment increases the grain size of the CdTe film, improves the quality of the grain boundaries and promotes the intermixing of CdS and CdTe layers at their interface. If this treatment is not performed, the short-circuit current of the solar cell is very low therefore the efficiency is very poor. The treatment is generally carried out by depositing a $CdCl_2$ film on top of the CdTe layer by evaporation or by dipping the CdTe layer in a solution of $CdCl_2$-methanol. The chlorine-treatment is completed keeping the $CdTe + CdCl_2$ system at a temperature in the range of (350–400) °C in air or in an inert gas such as Ar at atmospheric pressure [77, 78]. Recently, alternative Cl_2-containing compounds were proposed such as $MgCl_2$, NaCl and NH_4Cl [79–84]. Cd-free salts have the great advantage of being much more environmentally sustainable than $CdCl_2$. Moreover, another way to perform a Cd-free chlorine treatment is to use a halogen gas capable of releasing chlorine or chlorine-containing radicals at the processing temperature. These gases belong to the chlorine-based Freon family or to the same family of chlorine, such as HCl or Cl_2, but hydrochloric acid or chlorine are very aggressive gases making the CdTe treatment extremely critical [85, 86]. Since 2011, the industrial use of chlorinated Freon gases has been banned in Europe because they are ozone depleting substances (ODS). For this reason, other chlorinated hydrocarbons, effective in the treatment of CdTe and not considered ODS, were considered. Possible candidates have been identified among liquid chlorinated hydrocarbons (LCHY).

It is assumed that the reaction on the CdTe surface during the treatment is:

1. if Cl_2 is supplied by means of chlorine-based salt

$$CdTe(s) + CdCl_2 \Rightarrow 2Cd(g) + 1/2Te_2(g) + Cl_2(g) \Rightarrow CdCl_2(s) + CdTe(s) \quad (9.4)$$

2. if Cl_2 is supplied by means of chlorine-based gas

$$CdTe(s) + 2Cl_2(g) \Rightarrow CdCl_2(g) + TeCl_2(g) \Rightarrow 2Cl_2(g) + CdTe(s) \quad (9.5)$$

After the treatment the CdTe surface morphology is completely changed due to an increase in the size of the small grains (see Fig. 9.9a, b). The morphology of the surfaces shown in Fig. 9.9b presents the typical mesa-like structure resulting from the etching of the surface by chlorine and/or hydrogen chloride. The crystalline grains tend to coalescence, becoming more compact with narrow boundaries.

From a stoichiometric point of view, a CSS-deposited CdTe film shows a Te-rich surface if Cl_2 is supplied by means of chlorine-based gas. In this case, the peculiarity is that this treatment does not introduce Cd on the surface of the CdTe film from outside, leaving the Cd-vacancies not compensated; this fact cannot happen if a cadmium salt such as $CdCl_2$ is used. This important feature could be exploited by carrying out the back-contact directly on the surface of the CdTe layer since, to obtain good ohmic contact with p-type CdTe, a Te-rich surface is required.

9.3.6 The Back-Contact

In the production of high efficiency CdS/CdTe solar cells, the realization of a stable over time ohmic back-contact has long been critical. Commonly, many manufacturing processes use a Cu-containing compound, such as the Cu-Au alloy, Cu_2Te, ZnTe:Cu and Cu_2S [87–90]. Excluding the Cu-Au bilayer, the electrical contact is terminated with a sputtering deposition of a metallic layer such as Mo, Ni-V or with a graphite paste. The diffusion of copper reduces the resistivity of the CdTe layer, resulting effective in the formation of a very low resistance ohmic contact suitable for high-performance solar cells.

Devices made with non-Cu-containing contacts exhibit high series resistance, mainly due to the non-ohmic contact. For this reason, very high efficiency solar cells have been realized, exploiting the presence of a Cu-based compound in the back contact.

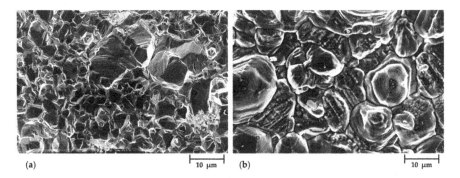

Fig. 9.9 SEM image of the surface morphology of a CSS-deposited CdTe film at a substrate temperature of 520 ° C. The thickness of the CdTe film is 6 μm. (**a**) as deposited, (**b**) after heat-treatment in LCHY atmosphere. (Reprinted from "A. Bosio, S. Pasini, N. Romeo, The History of Photovoltaics with Emphasis on CdTe Solar Cells and Modules, Coatings 2020, 10, 344". Published 2020 by MDPI as open access)

Typically, a chemical attack is performed in Br_2-methanol or in a mixture of HNO_3/HPO_3 acids (N-P), which leaves the CdTe surface Te-rich. A Te-rich surface favors the correct band alignment between the valence band of the CdTe and the working function of the metal contact and/or the electronic affinity of a degenerated semiconductor, through the formation of a low resistance tunneling barrier. Since copper is very reactive with Te, a thin layer of Cu_2Te is formed at the interface between CdTe and the back contact, which limits the diffusion of copper into the grain boundaries of the absorber film. Unfortunately, Cu_2Te is not a stable compound, triggering the usual diffusion of Cu atoms into the grain boundaries. To overcome this problem, the amount of copper is carefully controlled. Several solutions are proposed; one of the most reliable methods makes use of a very thin layer of Cu (1/2 nm thick) or by means of a buffer layer, which acts as a filter for the diffusion of Cu atoms. In both cases the formation of a layer of Cu_xTe is necessary to ensure a good contact able to stop the diffusion of Cu. It was discovered that Cu_xTe behaves as a stable material only if $x \leq 1.4$. This condition is obtained by placing a buffer layer, based on M_2Te_3, between the very thin layer of Cu and the surface of CdTe (where M = Sb, Bi, As) [91, 92]. Two different phenomena could take place:

(i) M atoms can bind tellurium excess, forming a M_2Te_3 or Cu_xTe interface, confirming the fundamental role of a Te-rich surface of the CdTe film.
(ii) A heat treatment in air of the whole system, carried out at a temperature of 200 °C for 15 min, favors the formation of an oxide compound into the grain boundaries of the CdTe layer, which contributes to the passivation of the grain boundaries by removing their ability to recombine charge carriers. A similar result is obtained if a ZnTe film is used as a buffer layer [93].

This is a general recipe; if respected, an efficient and stable over-time solar cell is regularly produced, which is the basis of an industrial production of large-sized modules based on CdTe.

9.4 Conclusion

Since its discovery, CdTe has immediately proved to be a very promising material for the realization of electro-optical devices thanks to the fact that good-quality single crystals can be easily obtained. Historically, many devices have been obtained thanks to the material transparency in the mid infrared also shown by a few cm thick monocrystalline sheets.

The greatest success of this material is due to its use as an absorber material in solar cells. What is most surprising is that solar cells totally made with thin film technology are much more performing than the same devices built with bulk CdTe single crystal. Recently, CdTe thin film solar cells achieved efficiencies of 22.1% by exploiting a sophisticated engineering of the energy gap of the absorber simply adding a little amount of Se at the interface with the *n*-type partner.

This is a very mature technology since fully automated in-line production machine could produce large-area, 19% efficiency module every 1 minute. Moreover, the main actors on the market assure that at their end-of- life these modules are completely recyclable implementing a virtuous circular loop.

References

1. Fabre M (1888) Tellurures métalliques cristallisés. Ann Chim Phys Ser 6:110–120
2. Frerichs R (1947) The photo-conductivity of "incomplete phosphors". Phys Rev 72:594
3. Schwarz E (1948) New photoconductive cells. Nature 162:614
4. Jenny DA, Bube RH (1954) Semiconducting cadmium telluride. Phys Rev 96:1190
5. De Nobel D (1959) Phase equilibria and semiconducting properties of cadmium telluride. Philips Res Repts 14:361-399 and 430-492
6. Lofersky JJ (1956) Theoretical considerations governing the choice of the optimum semiconductor for photovoltaic solar energy conversion. J Appl Phys 27:777
7. Kato Y, Fujimoto S, Kozawa M, Fujiwara H (2019) Maximum efficiencies and performance-limiting factors of inorganic and hybrid perovskite solar cells. Phys Rev Appl 12:024039
8. Werthen JG, Fahrenbruch AL, Bube RH, Zesch JC (1983) Surface preparation effects on efficient indium-tin-oxide-CdTe and CdS-CdTe heterojunction solar cells. J Appl Phys 54:2750
9. Aranovich JA, Golmayo D, Fahrenbruch AL, Bube RH (1980) Photovoltaic properties of ZnO/CdTe heterojunctions prepared by spray pyrolysis. J Appl Phys 51:4260
10. Courreges F, Farhenbruch AL, Bube RH (1980) Surface preparation effects on efficient indium-tin-oxide-CdTe and CdS-CdTe heterojunction solar cells. J Appl Phys 51:2175
11. Arroyo J, Marfaing Y, Cohen-Solal G, Triboulet R (1979) Electric and photovoltaic properties of CdTe p-n homojunctions. Sol Energy Mater 1(1–2):171–180
12. Nakazawa T, Takamizawa K, Ito K (1987) High efficiency indium oxide/cadmium telluride solar cells. Appl Phys Lett 50:279
13. Bonnet D, Rabenhorst H (1972) New results on the development of thin film p-CdTe/n-CdS heterojunction solar cell. In: Proceedings of 9th IEEE Photovoltaic Specialists Conference. Silver Spring, 2–4 May 1972, pp 129–132
14. Tyan YS, Albuerne EA (1982) Efficient thin-film CdS/CdTe solar cells. In: Proceedings of 16th IEEE Photovoltaic Specialist Conference. San Diego, September 27–30, 1982, pp 794–800
15. Ferekides C, Britt J, Ma Y, Killian L (1993) High efficiency CdTe solar cells by close spaced sublimation. In: Proceedings of 23rd IEEE photovoltaic specialists conference. Louisville, 10–14 May, 1993, pp 389–393
16. Wu X, Keane JC, Dhere RG, DeHart C, Albin DS, Duda A et al (2002) 16.5% Efficiency CdS/CdTe polycristalline thin-film solar cells. In: Proceedings of the 17th European Photovoltaic Solar Energy Conference and Exhibition. Munich, 22–26 October, 2002, pp 995–1000
17. NREL. [Online]. Available: https://www.nrel.gov/pv/cell-efficiency.html. Accessed 29 Dec 2021
18. Green MA, Dunlop ED, Hohl-Ebinger J, Yoshita M, Kopidakis N, Hao X (2020) Solar cell efficiency tables (version 56). Prog Photovolt Res Appl 28:629–638
19. Pollok DB (1969) Thermal expansion values for installation of an lrtran-6 window. Appl Opt 8(4):837–838
20. Taylor JB, Boland R, Gowac E, Stupik P, Tricard M (2007) Recent advances in high-performance window fabrication, "Defense and Security Symposium". In: Proceedings of window and dome technologies and materials XIII, 1–2 May 2013. Baltimore, pp 8708–8740
21. EdmundOptics. [Online]. Available: https://www.edmundoptics.com/f/cadmium-telluride-cdte-windows/39727/. Accessed 02 Jan 2022

22. Long X-C, Myers RA, Brueck SRJ (1994) Measurement of the linear electro-optic coefficient in poled amorphous silica. Opt Lett 19(22):1819–1821
23. Sugie M, Tada K (1976) Measurements of the linear Electrooptic coefficients and analysis of the nonlinear susceptibilities in cubic GaAs and hexagonal CdS. Jpn J Appl Phys 15(3):421
24. Chenault DB, Chipman RA, Lu S-Y (1994) Electro-optic coefficient spectrum of cadmium telluride. Appl Opt 33(31):7382–7389
25. Huang H, Lin ZS, Chen CT (2008) Mechanism of the linear electro-optic effect in potassium dihydrogen phosphate crystals. J Appl Phys 104:073116
26. Sliker TR, Burlage SR (1963) Some dielectric and optical properties of KD_2PO_4. J Appl Phys 34:1837
27. Onaka R, Ito H (1976) Pockels effect of KDP and ADP in the ultraviolet region. J Phys Soc Jpn 41:1303–1309
28. Haibeiflavor, 2022. [Online]. Available: https://www.haibeiflavor.com/material-factory-optical-linbo3-ln-saw-filter-q-switch-electro-optic-dimmer-single-crystal-substrate-manufacturer-from-china_p242.html. Accessed 26 Jan 2022
29. Kurt MZ (2003) Electrooptical properties of LiTaO3. Ferroelectrics 296:127–137
30. Lovett DR (2017) Tensor properties of crystals. Taylor & Francis Group, New York
31. Akitt DP, Johnson C, Coleman PD (1970) Nonlinear susceptibility of CdTe. IEEE J Quantun Electron QE-6(8):196–499
32. Wagnière GH, Woźniak S (2017) Nonlinear optical properties: noncentrosymmetric crystals. In: Lindon JC, Tranter GE, Koppenaal DW (eds) Encyclopedia of spectroscopy and spectrometry, London. Academic, pp 375–387
33. Olson J, Rodriguez Y, Yang L, Alers G, Carter S (2010) CdTe Schottky diodes from colloidal nanocrystals. Appl Phys Lett 96:242103
34. Baines T, Papageorgiou G, Hutter O, Bowen L, Durose K, Major J (2018) Self-catalyzed CdTe wires. Nano 8:274
35. Wang X, Wang J, Zhou M, Wang H, Xiao X, Li Q (2010) CdTe nanorods formation via nanoparticle self-assembly by thermal chemistry method. J Cryst Growth 312:2310–2314
36. Wang X, Xu Y, Zhu H, Liu R, Wang H, Li Q (2011) Crystalline Te nanotube and Te nanorods-on-CdTe nanotube arrays on ITO via a ZnO nanorod templating-reaction. CrystEngComm 13:2955–2959
37. Lee S, Yu Y, Perez O, Puscas S, Kosel T, Kuno M (2010) Bismuth-assisted CdSe and CdTe nanowire growth on plastics. Chem Mater 22:77–84
38. Kret S, Szuszkiewicz W, Dynowska E, Domagala J, Aleszkiewicz M, Baczewski L, Petroutchik A (2008) MBE growth and properties of ZnTe- and CdTe-based nanowires. J Korean Phys Soc 53:3055–3063
39. Kulkarni R, Rondiya S, Pawbake A, Waykar R, Jadhavar A, Jadkar V, Bhorde A, Date A, Pathan H, Jadkar S (2017) Structural and optical properties of CdTe thin films deposited using RF magnetron sputtering. Energy Procedia 110:188–195
40. Yang G, Jung Y, Chun S, Kim D, Kim J (2013) Catalytic growth of CdTe nanowires by closed space sublimation method. Thin Solid Films 546:375–378
41. Salim H, Patel V, Abbas A, Walls J, Dharmadasa I (2015) Electrodeposition of CdTe thin films using nitrate precursor for applications in solar cells. J Mater Sci Mater Electron 26:3119–3128
42. Consonni V, Rey G, Bonaim J, Karst N, Doisneau B, Roussel H, Renet S, Bellet D (2011) Synthesis and physical properties of ZnO/CdTe core shell nanowires grown by low-cost deposition methods. Appl Phys Lett 98:96–99
43. Dan Y, Zhao X, Chen K, Mesli A (2018) A photoconductor intrinsically has no gain. ACS Photonics 5(10):4111–4116
44. Xie X, Kwok S-Y, Lu Z, Liu Y, Cao Y, Luo L, Zapien JA, Bello I, Lee C-S, Lee S-T, Zhang W (2012) Visible–NIR photodetectors based on CdTe nanoribbons. Nanoscale 4:2914–2920
45. Shaygan M, Davami K, Kheirabi N, Baek CK, Cuniberti G, Meyyappand M, Lee J-S (2014) Single-crystalline CdTe nanowire field effect transistors as nanowire-based photodetector. PhysChemChemPhys 16:22687

46. Wei H, Fang Y, Yuan Y, Shen L, Huang J (2015) Trap engineering of CdTe nanoparticle for high gain, fast response, and low noise P3HT:CdTe nanocomposite photodetectors. Adv Mater 27:4975–4981
47. Park H, Yang G, Chun S, Kim D, Kim J (2013) CdTe microwire-based ultraviolet photodetectors aligned by a non-uniform electric field. Appl Phys Lett 103:051906
48. Tang M, Xu P, Wen Z, Chen X, Pang C, Xu X, Meng C, Liu X, Tian H, Raghavan N, Yang Q (2018) Fast response CdS-CdS$_x$Te$_{1-x}$CdTe core-shell nanobelt photodetector. Sci Bull 63:1118–1124
49. Lofersky JJ (1956) Theoretical considerations governing the choice of the optimum semiconductor for photovoltaic solar energy conversion. J Appl Phys 27(7):777–784
50. Britt J, Ferekides C (1993) Thin-film CdS/CdTe solar cell with 15.8% efficiency. Appl Phys Lett 62:2851–2852
51. Green M, Hishikawa I, Dunlop E, Levi D, Hohl-Ebinger J, Yoshit M, Ho-Baillie A (2019) Solar cell efficiency tables (version 54). Prog Photovolt Res 27:565–575
52. "First Solar Establishes New World Record for CdTe Efficiency," 23 February 2016. [Online]. Available: https://www.solarpowerworldonline.com/2016/02/24939/. Accessed 13 Mar 2022
53. Cunningham D, Rubcich M, Skinner D (2002) Cadmium telluride PV module manufacturing at BP Solar. Prog Photovolt Res Appl 10:159–168
54. Bosio A, Menossi D, Mazzamuto S, Romeo N (2011) Manufacturing of CdTe thin film photovoltaic modules. Thin Solid Films 519:7522–7525
55. FirstSolar.com, "FirstSolar.com, Series 6 Datasheet," October 2021. [Online]. Available: http://www.firstsolar.com/-/media/First-Solar/Technical-Documents/Series-6-Datasheets/Series-6-Datasheet.ashx. Accessed 13 March 2022
56. Bosio A, Rosa G (2018) Past present and future of the thin film CdTe/CdS solar cells. Sol Energy 175:31–43
57. Salavei A, Artegiani E, Piccinelli F, di Mare S, Menossi D, Bosio A, Romeo N, Romeo A (2015) Flexible CdTe solar cells on polyimide and flexible glass substrates. In: Proceedings of the 31st European Photovoltaic Solar Energy Conference. Hamburg, 14–18 September, 2015, pp 1356–1357
58. Moss T (1954) The interpretation of the properties of indium antimonide. Proc Phys Soc B 67:775
59. Wu X, Ribelin R, Dhere RG, Albin DS, Gessert TA, Asher S, Levi DH, Mason A, Moutinho HR, Sheldon P (2000) High-efficiency Cd$_2$SnO$_4$/Zn$_2$SnO$_4$/Zn$_x$Cd$_{1-x}$S/CdS/CdTe polycrystalline thin-film solar cells. In: Record of the Twenty-Eighth IEEE Photovoltaic Specialists Conference. Anchorage, 15–22 Sept. 2000, pp 470–474
60. Romeo N, Bosio A, Canevari V (2003) The role of CdS preparation method in the performance of CdTe/CdS thin film solar cell. In: Proceedings of the 3rd world conference on photovoltaic energy conversion, WCPEC-3. Osaka, 11–18 May, 2003, pp 469–470
61. Bosio A, Romeo A, Romeo N (2011) Polycrystalline CdTe thin films solar cells. In: Thin film solar cells: current status and future trends. Nova Science Publishers Inc., New York, pp 161–200
62. Wei S, Zhang S, Zunger A (2000) First-principles calculation of band offsets, optical bowings, and defects in CdS, CdSe, CdTe, and their alloys. J Appl Phys 87:1304–1311
63. Mia M, Swartz C, Paul S, Sohal S, Grice C, Yan Y, Holtz M, Li J (2018) Electrical and optical characterization of CdTe solar cells with CdS and CdSe buffers—a comparative study. J Vac Sci Technol B 36:052904
64. Kephart J, McCamy J, Ma Z, Ganjoo A, Alamgir F, Sampath W (2016) Band alignment of front contact layers for high-efficiency CdTe solar cells. Sol Energy Mater Sol Cells 157:266–275
65. Munshi A, Kephart J, Abbas A, Shimpi T, Barth K, Walls J, Sampath W (2018) Polycrystalline CdTe photovoltaics with efficiency over 18% through improved absorber passivation and current collection. Sol Energy Mater Sol Cells 176:9–18

66. Fiducia T, Mendis B, Li K, Grovenor C, Munshi A, Barth K, Sampath W, Wright L, Abbas A, Bowers J (2019) Understanding the role of selenium in defect passivation for highly efficient selenium-alloyed cadmium telluride solar cells. Nat Energy 4:504–511

67. Romeo A, Bätzner D, Zogg H, Vignali C, Tiwari A (2001) Influence of CdS growth process on structural and photovoltaic properties of CdTe/CdS solar cells. Sol Energy Mater Sol Cells 67:311–321

68. Moutinho H, Al-Jassim M, Abulfotuh F, Levi D, Dippo P, Dhere R, Kazmerski L (1997) Studies of recrystallization of CdTe thin films after CdCl₂ treatment. In: Proceedings of the twenty sixth IEEE photovoltaic specialists conference. Anaheim, September 29–October 3, 1997, pp 431–434

69. Danaher W, Lyons L (1978) Photoelectrochemical cell with cadmium telluride film. Nature 271:139

70. Kröger F (1978) Cathodic deposition and characterization of metallic or semiconducting binary alloys or compounds. J Electrochem Soc 125:2028–2034

71. Ortega-Borges R, Lincot D (1993) Mechanism of chemical bath deposition of cadmium sulfide thin films in the ammonia-thiourea system in situ kinetic study and modelization. J Electrochem Soc 140:3464–3473

72. Fulop G, Taylor R (1985) Electrodeposition of semiconductors. Ann Rev Mater Sci 15:197–210

73. Basol B (1984) High-efficiency electroplated heterojunction solar cell. J Appl Phys 55:601–603

74. Ojo A, Dharmadasa I (2016) 15.3% efficient graded bandgap solar cells fabricated using electroplated CdS and CdTe thin films. Sol Energy 136:10–14

75. Paudel N, Wieland K, Compaan A (2012) Ultrathin CdS/CdTe solar cells by sputtering. Sol Energy Mater Sol Cells 105:109–112

76. Compaan A, Gupta A, Lee S, Wang S, Drayton J (2004) High efficiency, magnetron sputtered CdS/CdTe solar cells. Sol Energy 77:815–822

77. Zanio K (1978) Cadmium telluride, semiconductors and semimetals. Academic, New York

78. Ferekides CS, Marinskiy D, Viswanathan V, Tetali B, Palekis V, Selvaraj P, Morel D (2000) High efficiency CSS CdTe solar cells. Thin Solid Films 361-362:520–526

79. Bayhan H (2004) Investigation of the effect of CdCl₂ processing on vacuum deposited CdS/CdTe thin film solar cells by DLTS. J Phys Chem Solids 65:1817–1822

80. Hiie J (2003) CdTe:CdCl₂:O₂ annealing process. Thin Solid Films 431-432:90–93

81. Niles D, Waters D, Rose D (1998) Chemical reactivity of CdCl2 wet-deposited on CdTe films studied by X-ray photoelectron spectroscopy. Appl Surf Sci 136:221–229

82. Williams L, Major J, Bowen L, Keuning W, Creatore M, Durose K (2015) A comparative study of the effects of nontoxic chloride treatments on CdTe solar cell microstructure and stoichiometry. Adv Energy Mater 5:1500554–1500563

83. Potlog T, Ghimpu L, Gashin P, Pudov A, Nagle T, Sites J (2003) Influence of annealing in different chlorides on the photovoltaic parameters of CdS/CdTe solar cells. Sol Energy Mater Sol Cells 80:327–334

84. Potter M, Halliday D, Cousins M, Durose K (2000) A study of the effects of varying cadmium chloride treatment on the luminescent properties of CdTe/CdS thin solar cells. Thin Solid Films 361-362:248–252

85. Zhou T, Reiter N, Powell R, Sasala R, Meyers P (1994) Vapor chloride treatment of polycrystalline CdTe/CdS films. In: Proceedings of the 1st IEEE photovoltaic specialists conference. Waikoloa, pp 103–106. https://doi.org/10.1109/WCPEC.1994.519818

86. Qu Y, Meyers P, Mc Candless B (1996) HCl vapor post-deposition heat treatment of CdTe/CdS films. In: Proceedings of the 25th IEEE photovoltaic specialists conference, Washington, DC, pp 1013–1016. https://doi.org/10.1109/PVSC.1996.564303

87. Albright S, Jordan J, Akerman B, Chamberlain R (1989) Developments on CdS/CdTe photovoltaic panels at photon energy. Inc Sol Cells 27:77

88. Gessert T, Mason A, Sheldon P, Swartzlander A, Niles D, Coutts T (1996) Development of Cu-doped ZnTe as back-contact interface layer for thin-film CdS/CdTe solar cells. J Vac Sci Technol 14:806

89. Uda H, Ikegami S, Sonomura H (1990) Compositional change of the Au–Cu2Te contact for thin-film CdS/CdTe solar cells. Jpn J Appl Phys 29:495
90. McCandless B, Qu Y, Birkmire RA (1994) Treatment to allow contacting CdTe with different conductors. In: Proceedings of the first world conference on photovoltaic energy conversion. Waikoloa, pp 107–110. https://doi.org/10.1109/WCPEC.1994.519819
91. Du M (2009) First-principles study of back-contact effects on CdTe thin-film solar cells. Phys Rev B 80:205322
92. Durose K, Boyle D, Abken A, Ottley C, Nollet P, Degrave S, Burghelman M, Wendt R, Bonnet D (2002) Key aspects of CdTe/CdS solar cells. Phys Status Solidi 229:1055–1064
93. Bosio A, Ciprian R, Lamperti A, Rago I, Ressel B, Rosa G, Stupard M, Weschke E (2018) Interface phenomena between CdTe and ZnTe:Cu back contact. Sol Energy 176:186–193

Chapter 10
CdSe – Based Photodetectors for Visible-NIR Spectral Region

Hemant Kumar and Satyabrata Jit

10.1 Introduction

The discovery of the photoelectric effect by Albert Einstein in 1905 was the first stepping stone toward the creation of the new and emerging area of research in optoelectronics and photonics [1]. The photodetectors belong to a class of optoelectronic devices which are used to convert optical signals of certain wavelengths into an electrical signal for various applications including optical communications and computing, optoelectronics integrated circuits and optical imaging systems. Photodetectors are designed to detect lights of wavelengths belonging to one or more regions of the electromagnetic spectrum namely the ultraviolet (UV) region (i.e. wavelengths below 400 nm), visible region (i.e. wavelengths in between 400 and 700 nm) and infrared (IR) region (i.e. wavelengths above 700 nm but below 1 mm) [2].

The photodetection mechanism primarily depends upon the accurate detection of the wavelength (or energy) of incident photons on the photodetectors. Based on the spectral width of the incident photons to be detected by the photodetectors, the photodetectors are broadly divided into two types: narrowband and wideband photodetectors. When the photodetectors are designed to detect the incident photons of a single wavelength, they are known as narrowband photodetectors. The photo response characteristics of such type of photodetectors possess a dominant peak

H. Kumar
Department of Electronics and Communication Engineering, Jaypee Institute of Information Technology, Noida, India
e-mail: hemant.kumar@jiit.ac.in

S. Jit (✉)
Department of Electronics Engineering, Indian Institute of Technology (Banaras Hindu University), Varanasi, India
e-mail: sjit.ece@iitbhu.ac.in

© The Author(s), under exclusive license to Springer Nature Switzerland AG 2023   231
G. Korotcenkov (ed.), *Handbook of II-VI Semiconductor-Based Sensors and Radiation Detectors*, https://doi.org/10.1007/978-3-031-20510-1_10

centered around the wavelength of the incident photons to be detected with the full-width at half-maximum (FWHM) <= 100 nm [3]. On the other hand, the wideband photodetectors are designed to detect the incident photons of different wavelengths, may be ranging from UV to IR. Such photodetectors have a larger FWFH value >100 nm. The wideband photodetectors thus can't discriminate the incident photons on the basis of their wavelength or energy. The wideband or narrowband nature of the photodetectors primarily depends on the properties of the photoactive materials used for the light detection in the devices. Traditionally, Si is used for wideband photodetectors for detecting all the photons with wavelengths ranging from 400 to 1100 nm [4]. Since it is not possible to distinguish two incident photons of different energies or wavelengths by such detectors, external optical filters are connected with such devices for the same [5].

It is well known that different materials have different types of absorption spectral characteristics. Thus, the absorption properties of photoactive materials used in the photodetectors play vital roles in determining the spectral characteristics of the detectors. This chapter is devoted to discuss some Cadmium Selenide (CdSe) material-based photodetectors for their operation in the visible – NIR region [6]. CdSe belongs to the II-VI group semiconductors which carries exciting properties for optoelectronic applications [6]. CdSe exists in three crystalline forms: wurtzite, zinc blend, and rock-salt [7]. However, the wurtzite structure of CdSe is widely explored for optoelectronic applications due to its higher stability than the other structures [6]. CdSe is inherently an n-type material. The p-type CdSe is not stable and very difficult to achieve due to the self-compensation effect [6]. The absorption band edge of the bulk CdSe lies at ~1.7 eV (~729 nm) for both the wurtzite and zinc blend structures at the room temperature [7]. The band gap of the bulk CdSe is suitable for photodetectors operating in the visible and near infrared (NIR) regions of the electromagnetic spectrum [8, 9]. It may be mentioned that photodetectors working in the visible region find multiple areas of applications such as in artificial eye [3], imaging [3], machine vision [3], forensic analysis [10], atmospheric contamination [10], criminology [10], biomedical and diagnostic imaging [3], civil investigations [10] and detecting heart rate using smartwatches [3]. Similarly, the photodetectors operating in the NIR region are widely used for medical diagnosis and biomedical imaging without tampering with the biological tissues [11].

Various CdSe nanostructures (2D, 1D, 0D) are projected as potential materials for developing filter-less photodetectors [11]. Luo et al. [12] have reported CdSe nano-ribbons (1D) based photodetectors with a photo response around 650 nm corresponding to the red region of the visible spectrum without using any external filter. They [12] have used hollow gold nanospheres (HGN) on CdSe nano-ribbons (NRs) to enhance the absorption of CdSe nano-ribbons by multiple folds at the red region by using localized surface plasmon resonance (LSPR) effect. The device structure and TEM image of HGN-on-CdSe NRs surface are shown in Fig. 10.1a, b, respectively. The I-V and photo response characteristics of the aforementioned CdSe NRs based photodetector with and without HGN decorations are compared in Fig. 10.1c, d. In a different work, Kumar et al. [13] have explored optical resonance mechanism in the Pd/CdSe QDs (0D) based photodetector to demonstrate an

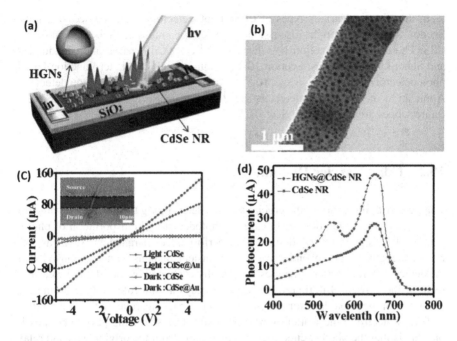

Fig. 10.1 (a) Device schematic, (b) TEM image of decorated HGN on CdSe NRs, (c) I-V response of the fabricated detector with and without HGN, and (d) spectral response of the device with or without HGNs. (Adapted from Ref. [12]. Published 2010 by Optica Publishing as open access)

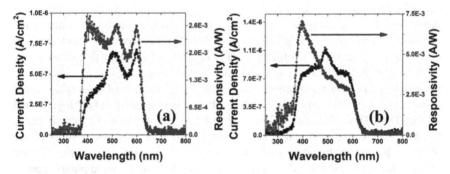

Fig. 10.2 Photo response of CdSe QD based Schottky device with (a) Au electrode and (b) Pd electrode. (Reprinted with permission from Ref. [13]. Copyright 2019: IEEE)

enhanced photo-response at ~400 nm corresponding to the blue region of the visible spectrum. The photo

response of the fabricated Schottky diode is shown in Fig. 10.2, where Fig. 10.2a shows the wide spectrum coverage (FWHM = 190 nm) of the device with Au electrode and Fig. 10.2b presents the improved photo response over blue region due to the resonance with Pd electrode thereby making detector a narrowband (FWHM = 61 nm) one.

In this chapter, various types of CdSe photodetectors are discussed in details. Section 10.2 focuses on the discussion of carrier dynamics and their applications in CdSe Photodetectors. Section 10.3 discusses different CdSe based photoconductors and Schottky photodiodes. Section 10.4 presents some CdSe heterojunction-based photodetectors while the Sect. 10.5 focuses on some hybrid photodetectors. Application of CdSe photodetectors are discussed in Sect. 10.6. Conclusion and future scopes are discussed in Sect. 10.7.

10.2 CdSe Carrier Dynamics

Charge carrier dynamics is the crucial part for determining the electrical characteristics of any electronic device. In case of photodetectors, transportation of photo-generated charge carriers defines the performance parameters such as the photoconductivity, photoconductive gain, transient response, and responsivity of the device. Charge carrier dynamics is generally evaluated by the lifetime measurement of the carriers. Lifetime of the carriers plays a vital role in determining the efficiency of the photodetectors. A photoexcited carrier can relax in the ground state and allow electron-hole scattering or electron-electron scattering [14]. Other possible way is that the photoexcited carrier may escape from the original site and relax on another site (trap site) or may be collected at the electrodes [14]. There should be one effective carrier lifetime (τ_{eff}) to describe the time elapsed before photoexcited carrier falls back into the ground state [14]. In photodetection operation, the photoexcited carriers must be collected by the electrodes within τ_{eff} period after their generation. Thus, the lifetime, τ_{eff}, of a carrier plays critical role for determining the responsivity and gain of the photodetectors. For the photoconductive detector one can write the relation of photocurrent (I_{ph}) in terms of incident flux density (ϕ), detector's quantum efficiency (η), and photoconductive gain (g) as $I_{ph} = e\phi\eta g$ [14] where g is the ratio of τ_{eff} to carrier transit time (τ_{tr}) or the ratio of the total collected carrier to total excited carriers, for both the thermally generated and photogenerated carriers [14]. Thus, one can easily notice that the improvement in τ_{eff} can improve the photoconductive gain and hence the photocurrent and the responsivity of the photoconductors or photodetectors. McGott et al. [18] estimated the carrier lifetime of $CdSe_xTe_{1-x}$ film on a $Mg_yZn_{1-y}O$ buffer layer using time resolved photoluminescence. They observed improvement in the carrier lifetime only after $x = 0.2$ mole fraction of Se [18]. They observed a response with a fast initial decay and a long-lived tail indicating the trap-dependent recombination in the film [18]. They further observed that the long-lived tail was absent for the samples prepared on Al_2O_3 instead of $Mg_yZn_{1-y}O$ due to lower density of surface trap states and better surface passivation [18]. Vietmeyer et al. [19] performed time-resolved lifetime (TCPSC) and transient differential absorption (TDA) analyses of CdSe nanowires to assess the carrier recombination process at 405 nm with 10 MHz repetition rate. They [19] observed that ~98% of the total signal constituted the short decay component (~100 ps). There were also other two components of ~450 ps and ~2.5 ns

constituting ~1.5% and ~0.5% of the signal, respectively. Mismatch was observed between the TCPSC (~100 *ps*) and TDA (~1.5 ns) kinetics due to the fast hole trapping and long-lived electrons in the CdSe nanowire conduction band [19]. The measurement results showed improvements in the photoconductive gain and photoresponsivity with the increase in the diameter of the CdSe nanowires [19]. Kung et al. [20] investigated the performance of CdSe nanowires based photoconductors with varying diameters. They observed the improvement in the photoconductive gain from $g = 0.017$ to $g = 4.9$ corresponding to the increase in the diameter of the nanowires from 10 nm to 100 nm. The improvement was also observed in the responsivity with the increase in the diameter of the CdSe nanowires. The results were consistent with the observations of Vietmeyer et al. [19]. However, they noted that the improvements in gain and responsivity were at the expense of deteriorating transient response of the device from 8 *μs* (at 10 *nm* diameter) to 8 s (*at* 100 *nm* diameter) [20].

Effect of doping on the carrier lifetime of CdSe QDs were investigated by Straus et al. [15]. They fabricated a FET by using a CdSe material-based channel deposited on an Indium film. The combination of CdSe QDs/Indium thin films was annealed at 300 °C to allow diffusion of Indium into the CdSe QDs. A schematic of the indium diffusion into the CdSe QDs film with annealing temperature is shown in Fig. 10.3a. The mobility and lifetime of the carriers were increased with the higher thickness of indium film due to enhanced diffusion of Indium into the CdSe QDs as indicated in Fig. 10.3b. Kongkanand et al. [16] assembled CdSe QDs over TiO_2 nanoparticles and nanaotubes for photovoltaic applications. The carrier lifetimes of 0.4 *ns* and 1.3 ns were observed in CdSe/TiO_2 nanoparticle and CdSe/TiO_2 nanotubes, respectively, which were smaller than 4.1 ns carrier lifetime of CdSe QDs [16]. The decrease in lifetime was attributed to the charge transfer from CdSe QDs

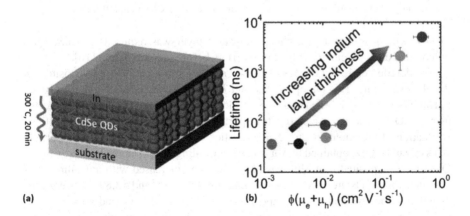

Fig. 10.3 (a) Schematic diagram indicating indium doping via annealing procedure and (b) Lifetime variation with varying thickness of Indium against quantum yield mobility product. (Reprinted with permission from Ref. [15]. Copyright 2015: American Chemical Society)

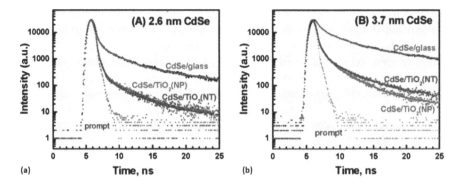

Fig. 10.4 Emission decay of CdSe quantum dots (**a**) for size 2.6 nm, and (**b**) for size 3.7 nm. The experiment also indicates the results with TiO_2 nanoparticles and TiO_2 nanotubes. (Reprinted with permission from Ref. [16]. Copyright 2008: American Chemical Society)

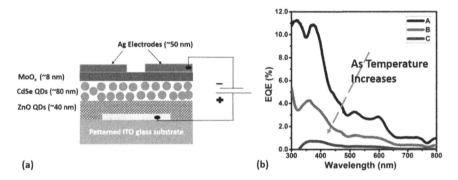

Fig. 10.5 (**a**) Schematic of fabricated CdSe QDs based photodetector with varying ZnO QDs size and (**b**) EQE of the detector showing effect of increased temperature on ZnO QDs acting as a charge transport layer. (Adapted with permission from Ref. [17]. Copyright 2017: IEEE)

to TiO_2 which implied that a charge transport layer was required for CdSe QDs based devices to efficiently collect the carriers before recombination [16].

The detailed emission decays of CdSe QDs of 2.6 nm and 3.7 nm deposited on the glass substrate, TiO_2 nanoparticles, and TiO_2 nanotubes are shown in Fig. 10.4a, b. Kumar et al. [17] fabricated a CdSe colloidal QDs (CQDs) based photodetector using ZnO CQDs as an electron transport layer (ETL) as shown in Fig. 10.5a. Evaluation of the photodetector characteristics was carried out for varying particle sizes of ZnO CQDs obtained by varying the annealing temperature. The size of ZnO QDs in the deposited thin film was observed to be increased with annealing temperature. ZnO QDs are reported to be 4.42 nm, 8.74 nm, and 13.89 nm were estimated at respective annealing temperatures of 250 °C, 350 °C, and 450 °C. They [17] observed that the transient response, contrast ratio, responsivity, detectivity, and external quantum efficiency (EQE) of the detector were deteriorated with the increased size of the ZnO QDs at higher annealing temperatures. The EQE characteristics of the photodetector with ZnO QDs ETL have been shown in Fig. 10.5b. Osedach et al. [21] used thermally evaporated spiro-TPD for the charge transport layer sensitized by CdSe QDs over a 10 μm finger width (or channel length). The photo response was shown to be very much dominated over the blue region of the spectrum, and the EQE followed a power square law function of the bias voltage due to the space charge limitation in spiro-TPD. The EQE peak observed at 400 nm confirmed the efficient dissociation of excitons at the interface of spiro-TPD and CdSe QDs.

10.3 Photoconductors and Schottky Photodiodes

Photoconductors are the simplest form of light-detecting devices which work on the principle of conductivity modulation by incident photons to be detected. The photoconductor contains a photoactive material with two electrodes placed on it to connect the device with an external circuit. When light is incident on the active material, its conductivity is increased (hence the resistivity is decreased) due to excess electron-hole pairs generated by absorption of photons in the photoactive material. When a potential is applied across the two electrodes of the device, the decreased resistivity increases the current flowing through the external circuit. On the other hand, the Schottky photodiodes are fabricated by forming rectifying contact on a semiconductor. The work function the metal must be larger (smaller) than that of an n-type (p-type) semiconductor for the Schottky contact formation. Since CdSe is inherently an n-type semiconductor, one needs to select a metal with work function larger than that of the CdSe for forming the Schottky junction based CdSe photodiodes [25].

The photoconductivity of the CdSe largely depends on the trap states which may lead to increased transient time. Skarman [26] carried out a detailed investigation on the photoconductivity of CdSe material with the trap states under a varying temperature. The effect of traps was found to be more dominating in single crystalline CdSe thin films than the polycrystalline films. However, the trap states were shown to be reduced by using controlled fabrication and activation techniques [26]. Margulis and Sibbett [27] fabricated an ultrafast CdSe based photoconductor using two Al electrodes placed with a separation distance of 40–100 μm and reported a transient response of 20 ps. Jiang et al. [22] fabricated a single crystalline CdSe nano-ribbons based photodetector using Ti/Au electrodes and observed photo response over 400–710 nm as shown in Fig. 10.6. Presence of various traps at different energy levels (shallow and deep in the bandgap) was observed. The device showed a transient response of <1 ms. A SiO_2 coating over the CdSe nano-ribbons was used to reduce the trap states and improve the transient response of the device at the expense of reduced sensitivity [22]. Researchers have also used variational technique for the efficient charge separation in CdSe. Xu et al. [23] used site-controlled growth of aligned CdS/CdSe core-shell nanowalls based photodetectors at wafer scale without post-growth transfer. The SEM image of the interdigitated

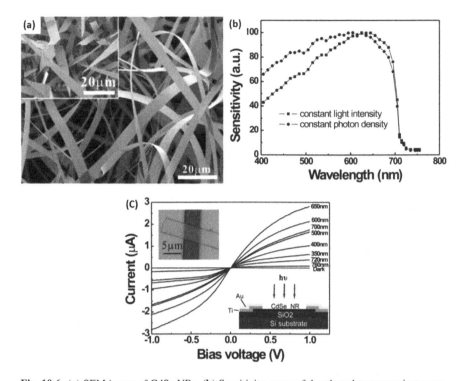

Fig. 10.6 (a) SEM image of CdSe NRs, (b) Sensitivity curve of the photodetector against wavelength, and (c) I-V characteristics of the device and structure of the device. (Adapted with permission from Ref. [22]. Copyright 2007: John Wiley and Sons)

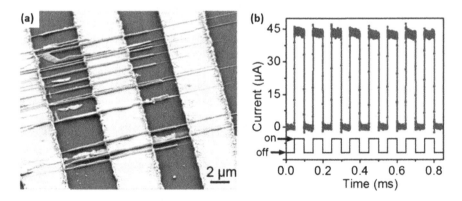

Fig. 10.7 (a) SEM image depicting the structure of the photodetector and (b) Transient response of the detector under 405 nm light at 10 V bias. (Adapted with permission from Ref. [23]. Copyright 2018: John Wiley and Sons)

detector structure is shown in Fig. 10.7a. The fabricated photoconductor showed a high gain of 3800 and photoresponsivity of 1200 A/W with an ultrafast transient response of 200 ns as shown in Fig. 10.7b [23]. The superior photo response and gain of the detector were attributed to enhanced charge-separation efficiency of core-shell nanowall geometry [23].

Some researchers have achieved spectral shift in CdSe photodetectors with efficient charge carrier separation techniques. Das et al. [28] fabricated a CuO/CdSe core-shell nanowire based photoconductor using Ag electrodes. Note that the bandgaps of CuO and CdSe are 1.2 eV and 1.7 eV respectively. However, the bandgap of CuO/CdSe nanowire structures was shown to be 3.96 eV which is much higher than the bandgaps of both the CdSe and CuO. The high bandgap of CuO/CdSe nanowire structure was explored for fabricating photodetector to work in deep UV region with a responsivity of 0.63 A/W at 254 nm wavelength [28]. The high bandgap was attributed to improved separation of photogenerated electron-hole pairs due to the Type-II band alignment of CuO/CdSe heterojunction formation.

It is well known that CdSe is inherently an n-type material. Wu et al. [24] studied the effect of impurity on the photoconductivity of intrinsic and n-type CdSe nanobelts (NB). They observed that the intrinsic CdSe NBs based device possessed a high gain while the n-type CdSe NBs based device showed a fast response. The above behavioral studies of the intrinsic and n-type CdSe NBs were carried out in both the Schottky Device (CdSe/Ni/Au) and Photoconductive device (CdSe/In/Au) with a channel length of 2.5 μm. The measured I-V characteristics of the devices are shown in Fig. 10.8a, while their transient responses are shown in Fig. 10.8b [24]. Ani et al. [29], introduced Cu impurity in CdSe thin films by vacuum annealing at 350 ° C. They observed an improved photoconductive gain due to the sensitizing effect of Cu centers in the CdSe lattice.

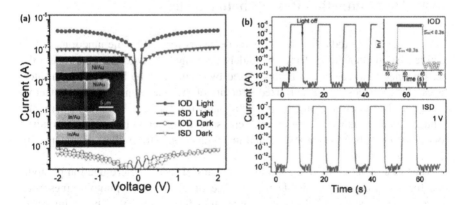

Fig. 10.8 (a) I-V relation of interdigitated Ohmic Device (IOD) and interdigitated Schottky device (ISD) under dark and under illumination of 633 nm light, inset indicates the FESEM image of the device, and (b) Transient response of both IOD and ISD device. (Adapted with permission from Ref. [24]. Copyright 2011: American Chemical Society)

The environmental conditions affect the photoconductivity of the CdSe based devices. Samanta et al. [8] studied the stabilization of CdSe thin films under air oxidation and observed stable photocurrent and dark current over a period of 12–25 days in room temperature depending on the thickness of the films. However, the CdSe thin films showed temperature-dependent quenching of photoconductivity at 265 °K [8]. When the graphite based Ohmic contact was formed on the films, photosensitivity of the device was increased with air exposure due to the reduction of free electrons owing to the oxygen diffusion along the grain boundaries of the CdSe in the films. The fabricated CdSe thin film based device also exhibited time varying photosensitivity due to slow recombination states and oxygen assisted conversion of selenium and cadmium vacancies [8]. An et al. [30] used one-step thermal evaporation method to synthesize hollow tubular CdSe nanotubes for fabricating Ag electrode based CdSe/Ag Schottky junction photodiodes. Their device showed a high photoresponsivity of 76 A/W with a high photoconductive gain of 190 at low operating voltage of −1 V. The performance of the device was observed to be strongly influenced by the presence of oxygen molecules in the atmosphere. Further, the dark current of the device was noted to be reduced with the increased content of oxygen molecules [30].

The photoconductive gain of the CdSe photodetector also depends of the metal electrodes used in the devices. Mehta et al. [25] achieved greater than unity photoconductive gain in Au/CdSe thin film Schottky photodiodes. It was attributed to the trapping of photogenerated holes in the Schottky junction which resulted in an injection of electrons in CdSe thin film by reducing the Schottky barrier. Photoconductive gain of 300 was measured in CdSe thin film-based Schottky device under an illumination of 800 nm wavelength.

10.4   Heterojunction Based Photodetectors

Heterojunction photodiodes are the photodetectors formed by the junctions of two different semiconductors. They are widely investigated photodetectors where the performance parameters can be optimized by modifying the heterojunction interfaces of the devices. The generalized structure of the heterojunction photodetectors is metal-electrode/semiconductor-1/semiconductor-2/metal-electrode where the semiconductor-1 and semiconductor-2 represent two different types of semiconductors of different band gap energies and metal electrodes form ohmic contacts with the semiconductors.

Yuan et al. [31] fabricated a CdSe nanoplates/2D MoS_2 heterojunction photodiode for operating in the visible region. The time resolution photoluminescence spectroscopies showed efficient charge transfer across the heterointerface. Suppression of photons within CdSe of the CdSe/MoS_2 heterojunction is demonstrated by their Streak camera images at 710 nm shown in Fig. 10.9a, b. The lifetime measurements of CdSe and CdSe/MoS_2 heterojunctions are shown in Fig. 10.9c. The lifetime of carriers in CdSe/MoS_2 is only 118.9 ps which is much smaller than

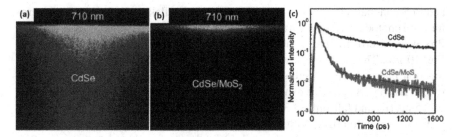

Fig. 10.9 Streak camera images recorded at 710 nm (**a**) CdSe only, (**b**) CdSe/MoS$_2$ heterostructure, and (**c**) Indicating the lifetime observed in 1 dimension indicating efficient charge separation in CdSe when in contact with MoS$_2$. (Adapted with permission from Ref. [31]. Copyright 2020: Elsevier)

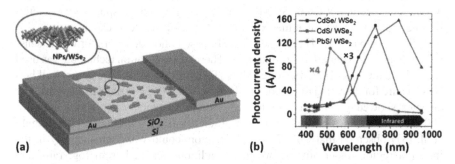

Fig. 10.10 (**a**) Schematic of the fabricated device consisting CdS, CdSe, and PbS NPs with *WSe$_2$* crystalline flakes, and (**b**) Photocurrent against wavelength indicating the spectrum selectivity and tunability with CdSe/*WSe$_2$*, CdS/*WSe$_2$*, and PbS/*WSe$_2$*. (Reprinted from Ref. [32]. Published 2022 by American Chemical Society as open access)

the lifetime of 1602 ps of the carriers of CdSe material. This shows that efficient charge separation in CdSe may take place when it is in contact with the MoS$_2$ [31]. Ren et al. [33] fabricated *CH$_3$NH$_3$PbI$_3$* nanowire arrays/single CdSe NB based heterojunction photodiode to operate in 400–800 nm with contrast ratio of 2.152×10^3, responsivity of 69.11 A/W, EQE of 11,000%, and detectivity of 8.6×10^{12} Jones. The device showed a rise time (τ_r) of 0.81 ms and fall time (τ_f) 0.77 ms. In order to improve the response, Medda et al. [34] prepared a heterojunction between 2D CdSe nanoplatelets (NPL) and *Au$_{25}$* nanoclusters (NCs) for photodetection application. The DFT based analysis of the CdSe NPL and *Au$_{25}$* NCs heterostructure showed the replacement of the conduction band of CdSe by Au. The device showed a detectivity of 2.5×10^{11} *Jones* and transient response of 200 ms. Luo et al. [35] explored a type-II heterojunction structure of CdSe/CdTe core-shell configuration to demonstrate a rise-time of 0.7 s (τ_r) and fall-time of 0.5 s (τ_f). The device showed a detection range extending from UV to NIR region. The optical properties of the device were improved by applying a compressive load over the device.

CdSe based heterostructures are employed to tune the spectrum coverage of the photodetectors. Ghods et al. [32] used the SILAR method to synthesize CdS, CdSe, and PbS nanoparticles directly over WSe_2 crystalline flakes to fabricate the hetero-junction photodetectors shown in Fig. 10.10a. The fabricated CdS/WSe_2, CdSe/WSe_2, PbS/WSe_2 heterostructures exhibited spectrum selective characteristics as shown in Fig. 10.10b. The response peaks of CdS/WSe_2, CdSe/WSe_2 and PbS/WSe_2 hetero-junction photodetectors were observed at 510 nm, 700 nm, and 810 nm, respectively. The reported device exhibited a quantum efficiency of 71% with a rise-time of 2.5 ms (τ_r) and fall-time of 3.5 ms (τ_f).

CdSe was introduced with other photoactive materials to improve the photo response of the photodetectors. Li et al. [36] fabricated a photodetector using PbI_2 nanosheet with CdSe NB (Nanobelt) to improve the detection capability of PbI_2 based photodetectors. The composite structure exhibited a broad spectral response over 400–730 nm. It was found the PbI_2 /CdSe NB interface improved the separation of the photogenerated e-h pairs, which in turn, improved the performance of the device. Introduction of the CdSe NB with PbI_2 improved the responsivity, detectivity, and EQE of the device at the expense of degraded transient response.

Villa-Angulo et al. [37] proposed CdSeTe based photodetectors for image sensing applications under adverse conditions. Thermal evaporation technique was used to fabricate a Glass/ITO/CdS/CdSe/CdSe$_{0.4}$Te$_{0.6}$/CdSe/Ag based quantum-well heterostructure for detecting of photons in the range of 500–1100 nm wavelengths. The detector showed good responsivity and specific detectivity for wavelengths of 500–1100 nm with a power conversion efficiency (PCE) of 10.4%. Villa-Angulo et al. [37] observed that CdSeTe based detectors could work satisfactorily in the infra-red region under adverse weather conditions resulted from fog, rain, and snow fall.

10.5   CdSe-Organic Hybrid Photodetectors

Organic semiconducting materials are extensively explored for fabricating large-area optoelectronic devices due to their flexible nature and cost-effective fabrication methods. However, the low light absorption coefficient and poor carrier mobility of the organic materials have restricted their use for optoelectronic device applications. Some researchers have explored hybrid heterojunctions formed between an organic semiconductor and an inorganic semiconductor to utilize the benefits of both the materials. Wang et al. [38] fabricated a hybrid photodetector using P3HT polymer and CdSe nanowires to explore low ionization potential of organic molecules of P3HT and high electron affinity of CdSe semiconductor. The I-V curve of the reported device is shown in Fig. 10.11a. They fabricated the photodetector on Si, PET, and printing paper to demonstrate the possibility of achieving a CdSe based flexible photodetector shown in the inset of Fig. 10.11a. The photodetector was shown to work over a broad spectral region of 400–710 nm as shown in Fig. 10.11b. The Si substrate-based detector showed a transient response of 0.1 s. The PET and printing paper substrate-based detectors showed transient response of about 10 ms. The latter two devices exhibited detection characteristics with 180° bending. They

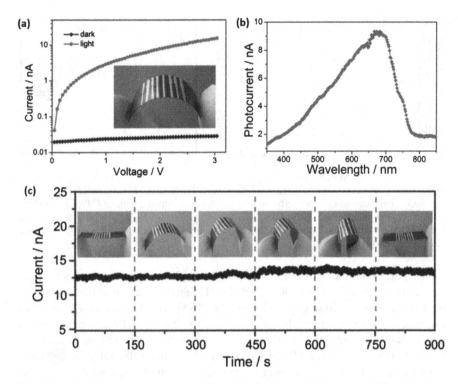

Fig. 10.11 (**a**) I-V characteristics of hybrid photodetector over PET substrate with and without illumination, inset shows the optical image of flexible detector, (**b**) Photo response of the flexible detector against different wavelengths, and (**c**) I-t curve of the flexible device under a constant bias of 3 V evaluating hybrid detector under different bending curvatures. (Adapted with permission from Ref. [38]. Copyright 2013: John Wiley and Sons)

observed a stable performance under different bending curvatures as shown in Fig. 10.11c.

Shih et al. [39] used CdSe colloidal quantum dots (CQDs) and rGO based heterojunction photodetector using low cost spin coating method. The fabricated detector was shown to have a very high ON/OFF current ratio of ~2195 for −1 V reverse bias voltage. The device showed a broadband spectral characteristic ranging from 350 nm (UV) to 900 nm (NIR). A huge factor of improvement in the figure of merit was achieved by decorating rGO with CdSe QDs. The EQE and detectivity of the photodetector are shown in Fig. 10.12a, b. The CdSe CQD based device is shown in the inset of the Fig. 10.12b.

Oertel et al. [40] fabricated the first hybrid structure photodetector using CdSe QDs and PEDOT:PSS. They used the solution processed CdSe QDs to grow a 200 nm film on the PEDOT:PSS polymer composite. The charge transport properties and exciton dissociation efficiency of the CdSe QDs thin film were improved by using n-butylamine. The result showed a blue shift in the band gap of the CdSe QDs thereby extending its photo response towards the visible region of the spectrum with

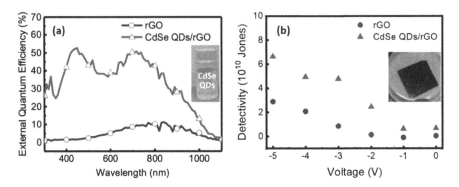

Fig. 10.12 (**a**) External Quantum efficiency of the detector with and without CdSe QDs and (**b**) Detectivity with and Without CdSe QDs, also shows synthesized CdSe QDs, and Spin Coated Large area rGO/CdSe QDs based photodetector in the inset. (Adapted from Ref. [39]. Published 2020 by IEEE as open access)

a cut-off at 600 nm. A contrast ratio of 100 of the hybrid heterojunction photodetector was observed at ~514 nm. Ramar et al. [41] fabricated a hybrid heterostructure photodetector using CdSe QDs and P3HT-PC71BM. The device showed a spectral response in 350–800 nm wavelengths. The P3HT:CdSe:PC71BM active layer based device showed a detectivity of 10 times higher than the device with the heterojunction of P3HT: PC71BM and P3HT:CdSe. This shows that CdSe QDs can be mixed with organic polymers to improve the spectral broadening, responsivity and detectivity of the hybrid photodetectors. Malik et al. [42] improved the photo response of a MEH-PPV and VOPcPhO heterojunction photodetector by mixing CdSe QDs in the materials. Authors observed that the introduction of CdSe QDs not only improved the absorption in the region of 400–560 nm but also improved the transient response two times faster. Apart from using CdSe QDs, other CdSe nanostructures can also be used. Dutta et al. [43] fabricated a hybrid heterojunction photodetector using 2D CdSe nanoplatelets (NPL) and phenothiazine (PTZ) heterostructure materials as the active layer in the device. They observed photoluminescence quenching and shortening of average decay time of CdSe QDs when PTZ was mixed with 2D CdSe NPL. Further, the recombination of excitons in the CdSe NPL. Photocurrent for the PTZ/CdSe NPL was shown to be ten-fold higher than that of the only CdSe NPL based photodetectors. The photodetector also showed an EQE of 40%, detectivity of 4×10^{11} *Jones*, and responsivity of 160 mA/W.

10.6 Application of CdSe Based Photodetectors

CdSe based photodetectors are actively used in polarization sensitive detectors. Polarization is used to distinguish the artificial objects in complicated environments. Polarization sensitive devices rarely operate in UV region due to difficulty in fabricating anisotropic materials and optical elements for polarization modulation. Ge et al. [46] developed a polarization sensitive device by combining an electron-multiplying charge-coupled device (EMCCD) with a polarized luminescence down-shifting material. They used CdSe@CdS-dot-in-rods (CdSe Dots in CdS rods) for the downscaling the UV light into visible light where the down converted visible light was allowed to be absorbed by an Si detector EMCCD (electron-multiplying charge-coupled device). Singh at al [47] also fabricated a polarization sensitive photodetector using solution processed CdSe nanowires with a bandgap of 1.75 eV. They observed a consistent polarization sensitivity in single as well as ensembled CdSe NWs. The fabricated detector showed a wide spectral response ranging from 350 to 720 nm. Though the authors claimed the device as a photoconductor but they did not provide the details of the metallic contacts used in the device [47].

CdSe based detectors integrated with Field Effect Transistors (FET) are used for imaging systems. Shalev et al. [44] synthesized guided CdSe nanowires on five different planes of sapphire (substrate planes) shown in Fig. 10.13a, b. They fabricated a CdSe nanowires based FET to examine the electronic and optoelectronic properties of the material as shown in Fig. 10.13c, d. Cr/Au metallic contacts and Al_2O_3

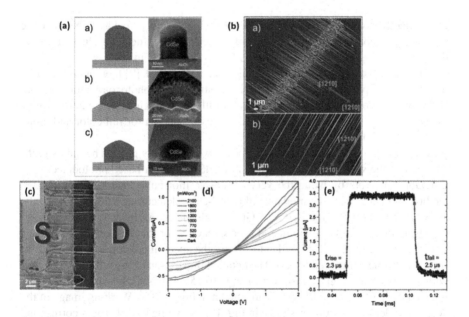

Fig. 10.13 (a) Three modes of guided growth with HRSEM image, (b) HRSEM image of Graphotepitaxial growth of CdSe NWs, (c) SEM image indicating the device with source, drain and CdSe NWs, (d) I-V curve of the device under different illumination at 473 nm, and (e) Transient response of the device (at 10 kHz, 2 V bias, and 520 mW/cm^2). (Adapted from Ref. [44]. Published 2017 by American Chemical Society as open access)

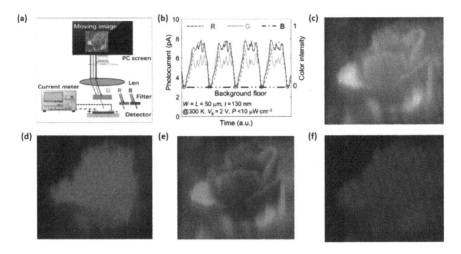

Fig. 10.14 (**a**) Experimental setup for capturing a image using external filter, (**b**) Photocurrent against time measured using CdSe/mica detector, (**c**) Full color image obtained using experimental setup, (**d**) Red mono color image, (**e**) Green mono color image, and (**f**) Blue mono color image. (Reprinted with permission from Ref. [9]. Copyright 2022: Springer Nature)

gate dielectric oxide were used in the FET to achieve 5 μm active area (between the interdigitated contacts) of the device [44]. The fabricated detector showed a rise-time of 2.3 μs (τ_r) and fall-time of 2.5 μs (τ_d) as shown in Fig. 10.13e.

To achieve a complete visible imaging system, Pan et al. [9] fabricated the CdSe thin film based photodetectors on flexible substrates. The photoconductor was analyzed over the complete visible spectrum for full-scale imaging using external filters as shown in Fig. 10.14a, b. The obtained results are shown for full color and mono color images in Fig. 10.14c–f.

The photo response and transient response were not very good. The only significant part of the work was the fabrication of CdSe thin film-based photodetector on a flexible substrate by using the very expensive molecular beam epitaxy (MBE) method. Kim et al. [45] reported a flexible full-color detector utilizing all QDs only. They used CdSe QDs for Red and Green light detectors by varying their sizes (bandgap tuning) from 7 to 5 nm and used CdS QDs for the Blue region also. They introduced PbS QDs in CdSe QDs to extend the spectral coverage of the detector from the visible to infrared region. The detailed illustration of the full-color detector with its absorption region is shown in Fig. 10.15a, b while the schematics of the infrared, red, green, and blue pixels are shown in Fig. 10.15c. Working image of the detector under illumination is shown in Fig. 10.15d, while Fig. 10.15e, f correspond to the cross-sectional area of the detector obtained from the HRSEM.

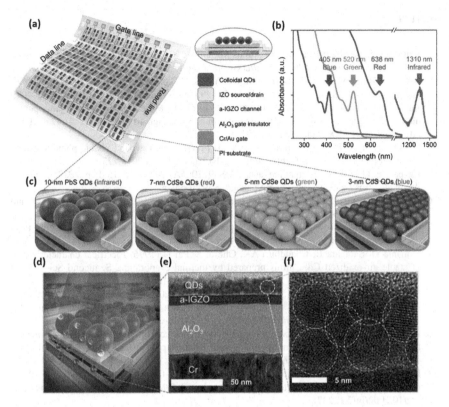

Fig. 10.15 (**a**) Full-color detector array fabricated using QDs and Structure of single-pixel, (**b**) Absorption characteristics of the QDs involved, and (**c**) Different devices for different regions of the spectrum are arranged as per the QDs size, (**d**) Working image of the detector under illumination, (**e**) and (**f**) corresponds to depicted cross-sectional area HRSEM. (Reprinted from Ref. [45]. Published 2019 by AAAS as open access)

10.7 Summary

The present chapter introduces various state-of-the-art developments in the area of CdSe based photodetectors operating in a wide range of wavelengths over visible to NIR region. It is observed that CdSe is a potential material for photodetection applications due its direct bandgap nature and easy synthesis routes. Various novel CdSe nanostructures of CdSe are believed to explored effectively for developing new generation photodetectors for imaging systems. Further, the solution-processed and low-temperature fabrication processes of CdSe based photodetectors are suitable for CdSe based flexible photodetectors.

References

1. Singh V (2007) Albert Einstein: his Annus Mirabilis 1905. arXiv:physics/0701240. Accessed 15 Feb 2022 [Online]. Available http://arxiv.org/abs/physics/0701240
2. Baeg K-J, Binda M, Natali D, Caironi M, Noh Y-Y (2013) Organic light detectors: photodiodes and phototransistors. Adv Mater 25(31):4267–4295. https://doi.org/10.1002/adma.201204979
3. Jansen-van Vuuren RD, Armin A, Pandey AK, Burn PL, Meredith P (2016) Organic photodiodes: the future of full color detection and image sensing. Adv Mater 28(24):4766–4802. https://doi.org/10.1002/adma.201505405
4. Lin Q, Armin A, Burn PL, Meredith P (2015) Filterless narrowband visible photodetectors. Nat Photonics 9(10):687–694. https://doi.org/10.1038/nphoton.2015.175
5. Bayer BE (1976) Color imaging array. US3971065A. Accessed 15 Feb 2022. [Online]. Available https://patents.google.com/patent/US3971065/en
6. Jin W, Hu L (2019) Review on quasi one-dimensional CdSe nanomaterials: synthesis and application in photodetectors. Nano 9(10):1359. https://doi.org/10.3390/nano9101359
7. Ninomiya S, Adachi S (1995) Optical properties of cubic and hexagonal CdSe. J Appl Phys 78(7):4681–4689. https://doi.org/10.1063/1.359815
8. Samanta D, Samanta B, Chaudhuri AK, Ghorai S, Pal U (1996) Electrical characterization of stable air-oxidized CdSe films prepared by thermal evaporation. Semicond Sci Technol 11(4):548–553. https://doi.org/10.1088/0268-1242/11/4/016
9. Pan W, Liu J, Zhang Z, Gu R, Suvorova A, Gain S et al (2022) Large area van der Waals epitaxy of II–VI CdSe thin films for flexible optoelectronics and full-color imaging. Nano Res 15(1):368–376. https://doi.org/10.1007/s12274-021-3485-x
10. Eyring MB, Martin P (2013) Spectroscopy in forensic science. In: Reference module in chemistry, molecular sciences and chemical engineering. Elsevier, p B978012409547205455X. https://doi.org/10.1016/B978-0-12-409547-2.05455-X
11. Li K, Lu Y, Fu XL, He J, Lin X, Zheng J et al (2021) Filter-free self-power CdSe/Sb$_2$(S$_{1-x}$,Se$_x$)$_3$ near infrared narrowband detection and imaging. InfoMat 3(10):1145–1153. https://doi.org/10.1002/inf2.12237
12. Luo L-B, Xie W-J, Zou Y-F, Yu Y-Q, Liang F-X, Huang Z-J, Zhou K-Y et al (2015) Surface plasmon propelled high-performance CdSe nanoribbons photodetector. Opt Express 23(10):12979. https://doi.org/10.1364/OE.23.012979
13. Kumar H, Kumar Y, Mukherjee B, Rawat G, Kumar C, Pal BN et al (2019) Effects of optical resonance on the performance of metal (Pd, au)/CdSe quantum dots (QDs)/ZnO QDs optical cavity based spectrum selective photodiodes. IEEE Trans Nanotechnol 18:365–373. https://doi.org/10.1109/TNANO.2019.2907529
14. Kochman B, Stiff-Roberts AD, Chakrabarti S, Phillips JD, Krishna S, Singh J et al (2003) Absorption, carrier lifetime, and gain in InAs-GaAs quantum-dot infrared photodetectors. IEEE J Quantum Electron 39(3):459–467. https://doi.org/10.1109/JQE.2002.808169
15. Straus DB, Goodwin ED, Gaulding EA, Muramoto S, Murray CB, Kagan CR (2015) Increased carrier mobility and lifetime in CdSe quantum dot thin films through surface trap passivation and doping. J Phys Chem Lett 6(22):4605–4609. https://doi.org/10.1021/acs.jpclett.5b02251
16. Kongkanand A, Tvrdy K, Takechi K, Kuno M, Kamat PV (2008) Quantum dot solar cells. Tuning photoresponse through size and shape control of CdSe−TiO$_2$ architecture. J Am Chem Soc 130(12):4007–4015. https://doi.org/10.1021/ja0782706
17. Kumar H, Kumar Y, Rawat G, Kumar C, Mukherjee B, Pal BN et al (2017) Heating effects of colloidal ZnO quantum dots (QDs) on ZnO QD/CdSe QD/MoOx photodetectors. IEEE Trans Nanotechnol 16(6):1073–1080. https://doi.org/10.1109/TNANO.2017.2761785
18. McGott DL, Good B, Fluegel B, Duenow JN, Wolden CA, Reese MO Carrier lifetime as a function of Se content for CdSe$_x$ Te$_{1-x}$ films grown on Al$_2$O$_3$ and MgZnO. In: 2021 IEEE 48th photovoltaic specialists conference (PVSC), Fort Lauderdale, FL, USA, pp 1301–1303. https://doi.org/10.1109/PVSC43889.2021.9518914

19. Vietmeyer F, Frantsuzov PA, Janko B, Kuno M (2011) Carrier recombination dynamics in individual CdSe nanowires. Phys Rev B 83(11):115319. https://doi.org/10.1103/PhysRevB.83.115319

20. Kung S-C, Xing W, van der Veer WE, Yang F, Donavan KC, Cheng M et al (2011) Tunable photoconduction sensitivity and bandwidth for lithographically patterned nanocrystalline cadmium selenide nanowires. ACS Nano 5(9):7627–7639. https://doi.org/10.1021/nn202728f

21. Osedach TP, Geyer SM, Ho JC, Arango AC, Bawendi MG, Bulović V (2009) Lateral heterojunction photodetector consisting of molecular organic and colloidal quantum dot thin films. Appl Phys Lett 94(4):043307. https://doi.org/10.1063/1.3075577

22. Jiang Y, Zhang WJ, Jie JS, Meng XM, Fan X, Lee S-T (2007) Photoresponse properties of CdSe single-nanoribbon photodetectors. Adv Funct Mater 17(11):1795–1800. https://doi.org/10.1002/adfm.200600351

23. Xu J, Rechav K, Popovitz-Biro R, Nevo I, Feldman Y, Joselevich E (2018) High-Gain 200 ns photodetectors from self-aligned CdS-CdSe core-shell nanowalls. Adv Mater 30(20):1800413. https://doi.org/10.1002/adma.201800413

24. Wu P, Dai Y, Sun T, Ye Y, Meng H, Fang X et al (2011) Impurity-dependent photoresponse properties in single CdSe nanobelt photodetectors. ACS Appl Mater Interfaces 3(6):1859–1864. https://doi.org/10.1021/am200043c

25. Mehta RR, Sharma BS (1973) Photoconductive gain greater than unity in CdSe films with Schottky barriers at the contacts. J Appl Phys 44(1):325–328. https://doi.org/10.1063/1.1661881

26. Skarman JS (1965) On the relationship between photocurrent decay time and trap distribution in CdS and CdSe photoconductors. Solid State Electron 8(1):17–29. https://doi.org/10.1016/0038-1101(65)90005-5

27. Margulis W, Sibbett W (1983) Picosecond CdSe photodetector. Appl Phys Lett 42(11):975–977. https://doi.org/10.1063/1.93820

28. Das B, Sa K, Mahakul PC, Subramanyam BVRS, Das S, Alam I et al (2018) Efficient ultraviolet photodetector device based on modulated wide band gap Type-II CuO/CdSe core-shell nanowires. Superlattice Microst 123:234–241. https://doi.org/10.1016/j.spmi.2018.08.021

29. Al-Ani SKJ, Mohammed HH, Al-Fwade EMN (2002) The optoelectronic properties of CdSe:Cu photoconductive detector. Renew Energy 25(4):585–590. https://doi.org/10.1016/S0960-1481(01)00088-X

30. An Q, Meng X, Xiong K, Qiu Y, Lin W (2017) One-step synthesis of CdSe nanotubes with novel hollow tubular structure as high-performance active material for photodetector. J Alloys Compd 726:214–220. https://doi.org/10.1016/j.jallcom.2017.07.336

31. Yuan Y, Zhang X, Liu H, Yang T, Zheng W, Zheng B et al (2020) Growth of CdSe/MoS$_2$ vertical heterostructures for fast visible-wavelength photodetectors. J Alloys Compd 815:152309. https://doi.org/10.1016/j.jallcom.2019.152309

32. Ghods S, Esfandiar A, Zad AI, Vardast S (2022) Enhanced photoresponse and wavelength selectivity by SILAR-coated quantum dots on two-dimensional WSe$_2$ crystals. ACS Omega 7(2):2091–2098. https://doi.org/10.1021/acsomega.1c05591

33. Ren W, Tan Q, Wang Q, Liu Y (2021) Hybrid organolead halide perovskite microwire arrays/single CdSe nanobelt for a high-performance photodetector. Chem Eng J 406:126779. https://doi.org/10.1016/j.cej.2020.126779

34. Medda A, Dutta A, Bain D, Mohanta MK, De Sarkar A, Patra A (2020) Electronic structure modulation of 2D colloidal CdSe nanoplatelets by au $_{25}$ clusters for high-performance photodetectors. J Phys Chem C 124(36):19793–19801. https://doi.org/10.1021/acs.jpcc.0c04774

35. Luo J, Zheng Z, Yan S, Morgan M, Zu X, Xiang X, Zhou W (2020) Photocurrent enhanced in UV-vis-NIR photodetector based on CdSe/CdTe core/shell nanowire arrays by piezo-phototronic effect. ACS Photonics 7(6):1461–1467. https://doi.org/10.1021/acsphotonics.0c00122

36. Li C, Li W, Cheng M, Yang W, Tan Q, Wang Q, Liu Y (2021) High sensitive and broadband photodetectors based on hybrid PbI$_2$ nanosheet/CdSe nanobelt. Adv Opt Mater 9(20):2100927. https://doi.org/10.1002/adom.202100927

37. Villa-Angulo C (2020) $CdSe_{0.4}Te_{0.6}$ quantum well-based photodetector toward imaging vision sensors. IEEE Sensors J 20(22):13357–13363. https://doi.org/10.1109/JSEN.2020.3006219

38. Wang X, Song W, Liu B, Chen G, Chen D, Zhou C, Shen G (2013) High-performance organic-inorganic hybrid photodetectors based on P3HT:CdSe nanowire heterojunctions on rigid and flexible substrates. Adv Funct Mater 23(9):1202–1209. https://doi.org/10.1002/adfm.201201786

39. Shih Y-H, Chen Y-L, Tan J-H, Chang SH, Uen W-Y, Chen S-L et al (2020) Low-power, large-area and high-performance CdSe quantum dots/reduced graphene oxide photodetectors. IEEE Access 8:95855–95863. https://doi.org/10.1109/ACCESS.2020.2995676

40. Oertel DC, Bawendi MG, Arango AC, Bulović V (2005) Photodetectors based on treated CdSe quantum-dot films. Appl Phys Lett 87(21):213505. https://doi.org/10.1063/1.2136227

41. Ramar M, Kajal S, Pal P, Srivastava R, Suman CK (2015) Study of binary and ternary organic hybrid CdSe quantum dot photodetector. Appl Phys A Mater Sci Process 120(3):1141–1148. https://doi.org/10.1007/s00339-015-9293-y

42. Ashraf Malik H, Aziz F, Asif CM, Raza E, Najeeb MA, Ahmad Z et al (2016) Enhancement of optical features and sensitivity of MEH-PPV/VOPcPhO photodetector using CdSe quantum dots. J Lumin 180:209–213. https://doi.org/10.1016/j.jlumin.2016.08.038

43. Dutta A et al (2020) Hybrid nanostructures of 2D CdSe nanoplatelets for high-performance photodetector using charge transfer process. ACS Appl Nano Mater 3(5):4717–4727. https://doi.org/10.1021/acsanm.0c00728

44. Shalev E, Oksenberg E, Rechav K, Popovitz-Biro R, Joselevich E (2017) Guided CdSe nanowires parallelly integrated into fast visible-range photodetectors. ACS Nano 11(1):213–220. https://doi.org/10.1021/acsnano.6b04469

45. Kim J, Kwon S-M, Kang YK, Kim Y-H, Lee M-A, Han K-J et al (2019) A skin-like two-dimensionally pixelized full-color quantum dot photodetector. Sci Adv 5(11):eaax8801. https://doi.org/10.1126/sciadv.aax8801

46. Ge Y, Zhang M, Meng L, Tang J, Chen Y, Wang L, Zhong H (2019) Polarization-sensitive ultraviolet detection from oriented-CdSe@CdS-dot-in-rods-integrated silicon photodetector. Adv Opt Mater 7(18):1900330. https://doi.org/10.1002/adom.201900330

47. Singh A, Li X, Protasenko V, Galantai G, Kuno M, Xing H, Jena D (2007) Polarization-sensitive nanowire photodetectors based on solution-synthesized CdSe quantum-wire solids. Nano Lett 7(10):2999–3006. https://doi.org/10.1021/nl0713023

Chapter 11
CdS-Based Photodetectors for Visible-UV Spectral Region

Nupur Saxena, Tania Kalsi, and Pragati Kumar

11.1 Introduction

Photodetector (PD) is a device that can convert light illuminations either directly (photoelectric effect) or indirectly (photothermal effect) into electric signals. In former case, the absorbed photons' energy generates electron-hole (e-h) pairs and are detected as change in resistance/current in external circuit like in high speed Si and GaAs photodetectors [1]. While, in other class of PDs, light induced temporal temperature gradient (dT/dt) is converted into an electrical signal like in pyroelectric detectors [2]. Further, PDs may also be classified in variety of ways viz.; on the basis of their basic structures & working mechanism, spectral sensing range & their applications, architecture etc. Figure 11.1 shows the cataloguing of photooelectric effect based PDs on the basis of their basic structure & working mechanism whereas Fig. 11.2 depicts spectral sensing range and field of applications of PDs.

The specific field of application require a PD that detects the light of particular spectral region. For example, ultra violet (UV) and UV-Vis. PDs are exploited in the monitoring of ozone layer to UV based skin therapy and missile warning systems to astronomy whereas Vis. and/or near infrared (NIR) PDs are employed in many commercial consumer electronics to imaging systems such as smart phones, digital camera, diagnostics, and remote sensing [9–11]. Further, the choice of materials, photosensitivity in diverse range of spectral wavelength are primarily depend on their band gaps. For instance, narrow band gap semiconductor materials like PbS, PbSe, InAs, and GaSb etc. are well suited for the fabrication of NIR to long

N. Saxena
Organisation for Science Innovations and Research, Garha Pachauri, India

T. Kalsi · P. Kumar (✉)
Nano-Materials and Device Lab, Department of Nanoscience and Materials, Central University of Jammu, Rahya-Suchani, Jammu & Kashmir, India
e-mail: pkumar.phy@gmail.com; pkumar.phy@cujammu.ac.in

© The Author(s), under exclusive license to Springer Nature Switzerland AG 2023   251
G. Korotcenkov (ed.), *Handbook of II-VI Semiconductor-Based Sensors and Radiation Detectors*, https://doi.org/10.1007/978-3-031-20510-1_11

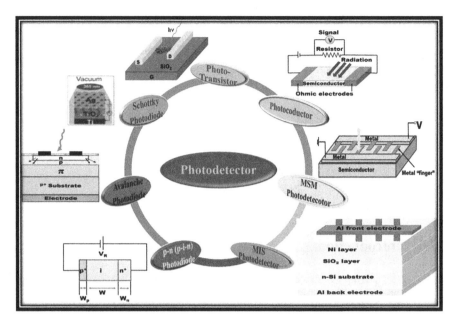

Fig. 11.1 Types of PDs on the basis of basic structures and working mechanism. (Photoconductor reprint from Ref. [3]. Published 2010 by MDPI as open access, M-S-M PD reprint from Ref. [4]. Published 2014 by Intechopen as open access, MIS PD adapted with the permission from Ref. [5]. Copyright 2015: Elsevier B.V., p-i-n photodiode. Reprinted from Ref. [6], Schottky photodiode adapted from Ref. [7]. Published 2019 by DE Gruyter as open access, and phototransitor adapted with the permission from Ref. [8]. Copyright 2014: WILEY-VCH Verlag GmbH & Co)

wavelength infrared (LWIR) PDs. Contrary, wide band gap semiconductor materials viz.: ZnO, ZnS, CdS, GaN, and SiC are usually sense the light in UV to NIR spectral region.

Among various functional materials CdS is one of the most extensively exciting and explored material due to its wide range of applications starting from photonic devices to diagnostic tools including PDs [12–24]. In particular, CdS is a phenomenal material for the development of UV-Vis.-NIR PDs because of its bulk band gap (~2.42 eV at room temperature) lying in visible region of spectrum which can be either tailored towards UV region by varying the particle size or NIR region by creating the intermediate energy states in between band gap via very familiar and easy strategy of impurity incorporation. In addition, solution processability, low cost synthesis, easy control on shape and size etc. are the other key factors for enormous use of CdS in PDs and other applications. Further, CdS based multispectral PDs may also be fabricated via synthesis of heterojunction, composite, decoration with other materials of interest. Till date, various nanostructures of CdS including quantum dots (QDs), nanowires, nanobelts, thin and thick films and their hybrids have been employed for the design of diverse configuration of PDs including photoconductive (ohmic), photodiodes (p-n junction, Schottky junction and p-i-n type photodiodes etc.), and phototransistor as illustrated in Fig. 11.1.

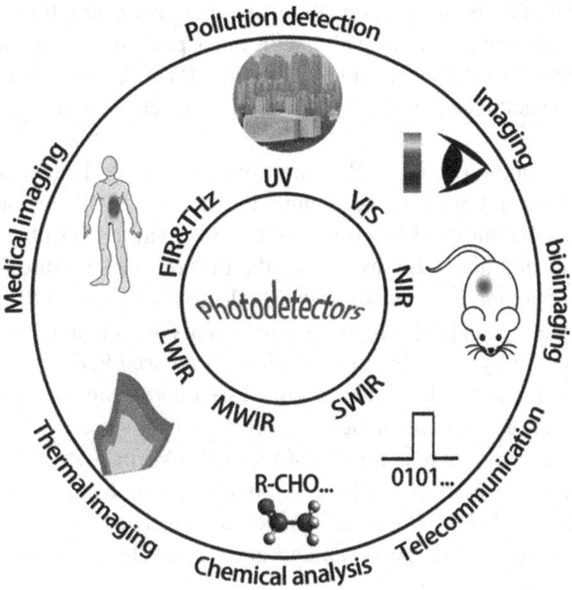

Fig. 11.2 Application areas of PDs for better living on the basis of their spectral range of detection. (Adapted with the permission from Ref. [1]. Copyright 2017: Wiley-VCH Verlag GmbH)

11.2 Conventional Photodetectors and Features of Their Functioning

A conventional photoconductor PD has a planar structure consist of photo-sensitive/active layer between two transverse metal electrodes often in ohmic contacts [25]. In spite of the advantages of the facile fabrication and high employment probability in flexible optoelectronic devices [11], large gain (G) [26], high external quantum efficiency (EQE) and responsivity (R) [25], application of such PDs are limited due to slow photoresponse, low photosensitivity and obvious dielectric hysteresis [11].

A conventional photodiode PD is a vertical structure (except M-S-M type) with rectification effect similar as photovoltaic devices. The photodiode PDs exhibit small dark current, quick response and high separation efficiency of e-h pairs especially at low bias. In specific cases, where high quantum efficiency of PDs is in demand, a large external bias is applied to trigger avalanche or carrier multiplication effects that leads to the generation of multiple carriers excited by a single photon [25].

A metal–semiconductor–metal (M-S-M) PD is a special kind of Schottky barrier detector with back-to-back Schottky contacts in planner geometry separated by semiconducting layer in its simplest structure, i.e., two transverse metallic electrodes on a semiconductor material, in contrast to a p–n junction as in a photodiode. However, the use of interdigited electrode structures (the finger spacing or ring-shaped) is a general practice for construction of M-S-M PDs [27]. A practically important aspect of M-S-M PDs is relatively simple planar structure that is suitable for monolithic integration in particular with other photonic components on semi-conductor integrated circuits. Besides, M-S-M PDs offer the advantages of fast response, low capacitance per unit area, low dark current, low-noise, and high sensitivity [28].

Conventional phototransistor PDs are three-terminal devices consisting of source, drain, gate, and photoactive channel in which gate voltage controls channel conductivity and encourage dark current to be lower and higher photocurrent [26]. Inclusion of gate and dielectric layers into the PDs device structure fallouts reduction in the noise signal, augmentation in the electrical signal, and finally boost up the responsivity and gain [11]. Figure 11.3 represents the schematic of architectures along with their energy band diagrams of above discussed PDs.

The different kinds of PDs structures discussed above are though based on different photodetection mechanism their performance is governed and assessed by the same key parameters known as figures of merit (FOM) or performance parameter. FOM include photosensitivity (S) [9, 29], photoresponsivity (R) [25, 26], external quantum efficiency (EQE) [25, 26, 29], specific detectivity (D^*) [11, 25, 29], gain (G) [11, 25], response time/speed (τ), and noise equivalent power (NEP) [11, 25] and are given by the expressions:

$$S = \frac{I_{Ph} - I_d}{I_d} \tag{11.1}$$

where, I_d and I_{ph} are the dark and photocurrent respectively.

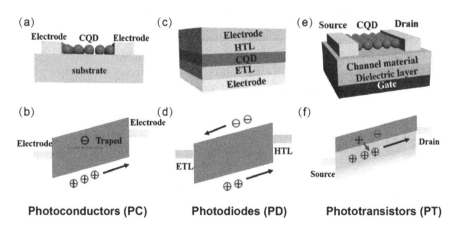

Fig. 11.3 Schematic of architectures and corresponding energy band diagrams of diverse PDs. (**a, b**) Photoconductors; (**c, d**) Photodiodes; (**e, f**) Phototransistors. (Adapted with the permission from Ref. [26]. Copyright 2020: Wiley-VCH GmbH)

$$R = \frac{I_{Ph}}{P_{in}} \quad (11.2)$$

where, P_{in} indicates the light intensity of the source. More precisely, spectral responsivity 'R_λ' is given by [11, 29];

$$R = \frac{I_{Ph} - I_d}{P_{in}} \quad (AW^{-1}) \quad (11.3)$$

EQE measures the number of collected charge carriers per unit incident photons and is closely related to R value. Therefore, it is the ratio of and can be expressed as;

$$EQE\left(\eta_{ex}\right) = \frac{N_c}{N_{ip}} = R_\lambda \frac{hc}{e\lambda} = R_\lambda \frac{1.24}{e\lambda} \quad (11.4)$$

where N_c is the number of photogenerated carriers, N_{ip} is the number of incident photons, h is Planck's constant, c is velocity of light, e is electronic charge, and λ is wavelength of incident light. Additionally, one can also estimate the internal quantum efficiency (IQE) by the equation [25];

$$IQE\left(\eta_{in}\right) = \frac{N_c}{N_a} = \frac{EQE}{\eta_a} \, or \, \frac{\eta_{ex}}{\eta_a} \quad (11.5)$$

Where η_a is the light absorption efficiency of photoactive material.

$$D^* = \frac{R_\lambda}{I_N} \sqrt{A \blacklozenge f} \quad \left(Jones \, or \, cmHz^{1/2}W^{-1}\right) \quad (11.6)$$

where A is effective area of device, Δf is electrical bandwidth, and I_N is noise current. Ordinarily, I_d is the dominating contribution to the noise signal, so the equation can be truncated as [10, 11, 29];

$$D^* = R_\lambda \sqrt{\frac{A}{2qI_d}} \quad \left(Jones \, or \, cmHz^{1/2}W^{-1}\right) \quad (11.7)$$

$$NEP = \frac{\sqrt{A \blacklozenge f}}{D^*} = \frac{I_N}{R_\lambda} \quad (11.8)$$

NEP may also be estimated as [26];

$$NEP = \frac{\sqrt{I_N^2}}{R_\lambda} \qquad (11.9)$$

Under these circumstance, electrons drifted to the external circuit may circulate several times in the channel and result in the photoconductive gain. 'G' is determined by [11, 25];

$$G = \frac{\tau_l}{\tau_t} = \frac{\tau_l}{L^2} \mu V_{DS} \qquad (11.10)$$

where τ_l is lifetime of trapped carrier say hole, τ_t is transit time of the other carrier say electron, L is the channel length, μ is the carrier mobility, and V_{DS} is the applied bias voltage.

Alternative critical performance parameter is 'τ' is defined as in [9–11, 25, 29] and can be estimated as [26];

$$f_c = \frac{1}{2\pi\tau} \qquad (11.11)$$

11.3   Fabrication of Photosensitive Devices

11.3.1   CdS PDs Based on Nanostructures

A PD device was fabricated with aligned CdS NTs arrays as active layer and Ag NWs network as transparent electrodes by An and Meng [30]. The schematic diagram depicting the fabrication process of their PD is shown in Fig. 11.4. The transparent and conducting Ag NWs networks with uniform distribution were produced by spraying the Ag NWs redispersed in ethanol onto the prepared quartz using micro syringes (Fig. 11.4a). Further, mechanically scratching of the substrate using a razor blade at a gap about 40 µm was carried out and was cut into two totally disconnected parts to use the Ag NWs network as electrodes (Fig. 11.4b). Next, the electrodes produced as aforementioned were turned over and placed on top of the aligned CdS NTs arrays substrate (Fig. 11.4c, d), producing PDs. The PD based on the single Sn doped CdS NW was fabricated by photo lithography on Si/SiO$_2$ substrate with In electrodes [11].

Jin et al. [31] fabricated PD with active layer of CdS flakes by a laser direct writing system and standard electron-beam lithography (Fig. 11.5a) followed by the deposition of Cr/Au (10 nm/50 nm) electrodes using the thermal evaporation. Nawaz et al. [32] fabricated M-S-M PDs of CdS NBs deposited over SiO$_2$/Si and polyimide (PI) substrates followed by the deposition of Ag electrodes (Fig. 11.5b), whereas PD with "L" shape electrodes of conductive Ag paste were designed onto

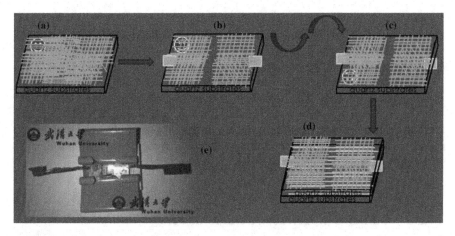

Fig. 11.4 Schematic diagram showing the fabrication process of PD. (**a**) Deposition of Ag NWs on quartz substrate, (**b**) Process of mechanically scratching of the substrate, (**c**) Turning the electrodes, (**d**) placing upside down on top of the aligned CdS NTs networks substrate, and (**e**) Photographs of final device. (Adapted with the permission from Ref. [30]. Copyright 2016: Springer)

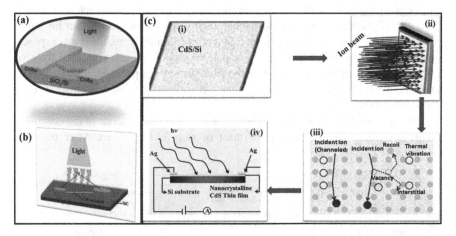

Fig. 11.5 (**a**) CdS NFs based PD fabricated by laser directed writing system. (Adopted with the permission from Ref. [31]. Copyright 2018: WILEY-VCH Verlag GmbH). (**b**) CdS NBs M-S-M PD. (Adapted with the permission from Ref. [32]. Copyright 2021: the Royal Society of Chemistry). (**c**) Schematic of (i) the pristine film, (ii) the film under exposure to ion beam, (iii) various promising phenomena as a consequence of passage of ion beam through the material and (iv) the fabricated PD. (Adapted with the permission from Ref. [9]. Copyright 2016: The Royal Society of Chemistry)

CdS over glass slides [33]. The PD of spray deposited Y:CdS TF on glass was fabricated by making Ag contacts in M-S-M geometry [3]. PD of CdS/ZnO core/shell NWs were fabricated by coating the paste of CdS/ZnO core/shell NWs in deionized water on a ceramic substrate (over which silver interdigitated electrodes with both finger-width and inter-finger spacing of about 200 μm was previously printed) [34]. An another core-shell PD was designed by deposition of two step CVD synthesized CdS-CdS$_x$Te$_{1-x}$ -CdTe core-shell NBs over SiO$_2$ substrate followed by the construction of electrodes using silver paste [35].

An entirely different strategy was adopted by Kumar et al. [9] to fabricate PDs in M-S-M geometry with Ni^{6+} ion beam irradiated CdS films over Si substrate (Fig. 11.5c (i & ii)). In this work, the CdS films were grown over Si via PLD and irradiated under different ion fluences. Ion beam irradiation resulted many processes (as shown in Fig. 11.5c (iii)), which modified various properties of the films. Finally, these films were used to design PD (Fig. 11.5c (iv)).

11.3.2 CdS-Based Heterostructures in PDs

Figure 11.6 shows process of PD fabrication and of final PD device developed by Chakrabarti et al. [36]. The spin coated PEDOT:PSS films on ITO coated glass substrate were patterned using water vapour assisted capillary force lithography. They have fabricated patterned stamps via Replica Molding of Sylgard 184 against the polycarbonate part of commercially available optical data discs such as CD and DVD. The details of using optical data storage discs for patterning can be found elsewhere [37]. To make the cross linked Sylgard 184 stamp wettable, it was exposed in a UV ozone chamber for 30 min. For pattern replication, the patterned Sylgard 184 stamp was carefully placed over the film surface (step 2, Fig. 11.6) and was subsequently transferred for 18 h in a water vapour's pre-saturated chamber at room temperature (step 3, Fig. 11.6).

The capillary driven rise of polymer meniscus along the contour of the confining stamp developed replica of pattern (inset of step 3, Fig. 11.6). After this, the

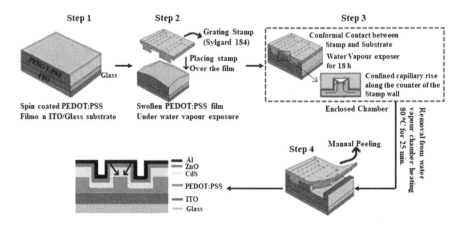

Fig. 11.6 Schematic diagram for imprinting of PEDOT:PSS film and final device. (Adapted with the permission from Ref. [36]. Copyright 2018: Institute of Physics)

assembly of patterned film and stamp was taken out from the chamber and heated under the application of uniform normal load of 4 KPa in a vacuum oven at 80 °C for 25 min to remove the residual water molecules in the film matrix. In the last step, stamp was gently peeled off after cooling (step 4, Fig. 11.6). The final architecture (ITO/PEDOT:PSS/CdS/ZnO/Al) of PD devices were achieved by growing 150 nm CdS TF over patterned and flat PEDOT:PSS films, followed by the deposition of ZnO TF over CdS TF layer via rf-sputtering and subsequent deposition of 150 nm thick layer of Al electrodes over ZnO TF by thermal evaporation method.

An architecture FTO/CdS NRs/SnS NFs/Au was designed by growing CdS NRs array via hydrothermal over FTO substrate followed by the deposition of SnS NFs layer and Au electrodes via thermal evaporation using a shadow mask [40]. Whereas, an architecture FTO/CdS NRs/perovskite/Spiro-OMeTAD/Ag was fabricated by post annealing of hydrothermally grown CdS NRs array on ITO in air at 200 °C for different time to make gradient O-CdS NRs array and then spin coated perovskite layer over gradient O-CdS NRs array followed by the coating of Spiro-OMeTAD layer. Finally, thermal evaporation was used to deposit the 100 nm thick Ag electrode [41]. Perpendicular CdS NWs array of 600 nm length and 50 nm dia was grown through a hydrothermal route on the flexible substrate of Si micropyramids (Fig. 11.7a, b (b2)) of bottom edge length ~3–7 μm that were synthesized by chemical etching process to form the p-Si/n-CdS heterojunction. Further, ITO layer as the transparent top-electrode and Al as bottom-electrode were deposited on the CdS NWs and the flexible p-Si substrate respectively to construct a flexible device structure ITO/CdS NWs/p- Si/Al [38]. An ordered and aligned CdSe@CdS dots in rods (DIRs) embedded in PVDF composite free standing films were synthesized by sol-gel and peel off method [42]. A PD of configuration FTO/TiO$_2$/CdS/Se/Ag was fabricated by growing 2 nm TiO$_2$ layer by atomic layer deposition (ALD) followed by microwave assisted hydrothermal growth of CdS layer and CVD growth of Se layer and 100 nm Ag electrode deposition via thermal evaporation [43]. An architecture of Al/CdS NPs-CdO/p-si/Al was fabricated by deposition of CdO thin film using sol-gel spin coating method onto p-Si, followed by CdS NPs structures using SILAR technique and Al contact by thermal evaporating using physical mask [44]. A position-sensitive array (PSA) PD was designed by the growth of comb-like Sn doped CdS NSs with cone-shape branched via in situ seeding CVD over Au coated Si substrate and depositing the Pt electrode using focused ion beam (FIB) technology by introducing precursor gas into the Ga ion beam as shown in Fig. 11.7c–h [39].

A PD of configuration FTO/CdS NFs/Ag was fabricated by annealing the as deposited thin film of CdS NFs over FTO at 300 °C in air for 30 min. Followed by the pattering of Ag electrodes of thickness 200 nm using mask in thermal evaporation and subsequently wiping out the part of CdS NFs film upon FTO to make the FTO another transport electrode as illustrated in Fig. 11.8a [45]. CdS core-Au/MXene-Based (Fig. 11.8b) PD was fabricated by Jiang et al. [46]. First, they dropped the solution of CdS core-Au NPs in ethanol onto fluorophlogopite substrate and natural air-drying at room temperature followed by the dropping of colloidal solution of MXene on both sides of CdS core-Au NPs as the electrodes and at last, the total device was annealed at 100 °C for 5 min in air. The architecture PET/ITO/

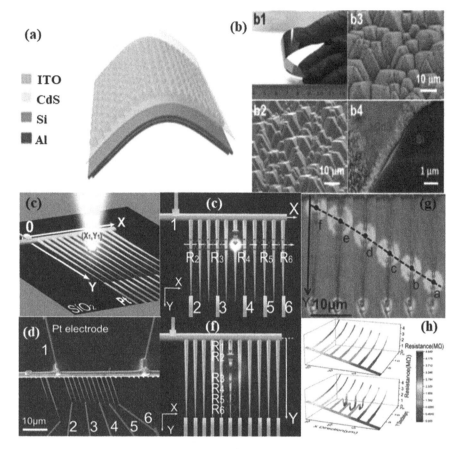

Fig. 11.7 (**a**) Schematic structure of the self-powered photodetector (SPPD). (**b**) Characterizations (b1) an optical image of flexible Si wafer after etching; (b2) SEM image of the etched Si wafer surface; (b3, b4) top view (b3) and side view (b4) of the CdS NWs array. (Adapted with the permission from ref. [38]. Copyright 2018: Wiley-VCH Verlag GmbH). (**c**) Schematic diagram of the CdS branched PD, (**d**) SEM image of the fabricated device, Schematic diagram to realize light position identification in the (**e**) X-direction, (**f**) Y-direction of the PD, (**g**) Optical image of the laser irradiating at different positions on a branch with wavelength 405 nm, and (**h**) Resistance distribution of the branched nanostructure. Above and below simulations correspond to the branched structures without and with light illumination, respectively. (Adapted with the permission from Ref. [39]. Copyright 2018: Wiley-VCH Verlag GmbH)

CdS/NiO/Al was developed by deposition of CdS TF onto PET/ITO substrate via photochemical method followed by the spin coating of hydrothermally synthesized NiO NPs and depositing the top Al metal contacts (120 nm) on the heterojunction using a shadow mask (Fig. 11.8c) [47].

The MoTe$_2$ flakes were exfoliated directly on a 300 nm SiO$_2$/Si substrate and contact electrode for flake was constructed using photolithography, followed by electron beam evaporation of a 50 nm Au electrode. Subsequently, the Ga-doped

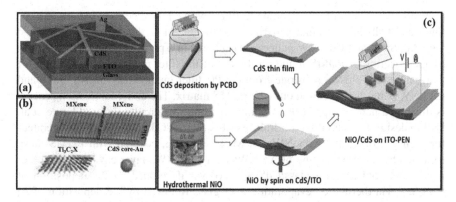

Fig. 11.8 Schematic illustrations (**a**) CdS NFs based PD. (Adopted with the permission from Ref. [45] © 2018 Elsevier GmbH). (**b**) Assembled CdS core-Au/MXene PD. (Adapted with the permission from Ref. [46]. Copyright 2020: Elsevier B.V). (**c**) the solution manufacturing process for CdS, NiO and PD construction. (Adapted with the permission from Ref. [47]. Copyright 2021: Elsevier B.V)

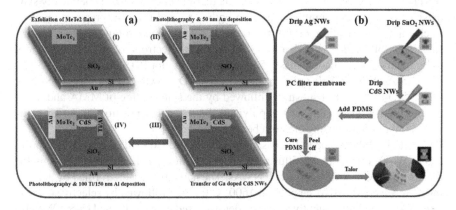

Fig. 11.9 (**a**) Experimental process for the fabrication of a 1D Ga-doped CdS NWs/2D MoTe$_2$ flake heterojunction PD. (Adapted with the permission from Ref. [48]. Copyright 2018: WILEY-VCH Verlag GmbH). (**b**) Fabrication process of the stretchable SnO$_2$-CdS interlaced NW PD and optical image of the as-fabricated stretchable device arrays. (Adapted with the permission from Ref. [49]. Copyright 2019: Springer Nature)

CdS NWs were placed using a manipulator onto a MoTe$_2$ flake followed by photolithography for the realization of (100 nm/150 nm) Ti/Au electrode to Ga-doped CdS NW to develop final structure of the p–n heterojunction PD (Fig. 11.9a) [48].

The PD based on stretchable SnO$_2$-CdS interlaced NWs was designed by Li et al. [49] They have used 3D printed two kinds of polydimethylsiloxane (PDMS) masks (one for Ag NW electrodes, and other for SnO$_2$ and CdS NW channels). Fabrication of the SnO$_2$-CdS interlaced NW PD involved the placement of the masks on a polycarbonate (PC) filter membrane to filter the Ag NW electrode pattern and the SnO$_2$ and CdS NW channel patterns respectively. After this PC filter membrane with Ag, SnO$_2$ and CdS NW patterns was placed in a plastic culture dish followed by the injection of PDMS liquid on the top of the filter membrane and peeled off thermally cured PDMS substrate from the PC filter membrane with Ag, SnO$_2$ and CdS NW patterns transferred onto the PDMS matrix. The entire fabrication process is illustrated here in Fig. 11.9b. A similar method was previously used by Pe et al. [50] to design PD with mixed cellulose esters (MCE)/CdS NWs/Ag NWs architecture. They drop casted CdS NWs on MCE substrate, covered with PDMS mask followed by the drop casting of Ag NWs over CdS NWs layer covered by another mask. The configuration glass/CdS$_x$Se$_{1-x}$/Al was developed by the deposition of CdS$_x$Se$_{1-x}$ via thermal co-evaporation over glass followed by the deposition of Al electrodes [51] and in subsequent year they fabricated Al/p-Si/CdS$_x$Se$_{1-x}$/Ag PD following the same process [52]. UV lithography followed by thermal evaporation and lift-off process were used to fabricate two terminal nanodevices of configuration Si/SiO$_2$/CdS NWs/(10 nm) Ti/(300 nm) Au [53] and n-Si/SiO$_2$/GaTe NFs/Sn:CdS NWs/(2 nm) Ti/(120 nm) Au [54]. An M-S-M PD based on NCs CdS-porous silicon (PS):p-Si heterostructure was developed via thermal evaporation of chemically synthesized CdS powder simultaneously on both glass and PS:p-Si (prepared by electrochemical anodization of (100) oriented boron doped p-type Silicon wafer) substrates followed by the deposition of Al metal electrode of ~20 nm thickness using interdigitated shadow mask [55]. A photoconductor PD with stacking Si/SiO$_2$/WS$_2$ NSs/CdS NWs/Cr/Au was designed via CVD growth of CdS NWs/WS$_2$ NSs over a SiO$_2$(300 nm)/Si substrate followed by the spin coating of MMA and polymethyl methacrylate (PMMA) copolymers layers and deposition of Au/Cr electrodes (50 nm/5 nm) using electron beam lithography [56].

Xu et al. [26] developed PD based on self-aligned CdS/CdSe core/shell nanowalls (NWAs). Figure 11.10a illustrates schematic of typical experimental steps for the growth of CdS/CdSe core/shell NWAs. The nanogroves were formed by annealing as-received M(1010) sapphire at 1600 °C for 10 h followed by selective deposition of Au NPs. Subsequently, CdS powder was used to grow aligned CdS NWAs by PVD followed by the covering of polycrystalline Al$_2$O$_3$ using ALD and defining the area to be etched via photolithography, etching in a buffered oxide etch (BOE) solvent, and lift-off the remained photoresist with acetone. Next, CdSe was grown via PVD on the sample with selective-etched area so that only the exposed NWs surfaces were coated with CdSe layers via a surface-epitaxial growth owing to the selective protection of the Al$_2$O$_3$ layer and remaining Al$_2$O$_3$ layer was removed by another etching in the BOE solvent. A PD is designed by sputtering p and n type Nd:TiO$_2$ and CdS layer onto FTO followed by the formation of connections in the circuit board as illustrated in Fig. 11.10b [58].

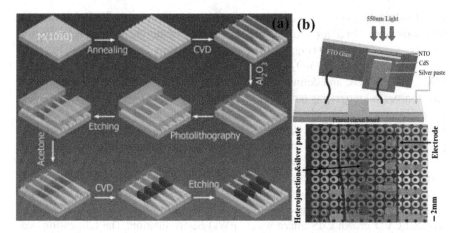

Fig. 11.10 Schematic illustrations (**a**) experimental steps for the growth of CdS/CdSe core/shell NWs. (Adapted with the permission from Ref. [57]. Copyright 2018: WILEY-VCH Verlag GmbH). (**b**) Device structure, the light is injected from the back of the glass (left) and the real devices with silver paste connection on the circuit board (in the red circles). (Adapted with the permission from Ref. [58]. Copyright 2020: IOP Publishing Ltd)

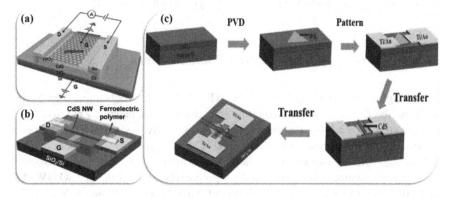

Fig. 11.11 Schematic illustrations (**a**) single CdS NB FET PD. (Adapted with the permission from Ref. [59]. Copyright 2019: American Institute of Physics). (**b**) Ferroelectric side-G single CdS NW FET PD. (Adapted with the permission from Ref. [60]. Copyright 2016: WILEY-VCH Verlag GmbH). (**c**) the fabrication process of the M-WS$_2$ /CdS/WS$_2$ NFs PD. (Adapted with the permission from Ref. [62]. Copyright 2021: the Royal Society of Chemistry)

11.3.3 CdS-Based Field Effect Transistor (FET) PDs

FET PDs of CVD grown single crystal CdS NBs on a SiO$_2$/Si substrate via electron-beam lithography process and metal deposition and lift-off process followed by the deposition of HfO$_2$ as a passivation layer using ALD (device-I) and exfoliation and transfer of a thin layer of transparent graphene onto HfO$_2$ as a top gate (G) electrode (device-II shown in Fig. 11.11a) [59]. The configuration Si/SiO$_2$/CdS/Ti-Au of PD was developed by syntheses of CdS TF on Si/SiO$_2$ substrate by CBD followed by the deposition of Ti(10 nm)/Au (90 nm) electrode by sputtering method. Ferroelectric polymer side-G FET PD (Fig. 11.11b) was fabricated by transferring CVD synthesized CdS NWs onto Si/SiO$_2$ (110 nm) substrate followed by the spin coating of MMA and PMMA, and defining the drain (D), source (S), and side-G patterns via electron-beam lithography technique. Subsequently, the Cr/Au (15 nm/50 nm) electrodes (S and D) were fabricated by metal evaporation and lift-off processes followed by spin coating of 200 nm of P(VDF-TrFE) (70:30 in mol%) ferroelectric polymer film on the NW channel to construct side-G electrode [60]. A similar process was used to fabricated single CdS NW based FET PD where, photoresist was spin coated onto Si/SiO$_2$/CdS NWs and photolithography was used to define S, D and G followed by the deposition of (10/100 nm) Ti/Au via e-beam evaporation [61]. A FET PD was designed based on PVD synthesized WS$_2$ NFs and SnO$_2$-assisted CVD grown CdS micro wires (μWs) heterojunction. The fabrication process is illustrated in Fig. 11.11c and involves first, spin coating of photoresist over Si/SiO$_2$/WS$_2$ NFs followed by photolithography to define S, D, and G and deposition of (10/60 nm) Ti/Au electrode via e-beam evaporation. Secondly a CdS μWs were transferred in the middle of the WS$_2$ NFs channel under the assistance of PMMA followed by mechanical exfoliation and transfer of multi-layered WS$_2$ NFs to the top of CdS μWs and WS$_2$ using polyvinyl alcohol (PVA) [62].

11.3.4 CdS-Based Self-Powered Photodetectors

The configuration FTO/ZnO NWs/G/CdS NPs of photoelectrochemical (PEC) type self-powered (SP) PDs was designed by spin coating seed layer of ZnO over FTO and growth of ZnO NWs via hydrothermal followed by the addition of graphene dispersion and subsequent annealing at 300 °C for 30 min and attachment of CdS NPs through SILAR method [63]. The PDs performance was analysed by electrochemical workstation using photoanode of polyacrylate encapsulated ZnO NWs/G/CdS NPs, electrolyte of aqueous solution of Na$_2$SO$_4$, and cathode of Pt wire. As another major breakthrough in order to design self-powered PDs, a relatively new phenomenon namely piezophototronic effect was harnessed. The PD device based on ZnO μWs-CdS NWs core-shell structure was fabricated by Zhang et al. [64] by fixing tightly the two ends of a single ZnO-CdS wire by silver paste (served as S and D electrodes) on a polystyrene (PS) substrate (flexible, and robust under repeated mechanical strains) followed by deposition of polydimethylsiloxane (PDMS) to package the device (Fig. 11.12a).

The above structure was modified further by the same group [65] to improve the device performance. This time, branched ZnO-CdS double-shell NW array was grown via a facile two-step hydrothermal method on the surface of a carbon fiber (CF/ZnO-CdS) to fabricate PD. In typical synthesis process, CF (black) was first coated with ITO (layer I-transparent) followed by coating of ZnO seed layer (blue) for a better adherence and conductivity and hydrothermal growth of highly dense

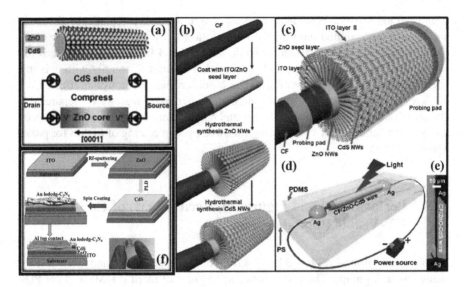

Fig. 11.12 Schematic illustrations (**a**) Zno-CdS core-shell NWs (top), the proposed sandwich model of the device, i.e. two back-to-back Schottky diodes connected to a ZnO core and CdS shell, respectively, and simulation of the piezopotential distribution in the ZnO core under compressive strain (bottom). (Adapted with the permission from Ref. [64]. Copyright 2012: American Chemical Society). (**b**) synthesis procedure of a branched ZnO-CdS double-shell NWs array on a CF, (**c, d**) fabrication of a single CF/ZnO-CdS wire based photodetector, (**e**) Optical microscopy image of a typical device. (Adapted with the permission from Ref. [65]. Copyright 2013: American Chemical Society). (**f**) Various steps involved in the fabrication of the hybrid heterojunctions of ACCZ along with the optical image of the device. (Adapted with the permission from Ref. [66]. Copyright 2018: Royal Society of Chemistry)

ZnO NW array (blue) perpendicular to the surface of the CF. Finally, the CdS NW array (yellow) was grown over ZnO NWs (Fig. 11.12b). Further, top transparent electrode (D) of ITO (layer II) was deposited for development of a photon sensitive device with S electrode of CF (Fig. 11.12c). Then two electrodes of a single CF/ZnO-CdS/ITO wire were fixed on a polystyrene (PS) tightly by silver paste (Fig. 11.12d) followed by packaging of the device by PDMS. Figure 11.12e illustrates microscopic image of the final device. Plasmonic Au-g-C$_3$N$_4$/CdS/ZnO (ACCZ) based hybrid heterojunction structure was used to design PD on a flexible platform [66]. The stepwise fabrication process is illustrated in Fig. 11.12f which involves deposition of rf-sputtered ZnO TF over ITO coated substrate (glass and PET) followed by growth of CdS TF via PLD, spin coating of Au-C$_3$N$_4$ TF and deposition of top electrode Al using thermal evaporation method. SPPD based on CdS N/μWs: Poly(3-hexylthiophene) (P3HT) μWs was fabricated by Yu et al. [67] In typical synthesis process, CVD grown an ultralong (10 mm) CdS N/μW was deposited on PS and fixed tightly by epoxy resin through one end followed by oxygen plasma treatment of device for 5 min and drop casting of P3HT. A flexible PD based on 2D WSe$_2$/1D CdS heterojunction was developed by Lin et al. [68] First, they transferred PVD grown CdS NWs on PET substrate by lift up and aligned stamp method and then WSe$_2$ nanosheet was synthesized via mechanical exfoliation and precisely positioned onto the CdS NW followed by the patterning of contact electrodes (Cr/Au (15 nm/50 nm)- Ohmic for CdS, and Pd/Au (15 nm/50 nm) for WSe$_2$) via electron beam lithography. In another attempt to comply with existing silicon technology and fabricating SPPD based on piezophototronic effect, CdS NWs were hydrothermally grown on Si micropyramids that were prepared by chemical etching of p-Si wafers. The top and bottom electrodes (Al and ITO respectively) were deposited by rf-sputtering followed by spin coating of PDMS for packaging the device [69].

11.4 Performance and Figures of Merit (FOM) of CdS Based UV-Visible Photodetectors

As discussed earlier, CdS based UV-Vis. PDs have been designed and fabricated by many research groups. The performance and figures of merit (FOM) of these PDs depend on various factors and they will be discussed in more detail in following sections.

11.4.1 Self-Powered Photodetectors

Among the most successful and less power consuming PDs are self-powered photodetectors (SPPDs). Their parameters are listed in the Table 11.1.

These are numerous effects and phenomena to realize a SPPD. A few of them are discussed here. High performing SPPDs were designed by perovskite/CdS heterostructure. The superior performance of device was driven by gradient energy band (Fig. 11.13a). It was found that photocurrent first increases with increase in gradient energy band and then a negative effect was noticed. The enhancement in photocurrent results from stronger extraction and separation efficiency of photogenerated e–h pairs at the interface due to large built-in potential (V_{bi} = 0.17 V) at CdS$_{10}$/perovskite interface than that of CdS$_x$/perovskite (0.07 V), which was estimated by M–S measurement under dark condition. Besides, the effect of perovskite layer on device performance was also analysed and found that the photocurrent mainly depends on perovskite layer rather than CdS. However, the presence of CdS layer enables to sense the broader range of spectral region starting from 350 to 800 nm [41]. The performance parameters in this study are summarized here in Table 11.1.

A visible SPPD of SnS NFs covered CdS NRs array was designed by coupling pyro-electricity and photo-electricity to optimize the cryogenic detecting performance. It was found that both the pyroelectric and photoelectric effects jointly contribute in the device performance and this dual effect becomes more significant with the reduction in temperature. Figure 11.13b, c show the current-time (I-t) response

Table 11.1 FOMs of different SPPDs based on CdS nanostructures (the bias voltage is 0)

Material	λ (nm)	S (%)/R (A/W)	EQE (%)/D (10^y Jones)	τ_r/τ_f	References
SnS/CdS	650	0.0104	3.56×10^{11}	30 ms	[40]
p-Si/n-CdS NWs	325–1550	0.00034	–	245 µs/277 µs	[38]
CdS NBs	484	1.54×10^6/0.036	2.36×10^{12}	40 ms/30 ms	[32]
ZnONAs/G/CdS/ electro-lyte	365; 475	.0273 .0043	–	5 ms	[63]
n-CdS/p-Se	500	9×10^3/0.040	1.3×10^{13}	2 µs/22 ms	[43]
Nb:TiO$_2$/CdS	550	0.125	–	~10 ms	[58]
CdS/ZnO/ PEDOT:PSS Flat Lp = 350 nm Lp = 750 nm	400–550	~2×10^2/0.017 4×10^3/0.038 ~1×10^2/0.032	7.10×10^{11} 5.76×10^{11} 2.65×10^{11}	0.14 s/0.16 s 0.12 s/0.16 s 0.14 s/0.17 s	[36]
ZnO/CdS core/shell	350	2.8×10^3	–	<20 ms/20 ms	[70]
NRs/PEDOT:PSS	470	1.07×10^3	–	<20 ms/20 ms	

λ wavelength, S sensitivity (I_{light}/I_{dark} or I_{light}-I_{dark}/I_{dark}), R responsivity, D detectivity, response time (τ_r-rise time, τ_f -fall time)

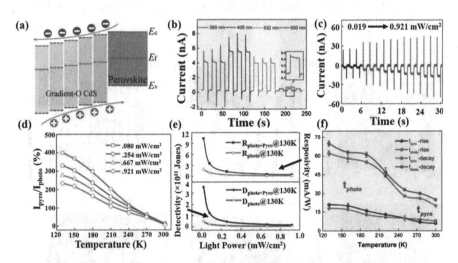

Fig. 11.13 Schematic illustrations (**a**) gradient energy levels, and carrier transport at the gradient-O CdS/perovskite interface. (Adapted with the permission from Ref. [41]. Copyright 2019: WILEY-VCH Verlag GmbH). (**b**) Time response curves under 365, 405, 532, and 650 nm light illumination at 0 V, (**c**) Photodetection capability and Pyroelectric current testing of SnS/CdS device under illumination of different light power density at temperatures 130 K, performance parameters of the SnS/CdS heterojunction PD (**d**) The ratio of pyroelectric current/photocurrent under different light power densities and temperatures, (**e**) Responsivity and detectivity at 130 K and (**f**) Response times of pyroelectric and photoelectric effect at different temperatures. (Adapted with the permission from Ref. [40]. Copyright 2020: WILEY-VCH Verlag GmbH)

curves for the SnS/CdS device under different light wavelengths without bias voltage and under illumination of different light power density at temperatures 130 K respectively. It was found that the amplitude of pyroelectric pulse signals increases with reduction in measurement temperature as one can see (Fig. 11.13c) the amplitude of pyroelectric pulse signals is significantly high (25 nA) with respect to photocurrent (just 3.5 nA) under the light irradiation of 0.019 mW cm^{-2}. Figure 11.13d illustrates the reduction in pyroelectric components in total photocurrent as a function of increasing temperature, whereas the contribution of this component in responsivity and detectivity of the sensor drastically fall with increase in power of illuminating light (Fig. 11.13e). In contrast to these, the response time majorly depends on the component of photoelectric effect induced photocurrent in all temperature range and become more prominent at cryogenic temperatures (Fig. 11.13f) [40]. The estimated performance parameters of the device are summarized the table I.

Chakrabarty et al. [36] developed high performance hybrid SPPDs on soft lithographically patterned organic platform and investigated the effect of line width (L_p), periodicity (λ_P) and the feature height (h_S) of the two grating patterned stamps used for patterning on the device performance. The morphology of PEDOT:PSS layer and final device (subsequent deposition of CdS, ZnO & Al) on (1) flat and (2) $L_p = 750$ nm; (3) $L_p = 350$ nm patterned films are illustrated in Fig. 11.14a (1–3) and 11.14b (1–3), respectively. The device fabricated on $L_p = 350$ nm patterned films demonstrated enhanced sensing parameters.

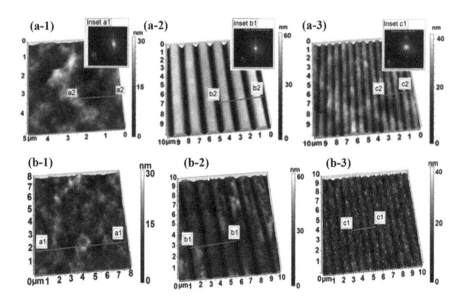

Fig. 11.14 AFM images (**a**) PEDOT:PSS films (1) flat and (2) $L_p = 750$ nm; (3) $L_p = 350$ nm patterned film. Inset shows the corresponding optical diffraction patterns, (**b**) devices (1) flat and patterned samples with line widths of $L_p = 750$ nm (2) and $L_p = 350$ nm (3), respectively. (Adapted with the permission from Ref. [36]. Copyright 2018: IOP Publishing Ltd)

11.4.2 Photodetectors Based on Nanostructures

The parameters estimated in their study are listed in Table 11.2. The effect of sulphur concentration on sensing characteristic of CdS based M-S-M device is studied by Halge and co-workers [71]. They tested the devices under the illumination of 360 nm, 550 nm and 700 nm light and observed that the photosensing parameters first improved with reducing Cd/S ratio, attained a maxima and then reduced with further reduction in Cd/S ratio. They achieved optimum sensing parameter (Table 11.2) for device S_2 (Cd/S = 1.24) under UV illumination. 2D microscale position-sensitive PDs were designed using highly ordered comb-like CdS NWs array with cone-shape branches. The position sensitivity of PDs was examined via variable resistance in different transportation routes and variable optical responses at different parts of the cone shape branches. They found that the gradient resistance rises from the trunk to the branch end due to the transportation route, lengthens along the X-direction and the photoresponse of cone shape branches varies along the Y-direction (Fig. 11.7h). The transportation route i.e. position/distance dependent resistance in trunk direction is found to decrease under the illumination of white light with respect to dark in both the cases either the connection position 1 or 6 (Fig. 11.7e) on the trunk to different terminals of branches (Fig. 11.15a, b), whereas position dependent resistance in branch direction i.e. along Y-axis (changes on one single conical branch Fig. 11.7f). The branch resistance under dark gradually decreases from bottom to top (Fig. 11.7g) and photocurrent increase linearly with increasing distance (Fig. 11.15c). The change in resistance per nm distance (resistance sensitivity) along X and Y direction was estimated ~85 K Ω and 58 Ω respectively [39].

The effect of dopant Sn [5, 53, 54], Cu [74] Y [78], Ce [79], Mg, Al and Mg:Al [55, 80] on the photo-sensing properties was studied by various researchers and it was found that single dopant may enhance the device performance up to a certain concentration of dopant and beyond that device performance may degrade, whereas at that critical concentration of single dopant introduction of other dopant element may results in degradation of device performance (Table 11.3). Lin et al. [72] developed a PD with picowatt sensitivity in UV region (larger than visible region) via deep ultraviolet (DUV) laser treatment on the surface of CdS. They investigated the device performance before and after irradiation with a single shot from a KrF laser at power densities of 0.7, 14, and 140 mJ cm^{-2} and found that the performance of devices was improved with increase irradiation in power density and improvement is maximum under 365 nm illumination (Fig. 11.16a). Figure 11.16b illustrates the detectivity of the laser-treated PD as a function of the illumination power density at a bias voltage of 1 mV. Channel lengths (CLs) specific broad spectral photoresponse properties were investigated by Sharma et al. [81]. They observed linear dependency of rise time with CLs up to 700 nm and non-linearity on further increase in CLs i.e. for CLs = 7 μm, 45 μm and 350 μm (Fig. 11.16c–f). Further, they were noticed increase in response time with decrease in CLs down to nanometers (300 nm).

Table 11.2 Comparison of the FOMs of different CdS nanostructure-based photodetectors

Material	Bias (V)	λ (nm)	S (%)/R (A/W)	EQE (%)/D (10^y Jones)	τ_r/τ_f	References
nc-CdS	5	470	1.02×10^3/82	19.5×10^3/5.05×10^{11}	183 ms/61 ms	[10]
CdS NCs	1	365	7.3×10^5	3.5×10^{16}	–	[72]
Al:CdS-PS:p-Si	–2	400	0.6	180/3.4×10^{13}	160 ms/350 ms	[55]
Au NPs/CdS QDs	10	390	$2/1.27 \times 10^{-4}$	1.42×10^9	–	[73]
CdS QDs/PVA	5	365	$20/9.5 \times 10^{-9}$	39.8×10^{-9}	–	[74]
CdS:Cu QDs/PVA			$24/15 \times 10^{-9}$	63×10^{-9}		
Graphene/CdSe/CdS/ZnS QDs	5	365	2/45.77	–	1 s/1 s	[75]
O-doped CdS/perovskite		700	0.48	–	–	[41]
CdS/ZnO core/shell NWs	4	367	0.0011	–	~26 ms/2.1 ms	[34]
		468	0.0013			
SnO$_2$-CdS interlaced NWs	5	370	7417	–	1.5 s/0.6 s	[49]
CdS/WS$_2$	5		~50	~10^{12}	–	[56]
M-WS$_2$/CdS/F-WS$_2$	1	405	1.5×10^4/~4.7	3.4×10^{12}	13.7 ms/15.8 ms	[62]
GaTe/Sn: CdS	2	White	100(f) & 3000(r)/607	–	260 ms/267 ms	[54]
CdS/ZnO	5	405	3.8×10^3	–	3.6 s/3.3 s	[76]
CdS-CdS$_x$Te$_{1-x}$-CdTe core-shell NB	5	355	4×10^3/1.5×10^3	4.7×10^5	11 μs/23 μs	[35]
ZnO NR/CdS	5	350	146/12.89	–	62.4 s/44.9 s	[77]
Ag/CdS/Ag	10	360	10^4	–	1.4 ms	[71]
Ag:Y:CdS/Ag	15	532	55.07/0.83	193.8/4.28×10^{11}	78 ms/87 ms	[78]
p-NiO/n-CdS	–	UV-vis.	5/0.04043	13.73/1.4×10^{10}	3.5 s	[47]
CdS–CdSe core–shell nanowalls	10	405	~100/1.2×10^3	–	250 ms/330 ms	[57]
CdS core-au/MXene	4	405	0.086	1.34×10^{11}	–	[46]

[a] f forward bias, r reverse bias

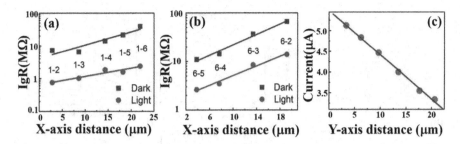

Fig. 11.15 The transportation route dependent resistance along X-direction (**a**) connection position 1 on trunk to different terminals of branches, (**b**) connection position 6 to different terminals 2–5 of branches, and (**c**) The light position dependent photocurrent on a branch along the Y-direction. (Adapted with the permission from Ref. [39]. Copyright 2018: Wiley-VCH Verlag GmbH & Co)

Table 11.3 Parameters of PDs based on doped CdS nanostructures

Material	Bias (V)	λ (nm)	S (%)/R (A/W)	EQE (%)/D (10^y Jones)	τ_r/τ_f	Referenes
Sn:CdS NWs	3	405	51.2	–	270 ms/310 ms	[11]
O-doped CdS/ perovskite		700	0.48	–	–	[41]
Sn-doped 1D CdS μWs/NSs	0.05	405 532 650	$30/1.46 \times 10^1$ $20/1.03 \times 10^1$ $8/2.66$	4.5×10^4 2.4×10^3 6.2×10^2	260 ms/260 ms	[53]
GaTe/Sn: CdS	2	White	$100(f)$ & $3000(r)/607$	–	260 ms/267 ms	[54]
Al:CdS-PS:p-Si	-2	400	0.6	$180/3.4 \times 10^{13}$	160 ms/350 ms	[55]
CdS QDs/PVA CdS:Cu QDs/ PVA	5	365	$20/9.5 \times 10^{-9}$ $24/15 \times 10^{-9}$	39.8×10^{-9} 63×10^{-9}	–	[74]
Ag/Y:CdS/Ag	15	532	55.1/0.83	$1948/4.3 \times 10^{11}$	78 ms/87 ms	[78]
CdS CdS:Mg(3%) CdS:Al(3%) CdS:Mg(3%) :Al(3%)	5	532	0.17 1.40 2.13 0.31	$41.4/7.1 \times 10^{10}$ $327.5/4.1 \times 10^{11}$ $497.3/5.2 \times 10^{11}$ $72.5/1.4 \times 10^{11}$	840 ms/910 ms 780 ms/880 ms 750 ms/850 ms 820 ms/890 ms	[80]
CdS:Ce (0%) CdS:Ce (10%) CdS:Ce (20%) CdS:Ce (30%)	–	460	560/0.0097 5523/0.1440 331/0.0245 84/0.0153	$2.51/3.5 \times 10^{10}$ $38.6/2.7 \times 10^{11}$ $6.2/1.3 \times 10^{11}$ $4.1/1.9 \times 10^{10}$	800 ms/860 ms	[79]

[a] *a* air, *v* vacuum

11.4.3 CdS Photodetectors Using Piezo-Phototronic Effect

A decade ago, it was found that the performance of PDs can be largely improved with the use of piezo-phototronic effect. This is a three-way coupling effect of piezoelectricity, semiconductor and photonic properties in piezoelectric

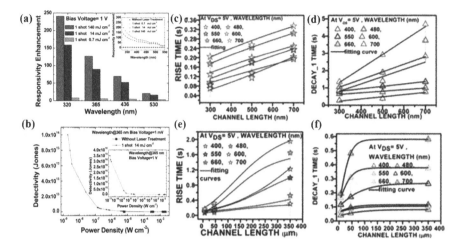

Fig. 11.16 (**a**) Responsivity enhancements at different wavelengths for laser-treated CdS devices, normalized with respect to the responsivity of the device prepared without laser treatment. (**b**) Detectivity of the laser-treated CdS device plotted with respect to the illumination power density at a bias voltage of 1 mV. Inset: Detectivity plotted with respect to the illumination power density at a bias voltage of 1 V. (Adapted with the permission from Ref. [72]. Copyright 2014: American Chemical Society). Variation in the rise and decay times as a function of channel lengths (**c**) and (**d**) for the short channels (nm) respectively and (**e**) and (**f**) for the long channels (μm) respectively. (Adapted with the permission from Ref. [81]. Copyright 2015: AIP)

(non-central symmetric) semiconductors like CdS, ZnO etc. The basic measurement setup for examine the piezo-phototronic effect and the device under various strain conditions are illustrated in Fig. 11.17a–d. There are various reports utilizing this effect to improve performance of many optical devices, here only few reports on PDs are discussed.

Zhang et al. [65] studied piezo-phototronic effect first for ZnO-CdS core-shell NWs [64] and subsequently for CF/ZnO-CdS core-shell NWs. Here, the various outcome of these studies are compared. Figure 11.17e, f illustrate the *I-V* characteristics of a single ZnO-CdS core-shell NW (device I) and CF/ZnO-CdS core-shell NW (device II) based device respectively under different tensile and compressive strains and 548 nm (1.43 mW/cm^2) illumination of light. One can see that as strain changes from compressive to tensile, photocurrent is increased continuously for both devices. However, the enhancement in photocurrent for device II is more than an order of 2 as compare to device I. Simultaneously, change of responsivity for device II under compressive strain and illumination of various wavelengths is much steeper than the device I as illustrated in Fig. 11.17g, h. It can also be noted that change in responsivity continuously increases with reduction in strain i.e. tensile to compressive strain. *I-t* responses were also studied by Yu et al. [67] for P3HT: CdS heterojunction based device under varying strain and different illuminations along with the various parameters that were studied in Refs. [64] and [65]. Such *I-t* response curves are illustrated in Fig. 11.17i, j under different strains for +ve and −ve

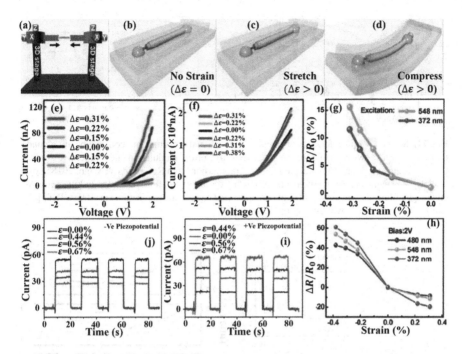

Fig. 11.17 Schematics of (**a**) experimental setup for studying the piezo-phototronic effect, (**b–d**) the device under various strain conditions. (Adapted with the permission from Ref. [65]. Copyright 2013: American Chemical Society). *I-V* characteristics of (**e**) device I and (**f**) device II under different tensile and compressive strains, excited green light of wavelength 548 nm and power 1.43 mW/cm². The change of responsivity (**g**) under compressive strains for device I and (**h**) tensile and compressive strains for device II excited by various wavelengths. (Adapted with the permission from Refs. [64] and [65]. Copyright 2012 and 2013: American Chemical Society). *I-t* response of P3HT: CdS heterojunction based device under 530 nm illumination with different (**i**) tensile strains and (**j**) compressive strain. (Adapted with the permission from Ref. [67]. Copyright 2017: Elsevier Ltd)

piezopotentials and 530 nm illumination respectively. It is evident from Fig. 11.17i, j, photocurrent increases as a function of increasing piezopotential and device shows good reproducibility and repeatability under various strains. Along with the above discussed studies Pal et al. [66] examined the *I–V* characteristics of the ZnO and ACCZ based flexible device under normal and concave bending conditions at an angle of ~32 ± 2° (Fig. 11.18a and inset of 11.18a). The enhancement in the current due to the development of a compressive strain induced piezopotential for both devices can be clearly seen. The enhancement in current is prominent in ACCZ based device than ZnO based device. Figure 11.18b illustrates the relative change in current between the bending and relaxed conditions at a constant bias of 2 V for ACCZ device. It is obvious that for each cycle of strain, the current initially increases sharply and becomes stable after reaching a peak value, which indicates good reproducibility and repeatability of device under various cycles of strain. In another study on p-Si/n-CdS heterojunction PD under 650 nm light illumination, Zhao et al. [69]

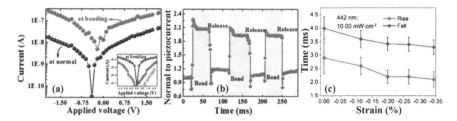

Fig. 11.18 (**a**) *I–V* characteristics of the bare ZnO TF under normal and concave bending conditions. The inset shows the same for the ACCZ and (**b**) Current modulation under repetitive application of bending cycles at 2 V. (Adapted with the permission from Ref. [66]. Copyright 2018: Royal Society of Chemistry). (**c**) Response time p-Si/n-CdS heterojunction PD under 442 nm (10 mW/cm^2) light illumination under different strains. (Adapted with the permission from Ref. [69]. Copyright 2019: The Royal Society of Chemistry)

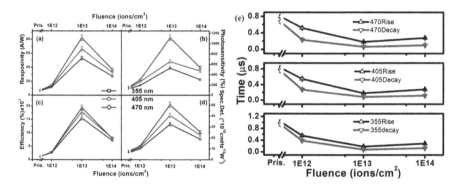

Fig. 11.19 Variation of (**a**) responsivity, (**b**) photosensitivity, (**c**) external quantum efficiency, (**d**) specific detectivity, and (**e**) response time with increase in ion irradiation influence. (Adapted with the permission from Ref. [10]. Copyright 2018: Elsevier B.V)

examined the effect of strain on response time of device as illustrated in Fig. 11.18c. It can be noticed that device becomes slower with increase in strain i.e. when stain changes from compressive to tensile.

An another tool for enhancing the PD performance was explored by Kumar et al. [4] by using ion beam modified CdS TFs for the fabrication of PD. Ion beam irradiation is a powerful technique to engineer the desired defects by multiple phenomena occurring simultaneously in the material. They investigated the photosensing characteristics of ion beam irradiated CdS TFs as a function of Ni^{6+} ion influence and found that up to a certain ion influence (1×10^{13} ion/cm^2) device performance was improved and then degraded for higher influences under illumination of both UV and visible light. Figure 11.19a–e shows the variation of different sensing parameters as a function of ion influence under illumination of various light [10].

11.5 Summary

In summary, CdS is enormously used as an active layer for photodetection in UV-Visible spectral region as well as in low frequency range. This chapter discussed CdS based PDs with various architectures, structures and working on different mechanisms. Various types of nanostructures, thin films, heterostructures, composites etc. have been utilized to fabricate high performing PDs. Besides, numerous approaches like doping with different elements, decoration with suitable materials, and ion beam irradiation etc. were proven to be fruitful in enhancing the performance and range of PDs fabricated. Moreover, self-powered PDs also have been realized using multifarious effects like piezoelectric, piezophototronic, pyroelectric, etc. in CdS and other materials used in architectures. In nut-shell, CdS is an integral part of UV-Visible PDs fabricated so far with sufficiently high FOM.

Acknowledgments The author TK is thankful to the University Grants Commission (UGC), New Delhi, India, for providing scholarship under NF-OBC scheme (NFO-2018-19-OBC-JAM-69666).

References

1. Zhuge F, Zheng Z, Luo P, Lv L, Huang Y, Li H, Zhai T (2017) Nanostructured materials and architectures for advanced infrared photodetection. Adv Mater Technol 2(8):1700005
2. Song K, Ma N, Mishra YK, Adelung R, Yang Y (2019) Achieving light-induced ultrahigh pyroelectric charge density toward self-powered UV light detection. Adv Electron Mater 5(1):1800413
3. Liu K, Sakurai M, Aono M (2010) ZnO-based ultraviolet photodetectors. Sensors (Basel) 10(9):8604–8634
4. Masouleh FF, Das N (2014) Application of metal-semiconductor-metal photodetector in high-speed optical communication systems. In: Das N (ed) Advances in optical communication. Intech, pp 87–114
5. Kim H, Kumar MD, Kim J (2015) Highly-performing Ni/SiO₂/Si MIS photodetector for NIR detecting applications. Sensors Actuators A Phys 233:290–294
6. https://www.electrical4u.com/p-i-n-photodiode-avalanche-photo-diode/. PIN photodiode. Accessed 08 Jan 2022
7. Gao XD, Fei GT, Xu SH, Zhong BN, Ouyang HM, Li XH, Zhang LD (2019) Porous Ag/TiO₂-Schottky-diode based plasmonic hot-electron photodetector with high detectivity and fast response. Nano 8(7):1247–1254
8. Park S, Kim SJ, Nam JH, Pitner G, Lee TH, Ayzner AL, Wang H, Fong SW, Vosgueritchian M, Park YJ, Brongersma ML, Bao Z (2015) Significant enhancement of infrared photodetector sensitivity using a semiconducting single-walled carbon nanotube/C60 phototransistor. Adv Mater 27(4):759–765
9. Kumar P, Saxena N, Dewan S, Singh F, Gupta V (2016) Giant UV-sensitivity of ion beam irradiated nanocrystalline CdS thin films. RSC Adv 6(5):3642–3649
10. Kumar P, Saxena N, Singh F, Gupta V (2018) Ion beam assisted fortification of photoconduction and photosensitivity. Sensors Actuators A Phys 279:343–350
11. Guo S, Wang L, Ding C, Li J, Chai K, Li W, Xin Y, Zou B, Liu R (2020) Tunable optical loss and multi-band photodetection based on tin doped CdS nanowire. J Alloys Compd 835:155330

12. Bao Q, Li W, Xu P, Zhang M, Dai D, Wang P, Guo X, Tong L (2020) On-chip single-mode CdS nanowire laser. Light Sci Appl 9:42

13. Wen Z, Liu P, Ma J, Jia S, Xiao X, Ding S, Tang H, Yang H, Zhang C, Qu X, Xu B, Wang K, Teo KL, Sun XW (2021) High-performance ultrapure green CdSe/CdS core/crown nanoplatelet light-emitting diodes by suppressing nonradiative energy transfer. Adv Electron Mater 7(7):2000965

14. Kumar P, Saxena N, Chandra R, Gao K, Zhou S, Agarwal A, Singh F, Gupta V, Kanjilal D (2014) SHI induced enhancement in green emission from nanocrystalline CdS thin films for photonic applications. J Lumin 147:184–189

15. Liu B, Guo J, Hao R, Wang L, Gu K, Sun S, Aierken A (2020) Effect of Na doping on the performance and the band alignment of CZTS/CdS thin film solar cell. Sol Energy 201:219–226

16. Saxena N, Kumar P, Gupta V, Kanjilal D (2018) Radiation stability of CBD grown nanocrystalline CdS films against ion beam irradiation for solar cell applications. J Mater Sci Mater 29(13):11013–11019

17. Li J-Y, Li Y-H, Qi M-Y, Lin Q, Tang Z-R, Xu Y-J (2020) Selective organic transformations over cadmium sulfide-based photocatalysts. ACS Catal 10(11):6262–6280

18. Xie T, Zhong X, Liu Z, Xie C (2020) Silica-anchored cadmium sulfide nanocrystals for the optical detection of copper(II). Mikrochim Acta 187(6):323

19. Faraz M, Abbasi A, Naqvi FK, Khare N, Prasad R, Barman I, Pandey R (2018) Polyindole/cadmium sulphide nanocomposite based turn-on, multi-ion fluorescence sensor for detection of Cr3+, Fe3+ and Sn2+ ions. Sensors Actuators B Chem 269:195–202

20. Saxena N, Kumar P, Gupta V (2019) CdS nanodroplets over silica microballs for efficient room-temperature LPG detection. Nanoscale Adv 1(6):2382–2391

21. Wang H, Ma J, Zhang J, Feng Y, Vijjapu MT, Yuvaraja S et al (2021) Gas sensing materials roadmap. J Phys Condens Matter 33(30):303001

22. Harish R, Nisha KD, Prabakaran S, Sridevi B, Harish S, Navaneethan M, Ponnusamy S, Hayakawa Y, Vinniee C, Ganesh MR (2020) Cytotoxicity assessment of chitosan coated CdS nanoparticles for bio-imaging applications. Appl Surf Sci 499:143817

23. Kim H-R, Bong J-H, Jung J, Sung JS, Kang M-J, Park J-G, Pyun J-C (2020) An on-chip chemiluminescent immunoassay for bacterial detection using in situ-synthesized cadmium sulfide nanowires with passivation layers. Biochip J 14(3):268–278

24. Saxena N, Kumar P, Gupta V (2015) CdS : SiO2 nanocomposite as a luminescence-based wide range temperature sensor. RSC Adv 5(90):73545–73551

25. Miao S, Cho Y (2021) Toward green optoelectronics: environmental-friendly colloidal quantum dots photodetectors. Front Energy Res 9:1–18

26. Xu K, Zhou W, Ning Z (2020) Integrated structure and device engineering for high performance and scalable quantum dot infrared photodetectors. Small 16(47):e2003397

27. Paschotta DR (2008) Metal–semiconductor–metal photodetectors. Laser Physics and Technology, Wiley-VCH

28. https://www.inup.cense.iisc.ac.in/msm-photodetectors. Fabrication Of ZnO Based MSM Photodetectors, 2021. https://www.inup.cense.iisc.ac.in/msm-photodetectors. 16 Oct 2021

29. Kalsi T, Kumar P (2021) Cd1-xMgxS CQD thin films for high performance and highly selective NIR photodetection. Dalton Trans 50(36):12708–12715

30. An Q, Meng X (2016) Aligned arrays of CdS nanotubes for high-performance fully nanostructured photodetector with higher photosensitivity. J Mater Sci Mater 27(11):11952–11960

31. Jin B, Huang P, Zhang Q, Zhou X, Zhang X, Li L, Su J, Li H, Zhai T (2018) Self-limited epitaxial growth of ultrathin nonlayered CdS flakes for high-performance photodetectors. Adv Funct Mater 28(20):1800181

32. Nawaz MZ, Xu L, Zhou X, Shah KH, Wang J, Wu B, Wang C (2021) CdS nanobelt-based self-powered flexible photodetectors with high photosensitivity. Mater Adv 2(18):6031–6038

33. Munde S, Shinde N, Khanzode P, Budrukkar M, Lahane P, Dadge J, Jejurikar S, Mahabole M, Khairnar R, Bogle K (2018) Nano-crystalline CdS thick films: a highly sensitive photodetector. Mater Res Exp 5:066203

34. Yang Z, Guo L, Zu B, Guo Y, Xu T, Dou X (2014) CdS/ZnO core/shell nanowire-built films for enhanced photodetecting and optoelectronic gas-sensing applications. Adv Opt Mater 2(8):738–745

35. Tang M, Xu P, Wen Z, Chen X, Pang C, Xu X, Meng C, Liu X, Tian H, Raghavan N, Yang Q (2018) Fast response CdS-CdS$_x$Te$_{1-x}$-CdTe core-shell nanobelt photodetector. Sci Bull 63(17):1118–1124

36. Chakrabarty P, Gogurla N, Bhandaru N, Ray SK, Mukherjee R (2018) Enhanced performance of hybrid self-biased heterojunction photodetector on soft-lithographically patterned organic platform. Nanotechnology 29(50):505301

37. Roy S, Bhandaru N, Das R, Harikrishnan G, Mukherjee R (2014) Thermally tailored gradient topography surface on elastomeric thin films. ACS Appl Mater Interfaces 6(9):6579–6588

38. Dai Y, Wang X, Peng W, Xu C, Wu C, Dong K, Liu R, Wang ZL (2018) Self-powered Si/CdS flexible photodetector with broadband response from 325 to 1550 nm based on pyro-phototronic effect: an approach for photosensing below bandgap energy. Adv Mater 30(9):1705893

39. Hao Y, Guo S, Weller D, Zhang M, Ding C, Chai K, Xie L, Liu R (2018) Position-sensitive array photodetector based on comb-like CdS nanostructure with cone-shape branches. Adv Funct Mater 29(1):1805967

40. Chang Y, Wang J, Wu F, Tian W, Zhai W (2020) Structural design and pyroelectric property of SnS/CdS heterojunctions contrived for low-temperature visible photodetectors. Adv Funct Mater 30(23):2001450

41. Cao F, Meng L, Wang M, Tian W, Li L (2019) Gradient energy band driven high-performance self-powered perovskite/CdS photodetector. Adv Mater 31(12):e1806725

42. Ge Y, Zhang M, Wang L, Meng L, Tang J, Chen Y, Wang L, Zhong H (2019) Polarization-sensitive ultraviolet detection from oriented-CdSe@CdS-dot-in-rods-integrated silicon photodetector. Adv Opt Mater 7(18):1900330

43. Wang J, Chang Y, Huang L, Jin K, Tian W (2018) Designing CdS/Se heterojunction as high-performance self-powered UV-visible broadband photodetector. APL Mater 6(7):076106

44. Gozeh BA, Karabulut A, Yildiz A, Dere A, Arif B, Yakuphanoglu F (2019) SILAR controlled CdS nanoparticles sensitized CdO diode based photodetectors. SILICON 12(7):1673–1681

45. Li J, Zhu Y, Li M, Cai H, Ding H, Pan N, Wang X (2018) One-step fabrication of CdS nano-flake arrays and its application for photodetector. Optik 169:190–195

46. Jiang T, Huang Y, Meng Z (2020) CdS core-Au/MXene-based photodetectors: positive deep-UV photoresponse and negative UV–vis-NIR photoresponse. Appl Surf Sci 513:145813

47. Reddy KCS, Selamneni V, Rao MGS, Meza-Arroyo J, Sahatiya P, Ramirez-Bon R (2021) All solution processed flexible p-NiO/n-CdS rectifying junction: applications towards broadband photodetector and human breath monitoring. Appl Surf Sci 568:150944

48. Lu MY, Chang YT, Chen HJ (2018) Efficient self-driven photodetectors featuring a mixed-dimensional van der Waals heterojunction formed from a CdS nanowire and a MoTe2 flake. Small 14(40):e1802302

49. Li L, Lou Z, Chen H, Shi R, Shen G (2019) Stretchable SnO2-CdS interlaced-nanowire film ultraviolet photodetectors. Sci China Mater 62(8):1139–1150

50. Pei Y, Pei R, Liang X, Wang Y, Liu L, Chen H, Liang J (2016) CdS-nanowires flexible photodetector with Ag-nanowires electrode based on non-transfer process. Sci Rep 6:21551

51. Moger SN, Mahesha MG (2020) Colour tunable co-evaporated CdS$_x$Se$_{1-x}$ ($0 \leq x \leq 1$) ternary chalcogenide thin films for photodetector applications. Mater Sci Semicond Process 120:105288

52. Moger SN, Mahesha MG (2021) Investigation on spectroscopic and electrical properties of p-Si/CdS$_x$Se$_{1-x}$ ($0 \leq x \leq 1$) heterostructures for photodetector applications. J Alloys Compd 870:159479

53. Zhou W, Peng Y, Yin Y, Zhou Y, Zhang Y, Tang D (2014) Broad spectral response photodetector based on individual tin-doped CdS nanowire. AIP Adv 4(12):123005

54. Zhou W, Zhou Y, Peng Y, Zhang Y, Yin Y, Tang D (2014) Ultrahigh sensitivity and gain white light photodetector based on GaTe/Sn:CdS nanoflake/nanowire heterostructures. Nanotechnology 25(44):445202

55. Das M, Sarmah S, Sarkar D (2019) Photo sensing property of nanostructured CdS-porous silicon (PS):p-Si based MSM hetero-structure. J Mater Sci Mater 30(12):11239–11249

56. Gong Y, Zhang X, Yang T, Huang W, Liu H, Liu H, Zheng B, Li D, Zhu X, Hu W, Pan A (2019) Vapor growth of CdS nanowires/WS$_2$ nanosheet heterostructures with sensitive photodetections. Nanotechnology 30(34):345603

57. Xu J, Rechav K, Popovitz-Biro R, Nevo I, Feldman Y, Joselevich E (2018) High-gain 200 ns photodetectors from self-aligned CdS-CdSe core-shell nanowalls. Adv Mater 30(20):e1800413

58. Wang D, Chen H, Min Y, Liang J, Pan F (2020) Tunable p- and n-type Nb:TiO$_2$ and performance optimizing of self-powered Nb:TiO2/CdS photodetectors. Semicond Sci Technol 35(7):075015

59. Peng M, Wu F, Wang Z, Wang P, Gong F, Long M, Chen C, Dai J, Hu W (2019) Enhancement-mode CdS nanobelts field effect transistors and phototransistors with HfO$_2$ passivation. Appl Phys Lett 114(11):111103

60. Zheng D, Fang H, Wang P, Luo W, Gong F, Ho JC, Chen X, Lu W, Liao L, Wang J, Hu W (2016) High-performance ferroelectric polymer side-gated CdS nanowire ultraviolet photodetectors. Adv Funct Mater 26(42):7690–7696

61. Zhao W, Liu L, Xu M, Wang X, Zhang T, Wang Y, Zhang Z, Qin S, Liu Z (2017) Single CdS nanorod for high responsivity UV-visible photodetector. Adv Opt Mater 5(12):1700159

62. Zhou Y, Zhang L, Gao W, Yang M, Lu J, Zheng Z, Zhao Y, Yao J, Li J (2021) A reasonably designed 2D WS2 and CdS microwire heterojunction for high performance photoresponse. Nanoscale 13(11):5660–5669

63. Huang G, Zhang P, Bai Z (2019) Self-powered UV–visible photodetectors based on ZnO/graphene/CdS/electrolyte heterojunctions. J Alloys Compd 776:346–352

64. Zhang F, Ding Y, Zhang Y, Zhang X, Wang ZL (2012) Piezo-phototronic E ff ect enhanced visible and ultraviolet photodetection using a ZnO-CdS core-shell micro/nanowire. ACS Nano 6(10):9229–9236

65. Zhang F, Niu S, Guo W, Zhu G, Liu Y, Zhang X, Wang ZL (2013) Piezo-phototronic E ffect enhanced visible/UV photodetector of a carbon-fiber/ZnO-CdS double-shell microwire. ACS Nano 7(5):4537–4544

66. Pal S, Bayan S, Ray SK (2018) Piezo-phototronic mediated enhanced photodetection characteristics of plasmonic Au-g-C3N4 /CdS/ZnO based hybrid heterojunctions on a fl exible platform. Nanoscale 10:19203–19211

67. Yu X-X, Yin H, Li H-X, Zhang W, Zhao H, Li C, Zhu M-Q (2017) Piezo-phototronic effect modulated self-powered UV/visible/near-infrared photodetectors based on CdS:P3HT microwires. Nano Energy 34:155–163

68. Lin P, Zhu L, Li D, Xu L, Wang ZL (2018) Tunable WSe$_2$-CdS mixed-dimensional van der Waals heterojunction with a piezo-phototronic effect for an enhanced flexible photodetector. Nanoscale 10(30):14472–14479

69. Zhao ZH, Dai Y (2019) Piezo-phototronic effect-modulated carrier transport behavior in different regions of a Si/CdS heterojunction photodetector under a Vis-NIR waveband. Phys Chem Chem Phys 21(18):9574–9580

70. Sarkar S, Basak D (2015) Self powered highly enhanced dual wavelength ZnO@CdS core-shell nanorod arrays photodetector: an intelligent pair. ACS Appl Mater Interfaces 7(30):16322–16329

71. Halge DI, Narwade VN, Khanzode PM, Begum S, Banerjee I, Dadge JW, Kovac J, Rana AS, Bogle KA (2021) Development of highly sensitive and ultra-fast visible-light photodetector using nano-CdS thin film. Appl Phys A Mater Sci Process 127(6):446

72. Lin KT, Chen HL, Lai YS, Liu YL, Tseng YC, Lin CH (2014) Nanocrystallized CdS beneath the surface of a photoconductor for detection of UV light with picowatt sensitivity. ACS Appl Mater Interfaces 6(22):19866–19875

73. Kan H, Liu S, Xie B, Zhang B, Jiang S (2017) The effect of Au nanocrystals applied in CdS colloidal quantum dots ultraviolet photodetectors. J Mater Sci Mater 28(13):9782–9787
74. Kakati J, Datta P (2015) Schottky junction UV photodetector based on CdS and visible photodetector based on CdS:Cu quantum dots. Optik 126(18):1656–1661
75. Al-Alwani AJK, Chumakov AS, Shinkarenko OA, Gorbachev IA, Pozharov MV, Venig S, Glukhovskoy EG (2017) Formation and optoelectronic properties of graphene sheets with CdSe/CdS/ZnS quantum dots monolayer formed by Langmuir-Schaefer hybrid method. Appl Surf Sci 424:222–227
76. Zhang C, Tian W, Xu Z, Wang X, Liu J, Li SL, Tang DM, Liu D, Liao M, Bando Y, Golberg D (2014) Photosensing performance of branched CdS/ZnO heterostructures as revealed by in situ TEM and photodetector tests. Nanoscale 6(14):8084–8090
77. Lam KT, Hsiao YJ, Ji LW, Fang TH, Hsiao KH, Chu TT (2017) High-sensitive ultraviolet photodetectors based on ZnO nanorods/CdS heterostructures. Nanoscale Res Lett 12(1):31
78. Shkir M, Khan ZR, Chandekar KV, Alshahrani T, Ashraf IM, Khan A, Marnadu R, Zargar RA, Mohanraj P, Revathy MS, Manthrammel MA, Sayed MA, Ali HE, Yahia IS, Yousef ES, Algarni H, AlFaify S, Sanaa MF (2021) Facile fabrication of Ag/Y:CdS/Ag thin films-based photodetectors with enhanced photodetection performance. Sensors Actuators A Phys 331:112890
79. Ibrahim IM, Safi AA, Al-Hardan NHM (2019) Enhancement the sensitivity of CdS nano structure by adding of rare earth materials. J Phys Conf Ser 1178:012013
80. Kumar KDA, Mele P, Golovynskyi S, Khan A, El-Toni AM, Ansari AA, Gupta RK, Ghaithan H, AlFaify S, Murahari P (2022) Insight into Al doping effect on photodetector performance of CdS and CdS:Mg films prepared by self-controlled nebulizer spray technique. J Alloys Compd 892:160801
81. Sharma A, Kaur M, Bhattacharyya B, Karuppiah S, Singh SP, Senguttuvan TD, Husale S (2015) Channel length specific broadspectral photosensitivity of robust chemically grown CdS photodetector. AIP Adv 5(4):047116

Chapter 12
ZnTe-Based Photodetectors for Visible-UV Spectral Region

Jiajia Ning

12.1 Introduction

In the past few decades, the photoelectronic industry has been well developed to change the life in the world. Various photoelectronic devices and structures were developed and utilized, one of them, possessing the ability of transformation from light to electrical signals, such as photodetectors, has attracted scientists' intensive attention based on their potential for light detection. Benefiting from the development of semiconductor industry, the current detector based on the different semiconductors can detect the light from ultraviolet (UV: 10–400 nm) to the visible (vis: 400–700 nm) to the near-infrared (NIR: 700–1000 nm) to the terahertz (0.1–10 THz), and transfer them to electrical signals [1]. Generally, the whole photodetection system mainly contains the two parts, the photodetectors and signal processing chips. The photodetector part is built on the basis of semiconductors [2].

Due to the different band gap in semiconductors, the photodetector can absorb the light in the different range and excite the electron to conduction band, finally, the excitons are transferred to electronic signals. The range of light detected is depended by the ban gap in semiconductors. So various semiconductors are used in photodetectors to cover the full range of light (UV to THz). In the whole range of light, the UV and visible range are more important and closer to our life, which can be used in flame sensing, ozone sensing, convert communications, air and water purification, environmental monitoring, video imaging, materials identification [3–11].

The band gap is an important parameter in photodetectors, depending the detected range of light. The general way to change the band gap is using semiconductors with different composition. To one semiconductor, the band gap is difficult

J. Ning (✉)
Key Laboratory of Physics and Technology for Advanced Batteries, Ministry of Education, College of Physics, Jilin University, Changchun, China
e-mail: jiajianing@jlu.edu.cn

© The Author(s), under exclusive license to Springer Nature Switzerland AG 2023
G. Korotcenkov (ed.), *Handbook of II-VI Semiconductor-Based Sensors and Radiation Detectors*, https://doi.org/10.1007/978-3-031-20510-1_12

to change. In 1980s, the quantum size confinement effect was found in the semiconductors [12–14], the band gap of semiconductors can be tuned via changing the size of semiconductors [15, 16], which provides a simple way to turn the band gap in semiconductors. The photodetectors with one semiconductor, covering a broad region of light, is possible on the basis of quantum size confinement effect. Such as the full visible region can be detected just used CdSe nanocrystals with the different size [17].

Based on the intrinsic band gap in semiconductors and the quantum size confinement effect, the developed photodetectors based on semiconductors can cover the region from violet to infrared range [2, 9]. Among the whole detected region of light, the photodetectors for visible region the most widely used in our life. The commercial reason of the photodetectors for the visible region promotes the investigation and development on visible region materials for photodetectors. Up to now, CdSe [18], CdS [19], CdTe [20], InP [21], Cd_3P_2 [22], and Si [23] based photodetectors were developed for the detection of visible region light. However, the toxic element of Cd limits the further applications of cadmium chalcogenide-based photodetector in our life. The Cd-free semiconductors-based photodetectors were attracted much attention for their potential in applications. As an important Cd-free semiconductor, ZnTe is an important candidate for visible region detectors. Herein, we review the development of ZnTe based photodetector for visible region from controlled synthesis, optical characterizations and the design of photodetectors.

12.2 Optical Properties of ZnTe

Zinc chalcogenide are the important members in II-VI semiconductors, exhibiting the widely band gap of 3.8 eV for ZnS, 2.8 eV for ZnSe and 2.2 eV for ZnTe [24]. The widely band gap in ZnS and ZnSe makes them to be used in ultraviolet to deep blue region [25]. Compared with ZnS and ZnSe, the smaller band gap of 2.2 eV in ZnTe can extend the optical properties to visible light. The absorption peak of ZnTe nanocrystals (NCs) shifts to 430 nm with the smaller diameter based on the quantum size confinement effect [26].

Due to the band gap of 2.2 eV in ZnTe, the fluorescence emission for ZnTe NCs is limited 563 nm. In order to extend the fluorescence emission of ZnTe NCs to the whole visible range, the alloyed ZnTe with Se to form ZnSeTe [27, 28] or the formation of type-II heterostructures [29, 30] with another compound are the available methods. The band gap in alloyed semiconductor can be tuned to smaller than each of them due to the bowing effect [31]. Alloyed ZnSeTe NCs showed the fluorescence emission at 540 nm with the size of 4.3 nm [32].

The fluorescence emission from ZnTe-based NCs was also widely tuned for the type-II heterostructures, such as type-II core/shell NCs, combining the charge in the core and shell for recombination [33]. The red shift and larger stokes-shift are observed in type-II core/shell NCs. On the basis of band gap alignment between

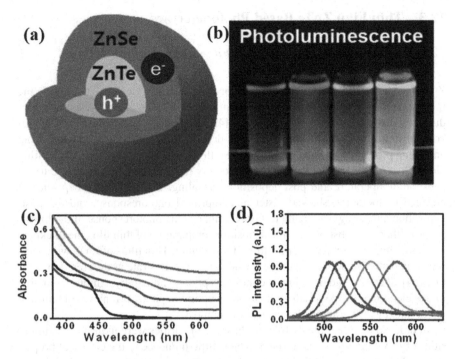

Fig. 12.1 (a) the schematic illustration for ZnTe/ZnSe core/shell NCs, (b) the optical properties for ZnTe/ZnSe core/shell NCs, (c) the absorption spectra of ZnTe/ZnSe core/shell NCs with different thickness of ZnSe shell, (d) the photoluminescence emission spectra of ZnTe/ZnSe core/shell NCs with different thickness of ZnSe shell. (Reproduced with permission from Ref. [34]. Copyright 2010: American Chemical Society)

ZnTe and shell materials, ZnTe/ZnSe core/shell NCs are one of typical type-II heterostructures (Fig. 12.1a), coupled with the growth of ZnSe on ZnTe core NCs, an obvious red shift can be observed in absorption spectra and photoluminescence emission spectra (Fig. 12.1c, d). The emission can be tuned from blue to orange region via changing the thickness of ZnSe shell (Fig. 12.1b) [34].

In semiconductor NCs, the optical properties and energy-level structure are also influenced by the shape and crystal structure of NCs [35]. 1D nanorods were developed for the special polarized photoluminescence emission [36], which have been widely used in display, solar energy conversion and optoelectronic devices [37–40]. Based on the 1D ZnTe nanorods, ZnSe dots were further grown on the two tips of nanorods to form nanodumbbells structure [41]. The formed dumbbells structured of ZnTe/ZnSe NCs exhibited the optical characters for type-II heterostructures, the fluorescence emission can tunned between 500 nm to 580 nm via changing the size of ZnSe dots on the tips of nanorods, and the PLQY in ZnTe/ZnSe nanodumbbells can reach 40%.

12.3   Thin Film ZnTe Based Photodetectors

12.3.1   ZnTe Thin Film with Vacuum Evaporation Method

ZnTe has been attracted much attention for optoelectronic device. Thin film is the most widely developed for device in industry. The thin film of ZnTe was early produced by the vacuum evaporation method [42]. ZnTe was deposited on glass or silicon substrate, the type of substrate, the deposit temperature and the following annealing procedure showed the important influence on the properties of ZnTe film. Gowrish et al. [43] found the photoconductivity of ZnTe thin film is related to the substrate temperature and post deposition annealing. The ZnTe film, deposited at elevated temperature, showed faster and improved photoresponse, and the post-deposition annealing was found to further enhance the photoresponse of the films.

More than the substrate and the deposition temperature of thin film, the thickness is another important parameter for thin film structure. Thin films of ZnTe with varying compositions and thickness are formed on glass substrate at different temperature [44]. The dark and illuminated conditions photoconductivity exhibits a function of wavelength of incident radiation, and the maximum photocurrent was obtained at about thickness of 500 nm irrespective of composition.

In order to modified the ZnTe thin films, the impurities were induced. The impurities of BaF_2 and $PbCl_2$ were found to have little influence to the electrical properties of ZnTe thin films [42]. Furtherly, hydrogen was induced to ZnTe thin film to passivate the surface of ZnTe film [45]. After the hydrogen passivation of ZnTe film via hydrogen plasma method, the initial PL emission peaks of 2.06 eV, 1.47 eV, 1.33 eV and 1.06 eV became one strong band edge green emission at 2.37 eV. The hydrogen passivated ZnTe film also showed an improved photoconductivity at higher temperature. The investigation on the mechanism of hydrogen treat showed the passivated deep defect energy level of ZnTe.

12.3.2   ZnTe Thin Film with Wet Chemical Method

Beside the vacuum evaporation method, the wet chemical methods were also developed to prepare ZnTe thin film. Hamdi and his co-workers [46] developed the modified Bridgman solvent method to produce ZnTe film at 1980, which showed the responsivity of 5×10^{-2}A/W with the light of 610 nm. At 2000, Bozzini et al. [47] further developed the electrodeposition method to synthesize ZnTe thin film (Fig. 12.2). The crystalline structure of the deposited ZnTe film occur the cubic zinc-blende structure under suitable electrochemical conditions.

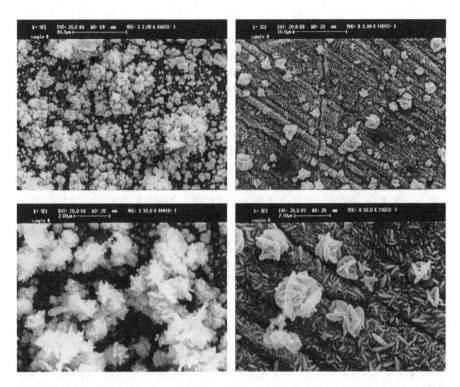

Fig. 12.2 SEM micrograph of an electrodeposit from the ZnTe-bath on Cu, deposited at 0 mV (Ag/AgCl) (fibrous appearance-left) and at −200 mV(Ag/AgCl) (rosette appearance-right). (Reproduced with permission from Ref. [47]. Copyright 2000: Elsevier)

12.4   ZnTe Nanostructures Based Photodetectors

Because of the difficulty in the preparation of dot-shape ZnTe NCs-based photodetectors, 1D or two-dimensional (2D) nanostructures are widely used for ZnTe based photodetectors. As the above discussion, ZnTe can be used for photodetectors between ultra-violet (UV) and green region. 1D nanowires and 2D nanosheets are produced by physical vapour deposition (PVD), chemical vapour deposition (CVD) and metal-organic chemical vapour deposition (MOCVD) method for scale-up [48–51].

12.4.1   One Dimensional ZnTe Based Photodetectors

1D nanostructures have been studied extensively for their novel properties and potential applications in nanoscale devices, such as ultra-sensitive nanosensors and photodetectors [52, 53]. The single ZnTe nanowire-based photodetectors was developed in 2011 [51]. The untreated ZnTe nanowire produced by PVD method exhibited the diameter of 60–400 nm with cubic zinc-blende crystal structure. One single ZnTe nanowire with length of 88 μm and width of 330 nm was selected to build the photodetectors on the SiO_2/Si substrate. Ti/Au (2 nm/60 nm) interdigitated electrodes with 3 μm separation was deposited on the ZnTe nanowires. Compared with dark condition, this ZnTe nanowire-based photodetector showed 2 times of photocurrent under the light of 500 nm.

In order to improve the performance of ZnTe nanowire-based photodetectors, the surface of ZnTe nanowires can be modified with other ligand or compound to passivate the defects [54, 55]. Growth of the inorganic compound as the shell is the most efficient way to passivate the surface. ZnO shell was grown on ZnTe nanowires to form ZnTe/ZnO core/shell nanowires [56]. Starting from the initial ZnTe nanowires, the thickness of ZnO shell can be tuned to increase the ratio of shell for 19% in core/shell nanowires. The performance of ZnTe nanowires-based photodetectors greatly improved by increasing the thickness of ZnO shell. The I_{pn}/I_{dark}, responsivity, and the photoconductive gain can be improved from 1.95 (A W^{-1}) and 8×10^2% to 199 (A W^{-1}) and 8.12×10^4% with increasing the thickness of ZnO shell, respectively.

A method to modify the intrinsic properties in semiconductor is inducing the impurities with controlling amount. The present impurities can increase the density of charge in semiconductor to form n-type or p-type semiconductor based on the different type of impurities [57, 58]. During the formation process of ZnTe structures, Cu, P, Sb and N can act as induced impurities in ZnTe [59–61]. The doped ZnTe can reveal the enhanced hole mobility or electron mobility. The performance of 1D ZnTe nanowires in optoelectronic devices can be promoted via the present impurities, such as Sb dope ZnTe nanowires and Cu doped ZnTe nanowires.

Ga doped ZnTe nanowires were developed via the simple thermal evaporation method [62]. The Ga content in the ZnTe nanowires can be tuned from 1.3 to 8.7%, and the hole mobility and hole concentration will increase from 0.0069 to 0.46 $cm^2V^{-1}S^{-1}$, respectively. The ZnTe:Ga nanowires based photodetectors also show the high sensitivity to visible light illumination, the responsivity and detectivity were estimated to be 4.17×10^3 AW^{-1} and 3.19×10^{13} $cmHz^{1/2}$ W^{-1}, higher than other undoped ZnTe nanostructures-based photodetectors.

12.4.2 Two Dimensional ZnTe Based Photodetectors

More than 1D nanostructures, 2D nanostructures are another important candidate for optoelectronic device. As the most famous 2D nanostructures, graphene has been widely investigated and used in optoelectronic devices [63–65]. Because of the quantum size confinement effect in 2D nanostructures, the electron can move in another two dimensions. Generally, 2D nanostructures exhibit the thickness dependent optical and optoelectronic properties [66, 67].

Nanosheet is one type of 2D nanostructures. Wang and his co-workers [68] synthesized ZnTe nanosheet with the cubic zinc-blende structure via PVD method (Fig. 12.3a, b). The ZnTe nanosheets with single crystal structure had the thickness of 65 nm and the length of micrometers (Fig. 12.3c), showing the emission band peaked at 550 nm near-bandgap emission of ZnTe. Figure 12.3d shows the I_{ds}–V_{ds} curves of the device under various incident power density, which indicating a good photo conducting behavior. The obtained net photocurrent increases as the incident power density increased, indicating more electrons and holes are generated under stronger light illumination condition (Fig. 12.3e). Larger bias voltage can further enhance the charge transport process. Furthermore, such ZnTe nanosheets based photodetector shows good reversible switching properties under different bias voltages (Fig. 12.3g), indicating the good stability of the ZnTe devices. The produced

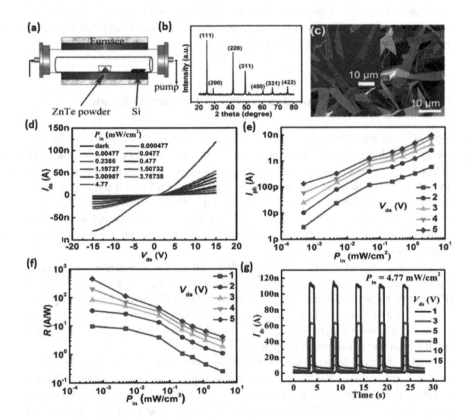

Fig. 12.3 (a) Schematic diagram of PVD system, (b) XRD pattern of ZnTe nanosheets, (c) SEM image of ZnTe nanosheets, (d) I_{ds}-V_{ds} curves of the photodetectors in the dark (black) and under 520 nm laser illumination at different incident power, (e, f) the net photocurrent and calculated photoresponsivity as a function of the incident power density under different cource-drain bias voltage, (g) time dependent I_{ds} of the ZnTe device with the laser (520 nm, 4.77 mW/cm²) switching on and off under a positive source-drain voltage V_{as} from 1 to 15 V. (Reproduced with permission from Ref. [68]. Copyright 2019: Elsevier)

ZnTe nanosheets have the p-type conductivity, and ZnTe nanosheet-based photode-tectors exhibit the high photoresponsivity of 453.9 A/W, excellent stability and reliability.

Recently, p-type 2D ZnTe nanostructures were developed by You et al. [69]. This ultrathin ZnTe nanoflakes were controllably synthesized by PVD method. The monolayer ZnTe flakes-based photodetectors showed a broadband response varying from the visible to near-infrared range under the dark and illumination with diverse light intensities (405 nm). The ZnTe based device gave a typical p-type semiconduc-tor behavior. The large and different photocurrent at the whole range of gate bias suggested the significant photoresponse. The photocurrent raises gradually with increase light intensity due to the enhancement of photogenerated carriers. The ZnTe flakes-based photodetectors also showed the stability after 3 months in air. This monolayer ZnTe nanoflake show photoresponse properties from the visible to near-infrared region and the responsivity of 18.3 A W^{-1} and detectivity of 2.89×10^9 Jones under 405 nm illumination in air, implying the potential application in elec-tronics and optoelectronics worked in harsh environment.

12.5   ZnTe Based Photodetectors for Terahertz Region

Most of ZnTe-based photodetectors are using the intrinsic band gap in ZnTe. The energy of light can be absorbed by ZnTe to produce the electron and holes, the charge can give the single via the transformation or recombination. Another type of photodetectors with band gap-independent properties can be built with the array of semiconductor, named as subwavelength structure (SWS), which can enhance ultra-broad band transmission [70, 71]. The working region of photodetectors is related to the surface of semiconductor-based SWS, covering the visible-near-infrared, infrared and terahertz region.

The SWSs on a ZnTe single crystal are designed via a modified reactive ion etch-ing method to increase the broadband transmission [72]. Large-area polystyrene (PS) nanoparticle nanolayers are spin-coated on the ZnTe surface by adopting oxy-gen plasma treatment to improve the surface wettability (Fig. 12.4a). After the etch-ing of Ch$_4$/H$_4$ and O$_2$ plasma, the SWSs based on etched ZnTe can be tuned via etching time and radio frequency power (Fig. 12.4b–d).

Finally, the well-define conical SWS arrays were fabricated on the ZnTe crystal by reactive ion etching over the PS monolayer template, with the size of SWS arrays customized by optimizing the etching process. The ultra-broadband antireflection on the surface structured ZnTe crystals in the visible-near-infrared, infrared, and terahertz regions with transmittance increase of 11.6%, 10.0% and 24.8%, which are attributed to the decrease of surface Fresnel reflection by SWS, and the transmit-tance can reach 70% in 0.2–1.0 terahertz (Fig. 12.5). This provides a new strategy to enhance the terahertz efficiency and detection sensitivity based on ZnTe crystals by surface engineering.

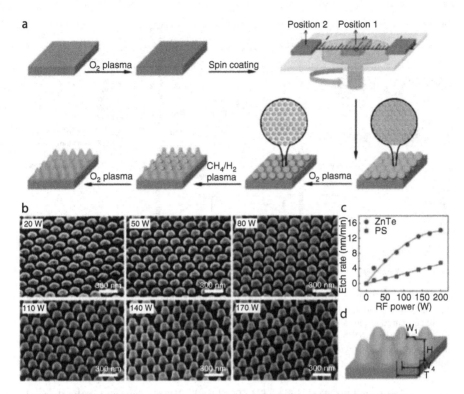

Fig. 12.4 (**a**) The schematic illustration of fabricating SWSs on ZnTe crystals, (**b**) SEM image of SWSs fabricated at a radio frequency power range of 20–170 W with an etching time of 15 min, (**c**) etch rates of ZnTe and PS nanoparticles as a function of radio frequency, (**d**) the schematic diagram of conical SWS morphology. (Reproduced with permission from Ref. [72]. Copyright 2021: American Chemical Society)

12.6   Heterostructured ZnTe Based Photodetectors

12.6.1   *ZnTe-Si Heterostructures Based Photodetectors*

As well-known, Si has response to broadband lights whose wavelength ranged from 400 to 1100 nm because of its narrow band gap of 1.12 eV. More and more scientific researchers have drawn much attention to silicon-based heterojunction materials because of two aspects: on the one hand, mature silicon microelectronic technology has laid a solid foundation for silicon-based heterojunction materials; on the other hand, the development of energy band engineering research, single atomic layer epitaxy and nanotechnology have provided deep theoretical basis and new technical means [73–75].

ZnTe could also be combined with other semiconductor to form heterostructures for photodetectors. ZnTe film can be formed via vacuum evaporation method on n-type or p-type Si substrate to build photodetectors with p-n junction structure. Lin

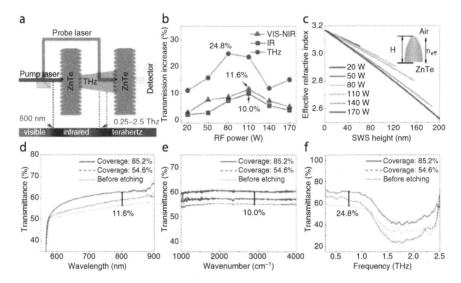

Fig. 12.5 (**a**) Schematic illustration of terahertz generation and detection of the ZnTe crystal by optical rectification and electro-optic sampling techniques, (**b**) transmission increase of the SWS ZnTe compared to the as-grown crystals from visible to terahertz region, (**c**) the effective refractive index profile with SWS height, (**d-f**) the optimal transmittance before and after etching in the visible-near infrared region (**d**), infrared region (**e**) and terahertz region (**f**). Reproduced with permission from Ref. [72]. (Copyright 2021: American Chemical Society)

et al. [76] reported the deposition of ZnTe film of silicon substrate via the thermal-furnace evaporation method. The crystallinity, charge mobility, carries concentration and sheet resistance in ZnTe film exhibited the dependent of argon pressure and deposition temperature. The higher annealing temperature and deposition temperature is better to increase the grain size of ZnTe on Si substrate. The highest carrier concentration of 1.9×10^{12} cm^{-3}, the lowest sheet resistance of 3180 ohm/square and the largest mobility of 5.1×10^3 cm^2V^{-1} S^{-1} were obtained at an argon pressure of 100 sccm and a deposition temperature of 580 °C, respectively.

In order to build the p-n junction heterostructure for optoelectronic devices, the doped ZnTe for p-type semiconductor was deposited on the n-type Si substrate by using thermal vacuum evaporation technique [77]. The deposited aluminum doped ZnTe film had the polycrystalline structure with cubic zinc-blende phase, and the roughness of film increased with increasing the doped amount of Al. With tuning the amount of impurities in ZnTe film, the optical band gap can be decreased from 2.24 eV to 1.86 eV. The increase of incident lighting intensity and doped Al can rise the illumination current of ZnTe-Si heterostructures. When the doped amount of Al is 0.2% in ZnTe films, the heterostructure-based device showed the best value of specific detectivity and quantum efficiency, originating from the height crystal quality of ZnTe films.

Furtherly, the high-quality ZnTe and TeO_2 composite was grown on an n-type silicon substrate via a modified metal-assisted chemical vapor deposition method [78]. Self-powered photodetector based on a ZnTe–TeO_2 composite/Si heterojunction with ultra-broadband and high responsivity is obtained. The photodetector shows ultra-broadband photoresponsivity from UV to NIR lights as ZnTe has a moderate, direct band gap of 2.26 eV, TeO_2 has a wide band gap of 4.0 eV, and Si has a narrow band gap of 1.12 eV. Upon exposure to 850 nm light at a zero-bias voltage, the detector shows a high responsivity of 75 mA/W, detectivity of 1.4×10^{13} cm $Hz^{1/2}$/W, fast response and recovery properties with response and recovery times both below 0.61 s, respectively.

12.6.2   Heterostructures Based on II-VI Semiconductors

ZnTe can form heterostructures with other numerous II-VI group semiconductor materials. Their hetero-structures were used to fabricate photodetector with high sensitivity, large photocurrent gain, good reliability and low response time. Among these, CdTe and ZnTe with the direct band gap of around 1.50 eV and 2.17 eV cover the visible and near IR regions, promising candidates for photodetector applications. ZnTe/Cd_xZn_{1-x}Te $(0.2 \leq x \leq 1.0)$ heterostructures were fabricated by thermal evaporation method by using CdTe and ZnTe as source materials [79]. ZnTe/Cd_xZn_{1-x} Te heterostructures with low barrier height and good photo response are explored in this work for the photodetector application. ZnTe/$Cd_{0.8}Zn_{0.2}$Te showed higher response in the visible region with improved response at longer wavelengths.

On the basis of band gap alignment of semiconductors in heterostructures, the heterostructures can be separated to type-I, type-II and quasi type-II. Generally, the type-II structure is good for charge separation and transfer in optoelectronic device. ZnTe and ZnSe can form the type-II heterostructures. Averin et al. reported the type-II ZnSe/ZnTe/GaAs superlattice for photodetectors [80]. The ZnSe/ZnTe/GaAs superlattice based-photodetector demonstrates very low dark current, high current sensitivity and external quantum efficiency. The maximum photoresponse of the ZnSe/ZnTe/GaAs superlattice based-detector at the wavelength 620 nm corresponds to current sensitivity 0.22 A/W and external quantum efficiency 44%. Photoresponse of the ZnSe/ZnTe/GaAs superlattice based-detector shows two peaks of response located at 620 nm and 870 nm. For the MSM-diode with finger width and gap of 3 μm and 100×100 μm^2 photosensitive area we have obtained dark current density 10–8 A/cm^2 at room temperature.

More than ZnSe, ZnTe can also combine with ZnO and CdSe to form type-II heterostructures. Rai et al. [81] reported the CdSe/ZnTe core/shell nanowire for broad band photodetector. CdSe nanowires arrays were first grown on the muscovite mica substrate. Then ZnTe was further deposited on CdSe nanowires to form CdSe/ZnTe core/shell nanowires. Photodetection is greatly enhanced by the piezo-phototronic effect. The photodetector performance under UV (385 nm), blue (465 nm), and green (520 nm) illumination infers a saturation free response with an intensity variation near two orders of magnitude, where the peak photocurrent (125 μA) is two orders higher at 0.25-kilogram force compared to no load (0.71 μA). The resulting (%) responsivity changed by four orders of magnitude. The significant increase in responsivity is believed to arise from: (1) the piezo-phototronic effect induced by a change in the Schottky barrier height at the Ag–ZnTe junction, and in the type-II band alignment at the CdSe–ZnTe interfaces, in conjugation with (2) a small lattice mismatch between the CdSe and ZnTe epitaxial layers, which lead to reduced charge carrier recombination.

With similar method, ZnO/ZnTe core/shell nanowires were developed by You et al. [82]. A self-powered core/shell photodetector was fabricated by sputtering a uniform p-type ZnTe layer on n-type ZnO nanorod array (Fig. 12.6). By integrating pyro-electric and photovoltaic effects, the photodetector realizes broadband detection from 325 nm ultraviolet to 1064 nm near infrared under zero bias. The maximum responsivity and detectivity reach 196.24 mA/W and 3.47×10^{12} cm $Hz^{1/2}$/W for 325 nm laser illumination with power density 2.13 mW/cm^2, respectively, which are improved ten-fold relating to the device responded to photovoltaic effect only. While the rise and fall time are drastically reduced from 1.2 ms to 62 μs and 1.6 ms to 109 μs, respectively (Fig. 12.7). Moreover, applied bias voltage and light power densities also play a significant impact in photovoltaic–pyroelectric coupled effects of device.

12.7   ZnTe-Based Materials for Solar Cells

Solar cells based on II-VI semiconductors (with II = Zn, Cd, Hg and VI = S, Se, Te) are the leading candidates for low-cost photovoltaic conversion of solar energy due to their high absorption coefficients and therefore the low material consumption for their production. Owing to its wide direct band gap of 2.23 ~ 2.28 eV at room temperature and low electron affinity 3.73 eV [83], zinc telluride, an important member of II-VI group semiconductors, is a suitable material for several applications, such as solar cells where it is used as a p-type doped window materials or a top layer for achieving low contact resistance in photovoltaic heterojunctions, switching devices and infrared and X-ray detectors [84, 85].

Tanaka et al. [86] developed the Al doped p-type ZnTe for solar cells. An open circuit voltage of approximately 0.9 V was obtained under 1 × sun AM1.5G condition in all solar cells, independent of diffusion times, while a short circuit current dropped down with increasing the diffusion time due to an increased light absorption in heavily defective Al-diffused layer. These fundamental results provide a basis for future development of intermediate band solar cells based on ZnTe materials.

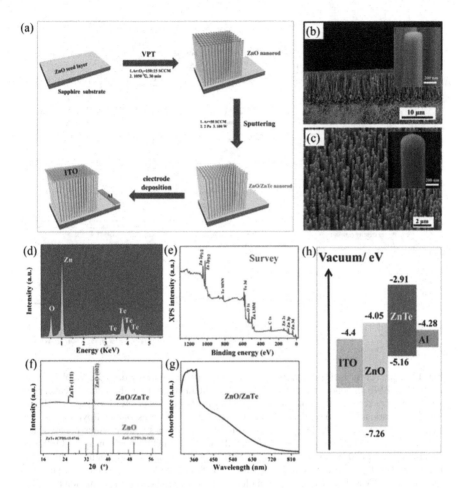

Fig. 12.6 (**a**) Schematic of synthesis process to fabricate vertically aligned ZnO/ZnTe core/shell nanorod array photodetector device. (**b, c**) Cross-sectional view SEM images of ZnO nanorods before and after the ZnTe layer coating, (inset b and c: high-magnification SEM of single nanorod). (**d**) EDS spectrum and (**e**) XPS measurement of ZnO/ZnTe core–shell nanorod array. (**f**) XRD patterns of bare ZnO and ZnO/ZnTe. (**g**) Optical absorption spectra of ZnO/ZnTe core–shell nanorods. (**h**) Energy band diagram of ITO/ZnO/ZnTe/Al. (Reproduced with permission from Ref. [82]. Copyright 2019: Elsevier)

Tang et al. [87] further doped ZnTe film with oxygen via ion implantation. The proper concentrations of oxygen ions attributed to the formation of the intermediate band which was approximately 1.88 eV above the valence band maximum. ZnTe with high crystalline quality and appropriate concentrations of oxygen ions led to the improvements of absorption efficiency of the intermediate band. The crystalline quality and the dose of oxygen concentration are important to achieve better ZnTe:O intermediate-band photovoltaic materials.

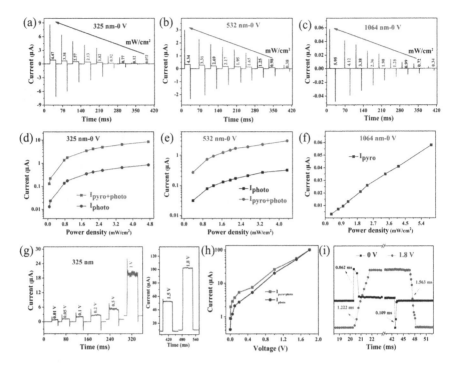

Fig. 12.7 Impacts of incident light power densities on photovoltaic–pyroelectric coupled effects on the ZnO/ZnTe photodetector under different laser illumination at zero bias under frequency of 20 Hz. I-t characteristics of the photodetector under (**a**) 325, (**b**) 532 and (**c**) 1064 nm laser illumination with various light power densities, respectively. (**d–f**) The corresponding I(pyro+photo) and I(photo) as a function of the light power densities. Impacts of bias voltage on the photovoltaic–pyroelectric coupled effects under 325 nm light illumination at light power density 2.13 mW/cm² under frequency of 20 Hz. (**g**) I–t characteristics of the photodetector at different bias voltages. (**h**) I$_{pyro}$+photo and I$_{photo}$ as a function of bias voltage. (**i**) Enlarged plot of a single output period as shown in a for 0 V and 1.8 V under 325 nm illuminations. (Reproduced with permission from Ref. [82]. Copyright 2019: Elsevier)

Beside as the absorber layer in photovoltaic device, ZnTe can also be used as back-contact in CdTe solar cell. With a cubic zinc-blende structure and a bandgap of ~2.26 eV, ZnTe has negligible valence band discontinuity with respect to CdTe (which would not impede hole transport), and a large conduction band offset, which can be beneficial for electron back reflection to CdTe and hence minimize minority carrier recombination related losses at the interface to the metal back contact [88, 89]. High p-type doping concentrations achievable with ZnTe can provide a low resistance Ohmic contact to the metal electrode. Cu-doped ZnTe back contact to CdTe have been credited with improvement in device efficiency [90, 91]. As an alternative to ZnTe:Cu, group V (N, As)-doped ZnTe can be used as a back contact (Fig. 12.8) [92], ZnTe:N has been recognized as a suitable back contact for CdTe thin film solar cells, showing improved device stability [93, 94].

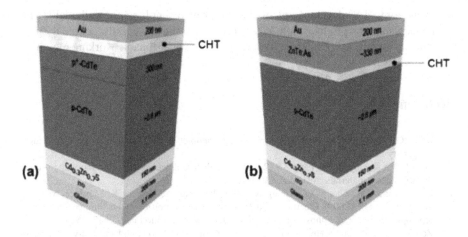

Fig. 12.8 Schematic of (a) baseline CdTe cell, (b) ZnTe:As back-contacted CdTe cell. (Reproduced from Ref. [92]. Published 2015 by MDPI as open access)

12.8   Summary and Outlook

In summary, recent advances of ZnTe based-photodetectors ranging from synthesis, device building and performance of device were discussed. On the basis of the band gap of 2.2 eV in ZnTe, the photodetectors covering the ultraviolet to visible region are developed, and the detected region can be further extended to near-infrared and terahertz region with the SWSs on a single ZnTe crystal. However, compared to other semiconductor-based photodetectors, the performance and investigation of ZnTe based photodetectors are limited. For example, ZnTe are not stable in air at room temperature, the shell growth of inorganic compound on the 1D ZnSe nanostructures can improve the stability and performance of photodetectors. There are several issues that remain unsolved for ZnTe based photodetectors. Some future aspects of ZnTe-based photodetectors might be explored in order to accomplish their widespread applications.

1. Semiconductor nanocrystals have the size and shape-dependent optical and optoelectronic properties. The precise controlled synthesis of semiconductor nanostructure can turn the band gap and optoelectronic properties based on the applications' requirement. 1D and 2D ZnTe nanocrystals via CVD or PVD method have the diameter or thickness of tens nanometers. The region of ultraviolet and blue light in ZnTe-based photodetectors are not be really achieved. Further controlled synthesis of ZnTe nanocrystals with tuned size to Bohr radius for strong quantum size confinement effect is necessary to develop.
2. Most of the developed ZnTe-based photodetectors are built on the PVD or CVD produced ZnTe nanostructures. The colloidal method produced nanocrystals have tunable shape and size. However, the photodetectors based on colloidal ZnTe nanocrystals are limited developed because of the complex ink-print

technology. The development of the ink-print technology for colloidal ZnTe nanocrystals can produce high-quality optoelectronic devices, the colloidal ZnTe nanocrystals-based photodetectors can give the wider tunable working region.

References

1. Rieke G (2003) Detection of light: from the ultraviolet to the Submilli-meter. Cambridge University Press, Cambridge
2. Teng F, Hu K, Ouyang W, Fang XS (2018) Photoelectric detectors based on inorganic p-type semiconductor materials. Adv Mater 30:1706262
3. Li PK, Liao ZM, Zhang XZ, Zhang XJ, Zhu HC, Gao JY, Laurent K, Leprince-Wang Y, Wang N, Yu DP (2009) Electrical and Photoresponse properties of an intramolecular p-n Homojunction in single phosphorus-doped ZnO nanowires. Nano Lett 9:2513–2518
4. Zou R, Zhang Z, Liu Q, Hu J, Sang L, Liao M, Zhang W (1848) High Detectivity solar-blind high-temperature deep-ultraviolet photodetector based on multi-layered (l00) facet-oriented β-Ga$_2$O$_3$ Nanobelts. Small 2014:10
5. Fang Y, Dong Q, Shao Y, Yuan Y, Huang J (2015) Highly narrowband perovskite single-crystal photodetectors enabled by surface-charge recombination. Nat Photonics 9:679–686
6. Han X, Du W, Yu R, Pan C, Wang ZL (2015) Piezo-Phototronic enhanced UV sensing based on a nanowire photodetector Array. Adv Mater 27:7963–7969
7. Lin Q, Armin A, Burn PL, Meredith P (2015) Filter less narrowband visible photodetectors. Nat Photonics 9:687–694
8. Youngblood N, Chen C, Koester SJ, Li M (2015) Waveguide-integrated black phosphorus photodetector with high responsivity and low dark current. Nat Photonics 9:247–252
9. Chen H, Liu H, Zhang Z, Hu K, Fang XS (2016) Nanostructured photodetectors: from ultraviolet to terahertz. Adv Mater 28:403–433
10. Yu X, Marks TJ, Facchetti A (2016) Metal oxides for optoelectronic applications. Nat Mater 15:383–396
11. Fang XS, Hu LF, Hui KF, Gao B, Zhao LJ, Liao MY, Chu PK, Bando Y, Golberg D (2011) New ultraviolet photodetector based on individual Nb$_2$O$_5$ Nanobelts. Adv Funct Mater 21:3907–3915
12. Ekimov AI, Onushchenko AA (1981) Quantum size effect in three-dimensional microscopic semiconductor crystals. JETP Lett 34:345–349
13. Efros AL, Efros AL (1982) Interband absorption of light in semiconductor sphere. Sov Phys Semicond 16:772–775
14. Brus LE (1983) A simple model for the ionization potential, electron affinity, and aqueous redox potentials of small semiconductor crystallites. J Chem Phys 79:5566–5571
15. Rossetti R, Nakahara S, Brus LE (1983) Quantum size effects in the redox potentials, resonance Raman spectra, and electronic spectra of CdS crystallites in aqueous solution. J Chem Phys 79:1086–1088
16. Brus LE (1984) Electron-electron and electron-hole interactions in small semiconductor crystallites: the size dependence of the lowest excited electronic state. J Chem Phys 80:4403–4409
17. Murray CB, Norris DJ, Bawendi MG (1993) Synthesis and characterization of nearly monodisperse CdE (E = S, Se, Te) semiconductor Nanocrystallites. J Am Chem Soc 115:8706–8715
18. Shalev E, Oksenberg E, Rechav K, Popovitz-Biro R, Joselevich E (2017) Guided CdSe nanowires Parallelly integrated into fast visible-range photodetectors. ACS Nano 11:213–220
19. Li LD, Lou Z, Shen GZ (2015) Hierarchical CdS nanowires based rigid and flexible photodetectors with ultrahigh sensitivity. ACS Appl Mater Interfaces 7:23507–23514
20. Xie X, Kwok S, Lu Z, Liu Y, Cao Y, Luo L, Zapien JA, Bello I, Lee CS, Lee ST, Zhang WJ (2012) Visible-NIR photodetectors based on CdTe nanoribbons. Nanoscale 4:2914–2919

21. Yan X, Li B, Wu Y, Zhang X, Ren XM (2016) A single crystalline InP nanowire photodetector. Appl Phys Lett 109:053109
22. Chen G, Liang B, Liu X, Liu Z, Yu G, Xie X, Luo T, Chen D, Zhu M, Shen G, Fan ZY (2014) High-performance hybrid phenyl-C61-butyric acid methyl Ester/Cd$_3$P$_2$ nanowire ultraviolet-visible-near infrared photodetectors. ACS Nano 8:787–796
23. Michel J, Liu J, Kimerling LC (2010) High-performance Ge-on-Si photodetectors. Nat Photonics 4:527–534
24. Langer DW, Vesely CJ (1970) Electronic Core levels of zinc chalcogenides. Phys Rev B 2:4885–4892
25. Katre A, Togo A, Tanaka I, Madsen GKH (2015) First principles study of thermal conductivity cross-over in nanostructured zinc-chalcogenides. J Appl Phys 117:045102
26. Zhang J, Sun K, Kumbhar A, Fang JY (2008) Shape-control of ZnTe nanocrystal growth in organic solution. J Phys Chem C 112:5454–5458
27. Jang E, Han C, Lim SW, Jo J, Jo DY, Lee SH, Yoon SY, Yang H (2019) Synthesis of alloyed ZnSeTe quantum dots as bright, color-pure blue emitters. ACS Appl Mater Interfaces 11:46062–46069
28. Lesnyak V, Dubavik A, Plotnikov A, Gaponik N, Eychmüller A (2010) One-step aqueous synthesis of blue-emitting glutathione-capped ZnSe$_{1-x}$Te$_x$ alloyed nanocrystals. Chem Commun 46:886–888
29. Fairclough SM, Tyrrell EJ, Graham DM, Lunt PJB, Hardman SJO, Pietzsch A, Hennies F, Moghal J, Flavell WR, Watt AAR, Smith SM (2012) Growth and characterization of strained and alloyed type-II ZnTe/ZnSe Core/Shell nanocrystals. J Phys Chem C 116:26898–26907
30. Jin S, Zhang J, Schaller RD, Rajh T, Wiederrecht GP (2012) Ultrafast charge separation from highly reductive ZnTe/CdSe type II quantum dots. J Phys Chem Lett 3:2052–2058
31. Smith AM, Nie SM (2010) Semiconductor nanocrystals: structure, properties and band gap engineering. Acc Chem Res 43:190–200
32. Asano H, Tsukuda S, Kita M, Fujimoto S, Omata T (2018) Colloidal Zn(Te, Se)/ZnS Core/Shell quantum dots exhibiting narrow-band and green photoluminescence. ACS Omega 3:6703–6709
33. Reiss P, Protiere M, Li L (2009) Core/Shell semiconductor nanocrystals. Small 5:154–168
34. Bang J, Park J, Lee JH, Won N, Nam J, Lim J, et al. (2010) ZnTe/ZnSe (Core/Shell) type-II quantum dots: their optical and photovoltaic properties. Chem Mater 22:233–240
35. Ngo CY, Yoon SF, Fan WJ, Chua SJ (2006) Effect of size and shape on electronic states of quantum dots. Phys Rev B 74:245331
36. Shabaev A, Efros AL (2004) 1D exciton spectroscopy of semiconductor nanorods. Nano Lett 4:1821–1825
37. Huynh WU, Dittmer JJ, Alivisatos AP (2002) Hybrid nanorod-polymer solar cells. Science 295:2425–2427
38. Nam S, Oh N, Zhai Y, Shim M (2015) High efficiency and optical anisotropy in double-heterojunction Nanorod light-emitting diodes. ACS Nano 9:878–885
39. Oh N, Kim BH, Cho SY, Nam S, Rogers SP, Jiang Y, et al. (2017) Double-heterojunction Nanorod light-responsive LEDs for display applications. Science 355:616–619
40. Pawar AA, Halivni S, Waiskopf N, Ben-Shahar Y, Soreni-Harari M, Bergbreiter S, Banin U, Magdassi S (2017) Rapid three-dimensional printing in water using semiconductor-metal hybrid nanoparticles as Photoinitiators. Nano Lett 17:4497–4501
41. Ji B, Panfil YE, Banin U (2017) Heavy-metal-free fluorescent ZnTe/ZnSe Nanodumbbells. ACS Nano 11:7312–7320
42. Pal U (1993) Dark-and photoconductivity in doped and Undoped zinc telluride films. Semicond Sic Technol 8:1331–1336
43. Rao GK, Bangera KV, Shivakumar GK (2010) Studies on the photoconductivity of vacuum deposited ZnTe thin films. Mater Res Bull 45:1357–1360

44. Shinde UP, Patil AV, Dighavkar CG, Patil SJ, Kapadnis KH, Borse RY, Nikam PS (2010) Photoconductivity study as a function of thickness and composition of Zn-Te thin films for different illumination conditions at room temperature. Optoelectron Adv Mater 4:291–294

45. Bhunia S, Pal D, Bose DN (1998) Photoluminescence and photoconductivity in hydrogen-passivated ZnTe. Semicond Sci Technol 13:1434–1438

46. Hamdi H, Valette S (1980) ZnTe extrinsic photodetector for visible integrated optics. J Appl Phys 51:4739–4741

47. Bozzini B, Baker MA, Cavallotti PL, Cerri E, Lenardi C (2000) Electrodeposition of ZnTe for photovoltaic cells. Thin Solid Films 361-362:388–395

48. Cui Y, Wei Q, Park H, Lieber CM (2001) Nanowire nanosensors for highly sensitive and selective detection of biological and chemical species. Science 293:1289–1292

49. Chen RS, Wang SW, Lan ZH, Tsai JT, Wu CT, Chen LC, et al. (2008) On-chip fabrication of well-aligned and contact-barrier-free GaN nanobridge devices with ultrahigh photocurrent responsivity. Small 4:925–929

50. Cao Q, Rogers JA (2009) Ultrathin films of single-walled carbon nanotubes for electronics and sensors: a review of fundamental and applied aspects. Adv Mater 21:29–53

51. Cao YL, Liu ZT, Chen LM, Tang YB, Luo LB, Jie JS, Zhang WJ, Lee ST, Lee CS (2011) Single-crystalline ZnTe nanowires for application as high-performance green/ultraviolet photodetector. Opt Exp 19:6100–6108

52. Lieber CM, Wang ZL (2007) Functional nanowires. MRS Bull 32:99–108

53. Kind H, Yan HQ, Messer B, Law M, Yang PD (2002) Nanowire ultraviolet photodetectors and optical switches. Adv Mater 14:158–160

54. Luo J, Zheng Z, Yan S, Morgan M, Zu X, Xiang X, Zhou W (2020) Photocurrent enhanced in UV-vis-NIR photodetector based on CdSe/CdTe Core/Shell nanowires arrays by piezo-Phototronic effect. ACS Photonics 7:1461–1467

55. Tang M, Xu P, Wen Z, Chen X, Pang C, Xu X, Meng C, Liu X, Tian H, Raghavan N, Yang Q (2018) Fast response CdS-CdS$_x$Te$_{1-x}$-CdTe Core-Shell Nanobelt photodetector. Sci Bull 63:1118–1124

56. Shaygan M, Davami K, Jin B, Gemming T, Lee J, Meyyappan M (2016) Highly sensitive photodetectors using ZnTe/ZnO Core/Shell nanowire field effect transistors with a tunable Core/Shell ratio. J Mater Chem C 4:2040–2046

57. Chattopadhyay D, Queisser HJ (1981) Electron scattering by ionized impurities in semiconductors. Rev Mod Phys 53:745–768

58. Grimmeiss HG (1977) Deep level impurities in semiconductors. Annu Rev Mater Sci 7:341–376

59. Luo L, Huang X, Wang M, Xie C, Wu C, Hu J, Wang L, Huang J (2014) The effect of Plasmonic nanoparticles on the optoelectronic characteristics of CdTe nanowires. Small 13:2645–2652

60. Zhang Q, Zhang J, Utama MIB, Peng B, de la Mata M, Arbiol J, Xiong Q (2012) Exciton-phonon coupling in individual ZnTe Nanorods studied by resonant Raman spectroscopy. Phys Rev B 85:085418

61. Li S, Jiang Y, Wu D, Wang L, Zhong H, Wu B, Lan X, Yu Y, Wang Z, Jie J (2010) Enhanced p-type conductivity of ZnTe nanoribbons by nitrogen doping. J Phys Chem C 114:7980–7985

62. Luo L, Zhang S, Lu R, Sun W, Fang Q, Wu C, Hu J, Wang L (2015) p-Type ZnTe:Ga Nanowires: controlled doping and optoelectronic device application. RSC Adv 5:13324–13330

63. Wan X, Huang Y, Chen Y (2012) Focusing on energy and optoelectronic applications: a journey for graphene and graphene oxide at large scale. ACS Chem Res 45:598–607

64. Chang H, Wu H (2013) Graphene-based nanomaterials: synthesis, properties, and optical and optoelectronic applications. Adv Funct Mater 23:1984–1997

65. Xie C, Wang Y, Zhang Z, Wang D, Luo L (2018) Graphene/semiconductor hybrid Heterostructures for optoelectronic device applications. Nano Today 19:41–83

66. Mak KF, Shan J (2016) Photonics and optoelectronic of 2D semiconductor transition metal dichalcogenides. Nat Photonics 10:216–226

67. Wang J, Ardelean J, Bai Y, Steinhoff A, Florian M, Jahnke F, Xu X, Kira M, Hone J, Zhu X (2019) Optical generation of high carrier densities in 2D semiconductor conductor heterobilayers. Sci Adv 5:0145
68. Wang Y, Li H, Yang T, Zou Z, Qi Z, Ma L, Chen J (2019) Space-confined physical vapour deposition of high quality ZnTe Nanosheets for optoelectronic application. Mater Lett 238:309–312
69. You S, Wu Z, Niu L, Chu X, She Y, Liu Z, Cai Y, Liu H, Zhang L, Zhang K, Luo Z, Huang S (2022) 2D ultrathin p-type ZnTe with high environmental stability. Adv Electron Mater 8:2101146
70. Li Y, Zhang J, Yang B (2010) Antirereflective surfaces based on biomimetic Nanopillared arrays. Nano Today 5:117–127
71. Choi HJ, Huh D, Jun J, Lee H (2019) A review on the fabrication and applications of subwavelength anti-reflective surfaces based on Biomimietics. Appl Spectrosc Rev 54:719–735
72. Sun H, Liu J, Zhou C, Yang W, Liu H, Zhang X, Li Z, Zhang B, Jie W, Xu Y (2021) Enhanced transmission from visible to terahertz in ZnTe crystals with scalable subwavelength structures. ACS Appl Mater Interfaces 13:16997–17005
73. Wang Z, Yu R, Wen X, Liu Y, Pan C, Wu W, Wang ZL (2014) Optimizing performance of silicon-based p-n junction photodetectors by the piezo-phototronic effect. ACS Nano 8:12866–12873
74. Avasthi S, Lee S, Loo YL, Sturm JC (2011) Role of majority and minority carrier barriers silicon/organic hybrid hetero-junction solar cells. Adv Mater 23:5762–5766
75. Masuko K, Shigematsu M, Hashiguchi T (2014) Achievement of more than 25% conversion efficiency with crystalline silicon heterojunction solar cell. IEEE J Photovolt 4:1433–1435
76. Lin J, Wei S, Yu Y, Hsu C, Kao W, Chen W, Tseng C, Lai C, Lu J, Ju S, Hsieh J (2013) Synthesis and characterization of ZnTe thin films on silicon by thermal-furnace evaporation. Proc SPIE 8913:89130k
77. Maki SA, Hassun HK (2018) Effect of aluminum on characterization of ZnTe/n-Si heterojunction photodetector. J Phys Conf Series 1003:012085
78. Song Z, Liu Y, Wang Q, Yuan S, Yang Y, Sun X, Xin Y, Liu M, Xia Z (2018) Self-powered photodetectors based on a ZnTe-TeO$_2$ composite/Si heterojunction with ultra-broadband and high responsivity. J Mater Sci 53:7562–7570
79. Moger SN, MG, M. (2020) Investigation on ZnTe/Cd$_x$Zn$_{1-x}$Te Heterostructure for photodetector applications. Sens Actuators A 315:112294
80. Averin SV, Kuznetsov PI, Zhtov VA, Zakharov LY, Kotov VM (2018) Multicolour photodetector based on a ZnSe/ZnTe/GaAs heterostructure. Quant Electron 48:675–678
81. Rai SC, Wang K, Chen J, Marmon JK, Bhatt M, Wozny S, Zhang Y, Zhou W (2015) Enhanced broad band Photodetection through piezo-Phototronic effect in CdSe/ZnTe Core/Shell nanowire Array. Adv Electron Mater 1:1400050
82. You D, Xu C, Zhang W, Zhao J, Qin F, Shi Z (2019) Photovoltaic-pyroelectric effect coupled broadband photodetector in self-powered ZnO/ZnTe Core/Shell Nanorod arrays. Nano Energy 62:310–318
83. Pistone A, Arico AS, Antonucci PL, Silvestro D, Antonucci V (1998) Preparation and characterization of thin film ZnCuTe semiconductors. Sol Energy Mater Sol Cell 53:255–267
84. Ernt K, Seiber I, Neumann-Spalart M, Lux-Steiner MC, Könenkamp R (2000) Characterization of II-VI compounds on porous substrates. Thin Solid Films 361:213–217
85. Winnewiser C, Jepsen PU, Schall M, Schiyja V, Helm H (1997) Electrooptic detection of THz radiation in LiTaO$_3$, LiNbO$_3$, and ZnTe. Appl Phys Lett 70:3069–3071
86. Tanaka T, Yu KM, Stone PR, Beeman JW, Dubon O, Reichertz LA, Kao VM, Nishio M, Walukiewicz W (2010) Demonstration of Homojunction ZnTe solar cells. J Appl Phys 108:024502
87. Tang N, Hu Q, Ren A, Li W, Liu C, Zhang J, Wu L, Li B, Zeng G, Hu S (2017) An approach to ZnTe:O intermediate-band photovoltaic materials. Sol Energy 157:707–712

88. Amin N, Yamada A, Konagai M (2002) Effect of ZnTe and CdZnTe alloys at the back contact of 1-μm-thick CdTe thin film solar cells. Jpn J Appl Phys 41:2834–2841
89. Späth B, Fritsche J, Klein A, Jaegermann W (2007) Nitrogen doping of ZnTe and its influence on CdTe/ZnTe interfaces. Appl Phys Lett 90:062112
90. Li J, Beach JD, Wolden CA (2014) Rapid thermal processing of ZnTe:cu contacted CdTe solar cells. In: Proceedings of the 40th IEEE photovoltaic specialist conference, Denver, CO, USA, 8–13 June 2014, pp 2360–2365
91. Uliˆcnâ S, Isherwood PJM, Kaminski PM, Walls JM, Li J, Wolden CA (2016) Development of ZnTe as back contact material for thin film cadmium telluride solar cells. Vacuum 139:159–163
92. Oklobia O, Kartopu G, Irvine SJC (2019) Properties of arsenic-doped ZnTe thin films as a back contact for CdTe solar cells. Materials 12:3706
93. Amin N, Yamada A, Konagai M (2000) ZnTe insertion at the back contact of 1 μm-CdTe thin film solar cells. In: Proceedings of the 28th IEEE photovoltaic specialists conference, Anchorage, AK, USA, 15–22 September 2000, pp 650–653
94. Makhratchev K, Price KJ, Ma X, Simmons DA, Drayton J, Ludwig K, Gupta A, Bohn RG, Compaan AD (2000) ZnTe:N back contacts to CdS/CdTe solar cells. In: Proceedings of the 28th IEEE photovoltaic specialists conference, Anchorage, AK, USA, 15–22 September 2000, pp 475–478

Chapter 13
ZnSe-Based Photodetectors

Ghenadii Korotcenkov

13.1 Introduction

ZnSe, as well as other direct-gap wide-gap semiconductors, is of great interest in the development of ultraviolet photodetectors (PD). Currently, UV photodiodes are being developed based on various materials such as silicon, gallium phosphide, silicon carbide, gallium nitride, zinc oxide, etc. [44]. The disadvantage of silicon photodiodes is the presence of photosensitivity outside the UV range, namely in the visible and infrared (IR) region of the spectrum. Photodiodes based on gallium phosphide have a low sensitivity in the region of 200–250 nm. Photodiodes based on silicon carbide are quite sensitive to radiation with a wavelength of about 300 nm. However, their sensitivity in the range of 200–250 nm and 350–400 nm is low. Gallium nitride photodiodes have good parameters, but thin-film GaN is quite expensive, which limits its wide application in general-purpose devices. Good parameters are also demonstrated by photodiodes based on zinc oxide. But, due to the large band gap (E_g = 3.37 eV), photodetectors have low sensitivity in the near UV region. The photocurrent begins to decrease after 370 nm, and it continues decreasing to zero at 400 nm. In this regard, the ZnSe compound has a number of advantages. The band gap of ZnSe is 2.67–2.82 eV, which makes it possible to develop sensors sensitive to radiation in a wider spectral range. Photodiodes based on ZnSe are sensitive in the range of 250–470 nm, which are necessary for the manufacture of dosimeters in the UV-A (320–400 nm) and UV-B (290–320 nm) spectral ranges. In addition, photodiodes are not sensitive to optical radiation in the visible and IR ranges. ZnSe also has a high breakdown electric field strength (1 MV cm^{-1}) [68], as well as high resistance to degradation under intense UV and X-ray radiation. In addition, the existing technology for the synthesis and deposition of the

G. Korotcenkov (✉)
Department of Physics and Engineering, Moldova State University, Chisinau, Moldova
e-mail: ghkoro@yahoo.com

© The Author(s), under exclusive license to Springer Nature Switzerland AG 2023   301
G. Korotcenkov (ed.), *Handbook of II-VI Semiconductor-Based Sensors and Radiation Detectors*, https://doi.org/10.1007/978-3-031-20510-1_13

ZnSe compound does not require large capital investments. It should also be taken into account that ZnSe, unlike many II-VI compounds, can have both n- and p-type conductivity, which makes it possible to develop all possible types of ZnSe-based photodetectors. To date, there have been reports of the development of ZnSe based photodetectors such as photoconductive photodetectors [26], Schottky barrier photodiodes [79], metal-semiconductor-metal (MSM) photodiodes [68], p–n [31], p–i–n [2] and avalanche [30] photodiodes, phototransistors [15], and photodetectors based on various heterostructures [60], one-dimensional nanomaterials [36] and hybrid materials [1]. A theoretical analysis of how these devices work can be found in [8, 44, 62, 75].

13.2   Photoconductive Photodetectors

The structure of photoconductive detector is usually a semiconductor with two contacts, which represents a simple and low-cost structural design. Ideally, the contacts should be Ohmic without any energy barrier between the metal and semiconductor, so that electrons and holes can be injected без каких-либо ограничений into the semiconductor through the contacts and recombine with the photo-carriers. Therefore, the forming good Ohmic contacts is crucial in the photoconductive detector fabrication.

ZnSe photoconductors have been investigated in detail by a number of groups, especially before the year 2000. These studies have been carried out using both single crystal samples [11] and polycrystalline materials [21]. The spectral response of ZnSe photoconductors, measured under continuous excitation, is shown in Fig. 13.1. Undoped ZnSe single crystals show maximum photosensitivity at about 462 nm (2.68 eV) [11], which corresponds to the absorption edge of ZnSe. In crystals with halide impurity, the maximum is shifted about 10 nm to longer wavelengths. In most crystals of ZnSe:Br:Cu, the maximum is shifted even further to longer wavelengths, by as much as 30 nm. At the same time, doping with selenium shifts the peak to shorter wavelengths [58]. For polycrystalline ZnSe film, the photoconductivity maximum lies in the same spectral region, $\lambda \sim 450$–460 nm [21, 61]. Dima and Vasiliu [21] also showed that heat treatment significantly increases the photoconductivity of films without a significant change in the position of the photoconductivity maximum.

The main feature of photoconductors is that (a) an external bias voltage is required to operate the device, and (b) they can have an external quantum yield greater than unity. In general, when photoconductive detectors absorb photons and generate electron-hole pairs, the majority carriers (electrons) have higher mobility and shorter transit time between electrodes than carrier lifetime; while the minority carriers (holes) move slowly and have a transit time longer than the lifetime of the carriers. In such conditions, the electrons are quickly swept out of the detector, which leads to the creation of an excess of holes. In this case, to maintain electrical neutrality, the other electrode must supply electrons. Due to this behavior, the electrons can move back and forth between the electrodes several times during the life

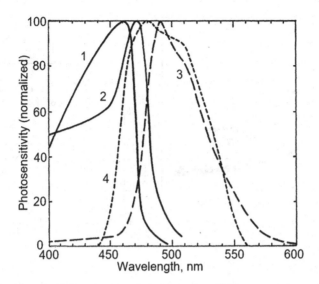

Fig. 13.1 Spectral response curves for photoconductivity in ZnSe crystals. (1) "Pure" ZnSe, (2) ZnSe with halide impurity, (3) ZnSe:Br:Cu, and (4) ZnSe:Br:Cu at −183 °C. (Reprinted with permission from Ref. [11]. Copyright 1958: American Physical Society)

of the carrier, and thus the photodetector can obtain a higher gain, which is usually understood as a persistent photoconductivity. The longer the carrier recombination time, the greater will be the responsivity of the photodetector. However, this mechanism will inevitably lead to an increase in the recovery time of the device after switching from on to the off state [33]. It is clear that to increase the carrier life time, any carrier quenching processes should be minimized. Therefore, a high quality semiconductor film is desired in the development of semiconductor photoconductors. Selecting a smaller distance between electrodes in a photoconductor can shorten the transit time, but shortening the distance also increases the dark current which affects the performance of device. A balance of these two factors should be considered in the device design.

The weakness of the photoconductive PD is the exposed optical receiving area to air, which may suffer from pollutants and surface effect. In addition, the devices display poor UV/visible contrast, hardly reaching a factor of ten. As a rule, ZnSe photoconductors exhibit increased sensitivity in the impurity region of the spectrum as well. For example, in most crystals of ZnSe:Br:Cu a second impurity-associated shoulder in the spectral response curve occurs at about 510 nm, with a long-wavelength tail extending out to about 600 nm [11].

However, the main drawback of photoconductive detectors is the presence of persistent photoconductivity (PPC), i.e., the photocurrent persists for a long time (up to hours) after the light is removed (see Fig. 13.2a). As a consequence of persistence, the measured responsivity depends drastically on the time that the sample has been kept in the dark. However, this effect is most pronounced for polycrystalline material. The rise and decay of monocryctalline samples are usually quite fast. For

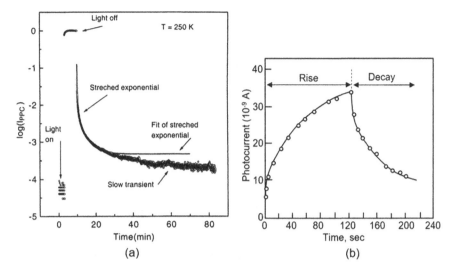

Fig. 13.2 (a) A typical logarithmic plot of the persistent photocurrent decay in MBE-grown p-ZnSe/GaAs structures at T = 250 K. The arrows indicate the moments at which the excitation light was turned on and off. Reprinted with permission from [65]. Copyright 2000: Elsevier; (b) Rise and decay curves of ZnSe single crystals when annealed in the presence of zinc. (Reprinted with permission from Ref. [9]. Copyright 1987: Springer)

example, Bube and Lind [11] reported that the insulating crystals had a measured decay time of 2 ms; the more conducting crystals had a decay time of 14 ms, which indicates the high structural quality of the crystals. If, however, the concentration of structural defects is increased, which can be done, for example, by annealing ZnSe in the presence of metal zinc [9], then the rise and decay become quite slow even for a single crystal material (see Fig. 13.2b).

In fact, the presence of PPC can be explained by the same mechanism responsible for the high gain in photoconductors [49]. The photoionization of electrons trapped on extended defects causes a local charge change and, consequently, a narrowing of the space charge region (SCRs) around them. Because of the band bending, holes are captured by defects, and when the light is off, the electrons have to overcome the potential barrier before recombination begins. Since the barrier height increases in proportion to the square of the charge on the defect, the decay of the photocurrent is non-exponential. This model can be applied if the dominant defects are located either in dislocations or grain boundaries, or at the ZnSe interfaces with air or substrate [64]. In a polycrystalline material, PPC manifests itself most clearly, since, on the one hand, an increased concentration of structural defects is characteristic of a polycrystalline material, and, on the other hand, grain boundaries act as trapping centers for photogenerated charge carriers.

If the photoconductors are not encapsulated, then oxygen chemisorbed on the surface of crystallites begins to play a significant role in PPC. In short, surface traps tend to chemisorb oxygen molecules from the air, which is accompanied by the capture of free electrons from an n-type semiconductor [49]. This leads to the formation of an electron-depleted region near the surface and band bending, which significantly reduces the conductivity of the device. The bending of the surface bands creates an internal electric field that spatially separates the photogenerated electron-hole pars, which leads to the suppression of photocarrier recombination and a significant increase in the carrier lifetime. These effects are especially noticeable in nanocrystalline films, where the surface area is large and depletion regions can spread throughout the crystallite. When illuminated with a photon energy greater than or equal to the band gap of the semiconductor, electron-hole pairs are generated. Holes that migrate to the surface along the potential gradient created by band bending either discharge the negatively charged chemisorbed oxygen ions, facilitating desorption of oxygen from the surfaces, or are effectively trapped on the surface of ZnSe crystallites. This leads to an increase in the concentration of free carriers and a decrease in the width of the depletion layer, and hence to an increase in the conductivity of the material. When the UV illumination is turned off, the residual accumulated holes recombine with unpaired electrons, and oxygen is gradually re-chemosorbed by the surface, which leads to a slow decay of the current. Considering the dependence of the process of oxygen chemisorption on many factors, such as pressure, temperature and humidity, it must be stated that in order to exclude the influence of external factors on the parameters of devices, photoconductive detectors must be encapsulated.

As follows from our consideration, the slowness of the decay process is mainly due to the presence of impurities, grain boundaries, and structural defects that play the role of charge carrier traps. This means that the optimization of the ZnSe synthesis technology is the main way to reduce the effect of PPC on the characteristics of photodetectors and thereby make photoconductive detectors faster. Research conducted by Rao et al. [61] and Sharma and Tripathi [64] support this claim. Sharma and Tripathi [66] showed that the photoresponse rate and photocurrent decay depend on the ZnSe film deposition conditions (see Fig. 13.3). They believe that the observed improvement in decay time is due to a decrease in the intergrain potential barrier and trap density. The same results were obtained by Rao et al. [61]. They found that elevated substrate temperature during ZnSe deposition reduces the density of traps in the films, and hence the rise and decay processes become much faster.

As for the most efficient photoconductive detector developed on the basis of ZnSe, such a device is the PD developed by Huang et al. [26]. They reported that a photoconductor fabricated from a ZnSe epitaxial layer grown on semi-insulating GaAs by molecular beam epitaxy (MBE) showed a sensitivity of 320 A/W (U = 10 V) and a response time of 2.3 ms. Ohmic contacts were fabricated by evaporating 0.2 μm Au and annealing at 450 °C for 5 min. In the interdigital electrodes, the fingers were 2 μm wide and 50 μm long with a distance of 2 μm between them. The responsivity increased linearly with the applied voltage up to 20 V. This behavior is typical for all photoconductive PCs. It is important to note that these parameters of the photodetector were achieved due to the high quality of the epitaxial layers and the use of interdigital contacts with a small distance between the electrodes. However, despite their high sensitivity, photoconductive detectors are unsuitable for some applications that require speed or certain spectral contrast.

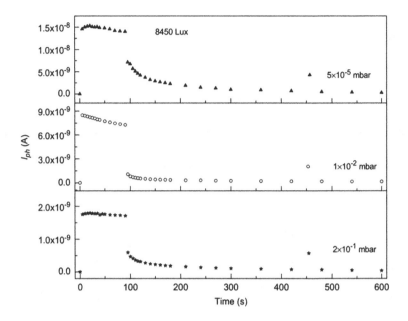

Fig. 13.3 Decay of photocurrent at 298 K for ZnSe thin films deposited at different pressures. ZnSe thin films have been prepared by inert gas condensation method. Thin films were grown in a conventional vacuum coating system on well degassed chemically cleaned Corning 7059 glass substrates. (Reprinted with permission from Ref. [66]. Copyright 2011. Elsevier)

13.3 The p–n Junction Photodiodes

As follows from the previous chapter, photoconductive detectors suffer from a poor time response, due to the long electron lifetime. In addition, it should be noted that such PDs are also characterized by significant shot noise from the high level of dark current. Both of these problems are remedied in the photodiode detector. In this device, light incident on the p-n junction of a semiconductor creates electron-hole pairs, which are swept out of the depletion region by the electric field there. Current flows in the external circuit only while charges are moving through this E field region, so the time response of the detector can be made quite fast. The p-n junction also provides a potential barrier for majority charge carriers, greatly reducing the amount of dark current and associated shot noise.

An example implementation of a ZnSe-based p-n junction photodiode would be the PD developed by Ishikura et al. [31]. A schematic diode structure and the spectral response of the external quantum efficiencies of these PDs are shown in Fig. 13.4a. The ZnSe p^+–n structure photodiodes were fabricated by conventional MBE growth on (100) GaAs substrates. This PD operates in the blue-ultraviolet optical region at room temperature. The diode fabricated has fairly good forward-bias characteristics with a built-in voltage $V_{bi} \sim 1.5$ V, and an ideal diode factor (n) ~ 1.3–1.4. The open-circuit photo-voltage (V_{oc}) for the blue incident light with an intensity of 10 μW is 1.55 V, which is about five times larger in magnitude than

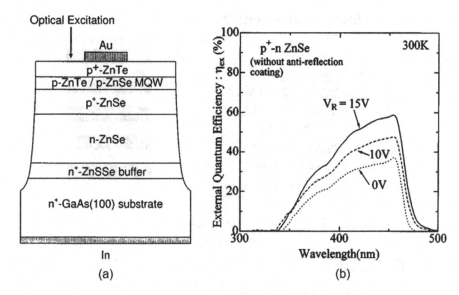

Fig. 13.4 (a) A schematic structure of a ZnSe p⁺– n avalanche photodiode grown on GaAs by MBE. The surface ohmic contact layer consists of ZnTe–ZnSe multi-quantum wells (with five periods) and a thin p⁺-ZnTe cap layer (10 nm); (b) Spectral response of an external quantum efficiency in the p⁺– n ZnSe photodiodes under zero-and reverse-bias conditions. (Reprinted with permission from Ref. [31]. Copyright 2000: AIP)

the case in conventional Si photodiodes. As it is seen in Fig. 13.4b, in zero-bias conditions the efficiency with the maximum value in the fundamental absorption-edge region is about 38%. The zero-bias efficiency is not so high because the active region (depletion region of the n-ZnSe layer) is small ~0.2 μm, compared to the thickness of the ZnSe layer. However, one can see a существенное increase in the efficiencies with applying reverse bias. In high reverse-bias condition of 15 V, the efficiencies are found to increase up to 60%. According to Ishikura et al. [31], reported parameters of the PD were achieved due to optimization of the MBE process, perfect lattice matching of the ZnS_xSe_{1-x} buffer (x = 5.6–6.0%) on GaAs substrates, high-quality initial preparation of the interface, and correctly chosen doping conditions. This made it possible to significantly improve the quality of the epitaxial layers and reduce the dislocation density below 1×10^5 cm⁻². Росту чувствительности способствовало также использование an extremely thin surface contact layer which consisted of ZnTe (well)-ZnSe (barrier) multi-quantum wells. The high quality of the structure allowed the developed PD to work in the multiplication mode with an avalanche gain of G = 60 under the electric field 8×10^5 V/cm.

Subsequently, based on the developed technology for growing ZnSe p⁺–n structures by MBE, Ishikura et al. [30] developed ZnSSe and ZnSe p⁺–n avalanche photodiodes (APDs) with back surface photo excitation (hole injection) (see Fig. 13.5). GaAs substrate removed by wet etching. APDs were characterized by a very stable

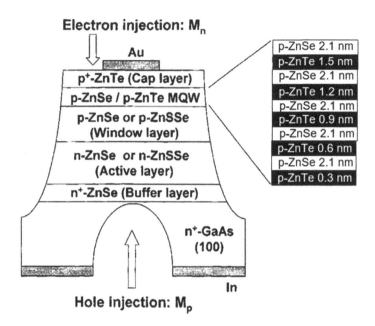

Fig. 13.5 Schematic structure of ZnSe (binary) or ZnSSe (ternary) APD with back excitation grown by MBE on n + -GaAs substrate. (Reprinted with permission from Ref. [30]. Copyright 2002: Wiley)

high field operation with extremely low dark current (~ 1 pA/mm^2). The ZnSSe APDs on GaAs had a large avalanche gain of G > 60 under the electric field of $\sim 1 \times 10^6$ V/cm at RT. ZnSe APDs had an avalanche gain of G $\sim$ 50.

13.4 The p-i-n Junction Photodiodes

As shown in the previous section, p-n junction PDs can have a fairly high an external quantum efficiency. However, such PDs suffer from the following disadvantages. First, an external quantum efficiency depends on the bias voltage, and high values are achieved with high reverse-bias voltages. Second, p-n junction PDs usually have an extended base (Fig. 13.6a) in which only diffusion, not drift, processes take place. As is known, diffusion proceeds much more slowly than drift. And this leads to the fact that PDs with a p-n junction are characterized by a large inertia and a reduced efficiency of carrier collection with a significant base thickness. The carriers generated in the base, upon reaching the space charge region (SCR), contribute to the photocurrent, but do so with a delay depending on their position relative to the SCR. Holes formed at different distances from the edge of the depletion region have different diffusion times, which leads to a diffusion "tail" in the response of the

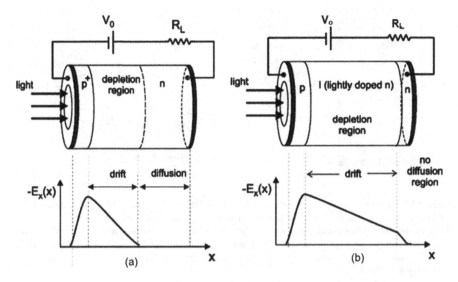

Fig. 13.6 Schematic diagram of (**a**) p-n and (**b**) p-i-n photodiodes. In a p-n junction photodiode, charge carriers may be created in a high-field drift region or a low-field diffusion region. In a PIN photodiode, charge carriers are mostly created in the high-field drift region, which extends almost to the far electrode. The lightly n-doped "intrinsic" region has a nearly constant E field, which sweeps charge carriers through the device without diffusion

photocurrent to a rectangular light pulse. This type of signal distortion is generally undesirable.

The solution of the problem of charge carrier diffusion is to simply eliminate the diffusion region. This can be achieved by decreasing the donor concentration in the n-region until the depletion region occupies almost the entire space between the electrodes. As shown in Fig. 13.6b, the E field extends almost the entire distance to the far electrode, so charge carriers generated anywhere in the material will drift rather than diffuse. As for the p^+ and n^+ layers necessary to create a good ohmic contact, in order to exclude diffusion processes in them, in the manufacture of photodiodes, they tend to make the p^+ and n^+ layers as thin as possible. Since the middle region is very lightly doped (almost intrinsic), it is labeled i and the device is termed a p-i-n photodiode.

It is important to note that the use of the p-i-n structure not only eliminates carrier diffusion, but also has the advantage that the depletion width d is fixed by the device geometry. The ability to adjust d according to design rather than applied voltage allows the photodiode performance to be optimized for specific applications. For example, increasing d increases the length of the light absorption path, which increases the efficiency of light absorption. This is especially important for wavelengths close to the semiconductor band gap. However, a larger value of d worsens the response time by increasing the transit time. Therefore, when developing p-i-n PD, a compromise is sought.

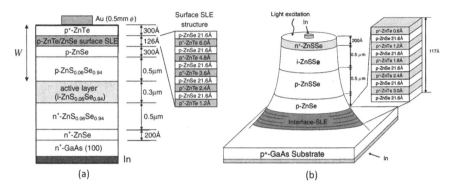

Fig. 13.7 (a) Conventional p-i-n structure APD on n-GaAs with surface SLE. The thickness of window layer (W) is 0.57 μm, including surface SLE. (Reprinted with permission from [4]. Copyright 2002: Wiley); (b) n⁺-i-p structure ZnSSe APD with a simple thin n⁺ window (W = 30 nm thickness) grown on p-GaAs substrate. Interface superlattice buffer layer (SLE) of total thickness of 11.7 nm is inserted between p-ZnSe and p-GaAs. (Reprinted with permission from [43]. Copyright 2006: Wiley)

Implementation examples of ZnSe-based p-i-n PDs are shown in Fig. 13.7. High gain and high sensitive blue-ultraviolet avalanche photodiodes (APDs) were developed using high quality ZnSSe n⁺-i-p heterostructure grown on p- and n-type GaAs substrates by molecular beam epitaxy (MBE) [2, 4, 32]. The same technology was used for fabrication of ZnSe-based p-n junction PDs [31]. An essential factor of the p–i–n diode is a high quality undoped i-layer (active layer) with high resistivity of over 10^6 Ω·cm. For n- and p-type conduction control, a chlorine (Cl) and an active nitrogen (N) by rf-radical doping [56] were used, respectively. The difference in two device structures is a position of superlattice buffer layer (SLE). In device on n-GaAs substrates, SLE of ZnTe-ZnSe MQWs is formed to obtain a hole ohmic contact in p-ZnSSe window layer and top metal (Au) electrode. This surface SLE has caused severe disadvantage such as incident light absorption loss and device instability under high electric field operation. The structure n⁺-i-p on p-GaAs has no surface SLE, in turn, very thin interface superlattice of 10 layers of ZnTe/ZnSe MQW (total MQW thickness = 11.7 nm) is formed, which is shown in the inset of the Fig. 13.7b. The role of this interface superlattice is to overcome energy barrier (~1 eV) for hole ohmic conduction between p-type GaAs and p-type ZnSe heterointerface. How important the presence of SLE at the interface with the substrate can be seen from the results shown in Fig. 13.8a. One can see a drastic improvement in the I–V characteristics in the photodiode with the interface SLE. The Al_2O_3 coating also increases the sensitivity of the PD [43].

Ando et al. [4], Abe et al. [2, 3] and Miki et al. [43] showed that the use of a p-type GaAs substrate and ZnS_xSe_{1-x} (x = 5.5–6%) instead of ZnSe gives a noticeable improvement in the PDs parameters. ZnS_xSe_{1-x} has a lattice parameter well matched to that of GaAs substrate. APDS fabricated on p-GaAs substrate operated in blue-violet-ultraviolet (450–300 nm) optical region (see Fig. 13.8c). The avalanche-gain reached G = 91.6 at 33 V (E = 6.5 × 10^5 V/cm). Abe et al. [2] also

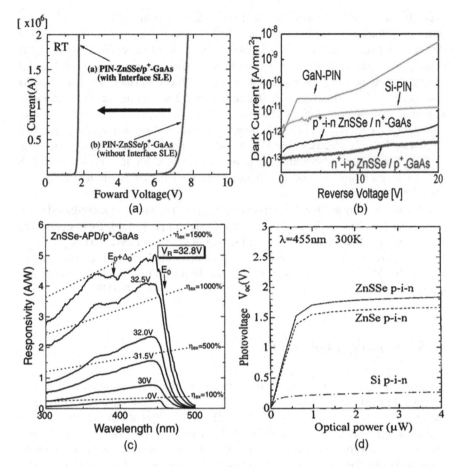

Fig. 13.8 (a) Current– voltage characteristics of ZnSSe PIN/p⁺-GaAs with interface SLE, in comparison with ZnSSe PIN/p⁺-GaAs without interface SLE. A built-in voltage of (a) is drastically improved below 2 V. (Reprinted with permission from Ref. [43]). Copyright 2006: Wiley; (**b**) Improved dark current characteristics of the n⁺-i-p structure on p⁺-GaAs in comparison with a conventional p⁺-i-n structure on n⁺-GaAs. For comparison, a GaN-based PIN and a Si-PIN are also shown. The dark current of ZnSSe p-i-n on p⁺-GaAs is below sub pA/mm². (Reprinted from Ref. [42]). Published 2008 by Korean Physical Society as open access; (**c**) Spectrum sensitivities of the n⁺-i-p structure ZnSSe APD in different reverse bias voltages. The highest sensitivities of 5 A/W for blue region (450 nm) and 3 A/W for ultraviolet region (300 nm) are obtained at reverse bias 32.8 V. (Reprinted with permission from Ref. [2]). Copyright 2005: Japan Society of Applied Physics; (**d**) Photo-voltages of ZnSe and ZnSSe p–i–n photodiodes vs. incident optical power. Here wavelength of 455 nm (blue region) is used. (Reprinted with permission from Ref. [4]). Copyright 2002: Wiley)

established that PDs had high sensitivities (~3 A/W) even for ultraviolet optical region ($\lambda = 300$ nm). According to Abe et al. [2], such sensitivity was achieved due to following two effects; (i) thin n⁺ window layer acting as semi-transparent window, and (ii) efficient hole impact ionization process in spin-orbit splitting valence band. It was found that such PDs had small dark current ($<10^{-9}$ A/ mm^2) under intrinsic avalanche breakdown region, which is one order smaller than in ZnSSe APDs on n-GaAs, and is smaller in two orders than in GaN based-APDs [2]. A noticeable decrease in the dark current was achieved by reducing the density of macro- and microscopic point defects in the developed devices. The same results were reported by Miki et al. [42]. These results are shown in Fig. 13.8b. Another important parameter of ZnSe–ZnSSe p–i–n devices is the large open-circuit photovoltages ($V_{oc} \sim 1.6$–1.7 V) [4]. V_{oc} is about eight times larger in magnitude than that of practical Si photodiodes (see Fig. 13.8d). Ando et al. [4] also suggested that ZnSSe p-i-n PDs are stable and can have a long device life in excess of 10,000 h because they did not observe any change in device parameters and film properties after 200 h of testing.

Also noteworthy are the developments of p-i-n PDs based on ZnSe-based compounds such as Zn(Mg)BeSe [79, 80] and ZnMgSSe [22]. ZnMgBeSe quaternary alloys with a lattice parameter matched to that of GaAs can have a band gap in the range of 2.75 to more than 3.8 eV. These compounds have high crystal quality, which should provide high sensitivity and detectability. The lattice mismatch between the ZnMgSSe layers and the GaAs substrate is only 0.03%.

13.5 Schottky Photodiodes

The metal-semiconductor contact, or Schottky barrier, is the equivalent of a p-i-n photodiode in the short wavelength region of the spectrum. Photons pass through a partially transparent metal layer (often gold) and are absorbed in the space charge region (SCR) of an n-type semiconductor. The metal film provides much less series resistance than a conventional p + −n junction, and the parasitic absorption of shortwave radiation in the metal film is less than in the p + region due to the large thickness difference. Another advantage of the Schottky photodiode compared to p-n (or p-i-n) structures is the improved kinetics of the photoresponse. Because it lacks a p-type layer, there is no residual "diffusion tail" arising from charge carriers generated in the p-type layer. This becomes especially important at short wavelengths, when, at a high absorption coefficient, a significant amount of light is absorbed in a p-n photodiode by a thin p-type layer. As in the case of a p-i-n photodiode, the "diffusion tail" in the response associated with the presence of a diffusion region in the n-type layer can be minimized by adjusting the concentration of donors in the n-region in such a way that the depletion region extends to the n$^+$-layer. Additional advantages of these devices are associated with their relative ease of manufacture compared to p-n and especially p-i-n photodiodes. Photodiodes with a Schottky barrier are technologically and physically compatible with integrated optics structures. In addition, Schottky barrier photodiodes can be fabricated on a wide variety of semiconductors, even those in which p-n junctions cannot be obtained.

It is important to note that photodiodes with a Schottky barrier are most effective in the blue-UV region of the spectrum, since at longer wavelengths, due to the reflection and absorption of light by the metal layer, their efficiency decreases. To reduce light reflection, it is necessary to apply an anti-reflection coating, which complicates the manufacture of such devices.

The greatest advances in the development of the Schottky barrier photodiode have been made at the Center National de la Recherche Scientifique (Valbonne, France) [10, 45, 79, 81, 82]. To fabricate the PD, they used undoped n-ZnSe layers grown by MBE on GaAs substrates. Two types of structures have been fabricated. The first ones are vertical geometry devices which consist of two layers: one nonintentionally doped (NID) layer which is "the absorption layer" and one Cl-doped n^+-type layer which is "the contact layer" (see Fig. 13.9a). The second ones are planar geometry devices which consist of a unique NID layer (see Fig. 13.9b). In theory, the vertical structure allows high-quality ohmic contacts (ohmic contact on a heavily doped layer) and low capacitance (Schottky contact on an undoped layer). In practice, the etching needed to realize that the mesa structure generates defects which might lead to bandwidth narrowing and noise increase. A semi-transparent Ni/Au layer (5 nm/5 nm) was used as a Schottky barrier. I-V characteristics had an

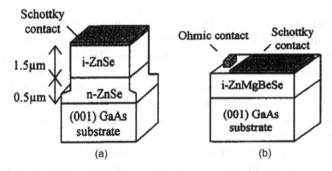

Fig. 13.9 A schematic representation of the (**a**) vertical and (**b**) planar geometry Schottky barrier photodiode. (Reprinted with permission from Ref. [77]. Copyright 2001: Springer)

Table 13.1 Performances of ZnSe-based Schottky barrier PDs

Device	Material	Wavelength cutoff (nm)	Bias (V)	Maximum responsivity (A/W)	Quantum efficiency	Rejection rate	Detectivity (mHz$^{1/2}$ W^{-1})
Schottky vertical	ZnSe	460	-2	0.1	27%	$3 \cdot 10^3$	-
			-3.5	0.115	31%	$3 \cdot 10^3$	$1.4 \cdot 10^{10}$
Schottky planar	ZnSe	460	-5	0.125	33%	10^3	-
			-10	0.127	34%	10^3	$1.3 \cdot 10^{11}$
Schottky planar	ZnMgBeSe	375	-2	0.1	31%	$5 \cdot 10^3$	$2 \cdot 10^{10}$
			-3	0.115	38%	$5 \cdot 10^3$	$1.5 \cdot 10^{10}$
			-10	0.145	57%	$5 \cdot 10^3$	-

Source: Data extracted from Vigué et al. [78, 80]

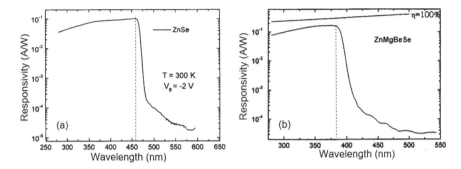

Fig. 13.10 (a) Spectral response of ZnSe- based Schottky photodiode at −2 V bias. (Reprinted with permission from Ref. [45]). Copyright 2000: AIP; (b) Spectral response of a ZnMgBeSe Schottky photodiode under 10-V bias. (Reprinted with permission from Ref. [78]. Copyright 2001: Springer)

ideality factor of 1.0. The ideality factor, very close to unity, underlines the high quality of the Schottky junction. The potential barrier height was 1.17–1.19 eV. Table 13.1 displays the results obtained on planar and vertical geometry Schottky devices based on ZnSe. Spectral characteristics are shown in Fig. 13.10.

A comparison of the parameters of the vertical and planar PDs showed that vertical devices are somewhat less sensitive than planar ones. The detectivity of a planar PD was ten times higher than that of a vertical one. This effect of Vigué et al. [79] explained on the basis that a vertical device had a much larger dark current due to leakage along the surface of the mesa structures. On the other hand, the rather low quality of ohmic contacts made on a high-resistance NID layer in planar structures makes it necessary to use a sufficiently high voltage to sufficiently bias the planar device.

Approximately the same parameters were demonstrated by Ni-ZnSe PDs developed by Naval et al. [50] and Voronkin [83]. These PDs were fabricated using single-crystal ZnSe substrate. Spectral sensitivity was in the range from 200 nm to 460 nm Voronkin [83]. A detector without a UV filter showed a maximum responsivity of about 0.11 A/W at 375 nm wavelength Naval et al. [50] and 0.1 A/W at 400 nm Voronkin [83]. The speed of the unfiltered detector was found to be about 300 kHz primarily limited by the RC time constant determined largely by the detector area [50].

Given the large surface leakage in ZnSe mesa structures, studies have been carried out to reduce them. In particular, Makhniy et al. [40, 41] proposed to treat the free ZnSe surface around mesa-structure with hydrogen peroxide to reduce the dark current in vertical ZnSe-based PDs. The effect of such treatment on the reverse I-V characteristics of Ni-ZnSe Schottky barriers is shown in Fig. 13.11. It can be seen that such treatment leads to a significant decrease in the reverse current (Fig. 13.11b) and an increase in the breakdown voltage (Fig. 13.11a). The most likely reason for the decrease in reverse current in surface-modified diodes is the formation of a

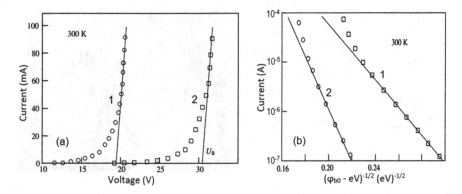

Fig. 13.11 (a) Reverse current–voltage characteristics of initial (1) and modified (2) diodes in the high current region; (b) Comparison of the reverse current–voltage characteristics of initial (1) and modified (2) diodes. (Reprinted with permission from Ref. [40]. Copyright 2003: Wiley)

high-resistance zinc oxide layer on the free surface of ZnSe, which acts as a passivating coating and leads to the suppression of leakage currents. The penetration depth of this layer in ZnSe is small and practically equal to the thickness of the zinc oxide layer, but it seems to be quite sufficient to reduce the possibility of surface breakdown.

Monroy et al. [45] and [79], have also shown that ZnMgBeS is also a very promising material for the development of Schottky barrier PDs. The spectral response of such PDs measured under − 2 V bias voltage is shown in Fig. 13.10b. It is seen that a very flat response is obtained at short wavelengths. A quantum efficiency over 50% is obtained from 380 to 315 nm, that is, over the whole UV-A range. Over the UV-B range (from 315 to 280 nm), a quantum efficiency over 35% is obtained. It is to be noted that a quantum efficiency of 57%, as obtained at 360 nm, is close to the theoretical limit taking into account the fact that this detector is not anti-reflection coated. Moreover, the cutoff at 380 nm is very sharp with a UV/visible contrast of more than three orders of magnitude: this is a consequence of the growth of a high-quality layer, quasi-lattice matched to the GaAs substrate.

13.6   Metal–Semiconductor–Metal Photodiodes

It should be noted that the metal–semiconductor–metal structure is most often used in the development of photodetectors based on ZnSe. A metal–semiconductor–metal photodetector (MSM PD) is a planar device containing two Schottky contacts, i.e., there are two metallic electrodes on a surface of semiconductor material. During operation, some electric voltage is applied to the electrodes, and therefore one Schottky barrier is turned on in the forward direction and the other in the reverse direction. When light impinges on the semiconductor between the electrodes, it

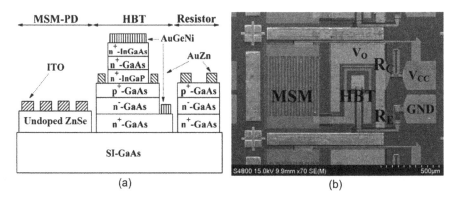

Fig. 13.12 (a) Cross-sectional diagram of the integrated MSM-PD/HBT photoreceiver; (b) Surface SEM of the fabricated MSM-HBT photoreceiver. (Reprinted with permission from [17]. Copyright 2009: IOP)

generates charge carriers (electrons and holes), which are collected by the electric field and thus can form a photocurrent.

MSM PD design typically uses an interdigital electrode structure (see Fig. 13.12) where the finger spacing can be as small as 1 μm. Light entering the device from the side of the electrodes is partially blocked by the electrodes, which, of course, reduces the quantum efficiency. For higher quantum efficiency, there are back-illuminated devices in which the light comes from the back side so that the electrodes do not obstruct it. Another possibility to reduce top exposure losses is to use extremely thin semi-transparent gold contacts. The advantage of overhead illumination is that the achievable detection bandwidth is usually larger as the charge carriers are generated closer to the contact.

The MSM PD design typically uses an interdigital electrode structure where the finger spacing can be as small as 1 μm. The light entering the device from the side of the electrodes is partially blocked by the electrodes, which naturally reduces the quantum efficiency. For higher quantum efficiency, there are back-illuminated devices in which the light comes from the back side so that the electrodes do not obstruct it. Another possibility to reduce top-irradiation losses is to use extremely thin semi-transparent gold contacts. The advantage of top-illumination is that the achievable detection bandwidth is usually larger as the charge carriers are generated closer to the contact.

An important advantage of MFM PD is a rather simple planar structure, which is convenient for monolithic integration with other components of photonic integrated circuits. There is also no need to form a low-resistance contact, which is a difficult task for II-VI compounds, during the MSM PD fabrication. In addition, MSM photodiodes can be made with a faster response than photodiodes. UV photodetectors based on MSM structures have a very low capacitance in the fF range and, therefore, can operate at high frequencies. Their detection bandwidth can reach hundreds of GHz (with an impulse response of less than 1 ps), making them suitable for high-speed fiber optic communications. As for ZnSe-based MSM PDs, their monolithic integration with InGaP/GaAs heterojunction bipolar transistors (HBTs) has been demonstrated by Chen and Chang [16–18]. Figure 13.12 shows one such device.

To develop ZnSe-based MSM PDs, undoped high-resistance ZnSe epitaxial layers grown on a GaAs substrate by molecular beam epitaxy (MBE) are commonly used. Due to the high quality structure and low concentration of charge carries in n-ZnSe, MSM photodetectors have a high resistivity, and a high bias voltage can be applied. This opens up the possibility to enhance the responsivity of ZnSe-based MSM UV photodetectors using internal gain mechanisms employing impact ionization (avalanche breakdown effect) due to the higher internal electric field strength [69].

As regards the performances of the MFM-PCs developed on the basis of ZnSe, they are summarized in Table 13.2. The MFM-PCs based on ZnSe have the same spectral characteristics as the Schottky barrier PDs discussed earlier. To achieve improved performance of UV photodetectors, a high Schottky barrier height at the metal-ZnSe interface is required. The high barrier height would result in low leakage current and high breakdown voltage, allowing operation at high bias voltage. This leads to an improvement in the sensitivity and response kinetics of such UV photodetectors, as well as a reduction in the noise level. Among different metals, the metals with high work functions like Ni, Cr, Pd, or Au are required to achieve a large Schottky barrier height on n-ZnSe (read Chap. 17, Vol. 1). In practice, a large group of metals such as Au [78], Ti/W [13], Pd [24], Cr/Au [69] and Ni/Au (25/140 nm) [13, 67] was used to fabricate the PD. Due to amphoteric properties [51], Au atoms may create both donor and acceptor centres during diffusion to ZnSe, while Cr and Ni create deep acceptor centres in ZnSe [59], which compensate for uncontrolled background donor impurities in ZnSe. Therefore, Sirkeli et al. [67] believe that Cr/Au and Ni/Au are attractive for the fabrication of the Schottky barrier structures with good characteristics. However, according to Sirkeli et al. [67], it is necessary to give preference to the use of Ni/Au contacts with a higher potential barrier height (Table 13.2). For the deposition of metals forming a Schottky barrier, either e-beam evaporation for refractory metals and thermal evaporation for gold are typically used.

Chang et al. [13] and Lin et al. [37] have shown that transparent conductive ITO (Indium Tin Oxide) layers can also be used for this purpose. In this case, it is the ITO layer that provides the maximum transparency of the contact (Fig. 13.13a), while the Ni/Au films have the minimum resistance [13]. However, although transparent ITO contact electrodes can provide detectors with greater photon absorption and higher photocurrent, the low height of the Schottky barrier between ITO and ZnSe results in a relatively large dark current (Fig. 13.13c).

During the development of ZnSe-based MSM PDs, the following regularities were also established:

- The breakdown voltage depends on the distance between the electrodes. For example, Hong and Anderson [24] reported an increase in breakdown voltage from 25 to 50 V with increasing distance from 2 to 4 μm.

Table 13.2 Performances of ZnSe-based MSM PDs with interdigital contacts

R, Ω·cm	ID width/space, μm	S, μm²	Contact	Φ_b, eV	Dark current	Responsivity, A/W	Response time	Bandwidth	Detectivity, 10^{10} cm Hz$^{1/2}$ W^{-1}	Ref.
MBE	(2–4)/(2–4)	$4 \cdot 10^4$	Pd	1.3	1 pA (10 V)	1.0 (380 nm, 5 V) 0.6 (450 nm, 5 V)		620–800 MHz	–	[24]
VP:~10^{12}	0.5/1.5	10^2	Cr/au	1.26	1.64 nA (5 V)	2.23 (325 nm, 15 V)	0.10–0.16 ms		9.9	[67]
			Ni/au	1.49	0.82 nA (5 V)	5.4 (325 nm, 15 V)	0.10–0.16 ms		33.7	
MBE	6/6		Au			0.13 (460 nm, 20 V)			1–2	[78]
VP: 10^{10}–10^{12}	0.5/1.5	10^2	Cr/Au	1.26	3.4 nA (20 V)	4.44 (325 nm, 20 V)	0.15 ms		14	[68]
MBE; 0.6	10/10	$4 \cdot 10^4$	ITO	0.78		0.13 (450 nm, 1 V)	0.04 ms			[37]
MBE	10/10	$5 \cdot 10^5$	ITO	0.66	1.5 nA (1 V)	0.128 (450 nm, 1 V)			93	[38]
MBE	10/10	$5 \cdot 10^5$	ITO		1.5 nA (1 V)	0.120 (450 nm, 1 V)			87	[13]
			Ti/W	0.695	0.35 nA (1 V)	0.051 (450 nm, 1 V)			41	
			Ni/au	0.715	0.20 nA (1 V)	0.028 (450 nm, 1 V)			77	
MBE	20/20	10^6	ITO		4 nA (5 V)	0.08 (440 nm, 5 V)				[18]

ID – interdigital; MBE- molecular beam epitaxy; R – resistivity; S – active area; VP-vapour phase method; Φ_b,- the height of potential barrier

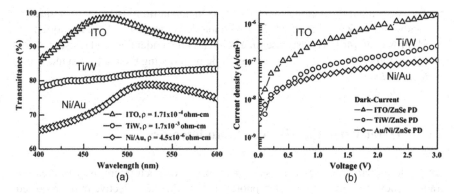

Fig. 13.13 (**a**) Normalized transmission spectral of ITO, TiW and Ni/Au; (**b**) Dark I–V characteristics of the photodetectors with ITO, TiW and Ni/Au contact electrodes. (Reprinted with permission from [13]. Copyright 2006: Elsevier)

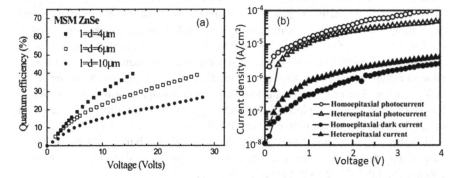

Fig. 13.14 (**a**) Variation of the quantum efficiency with the applied bias measured at 460 nm for a (6x6) MSM ZnSe photodetector. (Reprinted with permission from Ref. [78]). Copyright 2001: Springer; (**b**) I–V characteristics of the homoepitaxial and heteroepitaxial ZnSe MSM photodetectors with ITO transparent contact electrodes measured both in dark and under illumination. (Reprinted with permission from Ref. [38]. Copyright 2005: Elsevier)

- Devices with minimal electrode spacing have improved device speed due to the short carrier transit time between electrodes [24].
- Quantum efficiency increases as the characteristic size of ID contacts (width/space) in MSM PDs [78] decreases (see Fig. 13.14a).
- ZnSe-based MSM PDs based on homoepitaxial layers, due to better quality, have better performance compared to devices using heteroepitaxial layers [37, 38]. I-V characteristics of such PDs are shown in Fig. 13.14b. It is known that the slight lattice mismatch (0.27% at room temperature) between ZnSe and GaAs will still generate a huge amount of defects when we grow a very thick ZnSe epitaxial layer on top of the GaAs substrate [85]. The defects generated at the ZnSe–GaAs interface will significantly reduce the efficiency of the ZnSe-based photodetectors.

- The RC time of the measurement system is the main limiting factor of frequency bandwidth of ZnSe-based MSM UV photodetectors [67, 69]. As reported by Monroy et al. [46], the response time is linearly dependent on load resistance and can be reduced by 2–3 orders of magnitude by reducing the load resistance from $M\Omega$ to $k\Omega$-range.

13.7   Heterostructure-Based Photodetectors

As a rule, heterostructure-based structures in photodetectors are used to create a transparency window or to develop photodetectors that are selective in the required spectral region. To do this, a semiconductor with a larger band gap and low resistivity is deposited on the surface of the active photosensitive material. Such a wide-gap window transmits radiation in the active region without significant losses and, at the same time, is a contact layer with a low series resistance.

Processes in the active region of heterojunctions - absorption of radiation and generation, separation and accumulation of generated charge carriers - proceed in the same way as in Schottky barriers or a p-i-n structure. The difference lies in the fact that by choosing a suitable semiconductor for the photosensitive layer, it is possible to ensure complete absorption of radiation at a thickness of this layer of the order of 1 μm. The most important advantage of heterostructure-based photodetectors (HPDs) is also their physical and technological compatibility with integrated optics devices. The freedom to choose the HPD materials also makes it possible to achieve higher photo-emf values. However, HPDs are much more difficult to fabricate, require matching of the lattice parameters of the materials forming the heterojunction, and they have an increased noise level.

It is important to note that in HPDs containing ZnSe, ZnSe can act both as a photosensitive material [19, 74, 86] and as a wide-gap window [14, 34, 39], which cuts off the photoresponse at $\lambda < 450$ nm if the ZnSe layer is sufficiently thick. There are currently reports on the development of HPDs such as ZnSe/Si [14, 60], ZnSe/InSe and ZnSe/GaSe [34], ZnSe/ZnTe [60], $Zn_{1-x}Cd_xTe$ /ZnSe [76], ZnO/ZnSe [74, 86], ZnSTeSe/ZnSe/GaAs [19]. ZnSe/Si HPDs, whose schematic cross section and schematic band diagram are shown in Fig. 13.15, have a sensitivity range of 300–800 nm, while ZnSe/InSe and ZnSe/GaSe HPDs are sensitive in the range of 850–450 and 550–450 nm, respectively (see Fig. 13.16). InSe and GaSe are indirect band gap semiconductors with E_g equaled 1.25–1.4 eV and 2.0–2.1 eV, respectively. ZnSe/ZnTe HPDs developed by Rao [60] are sensitive in the range of 300–625 nm with one large maximum at around 455 nm, which corresponds to the band gap of ZnSe, and a second smaller peak at around 550 nm, which corresponds to energy bandgap of ZnTe (2.5 eV). At the same time, ZnSe/Si HPDs because of the narrow bandgap of silicon, exhibit considerably higher amount of photocurrent in longer wavelength up to 900 nm. However, large maximum lies at around 450 nm as in ZnSe/ZnTe HPDs. The measured dark current in ZnSe/Si p-i-n HPDs was as small as 0.2 nA/μm^2 at 8 V. The maximum photosensitivity of this device was 2.5

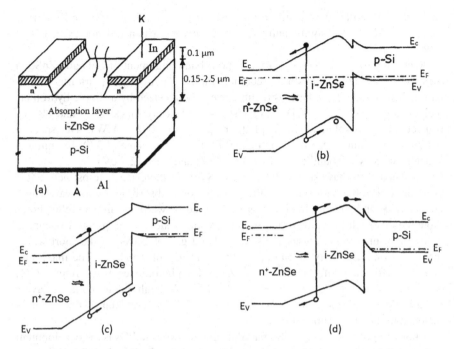

Fig. 13.15 (a) The schematic cross section of the ZnSe/Si photodiode. The PIN-like visible photodiode was grown on a (111)-oriented p-Si substrate using vapor phase epitaxy technique; (b-d) The schematic energy-band diagrams of the ZnSe/Si photodiode (b) at equilibrium, under (c) reverse and (d) forward biased conditions, respectively. (Reprinted with permission from Ref. [39]. Copyright 1996: Elsevier)

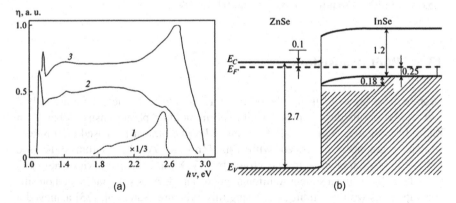

Fig. 13.16 (a) Spectral characteristics of the (1) n-ZnSe/p-GaSe heterojunction at 300 K, (2) n-ZnSe/p-InSe heterojunc tion at 300 K, and (3) n-ZnSe/p-InSe heterojunction at 120 K; (b) Energy band diagram for the n-ZnSe/p-InSe heterojunction. (Reprinted with permission from Ref. [34]. Copyright 2014: Springer)

A/W. ZnSe/Si HPDs developed on the basis of porous silicon (PSi) (n-ZnSe/PSi/p-Si structure) had a significantly lower sensitivity, which did not exceed 0.03 A/W [14].

According to Sunaina et al. [74], a Type II band alignment of ZnO and ZnSe in the heterostructure leads to slower recombination of electron-hole pairs after illumination. These ZnO-ZnSe heterostructures were synthesized by a simple hydrothermal method. ZnO-ZnSe HPDs gave fast rise and decay times of 23 and 80 ms, respectively, and relatively high photo responsivity (0.091 A/W). The spectral region of maximum sensitivity of such HPDs lies in the range of 380–460 nm and is determined by the band gaps of ZnO (3.2 eV) and ZnSe (2.7 eV).

Chen et al. [19] have shown that using ZnSTeSe/ZnSe/GaAs heterostructure it is possible to develop high sensitive PDs, which could absorb light with wavelength between 400 and 900 nm. In other words, the spectral width of this ZnSTeSe photodiode could reach 500 nm. In the long wavelength side, a sharp cutoff occurs at around 870 nm, which corresponds to the band gap of GaAs. In the short wavelength region, sensitivity is limited by the band gap of ZnSTeSe. The maximum quantum efficiency of the fabricated ZnSTeSe photodiodes was around 75%. Responsivity in this spectral range was 0.4–0.5 A/W. High quality epitaxial layers of ZnSe and ZnSTeSe were grown by MBE. ZnSTeSe epitaxial layers perfectly matched to the GaAs substrate used.

Another option for using ZnSe-based heterostructures in PDs is the development of PDs based on superlattices such as ZnSe/ZnTe [6, 7, 35], ZnSe/MgCdS [77], ZnSe/ZnS [5]. It was shown that with the use of such multilayer heterostructures it becomes possible to create selective multicolor photodetectors with high sensitivity. HPDs developed by Averin et al. [6, 7] and Kuznetzov et al. [35] had responsivity from 0.18 A/W to 0.28 A/W. As for ZnSe/MgCdS superlattice based PDs [77], no progress has been achieved here. The responsivity of such PDs in the region of maximum (400–450 nm) was only about 0.001 A/W.

13.8 Phototransistors

Despite the promise of using phototransistors in optoelectronics, there are only a few reports on the development of film and monolithic phototransistors based on ZnSe. In particular, Sun et al. [73] developed a phototransistor based on a hybrid structure: graphene decorated with ZnSe/ZnS core/shell quantum dots (see Fig. 13.17). Monolayer graphene grown by CVD was decorated with ZnSe/ZnS core/shell QDs using a simple solution method. The devices were fabricated on silicon substrates with a bottom gate. Using this structure, Sun et al. [73] achieved a high responsivity above 10^3 AW^{-1} for UV light. The photodetectors exhibited a selective photo responsivity for the UV light with the wavelength of 405 nm, confirming the main light absorption from QDs. According to Sun et al. [73], the

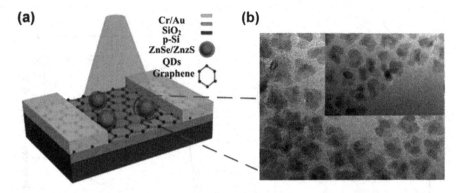

Fig. 13.17 (**a**) Schematic illustration of a graphene FET decorated with ZnSe/ZnS quantum dots under the illumination; (**b**) The TEM image of the ZnSe/ZnS quantum dots on the SiO₂ substrate with different scale bars of 10 nm and 5 nm, respectively. (Reprinted from Ref. [73]. Published 2018 by Nature as open access)

sensing mechanism is attributed to the photo-generated charge carriers in the QDs and the transfer process from QDs to graphene, which may directly modulate the Fermi Level of graphene and the conductance of graphene channel. This charge transfer results in a gate-controlled photosensitivity with a maximum value obtained at V_G of about 15 V. It is important to note that the phototransistors showed good stability during the 21 days test. During this time, responsivity decreased by only 13%. The response time of devices developed was about 0.5 s. Sun et al. [73] believe that the response time is limited by the ligand covering the surface of QDs. Therefore, ligand optimization is required to further improve the response time of phototransistors.

Chang and Wu [15] have proposed a different approach. They developed a phototransistor based on n-ZnSe/p-Si/n-Si heterojunction. The cross-sectional schematic of the studied ZnSe/Si heterojunction phototransistor is shown in Fig. 13.18a. The 0.45 μm n-ZnSe emitter epilayer was grown on a (111) Si substrate utilizing an ordinary infrared (IR) furnace CVD system and a two-step process to overcome the lattice constant mismatch of approximately 4.1%, existing between the ZnSe epilayer and the Si substrate. The resulting photo-current responsivity was approximately 50 A/W at a bias voltage of $V_G = 15$ V, with the highest photo-current responsivity occurring at a wavelength of 470 nm (Fig. 13.18b). It is seen that due to большой толщине n-ZnSe слоя, sensitivity is low in the spectral range $\lambda < 400$ nm. As the intensity of the incident light increases, the photosensitivity decreases. In addition, a strong photo-responsivity was also observed at a longer wavelength of 700 nm, due to the absorption of impinging photons in the Si substrate region. With a collector current density of 65 mA/ cm², the I-V current gain was approximately 28.

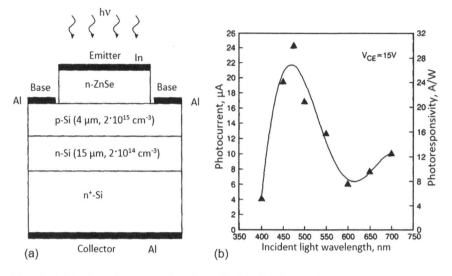

Fig. 13.18 (a) Schematic cross section of the n-ZnSc/p-Si/n-Si heterojunction phototransistor (b) The photocurrent and photo responsivity as a function of the wavelength of the incident light under the fixed incident intensity of 100 lux. (Reprinted with permission from Ref. [15]. Copyright 2000: The Institution of Engineering and Technology)

13.9   Nanowire-Based ZnSe Photodetectors

The Chap. 12 (Vol. 1) describes the main methods for synthesizing 1D nanostructures, and the Chap. 16 (Vol. 2) describes in detail all the approaches that can be used in the development of photodetectors based on 1D nanostructures of II-VI compounds. In the same Chap. 16 (Vol. 2) one can find an analysis of the photosensitivity mechanism of nanowires (NWs), as well as the advantages and disadvantages of their use in the development of photodetectors. In principle, all approaches discussed in the Chap. 16, Vol. 2 can also be used in the development of UV detectors based on ZnSe 1D structures. For example, PDs such as photoconductive PDs [63], field effect transistors [52], Schottky diodes based PDs [88], PDs with p-n junctions [52], PD with radial p-n heterojunctions [53], core-shell nanowire based PD [55], and axial nanowire photodetectors [47] have been developed based on ZnSe nanowires. In most studies, Cu/Au, Ti/Au or Au films were used as ohmic contacts.

There are other approaches to the development of 1D-based photodetectors. For example, Deng et al. [20] developed PDs based on ZnSe NWs decorated with PbSe nanoparticles (NPs). They showed that the formation of PbSe-ZnSe heterojunctions makes it possible to expand the spectral range of sensitivity in the visible and IR regions of the spectrum, and thereby increase the responsivity of the PDs from 5 A/W to 40 A/W (see Fig. 13.19). According to Deng et al. [20], electrons are transported from PbSe to ZnSe while holes are trapped at PbSe, and electrons are

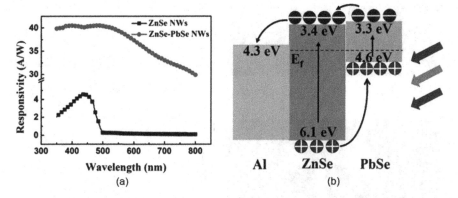

Fig. 13.19 (a) Wavelength-dependent responsivity of ZnSe and ZnSe-PbSe NW-based photodetectors. (b) Energy band alignment between ZnSe-PbSe heterojunction NWs and Al electrodes. (Reprinted from Ref. [20]. Published 2022 by Elsevier as open access)

Table 13.3 Performances of ZnSe nanostructure based photodetectors

Material	Responsivity (A/W)	Photoconductive gain (EQE)	Response/recovery times	Ref.
ZnSe:P/Si	1.1×10^5	2.9×10^5	74/153 µs	[71]
ZnSe:N/Si	2.4×10^4	6.5×10^4	0.4/0.5 s	[72]
ZnSe:Bi/Al	n.a.	$>10^3$	0.38/0.4 s	[88]
ZnSe:Cl/Si	4.1×10^2	1.1×10^3	2/2 s	[84]
ZnSe:Sb/ZnO	5.2×10^5	1.77×10^6	<1 s	[52]
ZnSe	4.2×105	1.3×10^6 (EQE)	0.55/0.6 s	[47]
ZnS$_{0.49}$Se$_{0.51}$/ ZnSe	6.3×105	2.08×10^6 (EQE)	23/50 ms	[47]

extracted to an external circuit by an Al electrode, resulting in a high gain of the photoconductor.

However, the most common variants of NW-based photodetectors are photoconductive 1D-based ZnSe photodetectors. In particular, in [36, 57, 88] one can find reports describing the results of testing such photodetectors. Photodetectors were fabricated both on the basis of single nanowires [23], and nanowires array [25]. Just as in the case of film photodetectors, doping can be used to control the parameters of nanowire-based PDs. Comparative characteristics of photodetectors based on ZnSe NWs doped with different dopants, is given in the Table 13.3. Functional principles of photoconductive 1D-based PDs in addition to Chap. 16 (Vol. 2) one can find in [70].

As can be seen from Table 13.3, photodetectors based on single 1D nanostructures demonstrate high sensitivity and rate of response. But it should be recognized that due to technological limitations (see Chap. 16, Vol. 2) it is unlikely that single NW-based PDs will appear on the market in the near future. Oksenberg et al. [53, 54] developed a technology for the horizontal growth of oriented ZnSe NWs to

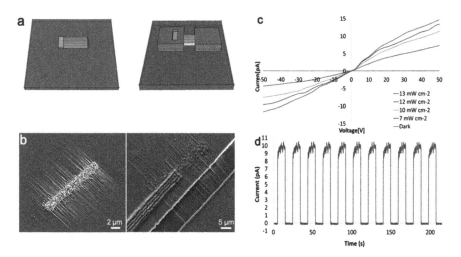

Fig. 13.20 Optoelectronic behavior of the guided ZnSe NWs: (**a**) A schematic illustration of the device fabrication process. Ti/Au electrodes are deposited over guided ZnSe NWs that grow from a patterned Au catalyst; (**b**) SEM images of the fabrication steps illustrated in (**a**); (**c**) I–V characteristics of a typical device under different 405 nm illumination intensities and under dark conditions; (**d**) The response of a typical photodetector set at a 30 V bias to on/off switching of a 405 nm laser illumination (13 mW cm^{-2}). (Reprinted with permission from Ref. [54]. Copyright 2015: Wiley)

solve this problem and fabricate photodetectors based on them (see Fig. 13.20). However, this technology is still far from perfect.

13.10 Photodetectors Based on Hybrid Structures

At present, two approaches have been used in the development of ZnSe-based photodetectors based on hybrid structures. Team from Tottori University (Japan) [27–29] used organic–ZnSe hybrid structures, while Chakrabortya et al. [12] and Xu et al. [87] developed photodetectors based on the ZnSe-graphene structure. Chakrabortya et al. [12] used a ZnSe-decorated reduced graphene oxide (RGO) nanocomposite and Xu et al. [87] used a two-layer graphene-ZnSe structure in which ZnSe films with a thickness of 60 nm were deposited on graphene by e-beam evaporation method. ZnSe-RGO nanocomposite for a large area thin film photodetector device was synthesized using a simple one step solvothermal reaction. Thin film of RGO-ZnSe composite was prepared on a pre-cleaned glass substrate by simple drop casting from dispersed solution in isopropyl alcohol (IPA). Chakrabortia et al. [12] reported that the photosensitivity of the RGO-ZnSe photodetector was 3 times higher than that of controlled ZnSe PD. Chakrabortia et al. [12] believe that this improvement is due to the specificity of the hybrid structure, in which ZnSe nanoparticles are evenly decorated onto the surface of RGO nanosheets. Photocurrent

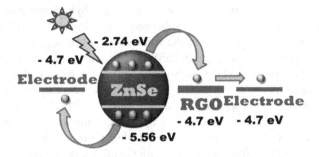

Fig. 13.21 Photocurrent generation mechanism in our thin film device. (Reprinted with permission from Ref. [12]. Copyright 2017: Royal Society of Chemistry)

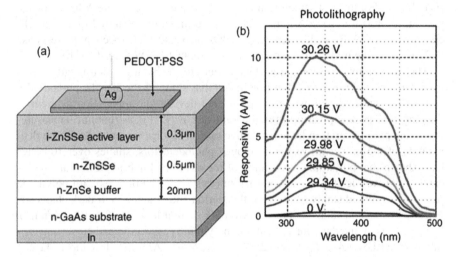

Fig. 13.22 (a) Schematic structure of organic–inorganic (PEDOT:PSS / ZnSSe) hybrid APD. (b) Responsivity spectra in the operation bias region of ZnSe-based hybrid APDs fabricated by photolithography. (Reprinted with permission from Ref. [27]. Copyright 2020: Springer)

generation mechanism for such structures is shown in Fig. 13.21. However, the responsivity of such PDs was small (2 ma/W), and the time constants of photoresponse were too large to use such PDs as detectors. Response and recovery times were about 29 and 41 seconds, respectively.

It should be noted that the photodetectors developed by Xu et al. [87] demonstrated a significantly higher responsivity, which sharply increased with decreasing channel width and reached 1.2×10^9 A/W for PD with 70 nm channel. At 5 μm channel the responsivity was three orders of magnitude lower. Obviously, the shorter channel allows the faster transport and more times recirculating of the charge carriers and thus produces larger photocurrent and responsivity. As a result, devices with a 70 nm channel had a response time of about 50 ms.

The approach based on the use of organic–inorganic hybrid structures, proposed by the team from Tottori University, is also very promising. A view of the structure of photodetectors developed within the framework of this approach is shown in Fig. 13.22a. ZnSe-based structure (i-ZnSSe ($<10^{15}$ cm^{-3}) - active layer/n$^+$-ZnSSe (4×10^{18} cm^{-3}) - contact layer) was grown by molecular beam epitaxy (MBE) on n-type GaAs substrates. The organic UV-transparent conducting polymer layer of poly 3,4-ethylenedioxythiophene:poly-styrenesulfonate (PEDOT:PSS) [48] with a thickness of 0.15 to 1.0 μm was formed by spin-coating and a photolithography technique. Smaller thickness provided lower absorption losses in the layer. The absorption loss of the PEDOT:PSS layer with the thickness of 1 μm is around 60% [29]. The energy barrier height of interface between organic (PEDOT: PSS) and inorganic (i-ZnSSe) was about 1.0 eV [28]. It was shown that the developed photodetectors are high-gain and highly sensitive ultraviolet avalanche photodiodes (UV-APDs). For the best samples, the leakage current before the breakdown voltage was suppressed to $<10^{-10}$ A/mm^2 [27], and the maximum external quantum efficiency reached 90% at 325 nm [28]. APDs also had high-speed photoresponse (3.25 ns) [1]. The maximum responsivity was about 10 A/W (Fig. 13.22b).

The maximum multiplication factor due to technology optimization was improved up to M = 3100 [27]. It is assumed that ultraviolet avalanche photodiodes (UV-APDs) under development are promising devices for application as medical imaging devices [e.g. positron emission tomography (PET)], astronomical measurement, and next-generation large capacitive optical storage systems. As for the degradation of the properties of the developed photodetectors, studies over 100 days have shown that polyimide passivation of the developed hybrid APDs can significantly improve the stability of their parameters [29]. The devices with polyimide passivation worked stably throughout this period and retained their dark characteristics with a steep avalanche breakdown (dark current less than 2×10^{-11} A/mm^2). For comparison, the dark current without polyimide passivation and N$_2$ sealing rapidly increased over 4–5 days from 2×10^{-11} to 2×10^{-6} A/mm^2 before the avalanche breakdown.

Acknowledgments This research was funded by the State Program of the Republic of Moldova, project 20.80009.5007.02.

References

1. Abe T, Inoue R, Fujimoto T, Tanaka K, Uchida S, Kasada H et al (2016) Development of ZnSe-based organic–inorganic hybrid UV-APDs array. Phys Status Solidi C 13:677–682
2. Abe T, Ando K, Ikumi K, Maeta H, Naruse J, Miki K, Ehara A, Kasada H (2005) High gain and high sensitive blue-ultraviolet avalanche photodiodes (APDs) of ZnSSe n$^+$-i-p structure molecular beam epitaxy (MBE) grown on p-type GaAs substrates. Jpn J Appl Phys 44(17):L 508–L 510
3. Abe T, Maeta H, Naruse J, Ikumi K, Kubota T, Fujiwara T, Kasada H, K. Ando K. (2004) New blue-ultraviolet PIN photodiodes of II-VI widegap compounds ZnSSe using p-type GaAs substrates grown by molecular beam epitaxy. phys stat sol a 1(4):1054–1057

4. Ando K, Ishikura H, Fukunaga Y, Kubota T, Maeta H, Abe T, Kasada H (2002) Highly efficient blue–ultraviolet photodetectors based on II–VI wide-bandgap compound semiconductors. Phys Stat Sol a 229(2):1065–1071

5. Averin SV, Kuznetzov PI, Zhitov VA, Zakharov LY, Kotov VM (2020) MSM-photodetector with ZnSe/ZnS/GaAs Bragg reflector. Opt Quant Electron 52:93

6. Averin SV, Kuznetzov PI, Zhitov VA, Zakharov LY, Kotov VM (2019) Electrical, optical and spectral characteristics of type-II ZnSe/ZnTe/GaAs superlattice and MSM-photodetector on their base. Opt Quant Electron 50:368

7. Averin SV, Kuznetzov PI, Zhitov VA, Zakharov LY, Kotov VM (2018) Multicolour photodetector based on a ZnSe/ZnTe/GaAs heterostructure. Quant Electron 48(7):675 –678

8. Bhattacharya P (1994) Semiconductor optoelectronic devices. Prentice Hall, New Jersey

9. Bhushan S (1987) Photoconductivity of ZnSe crystals. J Mater Sci Lett 6:591–592

10. Bouhdada A, Hanzaz M, Vigue F, Faurie JP (2003) Electrical and optical proprieties of photodiodes based on ZnSe material. Appl Phys Lett 83(1):171–173

11. Bube RH, Lind EL (1958) Photoconductivity of Zinc Selenide crystals and a correlation of donor and acceptor levels in II-VI photoconductors. Phys Rev 110(5):1040–1049

12. Chakraborty K, Chakrabarty S, Pal T, Ghosh S (2017) Synergistic effect of zinc selenide–reduced graphene oxide towards enhanced solar light-responsive photocurrent generation and photocatalytic 4-nitrophenol degradation. New J Chem 41:4662–4671

13. Chang SJ, Lin TK, Su YK, Chiou YZ, Wang CK, Chang SP et al (2006) Homoepitaxial ZnSe MSM photodetectors with various transparent electrodes. Mater Sci Eng B 127:164–168

14. Chang CC, Lee CH (2000) Study and fabrication of PIN photodiode by using ZnSe/PS/Si structure. IEEE Trans Electron Dev 47(1):50–54

15. Chang CC, Wu KT (2000) Fabrication of n-ZnSe/p-Si/n-Si heterojunction phototransistor using IR furnace chemical vapour deposition and its optical properties analysis. IEE Proc Optoelectron 147(2):104–108

16. Chen M-Y, Chang C-C (2009a) Comparison of performance of integrated photodetectors based on ZnS and ZnSe metal–semiconductor–metal photodiodes. Jpn Appl Phys 48:112201

17. Chen MY, Chang CC (2009b) Monolithic integration of a ZnSe MSM photodiode and an InGaP/GaAs HBT on a GaAs substrate. Semicond Sci Technol 24:045009

18. Chen M-Y, Chang C-C (2008) Monolithic photoreceiver constructed with a ZnSe MSM photodiode and an InGaP/GaAs HBT. IEEE Electron Dev Lett 29(11):1212–1214

19. Chen W-R, Meen T-H, Cheng Y-C, Lin W-J (2006) P-down ZnSTeSe/ZnSe/GaAs heterostructure photodiodes. IEEE Electron Dev Lett 27(5):347–349

20. Deng J, Lv W, Zhang P, Huang W (2022) Large-scale preparation of ultra-long ZnSe-PbSe heterojunction nanowires for flexible broadband photodetectors. J Sci: Adv Mater Dev 7:100396

21. Dima I, Vasiliu G (1967) Photoconductivity of ZnSe thin films. Phys Status Solidi 22:K79–K82

22. Ehinger M, Koch C, Korn M, Albert D, Nurnberger J, Hock V, Faschinger W, G. Landwehr G. (1998) High quantum efficiency II–VI photodetectors for the blue and blue-violet spectral range. Appl Phys Lett 73(24):3562–3564

23. Fang X, Xiong S, Zhai T, Bando Y, Liao M, Gautam UK et al (2009) High-performance blue/ultraviolet-light-sensitive ZnSe-nanobelt photodetectors. Adv Mater 21:5016–5021

24. Hong H, Anderson WA (1999) Cryogenic processed metal-semiconductor-metal (MSM) photodetectors on MBE grown ZnSe. IEEE Trans. Electron Dev. 46:1127–1133

25. Hsiao CH, Chang SJ, Wang SB, Chang SP, Li TC, Lin WJ et al (2009) ZnSe nanowire photodetector prepared on oxidized Silicon substrate by molecular-beam epitaxy. J Electrochem Soc 156(4):J73–J76

26. Huang ZC, Wie CK, Na I, Luo H, Mott DB, Shu PK (1996) High performance ZnSe photoconductors. Electron Lett 32(16):1507–1505

27. Ichikawa Y, Tanak K, Nakagawa K, Fujii Y, Yoshida K, Nakamura K et al (2020) High-gain ultraviolet avalanche photodiodes using a ZnSe-based organic-inorganic hybrid structure. J Electron Mater 49:4589–4593

28. Inagaki Y, Ebisu M, Otsuki M, Ayuni N, Shimizu T, Abe T et al (2012) New ultraviolet avalanche photodiodes (APDs) of organic (PEDOT: PSS)–inorganic (ZnSSe) hybrid structure. Phys Status Solidi A 209(8):1852–1855

29. Inoue R, Abe T, Fujimoto T, Ikadatsu N, Tanaka K, Uchida S et al (2015) ZnSe-based organic–inorganic hybrid structure ultraviolet avalanche photodiodes with long lifetime and its device integration. Appl Phys Express 8:022101

30. Ishikura H, Fukunaga Y, Kubota T, Maeta H, Adachi M, Abe T et al (2002) Blue-violet avalanche-photodiode (APD) and its ionization coefficients in II–VI wide bandgap compound grown by molecular beam epitaxy. Phys Stat Sol (b) 229(2):1085–1088

31. Ishikura H, Abe T, Fukuda N, Kasada H, Ando K (2000a) Stable avalanche-photodiode operation of ZnSe-based p+– n structure blue-ultraviolet photodetectors. Appl Phys Lett 76(8):1079–1071

32. Ishikura H, Fukuda N, Itoi M, Yasumoto K, Abe T, Kasada H, Ando K (2000b) High quantum efficiency blue-ultraviolet ZnSe pin photodiode grown by MBE. J Crystal Growth 214(215):1130–1133

33. Jia L, Zheng W, Huang F (2020) Vacuum-ultraviolet photodetectors. PhotoniX 1:22

34. Kudrynskyi ZR, Kovalyuk ZD (2014) Photosensitive anisotype n-ZnSe/p-InSe and n-ZnSe/p--GaSe heterojunctions. Technical Phys 59(9):1205–1208

35. Kuznetzov PI, Averin SV, Zhitov VA, Zakharov LY, Kotov VM (2017) MSM optical detector on the basis of II-type ZnSe/ZnTe superlattice. Semiconductors 51(2):249–253

36. Li S, Su Q, Zhao H (2013) Photoresponse properties of p-type ZnSe nanowire photodetectors. Micro Nano Lett 8(9):496–499

37. Lin TK, Chang SJ, Su YK, Chiou YZ, Wang CK, Chang CM, Huang BR (2005a) ZnSe homoepitaxial MSM photodetectors with transparent ITO contact electrodes. IEEE Trans Electron Dev 52(1):121–123

38. Lin TK, Chang SJ, Su YK, Chiou YZ, Wang CK, Chang SP, Chang CM, Tang JJ, Huang BR (2005b) ZnSe MSM photodetectors prepared on GaAs and ZnSe substrates. Mater Sci Eng B 119:202–205

39. Lour W-S, Chang C-C (1996) VPE growth ZnSe/Si PIN-like visible photodiodes. Solid State Electron 39(9):1295–1298

40. Makhniy VP, Melnik VV, Tkachenko IV, Gorley PN, Horváth ZJ, Horley PP, Vorobiev YV (2003) Electrical properties of UV detectors based on zinc selenide with modified surface barrier. Phys Status Solidi 0(3):1039–1043

41. Makhniy VP, Mel'nyk VV, Gorley PN, Horley PP, Sletov MM, Zhuo Z (2005) Detectors of UV and x-ray irradiation on the base of metal-zinc selenide contact. Proc SPIE 6024:60242J

42. Miki K, Oshita Y, Katada D, Nobe K, Nomura M, Abe T, Kasada H, Ando K (2008) High sensitivity ultraviolet PIN photodiodes of ZnSSe n+-i-p structure /p+-GaAs with an extremely thin n+-window layer grown by using MBE. J Korean Phys Soc 53(5):292–292

43. Miki K, Abe T, Naruse J, Ikumi K, Yamaguchi T, Kasada H, Ando K (2006) Highly sensitive ultraviolet PIN photodiodes of ZnSSe n+-i-p structure/p+-GaAs substrate grown by MBE. Phys Status Solidi 243(4):950–954

44. Monroy E, Omnes F, Calle F (2003) Wide-bandgap semiconductor ultraviolet photodetectors. Semicond Sci Technol 18:R33–R51

45. Monroy E, Vigue F, Calle F, Izpura JI, Munoz E, Faurie J-P (2000) Time response analysis of ZnSe-based Schottky barrier photodetectors. Appl Phys Lett 77(17):2761–2763

46. Monroy E, Calle F, Munoz E, Omnes F (1999) AlGaN metal-semiconductor-metal photodiodes. Appl Phys Lett 74:3401

47. Mu Z, Zheng Q, Liu R, Malik MWI, Tang D, Zhou W, Wan Q (2019) 1D ZnSSe-ZnSe axial heterostructure and its application for photodetectors. Adv Electron Mater 2019:1800770

48. Nakano M, Makino T, Tsukazaki A, Ueno K, Ohtomo A, Fukumura T et al (2008) Transparent polymer Schottky contact for a high performance visible-blind ultraviolet photodiode based on ZnO. Appl Phys Lett 93:123309

49. Nasiri N, Tricol A (2019) Nanomaterials-based UV photodetectors. In: Thomas S, Grohens Y, Pottathara YB (eds) Industrial Applications of Nanomaterials. Elsevier, New York, pp 123–149. https://doi.org/10.1016/B978-0-12-815749-7.00005-0

50. Naval V, Smith C, Ryzhikov V, Naydenov S, Alves F, Karunasiri G (2010) Zinc selenide-based Schottky barrier detectors for ultraviolet-A and ultraviolet-B detection. Adv OptoElectron 2010:61957

51. Nedeoglo ND, Nedeoglo DD, Sirkeli VP, Tiginyanu IM, Laiho R, Lähderanta E (2008) Shallow donor states induced in ZnSe:Au single crystals by lattice deformation. J Appl Phys 104:123717

52. Nie B, Luo L-B, Chen J-J, Hu J-G, Wu C-Y, Wang L et al (2013) Fabrication of p-type ZnSe:Sb nanowires for high-performance ultraviolet light photodetector application. Nanotechnology 24:095603

53. Oksenberg E, Martí-Sànchez S, Popovitz-Biro R, Arbiol J, Joselevich E (2017) Surface-guided core-shell ZnSe@ZnTe nanowires as radial p-n heterojunctions with photovoltaic behavior. ACS Nano 11(6):6155–6166

54. Oksenberg E, Popovitz-Biro R, Rechav K, Joselevich E (2015) Guided growth of horizontal ZnSe nanowires and their integration into high-performance blue–UV photodetectors. Adv Mater 27:3999–4005

55. Park S, Kim S, Sun G-J, Byeon DB, Hyun SK, Lee WI, Lee C (2016) ZnO-core/ZnSe-shell nanowire UV photodetector. J Alloys Comp 658:459–464

56. Park RM, Troffer MB, Rouleau CM, DePuydt JM, Haase MA (1990) p-type ZnSe by nitrogen atom beam doping during molecular beam epitaxial growth. Appl Phys Lett 57:2127

57. Philipose U, Ruda HE, Shik A, de Souza CF, Sun P (2006) Conductivity and photoconductivity in undoped ZnSe nanowire array. J Appl Phys 99:066106

58. Rabaco F, Martin JM, Vincent AB, Joshi NV (1991) Photoconductivity and optical absorption on the high energy side in an se treated ZnSe monocrystal. J Phys Chem Solids 52(4):575–578

59. Radevici I, Sushkevich K, Colibaba G, Sirkeli V, Huhtinen H, Nedeoglo N, Nedeoglo D, Paturi P (2013) Influence of chromium interaction with native and impurity defects on optical and luminescence properties of ZnSe:Cr crystals. J Appl Phys 114:203104

60. Rao GK (2017) Electrical and photoresponse properties of vacuum deposited Si/Al:ZnSe and Bi:ZnTe/Al:ZnSe photodiodes. Appl Phys A Mater Sci Process 123:224

61. Rao KG, Bangera KV, Shivakumar GK (2011) Photoconductivity and photo-detecting properties of vacuum deposited ZnSe thin films. Sol State Sci 13:1921–1925

62. Razeghi M, Rogalski A (1996) Semiconductor ultraviolet detectors. J Appl Phys 79:7433–7473

63. Salfi J, Philipose U, de Sousa CF, Aouba S, Ruda HE (2006) Electrical properties of Ohmic contacts to ZnSe nanowires and their application to nanowire-based photodetection. Appl Phys Lett 89:261112

64. Seghier D, Gislason HP (1999) The observation of persistent photoconductivity in N-doped p-type ZnSe/GaAs heterojunctions. J. Phys D: Appl Phys 32(4):369

65. Seghier D, Gislason HP (2000) Investigation of persistent photoconductivity in nitrogen-doped ZnSe/GaAs heterojunctions grown by MBE. J Crystal Growth 214(215):511–515

66. Sharma J, Tripathi SK (2011) Effect of deposition pressure on structural, optical and electrical properties of zinc selenide thin films. Physica B 406:1757–1762

67. Sirkeli VP, Nedeoglo ND, Nedeoglo DD, Yilmazoglu O, Hajo AS, Preu S et al (2021) ZnSe-based solar-blind ultraviolet photodetectors with different Schottky contact metals. Studia Universitatis Moldaviae 2(142):59–67

68. Sirkeli VP, Yilmazoglu O, Ong DS, Preu S, Küppers F, Hartnagel HL (2017a) Resonant tunneling and quantum cascading for optimum room-temperature generation of THz signals. IEEE Trans Electron Dev 64:3482–3488

69. Sirkeli VP, Yilmazoglu O, Hajo AS, Nedeoglo ND, Nedeoglo DD, Preu S et al (2017b) Enhanced responsivity of ZnSe-based metal–semiconductor–metal near-ultraviolet photodetector via impact ionization. Phys Status Sol RRL 12(2):1700418

70. Soci C, Zhang A, Bao X-Y, Kim H, Lo Y, Wang D (2010) Nanowire photodetectors. J Nanosci Nanotechnol 10(3):1430–1449

71. Su Q, Zhang Y, Li S, Du L, Zhao H, Liu X, Li X (2015) Synthesis of p-type phosphorus doped ZnSe nanowires and their applications in nanodevices. Mater Lett 139:487–490

72. Su Q, Li L, Li S, Zhao H (2013) Synthesis and optoelectronic properties of p-type nitrogen doped ZnSe nanobelts. Mater Lett 92:338–341
73. Sun Y-S, Xie D, Sun M-X, Teng C-J, Qian L, Chen R-S et al (2018) Hybrid graphene/cadmium free ZnSe/ZnS quantum dots phototransistors for UV detection. Sci Rep 8:5107
74. Sunaina, Ganguli AK, Mehta K (2022) High performance ZnSe sensitized ZnO heterostructures for photo-detection applications. J. Alloys Compounds 894:162263
75. Sze SM, Ng KK (2007) Physics of semiconductor devices. Wiley-Interscience, Hoboken
76. Terui Y, Yoshino M, Ogura M, Nakayama M, Yoneda M, Chikamura T et al (1981) A CCD imager using ZnSe-Zn$_{1-x}$ Cd$_x$Te heterojunction photoconductor, Jpn. J Appl Phys Sup 21–1:237–242
77. Ueno J, Ogura K, Ichiba A, Katsuta S, Kobayashi M, Onomitsu K, Horikoshi Y (2006) MBE growth of ZnSe/MgCdS and ZnCdS/MgCdS superlattices for UV-A sensors. Phys Stat Sol (c) 3(4):1225–1228
78. Vigue F, Faurie J-P (2001) Zn(MgBe)Se ultraviolet photodetectors. J Electron Mater 30(6):662–666
79. Vigué F, Tournié E, Faurie J-P (2001) Evaluation of the potential of ZnSe and Zn(Mg)BeSe compounds for ultraviolet photodetection. IEEE J Electron Dev 37(9):1146–1152
80. Vigué F, Tournié E, Faurie J-P (2000a) Zn(Mg)BeSe-based p-i-n photodiodes operating in the blue-violet and near-ultraviolet spectral range. Appl Phys Lett 76:242–244
81. Vigué F, de Mierry P, Faurie J-P, Monroy E, Calle F, Mufioz E (2000b) High detectivity ZnSe-based Schottky barrier photodetectors for blue and near-ultraviolet spectral range. Electron Lett 36(9):826–827
82. Vigué F, Tournié E, Faurie J-P (2000c) ZnSe-based Schottky barrier photodetectors. Electron Lett 36:352–354
83. Voronkin E (2013) Schottky diodes based on the Zinc Selenide semiconductor crystals. Functional Mater 20(4):534–537
84. Wang Z, Jie J, Li F, Wang L, Yan T, Luo L et al (2012) Chlorine-doped ZnSe nanoribbons with tunable n-type conductivity as high-gain and flexible blue/UV photodetectors. Chem Phys Chem 77:470–475
85. Yu Y-M, Nam SOB, Lee K-S, Yu PY, Lee J, Choi TD (2002) Strain effect in ZnSe epilayer grown on the GaAs substrate. J Crystal Growth 243:389–395
86. Xiao C, Wang Y, Yang T, Luo Y, Zhang M (2016) Preparation of ZnO/ZnSe heterostructure parallel arrays for photodetector application. Appl Phys Lett 109:043106
87. Xu Z, Wu C, Li A, Ma Y, Fei GT, Wang M (2018) Sub-100 nm channel ZnSe film/graphene hybrid-based photodetectors with an ultrahigh responsivity of 10^9 A/W. IEEE Electron Dev Lett 39(2):240–243
88. Zhang X, Jie J, Wang Z, Wu C, Wang L, Peng Q et al (2011) Surface induced negative photoconductivity in p-type ZnSe: Bi nanowires and their nano-optoelectronic applications. J Mater Chem 21:6736–6741

Chapter 14
ZnS-Based UV Detectors

Sema Ebrahimi, Benyamin Yarmand, and Nima Naderi

14.1 Introduction

The technology of photodetectors based on II-VI compound semiconductors has drawn much research attention over the past decades due to their optical properties, good transport properties, strong thermal stability, and high electron mobility [1, 2]. Zinc sulfide (ZnS), a prominent II-VI intrinsic semiconductor material, has been intensely studied because of its broad spectrum of diverse applications such as in photovoltaic, electroluminescence devices, bio-imaging devices, photodetectors, sensors, and solar cells for a long time [3–7].

When the size of these structures is comparable to their corresponding Bohr exciton radius, they show a quantum confinement effect with a considerable blue shift in absorption spectra, high surface-to-volume ratio, high aspect ratio, and large Debye length. On the contrary, engineering the composition of the materials is an alteration option to tuning the bandgap for specific applications [1–3, 8–12], in which the large bandgap of ZnS in UV regime and high quantum efficiency makes

S. Ebrahimi
Materials and Energy Research Center (MERC), Karaj, Iran

Light, Nanomaterials, Nanotechnologies (L2n) Laboratory, CNRS EMR7004, The University of Technology of Troyes, Troyes Cedex, France

Department of Physics and Mathematics, University of Hull, Cottingham Road, UK

G.W.Gray Centre for Advanced Materials, University of Hull, Cottingham Road, UK

B. Yarmand
Materials and Energy Research Center (MERC), Karaj, Iran

N. Naderi (✉)
Materials and Energy Research Center (MERC), Karaj, Iran

Photonics Research Centre, University of Malaya, Kuala Lumpur, Malaysia
e-mail: n.naderi@merc.ac.ir

© The Author(s), under exclusive license to Springer Nature Switzerland AG 2023   333
G. Korotcenkov (ed.), *Handbook of II-VI Semiconductor-Based Sensors and Radiation Detectors*, https://doi.org/10.1007/978-3-031-20510-1_14

it desirable for visible-blind UV detectors, revealing remarkable characteristics in widespread applications such as sensing, astronomy, communication, and medical instruments.

In the modern scientific area, the motivation for the fabrication of doped semiconductor nanostructures has attracted great interest due to their significance in basic scientific research and potential technological applications. To utilize semiconductor nanostructures as building blocks of photodetectors, it is significant to synthesize nanostructures that have remarkable optoelectrical properties [10, 12, 13]. In the case of doped materials, the addition of impurities into a wide-gap semiconductor can often induce dramatic changes in the optical and electrical properties as well as influence transition probabilities [11, 13, 14]. Since ZnS has a wide bandgap at room temperature, its nanocrystals are suitable to host material for doping elements including transition metal and/or rare-earth elements which are optically active. Therefore, doping a selective element such as Mn, Sn, Ni, W, and Graphene in ZnS nanostructures has been an important route for raising and controlling the bandgap and making it suitable to prepare UV-detectors with different cut-off wavelengths and high performance [2, 3, 6, 9, 15–22].

In this chapter, we focus on the most recent progress in UV-detectors based on the ZnS nanostructures. First, ZnS photodetectors including photoconductors, metal-semiconductor-metal (MSM) photodetectors, Schottky photodiodes, p-n junction photodiodes are discussed, followed by photodetectors based on the doped ZnS nanostructures. Finally, we conclude this review chapter with future perspectives in these brand-new fields.

14.2 ZnS-Based Photodetectors

This section discusses ZnS-based photodetectors, including photoconductors, metal-semiconductor-metal (MSM) photodetectors, and Schottky photodiodes.

Among the photons that strike a semiconductor, those with an energy greater than the semiconductor bandgap can give enough energy to the valance band electrons and move them into the conduction band. The photo-generated carriers increase the semiconductor conductivity, and this is the basic operating mechanism in photoconductors. The photoconductor can be simply placed in series with the load resistance. The change in the device conductivity due to the illumination changes the voltage across the load resistance, which can be detected by a high impedance voltmeter. The typical photoconductor usually contains a semiconductor layer sandwiched between two ohmic contacts, resulting in a high photoconductive gain and high responsivity and no need the external amplifying equipment. However, they show relatively large dark-current, low signal-to-noise ratio, and slow response speed. Many efforts have been taken to overcome these issues, such as using II–VI-based nanostructures [12, 20, 23] and, surface treatment [24] to increase the internal gain and other parameters on-demand.

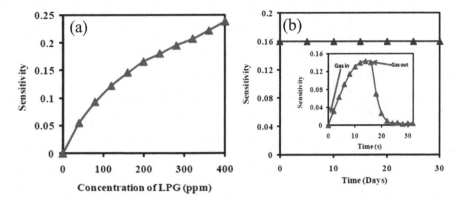

Fig. 14.1 LPG sensing response (**a**) at room temperature and (**b**) Stability response of ZnS nanoparticles against LPG and inset show the transient response at room temperature. (Reprinted with permission from [25]. Copyright 2021: Elsevier)

ZnS-based photoconductors have been investigated in detail by several groups. Different materials such as Au, Ag, Al, Pt, Al/Au, etc., have been used as electrodes. Nemade et al. [25] demonstrated UVC photoconductors based on the ZnS nanoparticles synthesized using flame-assisted spray pyrolysis method on a SiO$_2$ substrate. The highly conducting Ag was used as ohmic electrodes. In this report, the ZnS nanoparticles were applied for the liquefied petroleum gas (LPG) sensing, in which the response curve featured a decrease in the resistance of the nanoparticles suggesting the intrinsically n-type ZnS in the presence of LPG, as shown in Fig. 14.1a. The good response stability (Fig. 14.1b) has been recorded for the as-synthesized ZnS nanoparticles against LPG, with a fast response and recovery time of about 18 s and 11 s, respectively.

Hajimazdarani et al. [26] have fabricated UV detectors based on the ZnS nanoparticles using a chemical deposition method, followed by an annealing process. They focused on the effect of annealing temperature on the optoelectrical properties of the ZnS nanoparticles. According to this report, at the optimized annealing temperature of 500 °C, a phase change was applied to the ZnS lattice system from zinc blende to wurtzite, leading to a significant enhancement in the absorbance and photoluminescence intensity. The optoelectrical characteristics indicate the higher generation of electron carriers under the UV exposure for the annealed samples, resulting in higher photosensitivity and photoresponsivity compared to the as-synthesized sample as demonstrated in Fig. 14.2.

Prasad et al. [27] investigated a UV photoconductor based on the highly crystalline ZnS nanoparticles having cubic and wurtzite structure produced by a simple co-precipitation at room temperature and one-step hydrothermal method at 180 °C, respectively. They studied the influence of the crystalline structure of ZnS nanoparticles on the photoresponsivity behavior under UV illumination. Both fabricated devices showed ohmic behavior, which interestingly confirmed a better response of wurtzite lattice than the cubic one due to the high electron-phonon coupling. The

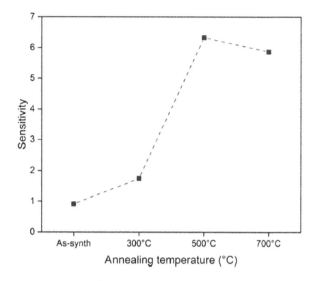

Fig. 14.2 The sensitivity of UV detectors based on detectors samples with different annealing temperature. (Reprinted with permission from [26]. Copyright 2021: Elsevier)

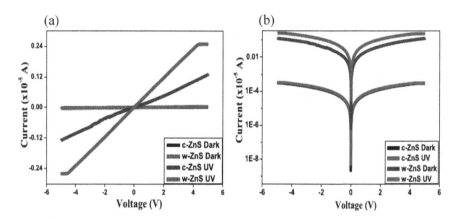

Fig. 14.3 (**a**) I–V characteristic curve of a photodetector with c-ZnS and w-ZnS nanoparticles as channel and Ag as electrode under dark and illumination of UV light and (**b**) Logarithmic plot of I–V characteristics of c-ZnS and w-ZnS nanoparticles-based UV photodetector. (Reprinted with permission from [27]. Copyright 2021: Elsevier)

I–V characteristic curves exhibited an enhanced photoresponsivity of the wurtzite ZnS nanoparticles-based UV detectors and high dependence of the performance on the dynamics of the phonon population illustrated in Fig. 14.3. It was then concluded that the fabricated device shows ohmic behavior and the wurtzite ZnS has a better optical response towards the UV illumination than the cubic type since a dominant enhancement in electron-phonon coupling has been achieved for wurtzite ZnS than for cubic-ZnS.

The Metal-Semiconductor-Metal (MSM) structure is the simplest type of photodetector. MSM photodetectors have a simple structure, easy to fabricate, have fast response, low capacitance per unit area, and large active area. MSM photodetectors are comprised of two back-to-back Schottky diodes using two interdigitated electrodes on an absorbing layer [28]. The main drawback of MSM photodetectors is their low photoresponsivity, due to the shadow effect arising from the metallization of electrodes on the active area. Under the incident optical power, electron-hole pairs are generated through the absorption of light and flow towards the gap between the two metal contacts, resulting in a photocurrent. For compound semiconductors, the light absorption layer is usually deposited on a semi-insulating substrate.

ZnS-based MSM photodiodes have been fabricated by different physical and chemical techniques such as sputtering [29], sol-gel [30], chemical vapor deposition [31], molecular beam epitaxy [32], atomic layer epitaxy [33], pulse electrochemical deposition [34], chemical bath deposition [35], spray pyrolysis [36] and hydrothermal/solvothermal method [37].

In the case of MSM photodiodes, Ebrahimi et al. [38] investigated the performance of UV detectors based on the different ZnS-based absorbing layers. First, they studied the optoelectrical properties of ZnS films deposited by a simple and low-cost spray pyrolytic method. In this report, the effect of the Zn: S molar ratio in the precursor solution on the structural, optical, and electrical properties of the ZnS layers has been investigated. For all the samples having a different molar ratio (Zn:S) of precursor, (1:1), (1:2), and (1:3), at the fixed substrate temperature of about 400 °C, a single-phase ZnS cubic with good crystallinity was obtained using a sulfur-rich solution. The band gap energy was found to be increased by the increment of sulfur concentration up to three times of zinc content in the ZnS lattice due to the improvement of crystallinity.

To continue, Ebrahimi et al. [36] reported the UV detectors based on the chalcogenide Mn-doped ZnS films deposited by the spray pyrolysis method. They studied the modification of the ZnS-based films by evaluating their optical properties including the optical density, optical band gap energy, Urbach energy, steepness parameter, electron-phonon interactions, and the relationship between them, to explore the photodetection and photoresponsivity of the UV detectors. From the optical properties in Fig. 14.4, by increasing the Mn^{2+} ions concentration from 0.05 to 0.15 mol, a slight redshift was observed in the absorption band-edge, leading to a decrease in the optical band gap energy of the samples from 3.98 eV to 3.86 eV. Otherwise, the Urbach energy corresponds to the extended defect levels and disorders introduced by Mn ions increased from 354 meV to 420 meV. Considering the linear relationship between the optical band gap and Urbach energies, two important phenomena were found. First, the value of the optical bandgap energy in the absence of defects and band tailing of about 4.61 eV. Second, is the inverse relevance between the steepness parameter and electron-phonon interactions. The PL spectra demonstrated two emission peaks located in UV and orange regions, corresponding to the band-to-band transition of host ZnS and $^4T_1$-$^6A_1$ inter-band transition of Mn^{2+} ions, respectively.

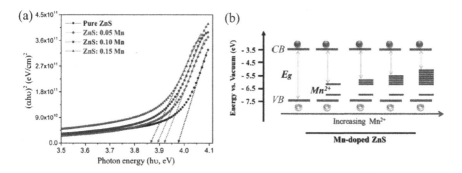

Fig. 14.4 (a) the variation of $(\alpha h\upsilon)^2$ versus $h\upsilon$, and (b) the schematic illustration of the energy diagram of pure ZnS and Mn-doped ZnS films. (Reprinted with permission from [36]. Copyright 2021: Elsevier).

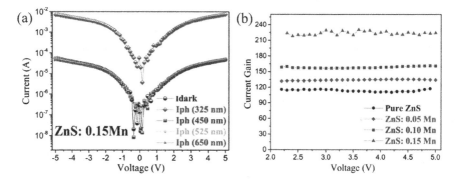

Fig. 14.5 The current-voltage curves of (**a**) ZnS: 0.15 Mn films-based UV detectors and (**b**) The photocurrent gain of Pure ZnS and Mn-doped ZnS films-based UV detectors. (Reprinted with permission from [36]. Copyright 2021: Elsevier)

The MSM UVA detectors based on the Mn-doped ZnS films were fabricated using two Schottky contacts of Au, as sketched in Fig. 14.5. Compared to the pure ZnS-based UV detectors, the as-fabricated devices revealed a high photocurrent gain of about 235.0 (Fig. 14.5b). The response and recovery times recorded the lowest values of about 9.9 ms and 13.2 ms, respectively, by which the results confirmed the performance of the fabricated visible-blind UV detectors based on Mn-doped ZnS films can be enhanced by introducing Mn^{2+} ions as dopants into the ZnS lattice.

As the results featured a strong dependence of the photodetection and sensing behavior on the Mn-content, they investigated structural, optical, and electrical properties of the composition-tunable ternary $Mn_xZn_{1-x}S$ thin films to use as UV detectors. At a high concentration of Mn^{2+}ions (>0.2 mol), a gradual decrease of the bandgap energy was found followed by an increase up to 4.05 eV. As Ebrahimi et al. [39] explained, when the Mn^{2+}ions are introduced into the ZnS lattice, they can occupy either substitutional or interstitial sites of Zn^{2+} ions, which caused the emerging of the Mn^{2+} impurity bands formed by the overlapped impurity states,

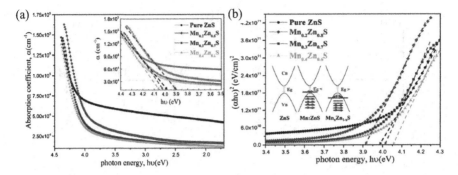

Fig. 14.6 (a) The variations of the optical absorption coefficient, and (b) $(\alpha h\upsilon)^2$ versus $h\upsilon$ for the pure ZnS and $Mn_xZn_{1-x}S$ thin films. (Reprinted with permission from [39]. Copyright 2021: Elsevier)

leading to a slight decrease of the ZnS bandgap energy. On the other hand, widening the bandgap energy at the higher content of Mn^{2+} ions is attributed to another phenomenological event named the Burstein-Moss shift theory, by which the blue shift of the bandgap energy is aroused by lying at the Fermi level (E_F) in the valence band for the ZnS structure, as depicted in Fig. 14.6. By increasing the Mn^{2+} content, the PL emission intensity located in the UV region was found to be enhanced which modifies the rate of electron-hole pair generation.

The performance of the UV detectors based on the $Mn_xZn_{1-x}S$ thin films showed an enhanced photoresponsivity and photoswitching behavior in the UVB region of the spectrum [39]. The current gain was improved over 4.5 times and the photoswitching behavior became faster over 5 times compared to the pure ZnS-based UV detectors (Fig. 14.7). The working mechanism of the Au/ $Mn_xZn_{1-x}S$ thin film/Au UV detectors can be described based on the energy band theory. When the Mn_xZn_{1-x} S alloys as a photo-sensing semiconductor connect to Au electrodes, due to a relatively lower work function of the Au contacts of about 5.31 eV than the $Mn_xZn_{1-x}S$ layer of about 5.43 eV, the electrons drift from the Au electrodes toward the Mn_xZn_{1-x} S thin films until their fermi levels are aligned. Under equilibrium conditions, a Helmholtz double-layer will be formed at the Au/$Mn_xZn_{1-x}S$ junction, where the Au is positively charged and $Mn_xZn_{1-x}S$ is negatively charged near their interface attributed to the electrostatic induction. Due to the electric field induced by the Helmholtz double-layer, a depletion layer is established by decreasing the free charge carrier concentration near the surface. Therefore, energy bands of $Mn_xZn_{1-x}S$ bend downward at the inter-face due to the electric field, which forms a Schottky barrier at the interface of Au and $Mn_xZn_{1-x}S$ thin film. When devices are illuminated by UV radiation with photon energy more than the optical bandgap of the samples, electron-hole pairs are generated. The generated carriers can be separated by a built-in potential or external electric field and then a photocurrent is produced. As well as the states caused by Mn impurities will function as effective trapping states. By trapping the minority charge carriers, the lifetime of the carriers is remarkably prolonged, which leads to enhancing the generated photocurrent which in turn decreases the SBHs

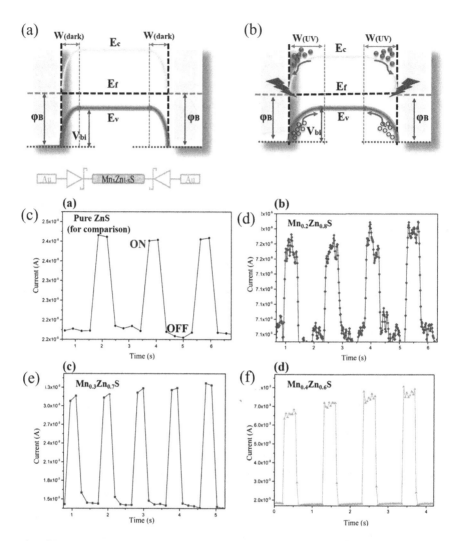

Fig. 14.7 The schematic energy band diagram of Au/Mn$_x$Zn$_{1-x}$S/Au based UV detector under (**a**) dark and (**b**) illumination conditions. The inset shows the circuitry of the back-to-back Schottky device. Time response of (**c**) pure ZnS, (**d**) Mn$_{0.2}$Zn$_{0.8}$S, (**e**) Mn$_{0.3}$Zn$_{0.7}$S and (**f**) Mn$_{0.4}$Zn$_{0.6}$S thin films-based UV detectors. (Reprinted with permission from [39]. Copyright 2021: Elsevier)

and depletion width across the junction. It can be concluded that the UV detector based on the Mn$_x$Zn$_{1-x}$S thin film is a high-speed performance device, which is much faster than many oxide and chalcogenide nanomaterial UV detectors. It can be attributed to the high surface-to-volume ratio, enhanced density of mobile charge carriers, optically active trap and defect states at the surface of the photo-detecting Mn$_x$Zn$_{1-x}$S layers, rearrangement and modification of the energy band structure as well as an efficient excitation transfer between the Mn^{2+} and ZnS energy levels, which drastically affects the photosensing and photoresponsivity of the devices. The results show that the visible-blind UV detector based on ternary $Mn_xZn_{1-x}S$ thin films is a promising candidate for future optoelectronic integration devices because it can offer remarkable characteristics for the high detection of the dangerous UV-B radiations.

Using different methods and ZnS-based compounds, designing simple and cost-effective self-powered UV detectors with high photoresponsivity, and photo-switching speed based on the ZnS nanostructures is still a noticeable challenge. An efficient and high-performance self-powered UV detector must satisfy 5S requirements, including high sensitivity, high spectral detectivity, high signal-to-dark current ratio, high stability, and high speed. Generally, the reported self-powered UV detectors suffer from complicated fabrication processes and high dark current. In this case, Ebrahimi et al. [37] aimed to study the self-powered UV detectors based on the ZnS thin films and their compounds through engineering the bandgap energy by doping and/or alloying processes. They developed a modified low-temperature solvothermal method using a capping agent of Ethylenediamine under short deposition time to synthesize the well-aligned $Sn_xZn_{1-x}S$ nanostructured thin films on the glass substrate for MSM UV detector applications. As depicted in Fig. 14.8, by incorporation of the Sn^{2+} ions into the ZnS lattice, a growth mechanism was

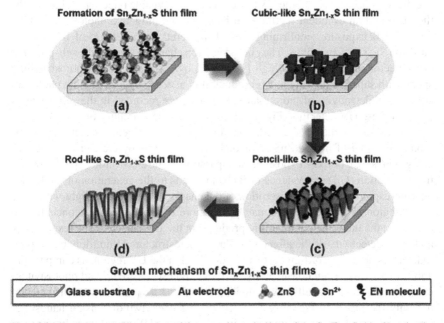

Fig. 14.8 The schematic illustration of the growth mechanism of the $Sn_xZn_{1-x}S$ thin films by the low-temperature solvothermal method; (**a**) the formation of the $Sn_xZn_{1-x}S$ thin film on the glass substrate in the presence of binary solvent of EN/water and (**b-d**) the formation of cubic-like, pencil-like and rod-like nanostructured $Sn_xZn_{1-x}S$ thin films developed by the variation of the Sn^{2+} doping content. (Reprinted with permission from [37]. Copyright 2021: Elsevier)

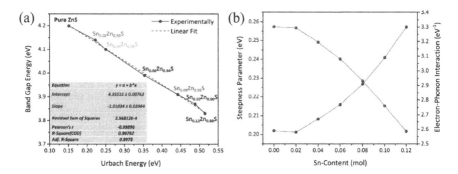

Fig. 14.9 (a) the relationship between the bandgap energy and Urbach energy, and (b) the variation of the steepness parameter and electron-phonon interaction of the pure ZnS and the $Sn_xZn_{1-x}S$ thin films. (Reprinted with permission from [37]. Copyright 2021: Elsevier).

proposed based on morphological evolution from aligned nanoflakes to cubic-like, pencil-like, and finally, rod-like nanostructures [37, 40, 41].

Considering the effect of the morphology on the optical properties, a linear relationship between the optical band gap energy and Urbach tails was found, detail shown in Fig. 14.9.

The fabricated UV detectors based on the $Sn_xZn_{1-x}S$ thin film junctions showed excellent photoresponse in the UVB region and high visible rejection. Additionally, they featured an excellent rectification behavior and attractive photovoltaic effect under UVB exposure, confirming the self-powered characteristic of the devices based on $Sn_xZn_{1-x}S$ thin films. The photodetection characteristics of the self-powered UV detectors demonstrated a high I_{on}/I_{off} ratio over 10^3 and a fast-photoswitching speed with superior stability and reproducibility at zero voltage. Notable photosensitivity and detectivity of more than ~20 times were recorded in the self-powered mode of these UV detectors (Fig. 14.10). The simplified circuitry of the UV detectors demonstrates that in the dark condition, the dark current emanates from the inbuilt potential of n-type ZnS semiconductor thin films by the cause of the free charge carriers. Upon UV exposure, the separation of the electron-hole pairs towards the conduction (CB) and valence (VB) bands leads to the generation of the photocurrent response due to the absorption of the energy of the incident light within the active region of the device. Ultimately, the more the charge carrier concentration increases, the more the photocurrent produces in the UV detector. The responsivity and photodetectivity of the aligned $Sn_xZn_{1-x}S$ nanostructured thin films-based photodetectors are the record-highest values obtained in the UV region, as compared in Table 14.1, which even considerably have the highest sensitivity and photoswitching performance at the self-power mode. The higher I_{on}/I_{off} ratio can be attributed to not only the larger surface-to-volume ratios and one−/two-dimension nanostructured $Sn_xZn_{1-x}S$ thin films, with which more absorbing UV and higher efficiency of electron-hole pair generation/separation are beneficial to produce larger photocurrent; but also Schottky junction formed between the thin films and Au contacts, which effectively hinders the transport of free charge careers and further gives rise to a lower dark current.

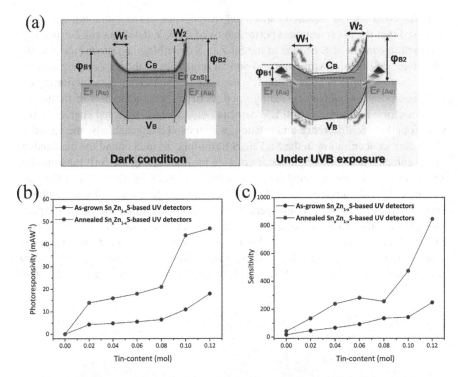

Fig. 14.10 (**a**) The schematic view of the n-type band energy diagram of the Schottky-type configuration, (**b**) photoresponsivity and (**c**) photosensitivity behavior vs. Sn^{2+} doping content for the as-grown and annealed $Sn_xZn_{1-x}S$ thin films-based MSM UV detectors at the self-power mode with zero bias voltage. (Reprinted with permission from [37]. Copyright 2021: Elsevier)

Table 14.1 Comparison of the important figures-of-merit of recently fabricated ZnS-based photodetectors

Photodetector	Bias (V)	Wavelength (nm)	Responsivity (AW^{-1})	Dark current (A)	Photo current (A)	I_{on}/I_{off}	References
ZnS nanowires	10	325/442	1.86	3×10^{-12}	24×10^{-12}	–	[42]
Sb-doped ZnS	1	254	–	$\sim 10^{-9}$	$\sim 10^{-7}$	$>10^2$	[43]
N-doped ZnS	5	254	1.4×10^5	–	–	5	[44]
Mn-doped ZnS	5	325	–	$\sim 10^{-8}$	$\sim 10^{-4}$	235	[36]
Sn-doped ZnS	0	280	$>10^{-2}$	$<9 \times 10^{-7}$	$>1 \times 10^{-3}$	$>10^3$	[37]
$Mg_xZn_{1-x}S$	5	400	1.523×10^{-6}	0.89×10^{-6}	1.52×10^{-6}	96.74	[45]
$Cd_xZn_{1-x}S$	0	325	130	$\sim 10^{-4}$	$\sim 10^{-2}$	–	[46]
$Mn_xZn_{1-x}S$	5	280	–	$\sim 10^{-6}$	$\sim 10^{-2}$	579.9	[39]
$Sn_xZn_{1-x}S$	0	280	212×10^{-3}	4.75×10^{-6}	6.98×10^{-3}	2406.1	[41]

In the other similar research carried out by the same group, Ebrahimi et al. [41] proposed that the photoresponse performance of the UV detectors can be tuned by varying the content of Sn^{2+} ions in the $Sn_xZn_{1-x}S$ absorbing layer, attributing to the advantages of the microstructural and optical properties of the $Sn_xZn_{1-x}S$ thin films. The Schottky barrier heights of the UV detectors were decreased by increasing the Sn^{2+} incorporation content, revealing good rectifying characteristics of the n-type UV detectors. Further, for each UV detector, superior self-powered characteristics were found in the difference and relation, which can be adjustable by the variation of the Sn^{2+} concentration in the $Sn_xZn_{1-x}S$ thin films. To understand the mechanism of the enhancement of UV photocurrent gain by introducing the Sn^{2+} ions into the ZnS network, the energy-band diagram in both junctions is illustrated in the dark and under UV illumination. From Fig. 14.11a, the orange arrows exhibit the photo-current in the $Sn_xZn_{1-x}S$ thin films generated by the motion of the electron-hole pairs under UV illumination. In the dark condition, the electrons flowing into the n-type $Sn_xZn_{1-x}S$ barrier layer accelerate due to the built-in electric field and/or applied external bias, resulting in a high photocurrent. Thus, the Au/$Sn_xZn_{1-x}S$ thin film/ Au devices exhibit a good rectification behavior in the dark current. On the other

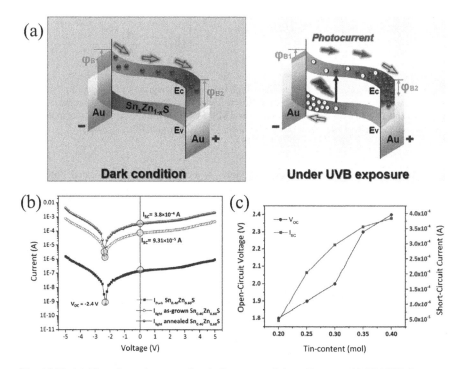

Fig. 14.11 (a) The schematic energy-band alignments of the self-powered MSM UV detectors based on ternary $Sn_xZn_{1-x}S$ alloy thin films in the dark and under UV illumination conditions. I–V curve of the self-powered UV detector based on the ternary (b) $Sn_{0.40}Zn_{0.60}S$ alloy thin films measured under the dark and UV illumination, and (c) the plot of VOC and ISC vs. Sn^{2+} content. (Reprinted with permission from [41]. Copyright 2021: Elsevier)

hand, upon UVB exposure, the electron-hole pairs are produced in the $Sn_xZn_{1-x}S$ nanostructures, in which the photogenerated holes migrate to the surface and discharge the negatively charged ions adsorbed at the surface of the photoactive layer. Thus, the high photoresponsivity and sensitivity are strongly dependent on effective photocurrent transport through the active $Sn_xZn_{1-x}S$ layer, and the introduction of the Sn^{2+} as donor states, resulting in the tunable self-powered $Sn_xZn_{1-x}S$ thin films-based UV detectors with tremendous photoresponsivity and photodetection performance. These UV detectors featured the highest photosensitivity of about 2406.1 (I_{on}/I_{off} ratio $\approx 8.47 \times 10^3$) and ultrafast photoresponse rise and decay times of 1.9 and 2.6 ms, respectively at the self-powered mode, which are the best results reported in the ZnS-based self-powered UV detectors until now. A clear comparison from Table 14.1 represents the important figures-of-merit of recently produced self-powered ZnS-based photodetectors. Generally, the data reveals the superiority of the self-powered UV detector based on the $Sn_xZn_{1-x}S$ nanostructured thin films with different Sn^{2+} content devices used in photodetection applications. It can be claimed that both the optical and electrical properties of the samples suggest the high photoresponsivity and excellent stability of the self-powered ultraviolet detectors.

References

1. Jiang D, Cao L, Su G, Liu W, Qu H, Sun Y et al (2009) Synthesis and luminescence properties of ZnS: Mn/ZnS core/shell nanorod structures. J Mater Sci 44(11):2792–2795
2. Mall M, Kumar L (2010) Optical studies of Cd^{2+} and Mn^{2+} Co-doped ZnS nanocrystals. J Lumin 130(4):660–665
3. Al-Rasoul KT, Abbas NK, Shanan ZJ (2013) Structural and optical characterization of Cu and Ni doped ZnS nanoparticles. Int J Electrochem Sci 8(4):5594–5604
4. Cao J, Yang J, Zhang Y, Wang Y, Yang L, Wang D et al (2010) XAFS analysis and luminescent properties of ZnS: Mn^{2+} nanoparticles and nanorods with cubic and hexagonal structure. Opt Mater 32(5):643–647
5. Salem JK, Hammad TM, Kuhn S, Draaz MA, Hejazy NK, Hempelmann R (2014) Structural and optical properties of Co-doped ZnS nanoparticles synthesized by a capping agent. J Mater Sci Mater Electron 25(5):2177–2182
6. Salem JK, Hammad TM, Kuhn S, Nahal I, Draaz MA, Hejazy NK et al (2014) Luminescence properties of Mn and Ni doped ZnS nanoparticles synthesized by capping agent. J Mater Sci Mater Electron 25(12):5188–5194
7. Zhang L, Qin D, Yang G, Zhang Q (2012) The investigation on synthesis and optical properties of ZnS: Co nanocrystals by using hydrothermal method. Chalcogenide Lett 9(3):93–98
8. Chen Y-C, Wang C-H, Lin H-Y, Li B-H, Chen W-T, Liu C-P (2010) Growth of Ga-doped ZnS nanowires constructed by self-assembled hexagonal platelets with excellent photocatalytic properties. Nanotechnology 21(45):455604
9. Dong L, Liu Y, Zhuo Y, Chu Y (2010) General route to the fabrication of ZnS and M-doped (M= Cd^{2+}, Mn^{2+}, Co^{2+}, Ni^{2+}, and Eu^{3+}) ZnS Nanoclews and a study of their properties. Wiley Online Library
10. Kang T, Sung J, Shim W, Moon H, Cho J, Jo Y et al (2009) Synthesis and magnetic properties of single-crystalline Mn/Fe-doped and Co-doped ZnS nanowires and nanobelts. J Phys Chem C 113(14):5352–5357

11. Zhu G, Zhang S, Xu Z, Ma J, Shen X (2011) Ultrathin ZnS single crystal nanowires: controlled synthesis and room-temperature ferromagnetism properties. J Am Chem Soc 133(39):15605–15612
12. Tian Y, Zhao Y, Tang H, Zhou W, Wang L, Zhang J (2015) Synthesis of ZnS ultrathin nanowires and photoluminescence with Mn2+ doping. Mater Lett 148:151–154
13. Shanmugam N, Cholan S, Kannadasan N, Sathishkumar K, Viruthagiri G (2014) Effect of polyvinylpyrrolidone as capping agent on Ce^{3+} doped flowerlike ZnS nanostructure. Solid State Sci 28:55–60
14. Cao J, Han D, Wang B, Fan L, Fu H, Wei M et al (2013) Low temperature synthesis, photoluminescence, magnetic properties of the transition metal doped wurtzite ZnS nanowires. J Solid State Chem 200:317–322
15. Karan NS, Sarkar S, Sarma D, Kundu P, Ravishankar N, Pradhan N (2011) Thermally controlled cyclic insertion/ejection of dopant ions and reversible zinc blende/wurtzite phase changes in ZnS nanostructures. J Am Chem Soc 133(6):1666–1669
16. Zhai T, Li L, Ma Y, Liao M, Wang X, Fang X et al (2011) One-dimensional inorganic nanostructures: synthesis, field-emission and photodetection. Chem Soc Rev 40(5):2986–3004
17. Azimi H, Ghoranneviss M, Elahi S, Yousefi R (2016) Photovoltaic and UV detector applications of ZnS/rGO nanocomposites synthesized by a green method. Ceram Int 42(12):14094–14099
18. Kumar V, Rawal I, Kumar V, Goyal PK (2019) Efficient UV photodetectors based on ni-doped ZnS nanoparticles prepared by facial chemical reduction method. Phys B Condens Matter 575:411690
19. Prasad N, Balasubramanian K (2018) Effect of morphology on optical and efficiently enhanced electrical properties of W-ZnS for UV sensor applications. J Appl Phys 124(4):045702
20. Sun Y-L, Xie D, Sun M-X, Teng C-J, Qian L, Chen R-S et al (2018) Hybrid graphene/cadmium-free ZnSe/ZnS quantum dots phototransistors for UV detection. Sci Rep 8(1):1–8
21. Song K-K, Lee S (2001) Highly luminescent (ZnSe) ZnS core-shell quantum dots for blue to UV emission: synthesis and characterization. Curr Appl Phys 1(2–3):169–173
22. Kim Y, Kim SJ, Cho S-P, Hong BH, Jang D-J (2015) High-performance ultraviolet photodetectors based on solution-grown ZnS nanobelts sandwiched between graphene layers. Sci Rep 5(1):1–8
23. Fang X, Zhai T, Gautam UK, Li L, Wu L, Bando Y et al (2011) ZnS nanostructures: from synthesis to applications. Prog Mater Sci 56(2):175–287
24. Tiwari A, Dhoble S (2017) Critical analysis of phase evolution, morphological control, growth mechanism and photophysical applications of ZnS nanostructures (zero-dimensional to three-dimensional): a review. Cryst Growth Des 17(1):381–407
25. Nemade K, Waghuley S (2016) Ultra-violet C absorption and LPG sensing study of zinc sulphide nanoparticles deposited by a flame-assisted spray pyrolysis method. J Taibah Univ Sci 10(3):437–441
26. Hajimazdarani M, Naderi N, Yarmand B (2019) Effect of temperature-dependent phase transformation on UV detection properties of zinc sulfide nanocrystals. Mater Res Express 6(8):085096
27. Prasad N, Karthikeyan B (2019) Phase-dependent structural, optical, phonon and UV sensing properties of ZnS nanoparticles. Nanotechnology 30(48):485702
28. Chatterjee A (2016) Performance analysis and optimization of MSM photodetectors using group-IV semiconductors. Project report, Kalyani Government Engineering College, Kalyani, India. https://www.researchgate.net/publication/304347238
29. Ghosh P, Ahmed SF, Jana S, Chattopadhyay K (2007) Photoluminescence and field emission properties of ZnS: Mn nanoparticles synthesized by rf-magnetron sputtering technique. Opt Mater 29(12):1584–1590
30. Tang W, Cameron D (1996) Electroluminescent zinc sulphide devices produced by sol-gel processing. Thin Solid Films 280(1–2):221–226
31. Lee EY, Tran NH, Russell JJ, Lamb RN (2003) Structure evolution in chemical vapor-deposited ZnS films. J Phys Chem B 107(22):5208–5211

32. Schön S, Chaichimansour M, Park W, Yang T, Wagner B, Summers C (1997) Homogeneous and δ-doped ZnS: Mn grown by MBE. J Cryst Growth 175:598–602
33. Torimoto T, Obayashi A, Kuwabata S, Yasuda H, Mori H, Yoneyama H (2000) Preparation of size-quantized ZnS thin films using electrochemical atomic layer epitaxy and their photoelectrochemical properties. Langmuir 16(13):5820–5824
34. Hennayaka H, Lee HS (2013) Structural and optical properties of ZnS thin film grown by pulsed electrodeposition. Thin Solid Films 548:86–90
35. Goudarzi A, Aval GM, Park SS, Choi M-C, Sahraei R, Ullah MH et al (2009) Low-temperature growth of nanocrystalline Mn-doped ZnS thin films prepared by chemical bath deposition and optical properties. Chem Mater 21(12):2375–2385
36. Ebrahimi S, Yarmand B, Naderi N (2019) Enhanced optoelectrical properties of Mn-doped ZnS films deposited by spray pyrolysis for ultraviolet detection applications. Thin Solid Films 676:31–41
37. Ebrahimi S, Yarmand B (2020) Solvothermal growth of aligned $Sn_xZn_{1-x}S$ thin films for tunable and highly response self-powered UV detectors. J Alloys Compd 827:154246
38. Ebrahimi S, Yarmand B, Naderi N (2017) Effect of the sulfur concentration on the optical band gap energy and Urbach Tail of spray-deposited ZnS films. Advanced Ceramics Progress 3(4):6–12
39. Ebrahimi S, Yarmand B, Naderi N (2020) High-performance UV-B detectors based on $MnxZn1-xS$ thin films modified by bandgap engineering. Sensors Actuators A Phys 303:111832
40. Ebrahimi S, Yarmand B (2019) Morphology engineering and growth mechanism of ZnS nanostructures synthesized by solvothermal process. J Nanopart Res 21(12):1–12
41. Ebrahimi S, Yarmand B (2021) Tunable and high-performance self-powered ultraviolet detectors using leaf-like nanostructural arrays in ternary tin zinc sulfide system. Microelectron J 116:105237
42. Liang Y, Liang H, Xiao X, Hark S (2012) The epitaxial growth of ZnS nanowire arrays and their applications in UV-light detection. J Mater Chem 22(3):1199–1205
43. Peng Q, Jie J, Xie C, Wang L, Zhang X, Wu D et al (2011) Nano-Schottky barrier diodes based on Sb-doped ZnS nanoribbons with controlled p-type conductivity. Appl Phys Lett 98(12):123117
44. Zhou Y, Chen G, Yu Y, Feng Y, Zheng Y, He F et al (2015) An efficient method to enhance the stability of sulphide semiconductor photocatalysts: a case study of N-doped ZnS. Phys Chem Chem Phys 17(3):1870–1876
45. Dive AS, Huse NP, Gattu KP, Sharma R (2017) Soft chemical growth of $Zn_{0.8}Mg_{0.2}S$ one dimensional nanorod thin films for efficient visible light photosensor. Sensors Actuators A Phys 266:36–45
46. El-Shazly O, Farag A, Rafea MA, Roushdy N, El-Wahidy E (2016) Light scattering and photosensitivity characteristics of nanocrystalline $Zn_{1-x}Cd_xS$ ($0 \leq x \leq 0.9$) films for photosensor diode application. Sensors Actuators A Phys 239:220–227

Chapter 15
Photodetectors Based on II-VI Multicomponent Alloys

Ghenadii Korotcenkov and Tetyana Semikina

15.1 Introduction

As was shown earlier in the (Chap. 2, Vol. 1), compounds II-VI have a specific set of parameters, which allows them to be widely used in the development of various devices (read Chaps. 1 and 2, Vol. 1). In particular, the band gap of these compounds can vary from -0.3 eV (HgTe) to 3.54 eV (ZnS). This means that photodetectors based on these compounds can cover the entire spectral region from infrared (IR) to ultraviolet (UV). However, in some cases it turns out that the band gap of binary compounds does not correspond to the optimal value required to achieve maximum efficiency.

The same situation arises when trying to develop monolithic optoelectronic devices in which it would be possible to combine the existing achievements of the well-established technology of III-V compounds and the properties of II-VI compounds, in other words, to develop technologies that allow fabrication on the same substrate both the III-V-based and II-VI-based high-quality optoelectronic devices. When using epitaxial methods for growing semiconductor compounds, the main obstacle to this is the lack of matching of the crystal lattice parameters of III-V and II-VI binary compounds.

However, it turned out that the above problems can be successfully solved. II-VI semiconductors can form a continuous series of solid solutions, which creates conditions for a smooth change in both the band gap and the crystal lattice parameter in a certain range of values determined by the materials used. It is these prospects that

G. Korotcenkov (✉)
Department of Physics and Engineering, Moldova State University,
Chisinau, Republic of Moldova
e-mail: ghkoro@yahoo.com

T. Semikina
Lashkarev Institute of Semiconductor Physics, National Academy of Science of Ukraine,
Kiev, Ukraine

© The Author(s), under exclusive license to Springer Nature Switzerland AG 2023
G. Korotcenkov (ed.), *Handbook of II-VI Semiconductor-Based Sensors and Radiation Detectors*, https://doi.org/10.1007/978-3-031-20510-1_15

are the main stimulus for the development of technology for the synthesis and study of multicomponent semiconductors based on II-VI compounds. Optimization of the electrophysical and physical properties of II-VI compounds, or giving them new properties not characteristic of binary compounds through the introduction of additional components, is another direction in the study of multicomponent II-VI compounds.

15.2 Photodetectors with Controlled Spectral Response

15.2.1 Solar Cells

The efficiency of any photodetector is determined by the consistency of the region of maximum sensitivity with the radiation wavelength. If this agreement does not exist, then high efficiency cannot be expected. This feature of photodetectors is most clearly manifested in solar cells. If the spectral distribution of the intensity of solar radiation is not taken into account, then the efficiency of using solar energy incident on the Earth will be insignificant. For example, the band gap of silicon ($E_g \sim 1.1$ eV, $\lambda = 1127$ nm), the material most widely used in solar cells, is not optimal for these applications. Most of the sunlight is not involved in the generation of photocarriers. In addition, due to the Si indirect band gap and the related low absorption coefficient, a thickness on the order of 100 μm are required for adequately absorb sunlight. Therefore, numerous studies are currently being carried out aimed at finding materials and configurations of solar cells that contribute to the most complete use of the energy of solar radiation (Fig. 15.1).

In particular, it has been shown that a II-VI compound such as CdTe with a band gap of 1.45–1.5 eV ($\lambda \sim 840$ nm) has more optimal properties for these applications

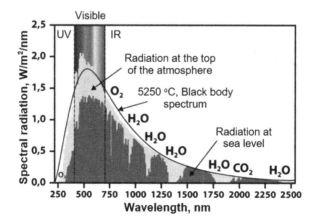

Fig. 15.1 Spectrum of solar radiation

[26]. CdTe has become a strong candidate for photovoltaic applications due to its optimum bandgap, high absorption coefficient, and ease of deposition. Currently the development of CdTe-based solar cells is one of the most promising thin film technologies that have already reached the commercial stage. At the research and development scale, solar submodules above 18% have been reported for this technology, while commercially available modules show overall area efficiencies in the range of 14–16%, depending on the manufacturer and the technology. But even the use of this material cannot solve the problem of the full use of solar energy.

One of the approaches being developed to solve these problems is the development of tandem solar cells, consisting of two cells of different bandgap semiconductors on the top of each other (1.7–1.8 eV ($\lambda \sim 708$ nm) on 1.1 eV). The Fig. 15.2 shows how effective the use of tandem solar cells can be to improve the efficiency of solar radiation conversion. Semiconducting alloys such as $Cd_{1-x}Zn_xTe$ and $Cd_{1-x}Mn_xTe$ are good candidates for the top cell since their bandgaps can be tailored between 1.45 eV (CdTe) and 2.26 eV (ZnTe) or 2.85 eV (MnTe) by adjusting the film composition. Studies have shown that these solid solutions have the required band gap at x = 0.42 and x = 0.16 for $Cd_{1-x}Zn_xTe$ and $Cd_{1-x}Mn_xTe$, respectively. However, it turned out that $Cd_{1-x}Zn_xTe$ films have a more uniform structure and a sharper interface, which ensured their predominant use in the development of various types of solar cells [6, 58]. $Cd_{1-x}Zn_xTe$ thin films have been deposited by several different techniques, and depending on the growing parameters used, whether epitaxial [54] or polycrystalline thin films with different grain sizes, they lead to variations of their optical and electrical properties [5]. This ternary system has been shown to be one of the semiconductor materials allowing tandem solar cell to absorb most of the energy in the solar spectrum.

Research has also shown that the CZT film can be more heavily doped with Cu than CdTe. This indicates that the Fermi level is closer to the valence band edge in CZT than in CdTe. The use of CZT:Cu/Te as the back buffer layers (BBLs) between CdTe and the back electrode, results in improved device performance by reducing recombination at the back of the device. Temperature dependent current-voltage

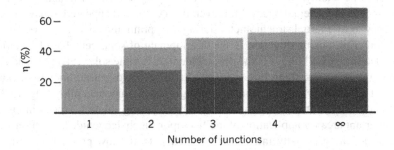

Fig. 15.2 The theoretical efficiency of solar cells can be improved by sequentially harvesting the Sun's constituent spectral components in tandem, using multi-junction cells that reduce losses associated with intraband relaxation. (Reprinted with permission from Ref. [18]. Copyright 2012: Springer Nature)

analysis shows that the valence band edge of $Cd_{0.5}Zn_{0.5}Te$ is close to the CdTe edge. Taken together, these results indicate that it should be possible to develop a high performance back contact with the desired Fermi level alignment at the back of the device [26] and upward band bending at the back of the device.

However, some developers believe that ternary $Cd_{1-x}Mn_xTe$ and $Cd_{1-x}Mg_xTe$ alloys still have some prospects for use in CdTe-based tandem solar cells [30]. To reach $E_g = 1.7$–1.8 eV, the Mn and Mg alloys require only about 15% alloying in contrast to more than 40% for Zn in $Cd_{1-x}Zn_xTe$. These two alloys also have smaller lattice contraction with x-value than does Zn. These two advantages are balanced somewhat by the greater sensitivity of the alloys of Mn and Mg to oxygen and water vapor. Therefore, the realization of the prospects of these alloys will depend on whether it is possible to reduce the change in their properties under the influence of the environment.

Another II-VI alloy promising for use in solar cells is $Cd_{1-x}Zn_xS$. Ternary cadmium-zinc sulfide combined with CdS is a good candidate for a wide bandgap window material for conventional CdTe/CdS cell. Its band gap can be tuned from 2.42 eV (CdS) to 3.6 eV (ZnS). Of course, ZnS can play the role of a wide band gap window, but the required efficiency of a solar cell can only be achieved if ZnS is heavily doped, which is difficult. At the same time, $Cd_{1-x}Zn_xS$, unlike ZnS, can have a much lower resistance. That is why the use of $Cd_{1-x}Zn_xS$ improves the spectral response of solar cells to wavelength without compromising the transport properties and shunt resistance of the cell. However, the use of a CdZnS film in a CdTe-based device is associated with some problems. First, as the band gap changes, the lattice constant also changes linearly. Second, as the Zn content in the compound increases, a significant increase in electrical resistivity is observed. Electrical resistivity increases from <1 Ω·cm to $>10^{10}$ Ω·cm as x increases from 0 to 1.0 [57]. This means that we have restrictions on the band gap engineering during forming transparency window of the solar cell.

Previously, we considered II-VI solid solutions as a material for wide-gap transparency windows or top cells in tandem solar cells. However, it turned out that some ternary alloys can also be used in bottom cells. Such material is $Hg_{1-x}Cd_xTe$ alloy [34, 51]. These alloys at x ~ 80% have a band gap Eg ≈ 1.1 eV, which is optimal for bottom cells in tandem devices. Moreover, this compound appeared to have a slight tendency to lose initial stoichiometry during the optimizing annealing process.

Another promising direction in the development of solar cells based on ternary and quaternary II-VI alloys is the development of flexible solar cells [21] and solar cells using as an absorbing layer a material with a bandgap that varies in thickness [31], which can be easily implemented using a multicomponent alloy (Fig. 15.3).

The development of cheap thin film solar cells using solution processing is also an important area for application of multicomponent chalcogenides. Solution based approaches are especially interesting for their potential low production costs and their easy scalability. It was found that for these purposes it is very promising Cu_2ZnSnS_4, a quaternary p-type semiconductor which presents great absorption coefficient ($>10^4$ cm^{-1}) and have tunable band gap (1.0–1.5 eV) (see Fig. 15.4) by varying the S/Se ratio [48]. The main advantage of this alloy is that it is formed by

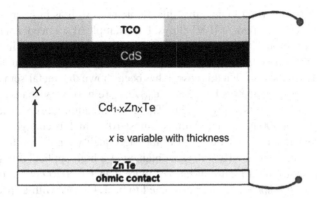

Fig. 15.3 Proposed structure for new solar cells based on CdZnTe with variable Zn content for which higher efficiencies than conventional CdTe solar cells are expected. (Adapted with permission from Ref. [31]. Copyright 2011: Elsevier)

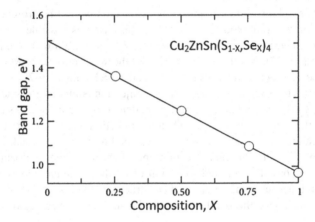

Fig. 15.4 The calculated band gap of $Cu_2ZnSn(S_{1-x}Se_x)_4$ at different composition (x). (Adapted with permission from Ref. [8]. Copyright 2011: American Physical Society)

abundant and nontoxic elements. Record solar cell efficiency for this material is 12.6% at the research scale denoting a high improvement potential [52].

15.2.2 Detectors for Visible Range

As we can see, solar energy converters require devices that are sensitive to the entire spectrum of solar radiation, which means that devices must respond to radiation in the entire spectral range from IR to UV (see Fig. 15.1). In the case of photodetectors, the requirements for devices may change radically, since in many cases it is not broadband photodetectors that are required, but devices characterized by spectral selectivity. It is important to note that this problem, as well as in solar cells, can be solved using multicomponent II-VI alloys. For example, if we need a photodetector for the visible region of the spectrum that is not sensitive to radiation in the near-IR region, then according to Ren et al. [36], photodetectors based on a single crystal of $Cd_{0.96}Zn_{0.04}Te$ can be useful in this case. It has been shown that metal-semiconductor-metal (MSM) photodetectors based on $Cd_{0.96}Zn_{0.04}Te$ have very sharp edges in the region of 900–800 nm (see Fig. 15.5a). The discrimination ratio between the near infrared region of 800 nm and 900 nm is almost 10^2, which is enough for the accurate spectra selectivity. Benefitting from the high-quality single crystallization, an ultra-low dark current of $\sim 10^{-10}$ A was obtained at a high applied voltage of 10 V, leading to a photo-to-dark-current ratio of more than 10^3 at 700 nm light illumination. The highest responsivity was estimated to be 1.43 A/W with a specific detectivity of 3.3×10^{12} Jones at -10 V at a relatively lower injection power density. In addition, the MSM photodetector also exhibited a fast response speed of ~ 800 μs (Fig. 15.5b), and extremely low persistent photoconductivity (PPC), while the PPC is inhibited at high temperatures.

If, however, solid solutions are used in the development of heterojunction, n-n+, or p-n junction-based photodetectors, then in this case it is possible to manufacture selective photodetectors for the visible region of the spectrum with a tunable sensitivity region [14, 27]. For the visible region of the spectrum, the most optimal and most studied material is CdSSe [30, 38] and CdSTe alloys [47]. For example, Fig. 15.6 shows the spectral characteristics of injection photodetectors developed on the basis of CdS_YSe_{1-Y} alloy [27]. The photodetectors had an n–n + structure and operated in the forward bias mode, when the injection amplification of the photocurrent is realized. Such photodiodes are designed to work with weak light fluxes. The mechanism of the internal injection amplification of the photocurrent under forward bias is based on the redistribution of the bias voltage between the high-resistance and low-resistance regions of the structure. The reason for this redistribution is the change in the conductivity of the high-resistance region when it is

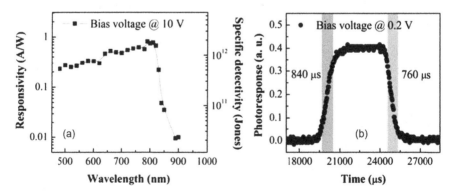

Fig. 15.5 (a) Spectral photoresponse/specific detectivity of fabricated $Cd_{0.96}Zn_{0.04}Te$ photodetector; (b) transient response time measurement at a bias voltage of 0.2 V with a chopping frequency of 100 Hz. (Reprinted from Ref. [36]. Published 2019 by Optica Publishing Group as open access)

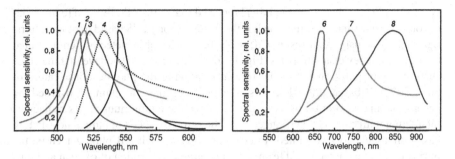

Fig. 15.6 Spectral characteristics of injection photodiodes based on CdS_YSe_{1-Y}: 1– y = 0.96, U_b = 2 V; 2 – y = 0.95, U_b = 1 V; 3 – y = 0.94, U_b = 0.5 V; 4 – y = 0.9, U_b = 4 V; 5 – y = 0.85, U_b = 0.5 V; 6 – y = 0.55, U_b = 4 V; 7 – y = 0.3, U_b = 4 V; 8 – y = 0.06, U_b = 6 V. (Data extracted from Ref. [27])

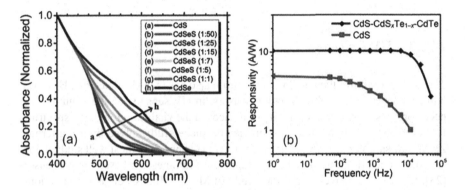

Fig. 15.7 (a) UV – visible absorption spectra of CdSeS NWs with varied Se:S ratio. Reprinted with permission from [22]. Copyright 2014: ACS; (b) Responsivity of CdS and CdS-CdS$_x$Te$_{1-x}$-CdTe-based photodetectors relative to modulation frequency under incident light density of 15.4 mW/cm^2 with a bias voltage of 5 V. (Reprinted with permission from Ref. [47]. Copyright 2018: Elsevier)

illuminated with light with a wavelength λ close to the absorption edge of the single crystal used, and the appearance of a positive current feedback in connection with this. This mechanism is well described in [50]. As can be seen, the positions of the photosensitivity maxima of the structures were in the wavelength range of 520–850 nm. The spectral sensitivity band at a level of 0.5 for most detectors was 15–30 nm, which indicates the high selectivity of the developed photodetectors.

Besides traditional photodetectors [38], CdSSe and CdSTe alloys are being used to develop nanowire- [22], and core-shell nanobelt-based photodetectors [47]. The steady-state absorption spectra of the CdSeS NWs with the absorption of the NWs normalized at 400 nm are shown in Fig. 15.7a. By varying the Se:S precursor ratio we were able to tune the optical band gap of the CdSeS NWs from 2.36 to 1.79 eV. At the same time, the measured response spectrum of CdS-CdSxTe$_{1-x}$-CdTe core-shell nanobelt photodetector covers a wide range from 355 nm (3.4 eV) to 785 nm

(1.58 eV) [47]. At that, Tang et al. [47] reported that CdS-CdS$_x$Te$_{1-x}$-CdTe photodetector had a rise time of 11 μs, which is the fastest among CdS based photodetectors, reported previously. As a result, CdS-CdS$_x$Te$_{1-x}$-CdTe photodetector demonstrated the best frequency characteristics (Fig. 15.7b). Tang et al. [47] believe that the core-shell structure plays an influential role in improving the photoresponse rate. For pure CdS NB-based photodetectors, the photoresponse is mainly governed by desorption and adsorption of oxygen [16]. Since the equilibrium of gas molecules interacting with carriers takes longer time, the response speed of pure nanostructure based photodetectors is limited. In contrast, in core-shell NB photodetectors, the intrinsic positive charge in CdTe shell will avoid the absorption of oxygen and thus operation of this kind NB photodetectors mainly depends on the built-in potential at the interface, which has a much faster carriers-separating speed. Because of the abovementioned mechanism, the responsivity and response speed of the CdS-CdS$_x$Te$_{1-x}$-CdTe core-shell NB photodetectors have been extraordinarily optimized.

15.2.3 UV Detectors

The need to optimize the absorption spectrum also exists in UV detectors. However, unlike solar cells or visible radiation sensors, in UV sensors, on the contrary, it is necessary to suppress the sensitivity of devices in the visible region of the spectrum. Research has shown that this problem can be successfully solved by using ternary Zn$_{1-x}$Mn$_x$S samples [13, 45]. Manganese (Mn^{2+}) is one of the most well studied dopant compounds II-VI, used in the development of luminescent-based optical sensors [23]. Due to the relatively close ionic radii of Mn and Zn, as well as the same ionic charge, Mn can fundamentally be introduced in the form of Mn^{2+} ions into various sites of the ZnS host lattice and change its structural and optoelectric properties. Ebrahimi et al. [13] found that all the deposited Zn$_{1-x}$Mn$_x$S showed a stable Zinc-Blende structure. The nanocrystalline, uniform and smooth surface of the alloys led to increasing the optical transmittance over 95% at the high-content of Mn^{2+} ions (see Fig. 15.8a). At a high concentration of Mn^{2+} ions (>0.2 mol), the values of bandgap energy increased dramatically from 3.91 eV to 4.05 eV, while the Urbach energy decreased considerably from 361 meV to 225 meV. Therefore, the bandgap energy in the absence of defects and band tailing and at the rearrangement of electronic localized states was obtained of about 4.28 eV. For comparison the E$_g$ of ZnS equals 3.54 eV.

The performances of UV detectors based on Zn$_{1-x}$Mn$_x$S thin films also showed a huge improvement in photosensitivity and photoswitching compared to ZnS devices. For example, the current gain of the detectors was improved from ~120 for pure ZnS to ~580 for the Mn$_{0.4}$Zn$_{0.6}$S sample. The photoresponse rate of UV detectors has also increased significantly. For Mn$_{0.4}$Zn$_{0.6}$S detectors, rise time (τ_r) and fall time (τ_d) decreased from 28.1 and 34.3 ms observed for ZnS devices to 5.3 and 6.1 ms, respectively. An important advantage of Zn$_{1-x}$Mn$_x$S-based detectors is also the expansion of the spectral region where these devices can be effectively used as

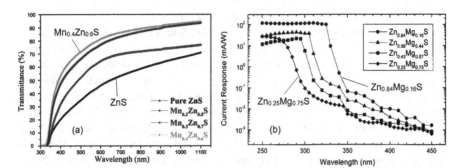

Fig. 15.8 (a) Optical transmittance of the pure ZnS and $Mn_xZn_{1-x}S$ thin films. Adapted with permission from [13]. Copyright 2020: Elsevier. (b) Photoresponse of four $Zn_{1-x}Mg_xS$ Schottky barrier photodiodes with long-wavelength cutoffs at 325, 305, 295, and 270 nm, respectively. (Adapted with permission from Ref. [45]. Copyright 2001: AIP Publishing)

selective UV detectors [45]. For example, with $Zn_{0.25}Mg_{0.75}S$ compositions, it is possible to manufacture a detector that will not only be insensitive to visible radiation, but also to UV radiation from the UV-A (320–400 nm) and UV-B (280–320 nm) regions (see Fig. 15.8b). For the device with the highest Mg composition of 75% and with a thickness of the active layer as thin as 30 nm, the external quantum efficiency still can be maintained around 15% in the plateau region. In particular, this device has a cutoff wavelength at 270 nm and offers a fall of almost three orders in responsivity at 300 nm relative to its peak, indicating very good solar–blind characteristics. The ability to manufacture sensors for different spectral ranges can be useful for pollution detection, medical applications, and as special purpose detectors. For example, such sensors can be used for development of flame sensing system. This system requires UV detectors that are insensitive to radiation above 280 nm [25].

Averin et al. [3] have shown that ZnCdS and ZnMgS can also be used in the development of metal-semiconductor-metal (MSM) photodetectors based on periodic heterostructures with ZnCdS quantum wells separated by ZnMgS and ZnS barrier layers. Heterostructures were grown on semi-insulating GaP substrates by MOVPE. It was established that these detectors exhibited very low dark currents and electrically tunable spectral response. At low bias detectors provide narrowband UV-response determined by a composition of ZnCdS quantum well. A shift of the peak position of the narrowband detector response to longer wavelengths is observed with increasing Cd content. A higher operating bias shifts the maximum sensitivity of the detector in the visible part of spectrum due to the penetration of external electric field down to the semi-insulating GaP substrate while a narrowband UV-response remains. Averin et al. [3] believe that developed highly selective two-color photodetector allows discriminating the optical channels and increasing dynamic range and noise immunity of optical informational and measuring systems.

Other ternary compounds, such as ZnSSe [44], ZnSeTe, ZnSTe [28, 44], CdMgTe [30], and CdS_xTe_{1-x} [47] have also been tried to control the spectral response region of UV detectors. However, the use of ternary compounds instead of binary ones did not always give the expected results. In particular, Sou et al. [44] studied ZnSSe and ZnSTe-based Schottky barrier UV detectors and concluded that of these compounds, only ZnSSe is a good candidate for an active material for UV detection applications that require excellent visible rejection and tunable turn-on wavelength capabilities. As seen in Fig. 15.9b, ZnSSe diodes have a sharper long wavelength limit compared to ZnSTe diodes. This is due to the fact that, unlike ZnSTe, ZnSSe does not contain isoelectronic traps responsible for the appearance of a photoresponse in the long-wavelength region of the spectrum in ZnSTe-based detectors (see Fig. 15.9a). In addition, good activation of dopant donor impurities up to 50% Se in the alloy is achieved for ZnSSe. In ZnSTe, this cannot be achieved, since high Te containing ZnSTe alloy cannot be n-doped. If we compare ZnSSe with CdS and ZnS, which are traditional materials for UV detectors, we will see that ZnSSe, unlike these semiconductors, can be used to detect UV radiation in the UV-A range (320–400 nm), where CdS and ZnS have very low sensitivity. This indicates that ZnSSe detectors can be used to measure UV radiation in all three spectral regions of UV radiation – UV-A (320–400 nm), UV-B (280–320 nm), and UV-C (100–280 nm).

The same results were obtained when using quadruple solid solutions, such as ZnMgBeSe [49]. Figure 15.10 shows the spectral response of a typical ZnMgBeSe-based detector measured at -10 V bias, with front-side illumination through the semitransparent Schottky contact.

It can be seen that the response is very flat above the band gap, due to the position of the depleted region on top of the structure: this is an advantage of Schottky barrier devices in comparison with p – n photodiodes. The cutoff at 380 nm is very sharp with a visible rejection rate of more than three orders of magnitude (5×10^3), which is a consequence of the high crystalline quality of the structure. A maximum responsivity of 0.17 A/W is obtained at 375 nm corresponding to a quantum efficiency of 54%. It is important to note that a quantum efficiency remains over 50% above the whole UV range from 380 to 315 nm, and drops off slightly as the wavelength decreases further. This is close to the theoretical limit. Moreover, due to the

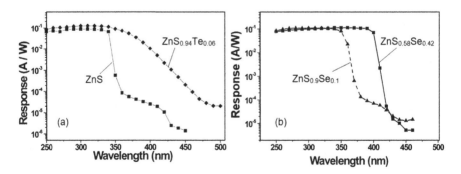

Fig. 15.9 Photoresponse curves of (**a**) ZnS and $ZnS_{0.94}Te_{0.06}$, and (**b**) $ZnS_{0.9}Se_{0.1}$ and $ZnS_{0.58}Se_{0.42}$ based Schottky barrier UV detectors. (Adapted with permission from Ref. [44]. Copyright 2000: Springer)

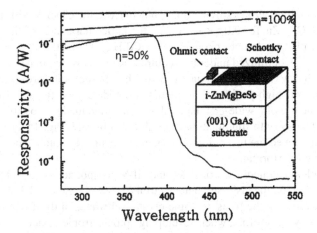

Fig. 15.10 Spectral response of a ZnMgBeSe Schottky photodiode at −10 V bias. Theoretical curves corresponding to different quantum efficiencies (η) are indicated. (Adapted with permission from Ref. [49]. Copyright 2001: AIP Publishing)

use of epitaxial films, no internal gain mechanism has been observed. And this means that a change in the state of the outer atmosphere will not affect the parameters of these detectors. A detectivity of 2×10^{10} mHz$^{1/2}$ W^{-1} has also been measured, highlighting the fact that these detectors are both sensitive and little noisy.

In addition, Vigué et al. [49] compared the fabricated detectors with detectors based on other materials and showed that the parameters of ZnMgBeSe detectors are superior to those of many analogues. ZnS- and ZnSSe-based structures have responsivities of 0.08 A/W at 335 nm and 0.09 A/W at 370 nm, respectively [43]. ZnSTe devices exhibit a high response of 0.13 A/W at 320 nm, but the tellurium incorporation leads to a significant decrease of the rejection rate and generates a large cutoff in the spectral response [28]. ZnMgBeSe photodetectors also compare advantageously with 6H-SiC (0.15 A/W) [2], GaN (0.10–0.18 A/W) [7, 35], and AlGaN (0.07 A/W) [33].

15.3 Solid Solutions Providing Lattice Matching of Contacting Semiconductor Materials

The most obvious example of a problem that has been solved through the use of II-VI ternary compounds is the growth of CdZnTe single crystals with lattice parameters matched to the lattice parameter of CdHgTe, the main material of IR technology. It turned out that for the manufacture of high-performance photodetectors array by epitaxial methods, large-diameter substrates (> 50 cm) with the same lattice parameter are needed. Unfortunately, at the moment there is no CdHgTe technology that makes it possible to grow single crystals of this diameter. At the same time,

CdZnTe technology allows solving this problem (read the Chap. 9, Vol. 1). With the composition $Cd_{1-x}Zn_xTe$ ($0.02 < x < 0.06$), the lattice parameter of this compound coincides with the lattice parameter of $Cd_xHg_{1-x}Te$ ($0.19 < x < 0.23$ and $0.27 < x < 0.32$), optimized under the required IR range, which makes it possible to grow high-quality CdHgTe epitaxial layers on their basis, which are required for the manufacture of high-performance matrix photodetectors (Chap. 15, Vol. 1). Moreover, the use of a wide-gap material as a substrate made it possible to develop IR photodetectors with illumination through the substrate (Fig. 15.11), which significantly facilitated the technology for manufacturing IR photodetector arrays and improved their performances.

The problem of integration of II-VI and III-V compounds is solved in a similar way. Depending on the tasks set, optoelectronic devices based on III-V compounds are manufactured on the basis of single-crystal substrates of these compounds, but mainly on GaAs. Therefore, when developing optoelectronic devices based on II-VI compounds integrated with devices based on III-V compounds, it is necessary to match the lattice parameters of GaAs and II-VI compounds in order to achieve the required device parameters. Matching the lattice parameters of the substrate and the epitaxial layers grown on their surface significantly reduces the mechanical stresses at the interface and the imperfection of the grown layer itself, which naturally affects the improvement of the operational parameters of the manufactured devices. The experiment showed that II-VI multicomponent solid solutions used as transition or buffer layers make it possible to solve this problem. In this case, the structures of II-VI-based photodiodes fabricated on a GaAs substrate may look as shown in Fig. 15.12.

Using this approach, it was possible to match the lattice parameters of AIIBVI compounds with the lattice parameters of other semiconductors, including Si [42],

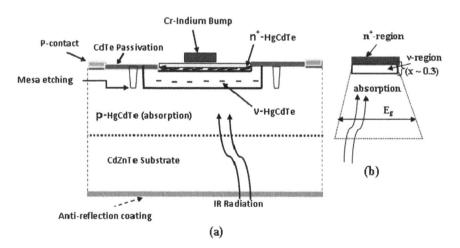

Fig. 15.11 Schematic diagram: (a) back side illuminated n+/ν/p + HgCdTe e-APD (b) bandgap grading in the device structure. (Reprinted with permission from Ref. [40]. Copyright 2018: Elsevier)

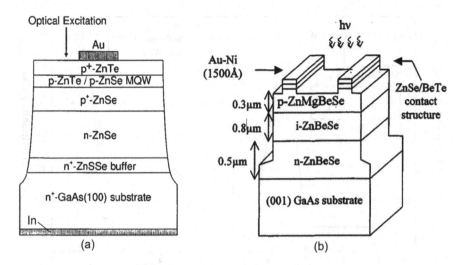

Fig. 15.12 (a) A schematic structure of a ZnSe p$^+$ – n avalanche photodiode grown on GaAs by MBE. The surface ohmic contact layer consists of ZnTe–ZnSe multiquantum wells (with five periods) and a thin ZnTe cap layer. Reprinted with permission from [20]. Copyright 2000: AIP Publishing. (b) Schematic representation of a typical ZnBeSe–ZnMgBeSe p-i-n photodiode. (Adapted with permission from Ref. [49]. Copyright 2001: AIP Publishing)

GaP [42], InSb [56], and on single-crystal substrates of these semiconductors to produce high-performance optoelectronic devices based on II-VI compounds, including lasers and photodetectors. In particular, using $ZnS_{1-x}Te_x$ solid solutions as buffer layers, it was possible to fabricate highly efficient UV detectors on GaP substrates [42]. The structure of these photodetectors and some parameters are shown in Fig. 15.13. Sou et al. [42] believe that a direct-gap semiconductor $ZnS_{1-x}Te_x$ alloy with a band gap covering 2.1–3.7 eV is a promising material for visible and ultraviolet optoelectronic applications. An important advantage of ZnSTe alloys is that their lattice can be matched with a number of commercially available substrates such as Si, GaAs, and GaP, which was confirmed experimentally. In particular, the 3%-Te alloy has a lattice-matched silicon substrate, and thus ZnSTe-based devices are compatible with advanced silicon integration technology. At the same time, Zhang and co-workers [24, 55] believe that for CdTe-based double heterostructures grown on InSb substrate, the $Mg_{0.46}Cd_{0.54}Te$ layer is an ideal barrier layer, which can effectively confine both electrons and holes and reduce the surface recombination.

As we can see, ternary alloys are used as a buffer layer in the given examples. However, it has been found that in addition to ternary alloys, more complex compounds can be used. For example, Vigué et al. [49] for such matching used the ZnMgBeSe solid solution, which can also have a lattice parameter equal to that of GaAs (Fig. 15.14). It is important to note that the use of a quadruple alloy instead of a ternary one has significant advantages, as it provides more opportunities for band gap engineering. If a ternary alloy has one composition with a fixed band gap, which ensures the equality of the lattice parameters with the substrate, then in a quaternary

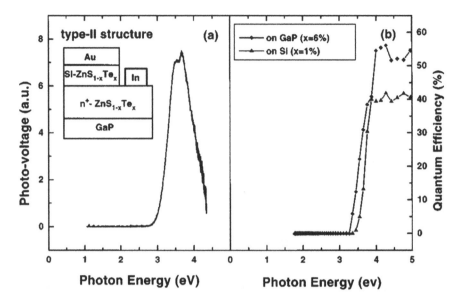

Fig. 15.13 (**a**) Photovoltage for Schottky barrier Au-ZnSTe structure grown on a GaP substrate; and (**b**) external quantum efficiency vs photon wavelength for Schottky barrier Au-ZnSTe structures grown on GaP and Si substrates. (Reprinted with permission from Ref. [42]. Copyright 1997: AIP Publishing)

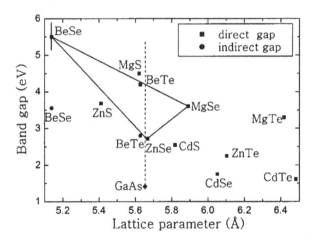

Fig. 15.14 Energy of wide-band gap II-VI semiconductors as a function of their lattice parameter. (Adapted with permission from Ref. [49]. Copyright 2001: AIP Publishing)

alloy there are much more such possibilities. As shown in Fig. 15.13, ZnMgBeSe quaternary alloys can be grown on lattice-matched GaAs substrates with a direct band gap ranging from 2.75 to greater than 4.2 eV. These compounds present high crystalline quality, which should lead to high sensitivity and detectivity.

15.4 Optimization of Electrophysical and Physical Properties of II-VI Compounds

Optimizing the parameters of optoelectronic devices is a multifactorial task that can be solved in various ways. Improving the electrical and physical properties of II-VI semiconductors through the complication of the composition is one of such approaches. Earlier, we noted that by using $Cd_{0.96}Zn_{0.04}Te$ alloys, it is possible to optimize the spectral sensitivity of photodetectors [36]. However, it turned out that the transition to solid solutions also makes it possible to get rid of some of the disadvantages inherent in CdTe [36]. Compared to the commercially used Si semiconductor, CdTe is much more sensitive to the light from visible to infrared region because it has a direct band gap and a relatively higher optical absorption coefficient. The higher carrier mobility also makes it promising for photovoltaic applications. However, polycrystalline CdTe photodetectors have been found to have a rather slow response rate due to the strong persistent photoconductivity (PPC), which was induced from abundant grain boundaries and surface state traps [53]. Single crystal CdTe has better parameters. However, even in this case, due to the specifics of the technology, the CdTe single crystal has a significant concentration of intrinsic point defects and residual impurities, which cause the appearance of localized levels in the band gap. This resulted in high leakage current, severe carrier trapping, and low responsivity of X-ray detectors. Another important issue in single crystal CdTe is polarization charges under bias, which can act as trapping centers affecting the space charge distribution and electric field profile [29]. Studies have shown that the use of single crystal $Cd_{1-x}Zn_xTe$ instead of CdTe can be a solution to the above problems. By increasing the Zn concentration, it is possible to adjust the band gap of alloys from 1.44 eV (for CdTe) to 2.2 eV (for ZnTe) and synthesize a material with high electrical resistivity ($\rho = (1-4) \times 10^{10}$ Ohm·cm) close to the theoretical maximum [46]. In addition, it was reported that the polarization effect can also be eliminated in $Cd_{1-x}Zn_xTe$ crystals [41].

A promising direction is also the doping of II-VI compounds with Mn, Cr, Co, and Fe [1]. Wide-gap II-VI compounds, which do not have magnetic properties, after doping with these elements, begin to exhibit magnetic and feromagnetic properties, which significantly expands the possible areas of their application. Such semiconductors belong to the class of so-called diluted magnetic semiconductors (DMS) [11, 12]. An important feature of dilute magnetic semiconductors is the ability to control the magnetic order through the concentration of carriers, which is easily changed by the gate voltage, and the ability to remagnetize layers using spin-polarized currents with a much lower density compared to traditional ferromagnetic metals. Due to their uniqueness, these materials are of extreme interest for both fundamental science and applied science. Diluted magnetic semiconductors are currently being considered as materials for spin polarized charge carrier injectors in semiconductor spin electronics devices [1], and for the development of a new generation of magnetic memory elements. Since ferromagnetic spin-spin interactions are mediated by holes in the valence band, changing the Fermi level using co-doping, electric fields, or light can directly affect magnetic ordering. Moreover, engineering the Fermi level position by co-doping makes it possible to modify solubility and self-compensation limits, affecting magnetic characteristics in a number of surprising ways. The Fermi energy can even control the aggregation of magnetic ions, providing a new route to self-organization of magnetic nanostructures in a semiconductor host [11, 12].

The most studied II-VI-based diluted magnetic semiconductors are $Zn_{1-x}Mn_xTe$, $Cd_{1-x}Mn_xTe$, $Zn_{1-x}Cr_xTe$, $Cd_{1-x}Cr_xTe$, $Cd_{1-x-y}Hg_xMn_yTe$, and ZnMnSe [1, 11, 12, 19, 32]. An exchange interaction between the sp-band electrons of the alloy and the localized d electrons associated with Mn^{2+} results in a strongly enhanced g-value for Zeeman splitting of electronic levels, which in turn causes a strong Faraday effect near the band gap of the DMS material. The large magneto-optical effects of single crystalline $Cd_{1-x}Mn_xTe$ and $Cd_{1-x-y}Hg_xMn_yTe$ are already find practical application as a Faraday rotator in optical isolators for the wavelength region in which magnetic garnet crystals are not applicable due to presence of strong absorption [19, 32]. However, the magnetic properties of most II-VI-based semiconductors are either paramagnetic or spin-glass state, which is not very attractive for practical applications. Recently, the development of a new II-VI semiconductor doping technology has made it possible to obtain p-type DMS materials based on II-VI allays. Theory predicts ferromagnetism in some of these DMS materials if they are heavily-doped p-type. Experiment confirmed that prediction [15]. Heavy p-doped $Zn_{1-x}Mn_xTe$ and $Cd_{1-x}Mn_xTe$ epilayers were synthesized [15].

II-VI compounds doped with indium, aluminum and copper can also be classified as multicomponent semiconductors, since the concentration of dopants reaches 5%. Moreover, doping is accompanied by a change not only in electrophysical parameters, but also in the optical properties of semiconductors, which affect the performances of optoelectronic devices. For example, Cui et al. [9] studied the photoelectron characteristics of ZnSe/ZnS/L-cys and Cu-doped ZnSe/ZnS/L-cys core shells and found that the surface photovoltaic property of Cu-doped ZnSe QDs is significantly improved when compared with ZnSe QDs. More specifically, the intensity of the surface photovoltaic (SPV) response of Cu-doped ZnSe QDs is approx. Twenty-one times stronger than that of un-doped ZnSe QDs. Cui et al. [9] believe that the excellent characteristics of Cu-doped ZnSe QDs may result from their unique electron structure and photogenerated charge-transfer (CT) transition behaviors. The new SPV response peaks of Cu-doped QDs can be ascribed to the defect-state levels that are closely related to the doped Cu^{2+} ions.

Doping with copper gives a positive result in the development of solar cells as well. It was established that the use of p-type ZnTe:Cu (III-V%) as a back contact in CdTe solar cells can significantly increase its efficiency [39]. A major challenge for CdTe solar cell technology is the formation of high quality ohmic back contacts [10]. p-CdTe has a high work function (~5.7 eV) which is higher than most common metals [4]. A direct metal contact with a low work function will form a negative Schottky barrier band bend in the back contact region [10], which is an important limiting factor for cell performance. Theory and experiment have shown that p-type ZnTe:Cu is an ideal back contact buffer material with CdTe absorber in both lattice matching and energy band matching. ZnTe has a high work function of 5.3–5.8 eV and a band gap of 2.2 eV, which can not only reduce the contact barrier with a small valence band gap, but also act as an electron reflector to suppress recombination at the interface [17, 37].

15.5 Summary

This chapter, which considers various types of photodetectors, has shown that ternary and quaternary solid solutions based on II-VI compounds are indeed promising materials that allow engineering of both band gap and lattice constant, and thereby develop devices with the required parameters.

Acknowledgments G. Korotcenkov is grateful to the State Program of the Republic of Moldova, project 20.80009.5007.02, for supporting his research.

References

1. Akai H, Ogura M (2006) Half-metallic diluted antiferromagnetic semiconductors. Phys Rev Lett 97:026401
2. Anikin M, Andreev AN, Pyatko SN, Rastegaeva MG, Savkina NS, Strelchuk AM et al (1992) UV photodetectors in 6H-SiC. Sens Actuators A 33:91–93
3. Averin SV, Kuznetzov PI, Zhitov VA, Zakharov LY, Kotov VM, Alkeev NV (2016) Wavelength selective UV/visible metal-semiconductormetal photodetectors. Opt Quant Electron 48:303
4. Bätzner DL, Romeo A, Zogg H, Wendt R, Tiwari AN (2001) Development of efficient and stable back contacts on CdTe/CdS solar cells. Thin Solid Films 387(1–2):151–154
5. Becerril M, Zelaya-Angel O, Fragoso-Soriano R, TiradoMejia L (2004) Band gap energy in Zn-rich $Zn_{1-x}Cd_xTe$ thin films grown by r.f. sputtering. Rev Mex Fis 50(6):588–593
6. Chander S, De AK, Dhaka MS (2018) Towards CdZnTe solar cells: an evolution to post-treatment annealing atmosphere. Solar Cells 174:757–761
7. Chen Q, Yang JW, Osinsky A, Gangopadhyay S, Lim B, Anwar MZ, Khan MA (1997) Schottky barrier detectors on GaN for visible–blind ultraviolet detection. Appl Phys Lett 70:2277
8. Chen S, Walsh A, Yang JH, Gong XG, Sun L, Yang P-X et al (2011) Compositional dependence of structural and electronic properties of $Cu_2ZnSn(S,Se)_4$ alloys for thin film solar cells. Phys Rev B 83:125201
9. Cui JY, Li KY, Ren L, Zhao J, Shen TD (2016) Photogenerated carriers enhancement in Cu-doped ZnSe/ZnS/L-cys self-assembled core-shell quantum dots. J Appl Phys 120:184302
10. Demtsu SH, Sites JR (2006) Effect of back-contact barrier on thin-film CdTe solar cells. Thin Solid Films 510(1–2):320–324
11. Dietl T, Ohno H (2003) Ferromagnetic III-V and II-VI Semiconductors. MRS Bull 28(11):714–719
12. Dietl T, Ohno H (2006) Engineering magnetism in semiconductors. Mater Today 9(11):18–26
13. Ebrahimi S, Yarmand B, Naderi N (2020) High-performance UV-B detectors based on $Mn_xZn_{1-x}S$ thin films modified by bandgap engineering. Sens Actuators A 303:111832
14. Faschinger W, Ehinger M, Schallenberg T (2000) High-sensitivity p-i-n-detectors for the visible spectral range based on wide-gap II-VI materials. J Crystal Growth 214(215):1138–1141

15. Ferrand D, Cibert J, Bourgognon C, Tatarenko S, Wasiela A, Fishman G et al (2000) Carrier induced ferromagnetic interactions in p-doped $Zn_{(1-x)}Mn_xTe$ epilayers. J Appl Phys 87:6451
16. Gao T, Li QH, Wang TH (2005) CdS nanobelts as photoconductors. Appl Phys Lett 86:173105
17. Gessert T, Mason A, Sheldon P, Swartzlander A, Niles D, Coutts T (1996) Development of Cu doped ZnTe as a back contact interface layer for thin film CdS/CdTe solar cells. J Vac Sci Technol A 14:806–812
18. Graetzel M, Janssen RAJ, Mitzi DB, Sargent EH (2012) Materials interface engineering for solution-processed photovoltaics. Nature 488:304–312
19. Imamura M, Tashima D, Kitagawa J, Asada H (2020) Magneto-optical properties of wider gap semiconductors ZnMnTe and ZnMnSe films prepared by MBE. J Electron Sci Technol 18(3):201–211
20. Ishikura H, Abe T, Fukuda N, Kasada H, Ando K (2000) Stable avalanche-photodiode operation of ZnSe-based p^+–n structure blue-ultraviolet photodetectors. Appl Phys Lett 76:1069–1071
21. Khalil MI, Bernasconi R, Lucotti A, Le Donne A, Mereu RA, Binetti S et al (2021) CZTS thin film solar cells on flexible molybdenum foil by electrodeposition-annealing route. J Appl Elecrochem 51:209–218
22. Kim J-P, Christians JA, Choi H, Krishnamurthy S, Kamat PV (2014) CdSeS nanowires: compositionally controlled band gap and Exciton dynamics. J Phys Chem Lett 5:1103–1109
23. Li D, Qin J, Xu Q, Yan G (2018) A room-temperature phosphorescence sensor for the detection of alkaline phosphatase activity based on Mn-doped ZnS quantum dots. Sensors Actuators B Chem 274:78–84
24. Liu S, Zhao X-H, Campbell CM, Lassise MB, Zhao Y, Zhang Y-H (2015) Carrier lifetimes and interface recombination velocities in $CdTe/Mg_xCd_{1-x}Te$ double heterostructures with different Mg compositions grown by molecular beam epitaxy. Appl Phys Lett 107:041120
25. Liu Y, Pang L-X, Liang J, Cheng M-K, Liang JJ, Chen JS, Lai Y-H, So I-K (2017) A compact solid-state UV flame sensing system based on wide-gap II-VI thin film materials. IEEE Trans Ind Elektron 65(3):2737–2744
26. Liyanage GK, Phillips AB, Alfadhili FK, Ellingson RJ, Heben MJ (2019) The role of Back buffer layers and absorber properties for >25% efficient CdTe solar cells. ACS Appl Energy Mater 2(8):5419–5426
27. Lubegin G, Onishchenko D, Guslyannikov V (2011) Injection photodiodes based on AIIBVI single crystals. Photonics (Russia) 1:34–37. (in Russian)
28. Ma ZH, Sou IK, Wong KS, Yang Z, Wong GKL (1998) ZnSTe-based Schottky barrier ultraviolet detectors with nanosecond response time. Appl Phys Lett 73:2251
29. Malm HL, Martini M (1974) Polarization phenomena in CdTe nuclear radiation detectors. IEEE Trans Nucl Sci 21(1):322–330
30. Mathew X, Drayton J, Parikh V, Compaan AD (2006) Sputtered $Cd_{1-x}Mg_xTe$ films for top cells in tandem devices. In: Proceedings of IEEE 4th world conference on PVEC, pp 327–331
31. Morales-Acevedo A (2011) Analytical model for the photocurrent of solar cells based on graded band-gap CdZnTe thin films. Solar Energy Mater Solar Cells 95:2837–2841
32. Onodera K, Matsumoto T, Kimura M (1994) 980 nm compact optical isolators using $Cd_{1-x-y}Hg_xMn_yTe$ single crystals for high power pumping laser diodes. Electron Lett 39:1954–1955
33. Osinsky A, Gangopadhyay S, Lim BW, Anwar MZ, Khan MA (1998) Schottky barrier photodetectors based on AlGaN. Appl Phys Lett 72:742
34. Parikh VY, Marsillac S, Collins RW, Chen J, Compaan AD (2007) $Hg_{1-x}Cd_xTe$ as the bottom cell material in tandem II-VI solar cells. MRS Proc 1012(Y12):37
35. Ravikiran L, Radhakrishnan K, Dharmarasu N, Agrawal M, Wang Z, Bruno A et al (2017) GaN Schottky metal–semiconductor–metal UV photodetectors on Si(111) grown by ammonia-MBE. IEEE Sensors J 17(1):72–77
36. Ren B, Zhang J, Liao M, Huang J, Sang L, Koide Y, Wang L (2019) High-performance visible to near-infrared photodetectors by using (Cd,Zn)Te single crystal. Opt Express 27(6):8935
37. Rioux D, Niles DW, Hochst H (1993) ZnTe: a potential interlayer to form low resistance back contacts in CdS/CdTe solar cells. J Appl Phys 73:8381–8385

38. Sahana NM, Mahesh MG (2020) Colour tunable co-evaporated CdS_xSe_{1-x} ($0 <x < 1$) ternary chalcogenide thin films for photodetector applications thin films for photodetector applications. Mater Sci Semicond Process 120:105288
39. Shen K, Wang X, Zhang Y, Zhu H, Chen Z, Huang C, Ma Y (2020) Insights into the role of interface modification in performance enhancement of ZnTe:Cu contacted CdTe thin film solar cells. Sol Energy 201:55–62
40. Singh A, Pal R (2018) Impulse response measurement in the HgCdTe avalanche photodiode. Solid State Electron 142:41–46
41. Sordo SD, Abbene L, Caroli E, Mancini AM, Zappettini A, Ubertini P (2009) Progress in the development of CdTe and CdZnTe semiconductor radiation detectors for astrophysical and medical applications. Sensors (Basel) 9(5):3491–3526
42. Sou IK, Man CL, Ma ZH, Yang Z, Wong GKL (1997) High performance ZnSTe photovoltaic visible-blind ultraviolet detectors. Appl Phys Lett 71:3847
43. Sou IK, Ma ZH, Zhang ZQ, Wong GKL (2000a) Temperature dependence of the responsivity of II-VI ultraviolet photodiodes. Appl Phys Lett 76:1098
44. Sou IK, Ma ZH, Wong GKL (2000b) ZnS-based visible-blind UV detectors: effects of isoelectronic traps. J Electron Mater 29(6):723–726
45. Sou IK, Wu MCW, Sun T, Wong KS, Wong GKL (2001) Molecular-beam-epitaxy-grown ZnMgS ultraviolet photodetectors. Appl Phys Lett 78:1811
46. Takahashi T, Watanabe S (2001) Recent progress in CdTe and CdZnTe detectors. IEEE Trans Nucl Sci 48(4):950–959
47. Tang M, Xu P, Wen Z, Chen X, Pang C, Xu X et al (2018) Fast response $CdS-CdS_xTe_{1-x}$-CdTe core-shell nanobelt photodetector. Sci Bull 63(17):1118–1124
48. Todorov TK, Reuter KB, Mitzi DB (2010) High-efficiency solar cell with earth-abundant liquid-processed absorber. Adv Mater 22:E156–E159
49. Vigué F, Tournié E, Faurie J-P, Monroy E, Calle F, Muñoz E (2001) Visible-blind ultraviolet photodetectors based on ZnMgBeSe Schottky barrier diodes. Appl Phys Lett 78:4190
50. Vikulin IM, Kurmashev SD, Stafeev VI (2008) Injection photodetectors. Phys Technol Semicond 42(1):113–127. (in Russian)
51. Wang SL, Lee SH, Gupta A, Compaan AD (2004) (2004) RF sputtered HgCdTe films for tandem cell applications. Phys Status Solidi C 1(4):1046–1049
52. Wang W, Winkler MT, Gunawan O, Gokmen T, Todorov TK, Zhu Y, Mitzi DB (2014) Device characteristics of CZTSSe thin-film solar cells with 12.6% efficiency. Adv Energy Mater 4(7):1301465
53. Yang G, Kim D, Kim J (2016) Photosensitive cadmium telluride thin-film field-effect transistors. Opt Express 24(4):3607–3612
54. Zhang ZZ, Shan DZ, Shan CX, Zhang JY, Lu YM, Liu YC, Fan XW (2003) The growth of the $Cd_xZn_{1-x}Te$ epilayers by low-pressure metalorganic vapor-phase epitaxy. Thin Solid Films 429:211–215
55. Zhao X-H, DiNezza J, Liu S, Campbell CM, Zhao Y, Zhao Y, Zhang Y-H (2014) Determination of CdTe bulk carrier lifetime and interface recombination velocity of CdTe/MgCdTe double heterostructures grown by molecular beam epitaxy. Appl Phys Lett 105:252101
56. Zhao X-H, Liu S, Zhao Y, Campbell CM, Lassise MB, Kuo Y-S, Zhang Y-H (2016) Electrical and optical properties of n-Type indium-doped $CdTe/Mg_{0.46}Cd_{0.54}Te$ double Heterostructures. IEEE J Photovoltaics 6(2):552–556
57. Zhou J, Wu X, Teeter G, To B, Yan Y, Dhere RG, Gessert TA (2004) $CBD-Cd_{1-x}Zn_xS$ thin films and their application in CdTe solar cells. Phys Status Solidi B 241:775–778
58. Zhou C, Chung H, Wang X, Bermel P (2016) Design of CdZnTe & crystalline silicon tandem junction solar cells. IEEE J Photovoltaics 6(1):301–308

Part III
New Trends in Development of II–VI Semiconductors–Based Photodetectors

Chapter 16
Nanowire-Based Photodetectors for Visible-UV Spectral Region

Ghenadii Korotcenkov and Victor V. Sysoev

16.1 Introduction

In recent decades, there has been a tremendous interest in the development of various devices based on (quasi-)1D nanostructures, such as nanowires (NWs), nanotubes (NTs), nanoribbons (NRs) and nanobelts (NBs) [17, 36, 54]. Semiconducting 1D nanostructures are quasi-one-dimensional nanocrystals, which normally exhibit a transverse width at the 10–100 nm range and length to be substantially longer, up to micrometers or even millimeters. NWs and nanobelts, which are most employed in the development of photodetectors, are so far the smallest structures, which could transport charge carriers efficiently. It has been shown that these 1D structures are promising nanostructures for the next generation of nanoscaled electrical and optical devices such as field-effect transistors, solar cells, diodes, LEDs, and lasers. The development of photodetectors for visual and UV range is no exception [8, 45, 65, 66]. This interest is explained by special shapes and specific fundamental chemical, physical and electro-optical properties of 1D structures, mostly grown as single crystals. In particular, it was found that the unique photonic features of 1D structures make them an excellent candidate for photodetectors. Due to the nanoscale size and matching the thickness or diameter of NWs and NBs to the Debye length, the electronic and electrophysical properties of 1D nanostructures strongly depend on the state of the surface and the processes occurring on it, that ensures appearing conditions to observe a superior sensitivity when compared to one of conventional thin-film photodetectors [64, 65]. The mechanism of this influence can be understood from the diagram presented in Fig. 16.1.

G. Korotcenkov (✉)
Department of Physics and Engineering, Moldova State University, Chisinau, Moldova
e-mail: ghkoro@yahoo.com

V. V. Sysoev
Yuri Gagarin State Technical University of Saratov, Saratov, Russia

© The Author(s), under exclusive license to Springer Nature Switzerland AG 2023
G. Korotcenkov (ed.), *Handbook of II-VI Semiconductor-Based Sensors and Radiation Detectors*, https://doi.org/10.1007/978-3-031-20510-1_16

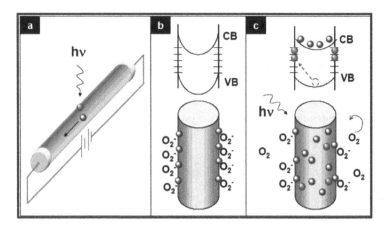

Fig. 16.1 (**a**) Schematic of a NW photodetector. (**b, c**) the energy band diagrams of a NW in dark and upon illumination, indicating band-bending and surface trap states. The bottom drawing shows oxygen molecules adsorbed at the NW surface. VB and CB are the valence and conduction band, respectively. (Reproduced with permission from Ref. [66]. Copyright 2007: American Chemical Society)

As illustrated in Fig. 16.1, oxygen molecules adsorbed on the NW surface under dark conditions, capture free electrons via $O_2(g) + e^- \rightarrow O_2^-(ad)$, and a low-conductivity depletion layer appears/extends near the surface (Fig. 16.1b). Once illuminated with the light at a photon energy above the energy gap ($h\nu > E_g$), electron-hole pairs are photogenerated [$h\nu \rightarrow e^- + h^+$] (Fig. 16.1a). The holes migrate to the surface, where they can be easily trapped at the surface state, leaving behind unpaired electrons, which results in the increase of the conductivity under an applied dc electric field. At the same time, the illumination stimulates the oxygen to desorb from the NW surface [$h^+ + O_2^-(ad) \rightarrow O_2(g)$] that causes a reducing the height of the potential barrier at the surface and the return of electrons captured by chemisorbed oxygen to the conduction band of the semiconductor, thereby enhancing the photoresponse (Fig. 16.1c). It is thought that the maximum effect is achieved at a certain ratio of the NW diameter (d) and Debye length (D_L). For example, if $d \gg D_L$, then no signal enhancement is expected. Therefore, it is very important to choose NWs of a suitable diameter while fabricating the photodetectors. A more detailed description of the mechanism of photosensitivity in NWs can be found somewhere else [61, 65, 66].

The low electrical cross-section of NWs implies a low electrical capacitance, but this comes without degradation of total light absorption due to antenna effects. With a certain configuration of highly ordered arrays of 1D structures, it is also possible to create conditions under which high absorption is achieved with a minimum volume of photosensitive material, required to achieve a high signal-to-noise ratio. Therefore, NW arrays can exhibit higher absorption than a thin film of the equivalent thickness [38]. In addition, the small footprint of NWs leads to an efficient relaxation of strain in lattice-mismatched heterostructures and facilitates heterogeneous integration of II-VI semiconducting NW devices with mass-scaled Si technologies. As a result, we get new degrees of design freedom to optimize electrical and photoelectrical performance of photodetectors. The advantages of nanostructures include also the possibility of growing one-dimensional nanostructures or converting them into flexible materials [46], which can be used to develop of wearable devices. In the following, we will mostly use the designation NWs, as the most common while discussing the properties of 1D nanostructures, for convenience.

16.2 Synthesis of NWs

Currently, high-quality II-VI semiconducting 1D nanostructures to be employed in photodetectors aimed mostly to visible and UV spectral range, can be rationally synthesized in a single-crystal phase whose key parameters such as chemical composition, shape, doping state, diameter, length, etc., are well controlled. For these purposes, a number of techniques have been developed. These methods can generally be classified into one of the following approaches: bottom-up or top-down. A more detailed discussion of the methods used to grow II-VI semiconducting NWs can be found in Chap. 12 (Vol. 1). Currently, the most common method to grow II-VI semiconducting 1D nanostructures is a thermal evaporation of II-VI compounds powders via a vapor–liquid–solid (VLS) process.

16.3 Fabrication Features of NW-Based Photodetectors

It should be noted that photodetectors based on individual NWs have a specific configuration and manufacturing technology, that is fundamentally different from the traditional one used to manufacture conventional photodetectors. The NW photodetectors can be fabricated under various designs (see Fig. 16.2). The most popular is the in-plane architecture where NWs are placed to lie on a substrate. This architecture is employed to fabricate photodetectors based on both single wires and their arrays. In this design, the NWs are either randomly dispersed or prepositioned using some alignment techniques, as dielectrophoresis, contact printing, and others.

Contacts to the NWs are usually made at the last step with the help of metal evaporation or sputtering by various methods and lithographic techniques. One can find the description of this approach in [36]. A typical example of photodetectors based on individual 1D structures and manufactured using this approach is shown in Fig. 16.2a. Another, a more progressive, approach to manufacture NW array-matured photodetectors employs a vertical architecture where NWs are stand generally on their native growth substrate (see Fig. 16.2c). The vertical configuration was mainly designed to demonstrate NW array devices using a large number of parallel-connected NWs as the active medium [8].

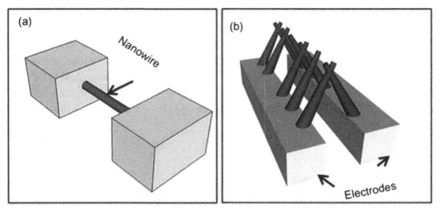

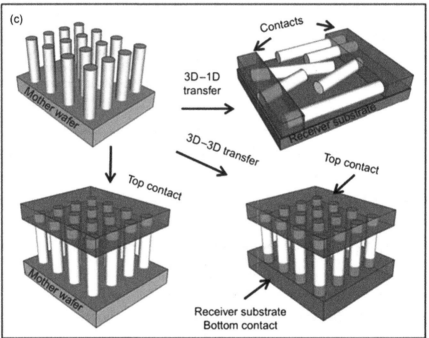

Fig. 16.2 Schematic of some of the approaches used to fabricate semiconducting NW-based photodetectors. (**a**) NWs can be directly grown between two electrodes using both top down or bottom up approaches. (**b**) NW bridges can form on isolated electrode contacts as a result of a serendipitous fusing phenomenon during their growth. (**c**) Vertically oriented NWs atop mother wafer can be fashioned into a PD by a lateral 3D-1D contact printing onto receiver substrates followed by deposition of electrical contacts or by a 3D-3D print transfer mechanism onto thermally pliable matrices followed by top contact deposition. Or whilst still atop the mother wafer, isolation and top contact layers can be deposited onto the array to create a device. (Reprinted with permission from Ref. [59]. Copyright 2016: Elsevier)

16.3.1 Direct NW Integration

If the deposition of the catalyst on the side walls of the platform used is acceptable, the semiconducting NWs can be most effectively integrated into the required sites of the device via direct growing to form a mechanically and electrically stable monolithic assembly. This process is shown schematically in Fig. 16.3. The positive output of this method depends on the proper making catalytic sites in the desired location. For this purpose, one can use technologies such as nanoimprint, electron beam lithography or evaporation followed by annealing. Many semiconducting NWs of Group IV [30], Group III-V [87], and Group II-VI [39] were integrated in frames of this technology. Unfortunately, due to the nucleation and migration of nanoparticles of the catalytic material upon annealing, it is difficult to control the catalyst diameter and its position. This leads to uncertainty in the position and diameter of the NWs [35].

It should be noted that the described integration method opens up new technological possibilities for fabricating NW-based photodetectors with advanced parameters. For large scale integrated circuits or embedded sensor applications, the "bridging" process can meet industry requirements for relatively low cost and high throughput. The *in situ* fabricated NW devices offer a massively parallel, self-assembling technique that provides a controlled connection of NWs between electrodes by means of just a coarse lithography [59].

16.4 Single NW-Based Photodetectors

16.4.1 Photoconductive Detectors Employing Single NWs

Single NWs are the simplest platform to develop photodetectors. Currently, such NW-based photoconductors are produced taking all wide-gap II-VI semiconducting compounds such as CdS, CdSe, ZnSe, CdTe, ZnTe and ZnS. They detect electromagnetic waves in a broad spectrum range, from infrared to ultraviolet region, due to their broad coverage of bandgap energies. CdS, ZnSe and ZnS NWs are the most

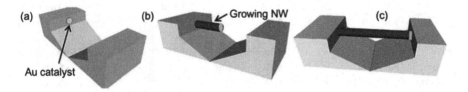

Fig. 16.3 (a) A schematic illustrating a lithography aided deposit of a gold catalyst onto the side of side wall of a trench. (b) Catalyst grown NW produces oriented NW directional growth. (c) Continued NW growth causes the NW to grow onto the opposing sidewall. (Reprinted with permission from Ref. [59]. Copyright 2016: Elsevier)

studied photoconductors in the group II–VI aimed to a visible and ultraviolet light detection [10]. As in the case of metal oxide NWs, the photoresponse of these low-dimensional structures is strongly affected by the photochemistry of surface oxygen, which can significantly change the relaxation dynamics of photocarriers [32, 33]. The characteristics of photodetectors developed on the basis of II-VI semiconducting NWs are shown in Table 16.1. The photoconductive detectors are highly sensitive ordinarily in the visible-UV spectral range (Fig. 16.4b, d) and have stable, fully reversible and periodic characteristics (Fig. 16.4c). Ideally, photoconductive detectors should have ohmic contacts (Fig. 16.4a). However, this condition is not met in many cases.

As shown in numerous works, the performance of NW-based photodetectors can be significantly improved through their doping. For instance, to optimize the parameters of CdS-based photodetectors, this material is doped with In [53], Ga [79], and Cl [83]. The observed effect to improve the parameters of photodetectors is explained by enhancing the concentration of charge carriers after doping. Typically, as-synthesized NWs exhibit a high resistance. Following doping, the resistance of NWs might drop by 3–4 orders of magnitude. For example, the photoconductive properties of CdSe nanobelts have been investigated in both intrinsic and n-doped structures [34]. Fast speed, response/recover times to be ~15/31 μs, and high photosensitivity, ca. 100, were observed in i-CdSe, while a high-gain is preserved in n-CdSe. These differences seem to mature from the traps created by impurities within the nanobelt [81]. In the visible-NIR range, of 400–800 nm, CdTe

Table 16.1 Performance of II-VI semiconducting NW-based photoconductive detectors

Mater.	Type	Min. size, nm	P_λ, mW/ cm^2	λ, nm	R_λ, AW^{-1}	I_{light}/I_{dark}	EQE, %	I_{dark}, pA	Res. time	Decay time	Ref.
CdS:Sn	NW	>300	16	405	14.6	~30	4.5 10^4	9 × 10^2	0.26 s	0.26 s	[105]
CdS	NB	–		405	4.1		1.3 10^3	~1	0.27 s	0.28 s	[105]
CdS:Er	NR	30–80	0.03	457	3.46 × 10^4	~10^3	9.3 10^4		–	–	[26]
CdS	NW	30	0.1	450		~6 × 10^2			–	–	[3]
CdS	NB	40	–		–	2 × 10^3		20	1 s	3 s	[16]
CdS	NW	>500	3.75	365	3.6 × 10^2	10^4	8.6 × 10^5	72	10 ms	10 ms	[18]
CdS	NR	~400	0.5	450	1.2 × 10^4	–	3.5 × 10^6		0.82 s	0.84 s	[104]
CdS	NB		0.21	484	0.036		1.5 10^6		30 ms	40 ms	[57]
CdSe	NB	60–80	3.4	650	3 10^3	1.7 × 10^4			20 μs	27 μs	[31]
ZnSe	NW				4.2 × 10^5		1.3 × 10^6		0.55 s	0.6 s	[56]
ZnSe	NB	40	–	400	0.12	>10^2	37.2	<10^{-2}	<0.3 s	<0.3 s	[14]
ZnSe:Mn	NW	3–4	0.02–9	365	0.13				<0.5 s	<0.4 s	[40]
ZnS$_{0.44}$Se$_{0.56}$	NW		–		1.5 × 10^6		4.5 × 10^6		0.52 s	0.93 s	[78]
ZnS	NB	20	–	300	0.18				<0.3 s	<0.3	[15]
ZnS	NT, array		40	350	50	1.6 10^5		~8	0.12	0.14	[2]

NW nanowires, *NB* nanobelts, *NR* nanoribbons, *NT* nanotubes

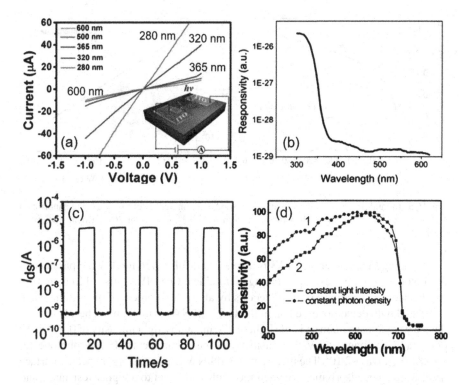

Fig. 16.4 (a) I–V curves ofs the ZnS:Ga nanoribbon-based PDs illuminated with varied wavelengths at a constant light intensity of 200 mW cm^{-2}. Inset: schematic diagram of the PCD for measurement. Reprinted with permission from [90]. Copyright 2014: RSC. (b) Spectral photoresponse measured from an individual ZnS nanobelt-based UV-light sensor for a bias of 10 V. (Reprinted with permission from Ref. [14]). Copyright 2009: Wiley. (c) Time response of CdS:Cl NW photodetector. (Reprinted with permission from Ref. [83]). Copyright 2010: IOP. (d) Spectral response of the CdSe nanoribbon. The curves represent the response spectra plotted at a constant light intensity (1) and a constant photon density (2), respectively. (Reprinted with permission from Ref. [31]. Copyright 2007: Wiley)

nanoribbons with p-type conductivity present a significant photoresponse to irradiation with high responsivity and gain of 7.8×10^2 AW^{-1} and 2.4×10^5%, respectively (Fig. 16.5).

An interesting effect was observed in CdS NWs doped with Sn [105]. It was found that a photodetector based on this material cannot only respond to light with energies exceeding the band gap, but can also respond to light with energies below the band gap, while maintaining excellent photoconductivity parameters. This phenomenon is distinctly different from that observed in pure CdS nanobelt photodetector, which can only response to light with energy exceeding the band gap. Zhou et al. [105] believed that both the trapped state induced by tin impurity and the optical whispering gallery mode microcavity effect in the Sn-doped CdS NWs contribute to improved conductivity and broad spectral response consistent with photoluminescence measurements.

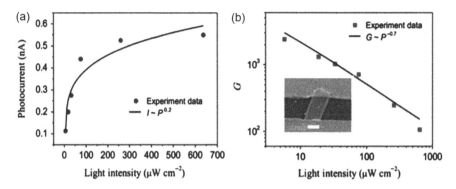

Fig. 16.5 (a) and (b) Dependence of gain and responsivity on light intensity of a CdTe nanoribbon ($\lambda = 400$ nm). (Reproduced with permission from Ref. [93]. Copyright 2012: Royal Society of Chemistry)

Changing the structure of NWs is also a powerful tool to modify the parameters of photodetectors. For example, taking hierarchical CdS NWs allowed Li et al. [41] to advance the sensitivity of the detectors by almost two orders of magnitude when compared with detectors based on individual CdS NWs without branches. When the device was exposed upon a visible light of 470 nm with a light intensity of 0.934 mW/cm^2 at a bias voltage of 5 V, the sensitivity of the hierarchical CdS NW photodetector, I_{light}/I_{dark}, was ~2 10^4. The hierarchical CdS NW has much larger specific surface area, which can adsorb more oxygen molecules and lead to a higher resistance and smaller dark current compared to conventional CdS NWs.

Another technique to optimize the parameters of NW-based photodetectors is based on a decoration of the crystal surface with plasmonic gold nanoparticles (PGNs). In particular, Luo et al. [48] showed that both sensitivity and detecting ability of CdSe NWs has been improved significantly after decorating their surface with PGNs.

It is important to note that nanostructures of II-VI ternary alloys are also utilized in the development of photodetectors for the visible-UV spectral range [49]. The most studied in this area are 1D structures of CdS$_x$Se$_{1-x}$. The photodetectors based on CdS$_x$Se$_{1-x}$ NWs were primarily reported by Choi et al. [7] who grown them by pulsed laser deposition on conductive electrodes. These devices have exhibited a high sensitivity to UV light with a reasonably fast response. An important advantage of photodetectors based on the specified II-VI ternary alloy, as shown by Choi et al. [7] and, further, by Lu et al. [50, 51], is the ability to control their spectral range of sensitivity by changing the stoichiometry of the materials. The spectral responsivity of CdS$_x$Se$_{1-x}$ NWs with different x value is shown in Fig. 16.6b. Lu et al. [50] also clarified that employing a modification of 1D structures via a focused laser could significantly improve the performance of CdS$_x$Se$_{1-x}$ nanoribbon photodetectors. In another study conducted by Guo et al. [21] it was found that the CdS$_x$Se$_{1-x}$ nanoribbons with lateral heterostructures could serve also as photodetectors with high photoresponsivity and quantum efficiency.

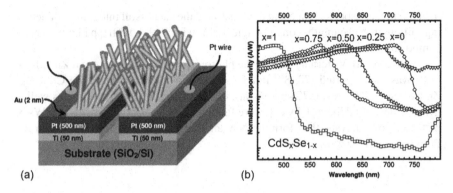

Fig. 16.6 (**a**) Schematic illustrating a network structure of CdS$_x$Se$_{1-x}$ NWs floating above the SiO$_2$/Si substrate. (**b**) Normalized spectral responsivity curves of CdS$_x$Se$_{1-x}$ NWs with a different x value. The applied bias to the device was 5 V. (Reproduced with permission from Ref. [7]. Copyright 2010: IOP Publishing Ltd)

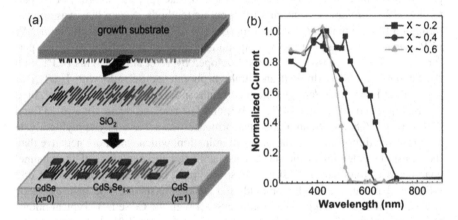

Fig. 16.7 (**a**) Schematic showing the contact printing method and the final device construction for CdSSe-based photodetectors. (**b**) The spectral response from different NW-array devices corresponding to S-rich, middle, and Se-rich regions. (Reprinted with permission from Ref. [68]. Copyright 2012: IOP Publishing Ltd)

Takahashi et al. [68] used another approach to fabricate photodetectors with II-VI ternary NWs. They designed wavelength-tunable detectors taking spatially composition graded CdS$_x$Se$_{1-x}$ NW arrays. A schematic illustration of their experiments is shown in Fig. 16.7a. CdS$_x$Se$_{1-x}$ NWs at varied compositions can be grown on the SiO$_2$/Si substrate taking an advantage to vary a distance to the vapour source. The as-synthesized NWs covered a wide range of visible light from 525 nm, S-rich area, to 650 nm, Se-rich area. The Ti/Au, of 5 nm/40 nm, thin film contact electrodes were finally deposited by electron beam evaporation to form an array of wavelength tunable photodetectors. As it is seen in Fig. 16.7b, NWs in different places on the substrate had different spectral responses, which contributed to an increase in the response range and demonstrated the successful integration of detecting different wavelengths on a single chip with NWs with spatially varying composition.

As for other II-VI ternary alloys, the photoresponse of ZnS_xSe_{1-x} and $Zn_xCd_{1-x}Se$ NWs was also reported [78, 88] in addition to photodetectors based on CdS_xSe_{1-x}. $Zn_xCd_{1-x}Se$ (0.31 < x < 0.72)-based variable-wavelength photodetectors, as well as the ones described in work by [7], were fabricated by a selective growth of NWs on patterned Au catalysts, thus forming NW air-bridges (see Fig. 16.6a) between two Pt pillar electrodes [88].

16.4.2 Phototransistors

Typically, NW field-effect phototransistors (FEPTs) have been fabricated via dispersing NWs on a dielectric-semiconductor substrate [22, 25, 85, 103] or by patterning NWs through conventional lithographic methods. A gate bias was applied through a lithographically patterned top gate, or a back gate. An example of NW FEPT made on the basis of II-VI compounds is given in Fig. 16.8a, b. A high performance CdS NW- based FEPT was developed by Ye et al. [85]. This FEPT with Au Schottky contact exhibited an ultrahigh photoresponse ratio with I_{light}/I_{dark} of 2.7×10^6 at $P_\lambda = 5.3$ mW/cm^2, the high current responsivity of 2.0×10^2 AW^{-1}, high external quantum efficiency of 5.2×10^2, and fast rise and decay time of 137 and 379 μs (Fig. 16.8c, d). Such a good performance was achieved by applying a gate shift near the threshold potential under illumination, which was more negative than the threshold potential in the dark. This facilitated the depletion of the channel charge carriers and led to a fast photoresponse [85]. Dai et al. [9] fabricated CdSe NBs-based metal-semiconductor field-effect phototransistors. The gate electrode was Au, which formed a good Schottky contact with the CdSe NB. Typical phototransistors exhibited a high responsivity, of ca. 1.4×10^3 A/W, high gain, of ca. 2.7×10^3, and fast response speed, of 35–60 μs, under a gate voltage of −1 V, which was attributed to the unique advantages of the high-performance MESFET structure.

In order to develop FEPTs based on ZnS:Ga nanoribbons, Yu et al. [90] used an ITO gate electrode, 80 nm thick, deposited by pulsed laser deposition. This study reported that the ZnS doping with Ga, combined with improved ohmic contact and the use of ITO as the contact, enabled the photodetectors to yield a high photoconductive gain under extremely low intensity of the light. Upon an ultralow operating voltage, of ~0.01 V, the device exhibited excellent photoresponse properties to $1 \cdot 10^{-14}$ W UV light such as a high gain of 10^6, and fast response speed. Response and decay times were equaled to 3.2 ms and 10 ms, respectively. These authors believe that doped ZnS NRs have a great potential as the building blocks for low-intensity UV phototransistors. The detectivity of the ZnS:Ga FEPT at 0.01 V was estimated to be 1.3×10^{19} cmHz $^{1/2}$ W^{-1} under a light intensity of 1.0 μW cm^{-2}, which is higher than the detectivity of traditional and most nanophotodetectors.

Phototransistors based on II-VI ternary alloys have also been developed. In particular, the results of testing FEPTs based on CdS_xSe_{1-x} are given in [43].

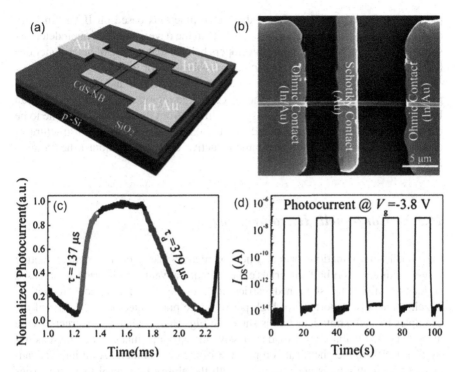

Fig. 16.8 (a) Schematic illustration of a single CdS nanobelt MESFET-based photodetector. (b) Typical FESEM image of a single CdS NB MESFET-based photodetector. (c) Transient photocurrent response of (A) at $V_G = 3.8$ V and $V_{DS} = 0.5$ V. (d) On/off photocurrent response of the CdS NB MESFET-based photodetector with $V_G = -3.8$ V as a function of time on an exponential scale. (Reprinted with permission from Ref. [85]. Copyright 2010: American Chemical Society)

As follows from the principle of FEPT operation, the transistor effect occurs due to the presence of a lateral electric field across the NW, which modulates the conductance in the channel. However, a similar effect is also present in NW photoconductors where a radial electric field is appeared, too, due to trapping charge carriers in deep trap states or surface defect states. The appearance of a charge on the surface is accompanied by band bending. This causes a separation of photogenerated carriers in the NW channel, which greatly extends the carrier recombination lifetime with enabling a much higher sensitivity. Thus, NW photoconductors can be viewed as phototransistors where the internal electric field in combination with light illumination acts as a photogate [8].

16.5 NWs-Based Heterostructures

Currently, heterojunctions are considered to be fundamental elements for modern electronic and optoelectronic devices such as light emitting diodes, laser diodes, photodetectors, and solar cells, which provide them with high performance parameters. Experiment has shown that heterojunctions based on II-VI semiconducting 1D nanostructures can also contribute to the development of photodetectors suitable for the market [95]. Many developers believe that the use of heterojunctions based on 1D nanostructures opens up new possibilities for improving the efficiency of photoconversion in photodetectors. For example, efficient and fast separation of photogenerated charge carriers in the space charge region of a heterojunction can lead to a fast photoresponse. This response differs from the slow- relaxed one to be associated with the behavior of adsorbates on the surface of nanostructuresas observed in 1D-structure-based photoconductive detectors without a heterojunction [37].

16.5.1 Core-Shell Heterojunctions

Core-shell heterojunctions or radial heterostructures based on 1D semiconducting nanostructures *versus* bulk materials exhibit unique advantages in terms of increased surface-to-bulk ratio, shortened carrier collection path, and reduced reflection, which are vitally important for high-performance photodetectors [96]. One of the examples of such photodetectors is shown in Fig. 16.9. Radial heterostructures are also interesting in that they provide a passivating envelope that reduces the sensitivity of the NW to the chemical composition of the environment. Additional advantages of such structures are associated with the ability to control their properties through changing the diameter and chemical composition of both the core and the shell.

Currently, a number of strategies have been developed to construct core-shell heterojunctions. These include chemical vapor deposition (CVD), atomic layer deposition (ALD), pulsed laser deposition (PLD), solution-based cation exchange reaction, sputtering, and molecular beam epitaxy (MBE). Core-shell

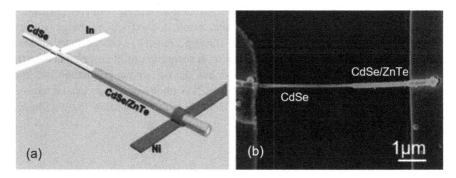

Fig. 16.9 (**a**) Schematics of a photovoltaic device based on an individual CdSe/ZnTe core-shell NW. The electrode materials for the core and shell are indium and nickel respectively. (**b**) A representative SEM image of a photodetector developed based on CdSe/ZnTe core-shell NW. The ZnTe shell was partially etched by a saturated, aqueous $FeCl_3$ solution. (Reprinted with permission from Ref. [75]. Copyright 2014: RSC)

Table 16.2 II-VI semiconducting based core-shell structure designed for photodetector application

Core-shell structure	Core material	Method of synthesis	Shell material	Method of synthesis	Type of PD	Ref.
Ge/CdS, NWs	Ge	PVD	CdS (NGrs)	ALD	Ind.	[67]
Si/CdS, NWs	p-Si	CE	CdS (NGrs)	PLD	Array	[55]
CdSe/ZnTe, NWs	CdSe	TE	ZnTe	PLD	Ind.	[75]
		CVD			Array	[62]
CdS/ZnTe, NWs	n-CdS:Ga	TE	p-ZnTe (NGr)	TE	Ind.	[74]
CdS/Cu$_2$S, NWs	CdS	CSM	Cu$_2$S	CSM	Ind.	[70]
CdS/CdTe, NBs	CdS	CVD	CdTe	CVD	Ind.	[69]
ZnSe/ZnO, NWs	p-ZnSe	Sputt.	n-ZnO (NPs)	Sputt.	Ind.	[97]
ZnS/CdS, NBs	ZnS	TE	CdS	TE	Ind.	[47]
ZnS/InP, NWs	ZnS	CVD	InP	CVD	Ind.	[95]
ZnS/ZnO, biaxial NBs	ZnS	TE	ZnO	TE	Ind.	[27]
ZnO/ZnSe, NWs	ZnO	TE	ZnSe	TE	Array	[60]
ZnO/ZnS, NRs	ZnO	CSM	ZnS	CSM	Array	[44]
ZnO/ZnS, NWs	ZnO	CVD	ZnS (NPs)	PLD	Array	[63]

ALD atomic layer deposition, *CE* chemical etching, *CSM* chemical solution methods, *CVD* chemical vapour deposition, *Ind.* individual, *NB* nanobelts, *NGr* nanograins, *NPs* nanoparticles, *NRs* nanorods, *NW* nanowire, *PD* photodetector, *PLD* pulsed laser deposition, *PVD* physical vapor deposition, *SC* solar cell, *Sputt.* spattering, *TE* thermal evaporation

heterojunctions employed in photodetector design are listed in the Table 16.2. For core-shell heterojunctions based on II-VI compound semiconductors, two-steps methods are usually applied as construction approaches. In this case, the NWs are first grown to build a core, and then shells are formed on the surface of these NWs in various ways. A schematic representation of a typical manufacturing process for a photodetector based on core-shell heterojunctions is shown in Fig. 16.10.

16.5.2   1D Axial Heterojunctions

The axial geometry of 1D-based heterojunctions is another option for II-VI semiconducting heterostructures to be promised for use in photodetectors. Numerous works have suggested that 1D axial heterojunction should demonstrate lower leakage currents and, therefore, they should possess a superior rectifying behavior compared to radial structures [52]. Currently, two approaches have been proposed for the implementation of such structures based on II-VI semiconductors. The first approach deals with modifying the composition of NWs during their growth by changing the precursors served as a source (Fig. 16.11a). In particular, Mu et al. [56]

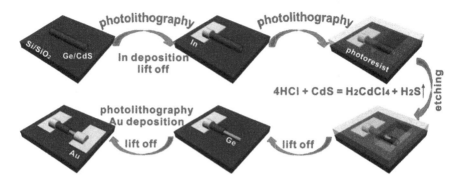

Fig. 16.10 Schematic of the fabrication process of the photodetector based on a Ge/CdS core–shell heterojunction NW. (Reprinted with permission from Ref. [67]. Copyright 2016: RSC)

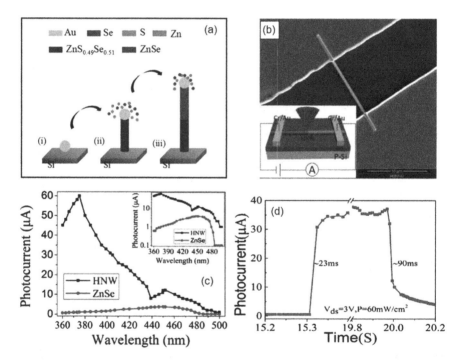

Fig. 16.11 (a) Schematic diagram of the growth process for $ZnS_{0.49}Se_{0.51}/ZnSe$ axial heterojunction. (b) SEM image of as-prepared HNW nanodevice. Inset is schematic model of photodetector based on single $ZnS_{0.49}Se_{0.51}/ZnSe$ HNW. (c) Wavelength-dependent photocurrent response of $ZnS_{0.49}Se_{0.51}/ZnSe$ heterojunction and pure ZnSe NW photodetectors. Inset is corresponding semi-logarithmic plot. (d) Experimental and fitted photocurrent versus power plot under 375 nm illumination and V_{ds} = 3 V. (e) Time-resolved photocurrent of the $ZnS_{0.49}Se_{0.51}/ZnSe$ heterojunction photodetector in response to 375 nm laser on/off at V_{ds} = 3 V. (Reprinted with permission from Ref. [56]. Copyright 2019: Wiley)

employed this approach to grow 1D ZnSSe-ZnSe axial heterostructures and to fabricate the related photodetectors. SEM image and characteristics of these photodetectors are presented in Fig. 16.11b, c, d. Zhang et al. [101] also applied this method with a variation. They constructed a kind of p-n heterojunction arrays by direct growing the p-type ZnSe NRs on highly-aligned n-type Si NWs arrays. Cross-section SEM images showed that these quasi-aligned nanoribbons were just grown as an array on the top of Si NWs array.

Another approach to employ the local oxidation of NWs of II-VI compounds allows one preparing single 1D-axial heterostructures. Thus, Zhang et al. [99], fabricated a single ZnSe-ZnO NW axial p-n junction following such a technique. For this, as-synthesized p-type ZnSe NWs were transferred onto the SiO_2/Si substrate by a sliding transfer process. Then, a Si_3N_4 protection layer was applied to one half of the NW with further oxidizing the exposed part via placing the sample into a fast heat treatment system. After the thermal oxidation, the Si_3N_4 protective layer was removed by reactive ion etching. As a result, the authors could get a NW whose one half consists of unoxidized ZnSe, and another one consists of post-oxidized ZnO. This process is illustrated in Fig. 16.12.

16.5.3   Crossed NW Heterojunctions

As previously demonstrated, fabricating core-shell heterojunctions and 1D axial heterojunctions is always a complicated multi-step preparation process [95]. Therefore, the search for simpler technological solutions is constantly being carried out. In this regard, using the crossed NWs to make heterojunctions is such a solution. The crossed NW heterojunction is a kind of post-assembly structure. According to Huang et al. [29], such a configuration of heterojunctions presents many advantages, including the ability: (1) to flexibly choose component materials; (2) to independently tune the dopant concentration of the component materials; (3) to define abrupt, nanoscale junctions that are ideal for high spatial resolution; and (4) to assemble arrays for integrated nano-optoelectronic devices. The manufacturing process of crossed NW heterojunctions is shown schematically in Fig. 16.13.

16.5.4   1D Nanostructure/Thin Film or Si
Substrate Heterojunctions

Coupling of II-VI group semiconducting 1D nanostructures and Si substrates is another approach to design heterojunctions. Moreover, this method to create heterojunctions based on 1D nanostructures is the simplest one. This approach has the following advantages [95]: (1) this could have avoided the complex multi-step process of preparing the device for fabrication; (2) it is compatible with conventional

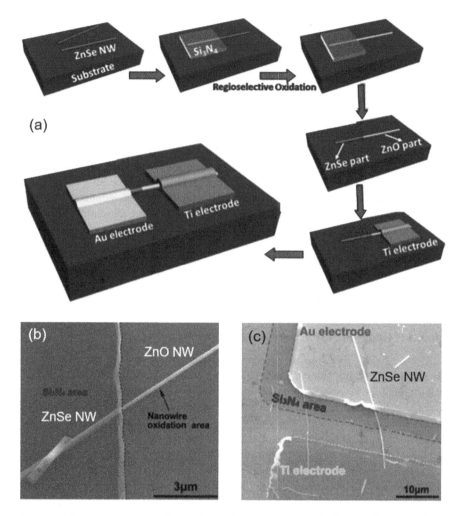

Fig. 16.12 (a) Schematic illustration of the construction process of the ZnSe-ZnO axial p-n junction. (b) The SEM image of a NW whose bare part has already been oxidized. (c) The SEM image of a prepared ZnSe-ZnO NW axial p-n junction. (Reproduced with permission from Ref. [99], Copyright 2015: Elsevier B.V)

microelectronic technology; and (3) it is suitable for developing aligned plane array integrated devices.

Ordinarily, the manufacturing process of such heterojunctions composes of the following operations [98] (see Fig. 16.14): (i) the formation of SiO_2 insulating pads on a n-Si substrate; (ii) the transfer of NWs from the growth substrate onto the patterned SiO_2/Si substrate; (iii) after dispersing, some NWs would cross on the edges of the SiO_2 pads and partially contact with the underlying n-Si substrate, where the heterojunction is formed; and (iv) finally, a metal film (usually, Au) is applied on the SiO_2 pads to provide the ohmic contacts to the NWs. Using this approach, such

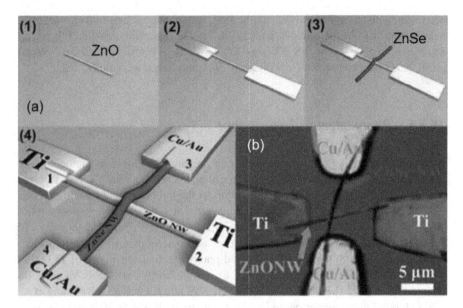

Fig. 16.13 (a) Schematic illustration of the step-wise process (1,2,3,4) for the fabrication of a ZnSe NW/ZnO NW p-n junction. (b) SEM image of ZnSe/ZnO crossed NW heterostructure. (Reproduced with permission from Ref. [58]. Copyright 2013: Institute of Physics)

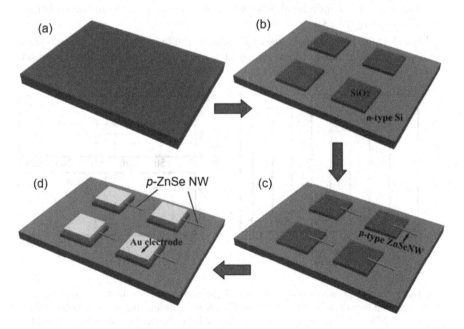

Fig. 16.14 Schematic illustration of the fabrication process of ZnSeNW/Si p-n junction. (Reproduced with permission from Ref. [98]. Copyright 2016, Elsevier B.V)

heterojunctions as p-ZnSe NW / n-Si [98], ZnSe nanoribbon/Si [73], CdS:Ga nanoribbon/Si [82], n-CdSe NWs/p$^+$-Si [28], p-ZnS nanoribbon/n-Si [91], p-CdS nanoribbon/n-Si [92], n-CdS NWs/p+-Si [4], ZnSe NWs/Si [76], ZnSe NWs/Si [100], CdTe nanoribbons/ n-Si NWs array [94], and so on, were fabricated.

16.5.5　Photodetector Performance

It is important to note that heterojunctions prepared by different methods were tested and showed acceptable photoelectrical characteristics. So, CdS-ZnTe core-shell heterojunction photodetectors tested under the light illumination with a wavelength of 638 nm and light intensity of 2 mWcm^{-2} at 1 V, exhibited responsivity of 1.55 × 10^3 AW^{-1}, conductive gain of 3.3 × 10^3, and detectivity of 8.7 × 10^{12} cm·Hz$^{1/2}$ W^{-1} [74]. Ge-CdS core-shell heterojunction NW photodetectors, developed by Sun et al. [67], showed photodetection sensitivity of 18,000%, which is remarkably much higher than sensitivity of arrays of CdS NWs and CdS quantum dots [1, 24]. CdSe/ZnTe core/shell NW array, fabricated by Wang et al. [75] exhibited even greater photodetection sensitivity. Rai et al. [63] and Lin et al. [44] developed an efficient and highly sensitive broad band UV/VIS photodetector based on wide band gap ZnO/ZnS heterojunction 3D core/shell NW array. It is worth noting that the photoresponse of heterojunction was always greater and faster than the response of individual NWs which are employed to form such a heterojunction (see Fig. 16.15).

Photodetectors based on a crossed ZnSe/ZnO p-n heterojunction and ZnSe-ZnO axial p-n junction, designed by Zhang et al. [95, 99], exhibited excellent diode

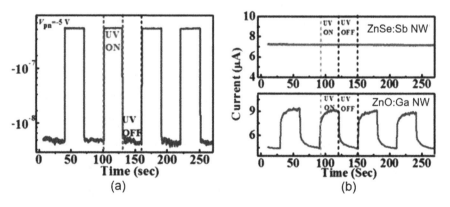

Fig. 16.15 (a) The photoresponse of the p-ZnSe:Sb/n-ZnO:Ga core shell NW junction diode under illumination with 365 nm UV light. (b) The photoresponse of the single ZnO:Ga NW and the ZnSe:Sb NW under illumination with 365 nm UV light. During the photodetection measurement, the light intensity and the bias voltage were kept at 100 μW cm^{-2} and − 5 V, respectively. The UV light was switched on and off manually. (Reprinted with permission from Ref. [58]. Copyright 2013: IOP)

behaviors and high sensitivity to ultraviolet light illumination with a good reproducibility and quick photoresponse. Both photodetectors showed much higher sensitivity than the detectors based on ZnO, ZnTe, and CdSe NWs [5, 6, 80]. CdS/CdS$_x$Se$_{1-x}$ axial heterostructure NWs, fabricated via a temperature-controlled chemical vapor deposition method [20] also demonstrated an appropriate photodetector performance. The photodetector based on the single CdS/CdS$_x$Se$_{1-x}$ NWs showed higher responsivity (1.2×10^2 A / W), faster response speed (rise ~68 µs, decay ~137 µs), higher I$_{light}$/I$_{dark}$ ratio (10^5), higher EQE (3.1×10^4%), and broader detection range (350–650 nm) at room temperature *versus* the photodetector based on single nanostructures. The performance of several photodetectors based on II-VI semiconducting NW heterojunctions is shown in Fig. 16.16.

Based on crossed CdS/p-Si NW heterojunction, Hayden et al. [23] designed avalanche photodiodes (APDs), which exhibited a photocurrent response (I$_{PC}$/I$_{dark}$) to be approx. 10^4 times higher than that in individual n-CdS or p-Si NW photoconductors. This effect was observed due to avalanche multiplication at the p-n crossed NW

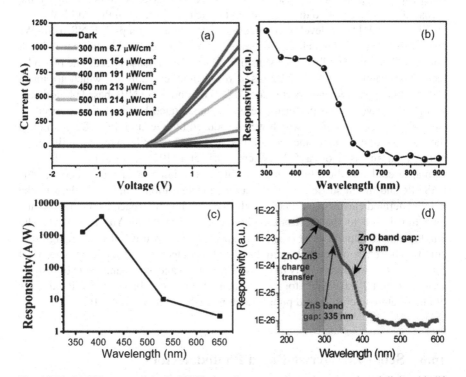

Fig. 16.16 (a) I-V curves of the ZnS/CdS photodetector under various wavelength light with different intensities and in the dark; (b) Spectroscopic photoresponses of the ZnS/CdS detector measured at different wavelengths ranging from 300 to 900 nm at a bias of 2 V. (Reprinted with permission from Ref. [47]. Copyright 2016: RSC). (c) Spectral photoresponse of CdS/CdS:SnS$_2$ superlattice NW. (Reprinted with permission from Ref. [18]. Copyright 2016: RSC). (d) Spectral photoresponse of ZnS/ZnO biaxial nanobelt with Cr/Au electrodes measured at a bias of 5.0 V. (Reprinted with permission from Ref. [27]. Copyright 2012: Wiley)

Table 16.3 Comparison of the spectral responsivity, external quantum efficiency, and response time of II-VI semiconducting NW heterostructure-based photodetectors

Materials	Responsivity, $A \cdot W^{-1}$	I_{ligh}/I_{dark}	EQE, %	Rise time	Decay time	Ref.
CdS/CdSSe	1.18×10^2		3.1×10^2	68 µs	137 µs	[20]
ZnS/ZnO	–		–	0.77 s	0.73 s	[71]
ZnO/ZnSe	8.6		–	–	–	[84]
CdS/CdSSe	1.2×10^3		10^6	280 ms	550 ms	[21]
$ZnS_{0.49}Se_{0.51}$/ZnSe	6.3×10^5		2.1×10^6	23 ms	90 ms	[56]
CdS/ZnTe	1.6×10^3		3.3×10^3	–	–	[74]
CdS/CdS:SnS_2	2.5×10^3	$\sim 10^5$	8.6×10^5	10 ms	10 ms	[18]
ZnSe:Sb/ZnO	5.2×10^5	1.8×10^6	–	–	–	[58]
ZnS/InP	–	$5 \ 10^3$	1.1×10^3	0.75 s	0.5 s	[96]

EQE External quantum efficiency

junction. A detection limit of about 75 photons was estimated for these devices. Polarization dependence of the photoresponse has also been observed in the "crossed" structure due to the predominant optical absorption in the CdS NW.

Xie et al. [94] have developed a photodetector, based on a p-CdTeNR/n-SiNWs array heterojunction, which can operate in the light, of visible to near-infrared range, with a good stability, high sensitivity, and fast response speed. The photodetector had small rise time, ca. 1.2 ms, and fall time, ca. 1.58 ms, which are much faster than the parameters reported for CdTe NWs and CdTe NRs-based photodetectors [1, 86]. Even better performance was demonstrated by photodiodes based on CdS NR/Si heterojunctions which were produced by Xie et al. [92]. Detectors had the rise time of ca. 300 µs and the fall time of ca. 740 µs under white light illumination with an intensity of ca. 5.3 mW cm^{-2}. Xie et al. [92] believed that such a performance could attribute to the high-quality of p-n junction appeared between the CdS NR and the Si substrate. Due to a low concentration of interface defects, the photogenerated carriers could be quickly separated by the space charge region and then transferred to the electrodes. Photodetectors made by Yu et al. [91], who employed heterojunctions of p-ZnS nanoribbon/n-Si substrate, also had a high sensitivity. The measurements showed that the photodetectors had stable optoelectrical properties with high sensitivity to UV-VIS-NIR light and an enhancement of responsivities of ca. 1.1×10^3 AW^{-1} for $\lambda = 254$ nm under a reverse bias of 0.5 V. Parameters of other heterostucture-based photodetectors are listed in the Table 16.3

16.6 Schottky Barrier-Based Photodetectors

Schottky photodiodes based on metal/semiconductor junction exploit such an interface to separate and collect the photogenerated carriers. Electron-holes pairs (EHP) generated in the vicinity of the depletion region are swept out by the built-in electric field of the Schottky junction. It is believed [8, 81, 106] that one of the major advantages of Schottky photodiodes relative to photoconductive structures with Ohmic

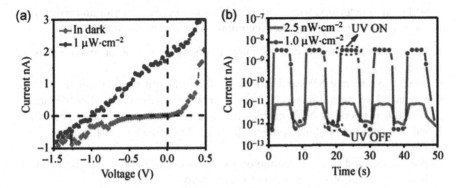

Fig. 16.17 (a) I-V curves of the Schottky barrier-based detectors (ITO-p-ZnS) measured in dark and under UV light intensity of ~1 μWcm^{-2}. (b) Time response of the ITO-p-ZnS detectors measured under UV light intensity of 1 μWcm^{-2} and 2.5 nWcm^{-2}, respectively. The external bias voltage is fixed at 0.01 V. (Reprinted with permission from Ref. [89]. Copyright 2015: Springer)

contacts in addition to the simplicity of the manufacturing technology are the fast response rate due to the high electric field (hence, the short transit time of charge carriers) through the junction under reverse bias, lower dark currents, and higher photoresponse. For example, Wu et al. [79] fabricated CdS:Ga NRs/Au Schottky barrier diodes, which demonstrated a high light sensitivity and fast response speed, response/decay times of ca. 95/290 μs, at a zero bias. A CdS single-NB Schottky contact photodetectors fabricated by Li et al. [42] were also characterized by a high sensitivity and short decay time. The photosensitivity, dark current and the decay time of the sensor were ca. 4×10^4, 0.2 pA, and 31 ms, respectively. Yu et al. [89] suggested using the ITO-p-ZnS contact as a Schottky barrier-based photodetectors. The device exhibited a pseudo-photovoltaic behavior which can allow one detecting UV light irradiation with incident power of ca. 6×10^{-17} W, or ~ 85 photons/s on the NR, at room temperature with excellent reproducibility and stability. The corresponding detectivity and photoconductive gain were calculated to be ca. 3.1×10^{20} cm·Hz$^{1/2}$·W^{-1} and 6.6×10^5, respectively (see Fig. 16.17).

More results from the Schottky photodiodes study based on CdS, and CdSe NWs can be found elsewhere [11, 12, 19, 77]. During these studies, it was shown that the photocurrent – voltage characteristics of such structures were typically asymmetric and photocurrent could be strongly localized near the metal electrode-NW contact in dependence on the biasing conditions.

16.7 Summary

In this chapter, we have described the various approaches utilized to fabricate quasi-1D NW-based photodetectors. Based on this consideration and analysis of the results presented in numerous articles, it can be concluded that photodetectors based on II-VI semiconductor-based 1D structures really have advantages over bulk and thin-film photodetectors, which can lead to cost-effective, superior and novel functionalities. The major advantages, according to [8], are the following:

- Large photosensitivity due to internal photoconductive or phototransistive gain [102];
- Ability to integrate NWs and substrates of different materials, including silicon;
- Enhanced light absorption with a small amount of active material in NW arrays. A dense network of NWs enables multiple reflections and scatterings, which enhance incident photon trapping, a feature, which does not exist in conventional PDs [102];
- Ability to build devices, as photoresistor, photodiode, phototransistor, on the basis of single NWs with different functionality;
- Opportunity for dense-packed device integration.

At the same time, along with advantages, the technologies dealing with NWs also has a number of significant disadvantages [34, 36, 59]. Primarily, in order to implement the most progressive ideas based on NW integration with silicon technology, a technology of direct growth and integration of II-VI semiconducting NW-based devices onto large Si wafers is required. Unfortunately, there are some challenges such as lattice constant and thermal mismatch, lack of good control over atomic and crystallographic structures, troubles with NWs assembly into functional devices, a complicated forming of ternary and quaternary NW alloys, high contact resistance and increased noise due to this fact [45], the lack of a well-controlled doping technology to develop sharp homo- and heterojunctions, catalyst-induced pollution, etc. makes it difficult to grow and design NW devices on Si or any other substrates which is a failure in the industrial application of NW PDs. Therefore, the technology for integrating II-VI NWs with Si-integrated circuits is still at an infancy stage despite on the significant research progress ongoing in this direction for recent years. There still exists a a large gap to overcome towards the practical integration and application of photodetectors based on 1D nanomaterials [34, 36]. Two issues should be resolved. First, we do not have an inexpensive, simple and reproducible method to synthesize one-dimensional nanomaterials with precisely managed geometric and physical properties, such as diameter, length, orientation, crystal quality, and doping concentration. Second, the controlled and scalable post-growth transfer and assembly of one-dimensional nanomaterials over target substrates is rather difficult in most cases to manufacture such photodetectors. Some methods have been proposed so far, including the use of fluid-assisted alignment, the Langmuir-Blodgett technique, the dielectrophoretic method, contact printing, etc. [36]. However, their ability to precisely control the position and orientation of each one-dimensional nanomaterial is still insufficient. Therefore, one cannot expect that there will be monolithic devices integrating II-VI semiconducting NWs and Si integrated circuits in the coming years. Most likely, hybrid approaches will be used at the initial stage, which would not require drastic changes in the current manufacturing process of Si-integrated circuits.

Another limitation to use the NWs-based photodetectors is their rather low quantum efficiency. While the most conventional semiconductor detectors operate at almost 100% quantum efficiency, then NW-based ones' efficiencies are much lower [59]. Although II-VI semiconducting 1D nanomaterials have a high absorption coefficient near the band edge, they absorb a little light due to thin diameters. With purpose to enhance the light absorption in 1D nanostructures, Wang et al. [72] and Fang and Hu [13] analyzed six routes to realize localized fields enhancement in two-dimensional material-based photodetectors. They considered ferroelectric field, photogating electric field, floating gate-induced electric field, interlayer built-in field, localized optical field, and photo-induced temperature gradient field. It is proved that localized fields substantially enhance the optical absorption, suppress background noise, and advance the efficiency of separation of electrons and holes. Wang et al. [72] and Fang and Hu [13] believe that combining 1D II-VI nanomaterial-based photodetectors with the idea of localized fields enhancement might be an efficient method to further improve the performance of photodetectors in the future.

There are also challenges associated with the scatter of parameters and the negative impact of surface traps on the photoelectrical properties of semiconducting NWs. Semiconducting NWs, especially those obtained by the bottom-up method, have stochastic deviations from each other due to their defects, which affect the charge capture and carrier mobility. Although tighter control of NWs growth and the use of high purity chemical precursors eliminate some of these defects, it is virtually impossible to prepare identical NWs with bottom-up approaches due to other physical and thermodynamic factors that arise during NWs growth. These problems were discussed in sufficient detail in [36, 45].

As we indicated earlier, involving surface effects in the photoresponse allows one to achieve very high sensitivityin the corresponding detectors. However, the strong dependence of the NW resistance on the state of the surface also has negative consequences, since the photoresponse becomes dependent on the state of the surrounding atmosphere, namely, on air humidity, temperature, and the presence of toxic and reducing gases. This means that the parameters of the photodetectors will alter when these factors vary, and this change can be significant. These effects have been detailed in Chap. 2 (Vol. 1). The encapsulation of NWs eliminates the influence of the surrounding atmosphere on the response of photoconductive detectors. In this case, the internal photoconductive or phototransistor gain will also disappear. In this regard, photodetectors based on Schottky barriers and core-shells structures as well as phototransistors may be more preferable when employed in an uncontrolled atmosphere.

There are also problems related to the mechanical stability of photodetectors based on 1D structures. A vibration might cause displacing NWs relative to each other or to the substrate, which should undoubtedly be accompanied by a change in the properties of homo- and heterojunctions having such contacts. The operational reliability of NW-based photodetectors against to Joule self-heating is also an issue for single nanostructures that needs to be addressed in high-power NW photodetectors [36, 45].

Acknowledgments G. Korotcenkov is grateful to the State Program of the Republic of Moldova, project 20.80009.5007.02, for supporting his research. V. Sysoev thanks the Russian Science Foundation, project No. 22-29-00793 (https://rscf.ru/project/22-29-00793/), for a partial support.

References

1. Amos FF, Morin SA, Streifer JA, Hamers RJ, Jin S (2007) Photodetector arrays directly assembled onto polymer substrates from aqueous solution. J Am Chem Soc 129:14296–14302
2. An Q, Meng X, Xiong K, Qiu Y, Lin W (2017) One-step fabrication of single-crystalline ZnS nanotubes with a novel hollow structure and large surface area for photodetector devices. Nanotechnology 28:105502
3. An B-G, Chang YW, Kim H-R, Lee G, Kang M-J, Park J-K, Pyun J-C (2015) Highly sensitive photosensor based on insitu synthesized CdS nanowires. Sensors Actuators B Chem 221:884–890
4. Cai J, Jie J, Jiang P, Wu D, Xie C, Wu C et al (2011) Tuning the electrical transport properties of n-type CdS nanowires via Ga doping and their nano-optoelectronic applications. Phys Chem Chem Phys 13:14663–14667
5. Cao YL, Liu ZT, Chen LM, Tang YB, Luo LB, Jie JS et al (2011) Single-crystalline ZnTe nanowires for application as high-performance green/ultraviolet photodetector. Opt Express 19:6100–6108
6. Chang SP, Lu CY, Chang SJ, Chiou YZ, Hsueh TJ, Hsu CL (2011) Electrical and optical characteristics of UV photodetector with interlaced ZnO nanowires. IEEE J Sel Top Quantum 17:990–995
7. Choi Y-J, Park K-S, Park J-G (2010) Network-bridge structure of CdS$_x$Se$_{1-x}$ nanowire-based optical sensors. Nanotechnology 21:505605
8. Dai X, Tchernycheva M, Soci C (2015) Compound semiconductor nanowire photodetectors. In: Semiconductors and semimetals, vol 94. Elsevier, Cambridge, MA, pp 75–107
9. Dai Y, Yu B, Ye Y, Wu P, Meng H, Dai L, Qin G (2012) High-performance CdSe nanobelt based MESFETs and their application in photodetection. J Mater Chem 22:18442–18446
10. Deng K, Li L (2014) CdS nanoscale photodetectors. Adv Mater 26(17):2619–2635
11. Doh Y-J, Maher KN, Ouyang L, Yu CL, Park H, Park J (2008) Electrically driven light emission from individual CdSe nanowires. Nano Lett 8:4552–4556
12. Dufaux T, Burghard M, Kern K (2012) Efficient charge extraction out of nanoscale Schottky contacts to CdS nanowires. Nano Lett 12:2705–2709
13. Fang H, Hu W (2017) Photogating in low dimensional photodetectors. Adv Sci 4:1700323
14. Fang X, Xiong S, Zhai T, Bando Y, Liao M, Ujjal K, Gautam UK et al (2009) High-performance blue/ultraviolet-light-sensitive ZnSe-nanobelt photodetectors. Adv Mater 21:5016–5021
15. Fang X, Bando Y, Liao M, Gautam UK, Zhi C, Dierre B et al (2009b) Single-crystalline ZnS nanobelts as ultraviolet-light sensors. Adv Mater 21:2034–2039
16. Gao T, Li QH, Wang TH (2005) CdS nanobelts as photoconductors. Appl Phys Lett 86:173105–173107
17. Garnett E, Mai L, Yang P (2019) Introduction: 1D nanomaterials/nanowires. Chem Rev 119(15):8955–8957
18. Gou G, Dai G, Qian C, Liu Y, Fu Y, Tian Z et al (2016) High-performance ultraviolet photodetectors based on CdS/CdS:SnS$_2$ superlattice nanowires. Nanoscale 8:14580–14586
19. Gu Y, Romankiewicz JP, David JK, Lensch JL, Lauhon LJ, Kwak ES, Odom TW (2006) Local photocarrent mapping as a probe of contact effects and charge carrier transport in semiconductor nanowire devices. J Vac Sci Technol B 24:2172–2177
20. Guo P, Xu J, Gong K, Shen X, Lu Y, Qiu Y et al (2016) On-nanowire axial heterojunction design for high-performance photodetectors. ACS Nano 10:8474–8481
21. Guo P, Hu W, Zhang Q, Zhuang X, Zhu X, Zhou H et al (2014) Semiconductor alloy nanoribbon lateral heterostructures for high-performance photodetectors. Adv Mater 26:2844–2849
22. Han S, Jin W, Zhang DH, Tang T, Li C, Liu XL et al (2004) Photoconduction studies on GaN nanowire transistors under UV and polarized UV illumination. Chem Phys Lett 389:176–180
23. Hayden O, Agarwal R, Lieber CM (2006) Nanoscale avalanche photodiodes for highly sensitive and spatially resolved photon detection. Nat Mater 5:352–356

24. He J, Chen J, Yu Y, Zhang L, Zhang G, Jiang S et al (2014) Effect of ligand passivation on morphology, optical and photoresponse properties of CdS colloidal quantum dots thin film. J Mater Sci Mater Electron 25:1499–1504
25. Heo YW, Tien LC, Kwon Y, Norton DP, Pearton SJ, Kang BS, Ren F (2004) Depletion-mode ZnO nanowire field-effect transistor. Appl Phys Lett 85:2274–2276
26. Hou D, Liu Y-K, Yu D-P (2015) Multicolor photodetector of a single Er^{3+}-doped CdS nanoribbon. Nanoscale Res Lett 10:285
27. Hu L, Yan J, Liao M, Xiang H, Gong X, Zhang L, Fang X (2012) An optimized ultraviolet-a light photodetector with wide-range photoresponse based on ZnS/ZnO biaxial nanobelt. Adv Mater 24:2305–2309
28. Hu Z, Zhang X, Xie C, Wu C, Zhang X, Bian L et al (2011) Doping dependent crystal structures and optoelectronic properties of n-type CdSe:Ga nanowries. Nanoscale 3:4798–4803
29. Huang Y, Duan X, Wei Q, Lieber CM (2001) Directed assembly of one-dimensional nano-structures into functional networks. Science 291:630–633
30. Islam MS, Sharma S, Kamins TI, Williams RS (2004) Ultrahigh-density silicon nano-bridges formed between two vertical silicon surfaces. Nanotechnology 15:L5–L8
31. Jiang Y, Zhang WJ, Jie JS, Meng XM, Fan X, Lee ST (2007) Photoresponse properties of CdSe single-nanoribbon photodetectors. Adv Funct Mater 17:1795–1800
32. Jie JS, Zhang WJ, Jiang Y, Meng XM, Li YQ, Lee ST (2006a) Photoconductive characteristics of single-crystal CdS nanoribbons. Nano Lett 6:1887–1892
33. Jie JS, Zhang WJ, Jiang Y, Lee ST (2006b) Single-crystal CdSe nanoribbon field-effect transistors and photoelectric applications. Appl Phys Lett 89:133118
34. Jin W, Hu L (2019) Review on quasi one-dimensional CdSe nanomaterials: synthesis and application in photodetectors. Nano 9:1359
35. Kayes BM, Filler MA, Putnam MC, Kelzenberg MD, Lewis NS, Atwater HA (2007) Growth of vertical aligned Si wire arrays over large areas (> 1 cm^2) with au and cu catalysis. Appl Phys Lett 91:10310–103113
36. Korotcenkov G (2020) Current trends in nanomaterials for metal oxide-based conductometric gas sensors: advantages and limitations. Part 1: 1D and 2D nanostructures. Nano 10:1392
37. Kum MC, Jung H, Chartuprayoon N, Chen W, Mulchandani A, Myung NV (2012) Tuning electrical and optoelectronic properties of single cadmium telluride nanoribbon. J Phys Chem C 116:9202–9208
38. Lähnemann J, Browne DA, Ajay A, Jeannin M, Vasanelli A, Thomassin J-L, BelletAmalric E, Monroy E (2019) Near- and mid-infrared intersubband absorption in topdown GaN/AlN nano- and micro-pillars. Nanotechnology 30:054002
39. Lee JS, Islam MS, Kim S (2006) Direct formation of catalyst-free ZnO nanobridge devices on an etched Si substrate using a thermal evaporation method. Nano Lett 6:1487–1490
40. Li D, Xing G, Tang S, Li X, Fan L, Li Y (2017) Ultrathin ZnSe nanowires: one-pot synthesis via a heat-triggered precursor slow releasing route, controllable Mn doping and application in UV and near-visible light detection. Nanoscale 9:15044–15055
41. Li L, Lou Z, Shen G (2015) Hierarchical CdS nanowires based rigid and flexible photodetectors with ultrahigh sensitivity. ACS Appl Mater Interfaces 7(42):23507–23514
42. Li L, Yang S, Han F, Wang L, Zhang X, Jiang Z, Pan A (2014) Optical sensor based on a single CdS nanobelt. Sensors 14:7332–7341
43. Li L, Lu H, Yang ZY, Tong LM, Bando Y, Golberg D (2013) Bandgap-graded CdS$_x$Se$_{1-x}$ nanowires for high-performance field-effect transistors and solar cells. Adv Mater 25:1109–1113
44. Lin H, Wei L, Wu C, Chen Y, Yan S, Mei L, Jiao J (2016) High-performance self-powered photodetectors based on ZnO/ZnS core-shell nanorod arrays. Nanoscale Res Lett 11:420
45. Logeeswaran VJ, Oh J, Nayak AP, Katzenmeyer AM, Gilchrist KH, Grego S et al (2011) A perspective on nanowire photodetectors: current status, future challenges, and opportunities. IEEE J Sel Topics Quant Electron 17(4):1002–1032
46. Lou Z, Shen G (2016) Flexible photodetectors based on 1D inorganic nanostructures. AdvSci 3:1500287

47. Lou Z, Li L, Shen G (2016) Ultraviolet/visible photodetectors with ultrafast, high photosensitivity based on 1D ZnS/CdS heterostructures. Nanoscale 8:5219–5225

48. Luo L-B, Xie W-J, Zou Y-F, Yu Y-Q, Liang F-X, Huang Z-J, Zhou K-Y (2015) Surface plasmon propelled high-performance CdSe nanoribbons photodetector. Opt Express 23(10):12979

49. Lu J, Liu H, Zhang X, Sow CH (2018) One-dimensional nanostructures of II-VI ternary alloys: synthesis, optical properties, and applications. Nanoscale 10:17456–17476

50. Lu J, Lim X, Zheng M, Mhaisalkar SG, Sow C-H (2012) Direct laser pruning of CdS_xSe_{1-x} nanobelts en route to a multicolored pattern with controlled functionalities. ACS Nano 6:8298–8307

51. Lu J, Sun C, Zheng M, Nripan M, Liu H, Chen GS et al (2011) Facile one step synthesis of CdS_xSe_{1-x} nanobelts with uniform and controllable stoichiometry. J Phys Chem C 115:19538–19545

52. Lysov A, Vinaji S, Offer M, Gutsche C, Regolin I, Mertin W et al (2011) Spatially resolved photoelectric performance of axial GaAs nanowire pn-diodes. Nano Res 4:987–995

53. Ma RM, Dai L, Huo HB, Yang WQ, Qin GG, Tan PH et al (2006) Synthesis of high quality n-type CdS nanobelts and their applications in nanodevices. Appl Phys Lett 89:203120

54. Machín A, Fontánez K, Arango JC, Ortiz D, De León J, Pinilla S et al (2021) One-dimensional (1D) nanostructured materials for energy applications. Materials 14:2609

55. Manna S, Das S, Mondal SP, Singha R, Ray SK (2012) High efficiency Si/CdS radial nanowire heterojunction photodetectors using etched Si nanowire templates. J Phys Chem C 116:7126–7133

56. Mu Z, Zheng Q, Liu R, Iqbal Malik MW, Tang D, Zhou W, Wan Q (2019) 1D ZnSSe-ZnSe axial heterostructure and its application for photodetectors. Adv Electron Mater 2019:1800770

57. Nawaz MZ, Xu L, Zhou X, Shah KH, Wang J, Wu B, Wang C (2021) CdS nanobelt-based self-powered flexible photodetectors with high photosensitivity. Mater Adv 2:6031–6038

58. Nie B, Luo LB, Chen JJ, Hu JG, Wu CY, Wang L et al (2013) Fabrication of p-type ZnSe:Sb nanowires for high-performance ultraviolet light photodetector application. Nanotechnology 24:095603

59. Ombaba MM, Karaagac H, Polat KG, Islam MS (2016) Nanowire enabled photodetection. In: Nabet B (ed) Photodetectors: Materials, Devices and Applications. Elsevier, pp 87–120

60. Park S, Kim S, Sun G-J, Byeon DB, Hyun SK, Lee WI, Lee C (2016) ZnO-core/ZnSe-shell nanowire UV photodetector. J Alloys Comp 658:459–464

61. Prades JD, Jimenez-Diaz R, Hernandez-Ramirez F et al (2008) Toward a systematic understanding of photodetectors based on individual metal oxide nanowires. J Phys Chem C 112:14639–14644

62. Rai SC, Wang K, Chen J, Marmon JK, Bhatt M, Wozny S et al (2015a) Enhanced broad band photodetection through piezo-phototronic effect in CdSe/ZnTe core/shell nanowire array. Adv Electron Mater 1:1400050

63. Rai SC, Wang K, Ding Y, Marmon JK, Bhatt M, Zhang Y et al (2015b) Piezo-phototronic effect enhanced UV/visible photodetector based on fully wide band gap type-II ZnO/ZnS core/shell nanowire array. ACS Nano 9:6419–6427

64. Shen G, Chen D (2010) One-dimensional nanostructures for photodetectors. Recent Patents Nanotechnol 4:20–31

65. Soci C, Zhang A, Bao X-Y, Kim H, Lo Y, Wang D (2010) Nanowire photodetectors. J Nanosci Nanotechnol 10:1430–1449

66. Soci C, Zhang A, Xiang B, Dayeh SA, Aplin DPR, Park J et al (2007) ZnO nanowire UV photodetectors with high internal gain. Nano Lett 7:1003–1009

67. Sun Z, Shao Z, Wu X, Jiang T, Zheng N, Jie J (2016) High-sensitivity and self-driven photodetectors based on Ge-CdS core–shell heterojunction nanowires via atomic layer deposition. Cryst Eng Comm 18:3919–3924

68. Takahashi T, Nichols P, Takei K, Ford AC, Jamshidi A, Wu MC et al (2012) Contact printing of compositionally graded CdSxSe1-x nanowire parallel arrays for tunable photodetectors. Nanotechnology 23:045201

69. Tang M, Xu P, Wen Z, Chen X, Pang C, Xu X et al (2018) Fast response CdS-CdS$_x$Te$_{1-x}$-CdTe core-shell nanobelt photodetector. Sci Bull 63(17):1118–1124

70. Tang J, Huo Z, Brittman S, Cao H, Yang P (2011) Solution-processed core–shell nanowires for efficient photovoltaic cells. Nat Nanotechnol 6:568–572

71. Tian W, Zhang C, Zhai T, Li SL, Wang X, Liu J et al (2014) Flexible ultraviolet photodetectors with broad photoresponse based on branched ZnS-ZnO heterostructure nanofilms. Adv Mater 26:3088–3093

72. Wang J, Fang H, Wang X, Chen X, Lu W, Hu W (2017) Recent progress on localized field enhanced two-dimensional material photodetectors from ultraviolet—visible to infrared. Small 13:1700894

73. Wang L, Chen R, Ren ZF, Ge CW, Liu ZX, He SJ et al (2016) Plasmonic silver nanosphere enhanced ZnSe nanoribbon/Si heterojunction optoelectronic devices. Nanotechnology 27:215202

74. Wang L, Song HW, Liu ZX, Ma X, Chen R, Yu YQ et al (2015) Core–shell CdS:Ga-ZnTe:Sb p-n nano-heterojunctions: fabrication and optoelectronic characteristics. J Mater Chem C 3:2933–2939

75. Wang K, Rai SC, Marmon J, Chen J, Yao K, Wozny S et al (2014) Nearly lattice matched all wurtzite CdSe/ZnTe type II core–shell nanowires with epitaxial interfaces for photovoltaics. Nanoscale 6:3679–3685

76. Wang L, Lu M, Wang X, Wu D, Xie C, Wu C et al (2013) Tuning the p-type conductivity of ZnSe nanowires via silver doping for rectifying and photovoltaic device applications. J Mater Chem A 1:1148–1154

77. Wei T-Y, Huang C-T, Hansen BJ, Lin Y-F, Chen L-J, Lu S-Y, Wang ZL (2010) Large enhancement in photon detection sensitivity via Schottky-gated CdS nanowire nanosensors. Appl Phys Lett 96:013508

78. Wu D, Chang Y, Lou Z, Xu T, Xu J, Shi Z, Tian Y, Li X (2017) Controllable synthesis of ternary ZnSxSe1-x nanowires with tunable band-gaps for optoelectronic applications. J Alloys Compd 708:623–627

79. Wu D, Jiang Y, Zhang Y, Yu Y, Zhu Z, Lan X et al (2012) Self-powered and fast-speed photodetectors based on CdS:Ga nanoribbon/Au Schottky diodes. J Mater Chem 22:23272

80. Wu D, Jiang Y, Zhang Y, Li J, Yu Y, Zhang Y et al (2012b) Device structure-dependent field-effect and photoresponse performances of p-type ZnTe: Sb nanoribbons. J Mater Chem 22:6206–6212

81. Wu P, Dai Y, Sun T, Meng H, Fang X, Yu B, Dai L (2011a) Impurity-dependent photoresponse properties in single CdSe nanobelt photodetectors. ACS Appl Mater Interfaces 3:1859–1864

82. Wu D, Jiang Y, Li S, Li F, Li J, Lan X et al (2011b) Construction of high-quality CdS:Ga nanoribbon/silicon heterojunctions and their nano-optoelectronic applications. Nanotechnology 22:405201

83. Wu CY, Jie JS, Wang L, Yu YQ, Peng Q, Zhang XW et al (2010) Chlorine-doped n-type CdS nanowires with enhanced photoconductivity. Nanotechnology 21:505203

84. Yan S, Rai SC, Zheng Z, Alqarni F, Bhatt M, Retana MA, Zhou W (2016) Piezophototronic effect enhanced UV/visible photodetector based on ZnO/ZnSe heterostructure core/shell nanowire array and its self-powered performance. Adv Electron Mater 2:1600242

85. Ye Y, Dai L, Wen X, Wu P, Pen R, Qin G (2010) High–performance single CdS nanobelt metal-semiconductor field-effect transistor-based photodetector. ACS Appl Mater Interfaces 2:2724–2727

86. Ye Y, Dai L, Sun T, You LP, Zhu R, Gao JY et al (2010b) High-quality CdTe nanowires: synthesis, characterization, and application in photoresponse devices. J Appl Phys 108:044301

87. Yi SS, Girolami G, Amano J, Islam MS, Sharma S, Kamins TI et al (2006) InP nanobridges epitaxially formed between two vertical Si surface by metal-catalyzed chemical vapor deposition. Appl Phys Lett 89:133121

88. Yoon Y-J, Park K-S, Heo J-H, Park J-G, Nahm S, Choi KJ (2010) Synthesis of Zn$_x$Cd$_{1-x}$Se (0 < x <1) alloyed nanowires for variable-wavelength photodetectors. J Mater Chem 20:2386–2390

89. Yu Y, Luo L, Wang M, Wang B, Zeng L, Chunyan WC et al (2015) Interfacial states induced ultrasensitive ultraviolet light photodetector with resolved flux down to 85 photons per second. Nano Res 8(4):1098–1107

90. Yu Y, Jiang Y, Zheng K, Zhu Z, Lan XZ, Yan ZY et al (2014) Ultralow-voltage and high gain photoconductor based on ZnS:Ga nanoribbons for the detection of low-intensity ultraviolet light. J Mater Chem C 2:3583–3588

91. Yu YQ, Luo LB, Zhu ZF, Nie B, Zhang YG, Zeng LH et al (2013) High-speed ultraviolet-visible-near infrared photodiodes based on p-ZnS nanoribbon-n-silicon heterojunction. Cryst Eng Comm 15:1635–1642

92. Xie C, Li F, Zeng L, Luo L, Wang L, Wu C, Jie J (2015) Surface charge transfer induced p-CdS nanoribbon/n-Si heterojunctions as fast-speed self-driven photodetectors. J Mater Chem C 3:6307–6313

93. Xie X, Kwok SY, Lu Z, Liu Y, Cao Y, Luo L et al (2012a) Visible–NIR photodetectors based on CdTe nanoribbons. Nanoscale 4:2914–2919

94. Xie C, Luo LB, Zeng LH, Zhu L, Chen JJ, Nie B et al (2012b) p-CdTe nanoribbon/n-silicon nanowires array heterojunctions: photovoltaic devices and zero-power photodetectors. Cryst Eng Comm 14:7222–7228

95. Zhang X, Wu D, Geng H (2017) Heterojunctions based on II-VI compound semiconductor one-dimensional nanostructures and their optoelectronic applications. Crystals 7:307

96. Zhang K, Ding J, Lou Z, Chai R, Zhong M, Shen G (2017b) Heterostructured ZnS/InP nanowires for rigid/flexible ultraviolet photodetectors with enhanced performance. Nanoscale 9:15416–15422

97. Zhang X, Meng D, Hu D, Tang Z, Niu X, Yu F, Ju L (2016) Construction of coaxial ZnSe/ZnO p-n junctions and their photovoltaic applications. Appl Phys Express 9:025201

98. Zhang X, Tang Z, Hu D, Meng D, Jia S (2016b) Nanoscale p-n junctions based on p-type ZnSe nanowires and their optoelectronic applications. Mater Lett 168:121–124

99. Zhang X, Hu D, Tang Z, Ma D (2015) Construction of ZnSe-ZnO axial p-n junctions via regioselective oxidation process and their photo-detection applications. Appl Surf Sci 357:1939–1943

100. Zhang X, Zhang X, Wang L, Wu Y, Wang Y, Gao P et al (2013) ZnSe nanowire/Si p-n heterojunctions: device construction and optoelectronic applications. Nanotechnology 24:395201

101. Zhang X, Zhang X, Zhang X, Zhang Y, Bian L, Wu Y et al (2012) ZnSe nanoribbon/Si nanowire p-n heterojunction arrays and their photovoltaic application with graphene transparent electrodes. J Mater Chem 22:22873–22880

102. Zhang A, You SF, Soci C, Liu YS, Wang DL, Lo YH (2008) Silicon nanowire detectors showing phototransistive gain. Appl Phys Lett 93:121110

103. Zhang D, Li C, Han S, Liu X, Tang T, Jin W, Zhou C (2003) Ultraviolet photo-detection properties of indium oxide nanowires. Appl Phys A Mater Sci Process 77:163–166

104. Zhao W, Liu L, Xu M, Wang X, Zhang T, Wang Y et al (2017) Single CdS nanorod for high responsivity UV–visible photodetector. Adv Optical Mater 5(12):1700159

105. Zhou W, Peng Y, Yin Y, Zhou Y, Zhang Y, Tan D (2014) Broad spectral response photodetector based on individual tin-doped CdS nanowire. AIP Adv 4:123005

106. Zhou J, Gu Y, Hu Y, Mai W, Yeh PH, Bao G et al (2009) Gigantic enhancement in response and reset time of ZnO UV nanosensor by utilizing Schottky contact and surface functionalization. Appl Phys Lett 94:191103

Chapter 17
QDs of Wide Band Gap II–VI Semiconductors Luminescent Properties and Photodetector Applications

M. Abdullah, Baqer O. Al-Nashy, Ghenadii Korotcenkov, and Amin H. Al-Khursan

17.1 Introduction

The optoelectronic device that converts the incident light into an electric signal is the photodetector (PD). The principle of the photosensitive semiconductor devices depends mainly on the electron transition under high energy to the conduction band (CB), leaving a hole behind. Ideally, each photon creates a single electron-hole pair. This pair can recombine at a characteristic time (lifetime), releasing additional energy in the form of heat. In an applied electric field, the drift of electrons and holes creates an electric current, and the way the field is applied determines the type of detector. The basics of PDs are explained well in the textbooks; see, for example [18, 75], and the reviewed paper [45, 89]. In semiconductor PDs, three types of mechanisms convert light into current: photoconductive effect, photogate effect, and photovoltaic effect [45]. In addition, PDs can be divided into photoconductive PDs and photodiode PDs. The photodiode accumulates a charge at the junction of n- and p-type semiconductors or between a semiconductor and a metal contact (Schottky junction). At the same time, the photoconductors have a semiconductor

M. Abdullah
Nasiriya Nanotechnology Research Laboratory (NNRL), College of Science, University of Thi-Qar, Nasiriya, Iraq

Department of Physics, College of Science, University of Thi-Qar, Nasiriya, Iraq

B. O. Al-Nashy
Department of Physics, College of Science, University of Misan, Omarah, Iraq

G. Korotcenkov
Department of Physics and Engineering, Moldova State University, Chisinau, Moldova
e-mail: ghkoro@yahoo.com

A. H. Al-Khursan (✉)
Nasiriya Nanotechnology Research Laboratory (NNRL), College of Science, University of Thi-Qar, Nasiriya, Iraq

© The Author(s), under exclusive license to Springer Nature Switzerland AG 2023   399
G. Korotcenkov (ed.), *Handbook of II-VI Semiconductor-Based Sensors and Radiation Detectors*, https://doi.org/10.1007/978-3-031-20510-1_17

crystal with ohmic contacts. The phototransistor is a photoconductor-type with an additional gate insulated by a dielectric material. The applied gate voltage modifies the conductivity and reduces noise through the dark current suppression.

The required PD characteristics are the high sensitivity, the spectral selectivity, the high signal-to-noise ratio, the fast photoresponse, the remarkable stability, and the simple manufacturing process [88]. PDs with high sensitivity in various spectral ranges are in demand in a wide range of applications, such as defense, industry, healthcare, imaging, biomedicine, object discrimination, advanced optical wireless communication systems, chemical analysis, and scientific research in various fields [2, 10]. PDs can be fabricated using single crystalline materials, epitaxial layers and polycrystalline films deposited by various methods. In recent decades, photodetectors have also been developed on the basis of nanomaterials such as 1D, 2D and quantum dots (QDs). Studies have shown that the use of these materials, especially QDs, can improve the performance of photodetectors [56, 85, 88].

Quantum dots (QDs), artificial atoms, particles of 2–20 nm have discrete atomic-like energy levels due to the size quantization. Such QDs mimic the atomic solids. One of the main characteristics of QDs is the tuning of their energy states by changing the QD size. The energy configuration in QDs is different for the same QD semiconductor crystal, depending on the QD size. This tuning in electronic properties is accompanied by a change in their optical characteristics [4]. Thus, the advantages of QDs lie in their tunable bandgap, as well as their superior electronic, surface and optical properties. Their surface-to-volume ratio, in addition to a short diffusion lengths, provides excellent photogeneration of charged carriers and then best used for PD operation. Therefore, QDs are used in various sciences including physics, material science, chemistry, energy technology, chemical and biosensing, medicine and biology [5, 26, 46, 48, 58, 87, 93]. In bio-detection, QDs are used in bio-imaging, bio-labeling, and bio-molecular detection [86]. In particular, in chemical sensors, QDs are used as probes for ions, such as Cu^{2+}, K^+, Na^+, Hg^{2+}, and small molecules, like glucose and aromatic hydrocarbons. The QDs as molecular detectors exhibit simplicity, selectivity, and cost-effectiveness [20, 62].

Experiment has shown that colloidal QD-based PDs fabricated by solution processing technology are cheaper than thin-film devices fabricated by vacuum deposition technology [49, 60, 95]. QD-based photodetectors are now being manufactured as self-powered devices, that generate potential when illuminated to operate in short-circuit or open-circuit mode [10, 50, 57]. However, a more promising direction is the development of flexible and UV photodetectors based on QDs. However, a more promising direction is the development of flexible and UV photodetectors based on QDs. Currently, there is a great need for flexible PDs, since wearable electronics, in contrast to traditional "hard" PDs, require both high sensitivity and flexibility. Wearable electronics, the milestone in future development of optoelecttronic devices intended for applications in the fields of biomedicine, sports and disease diagnostics. As far as UV photodetectors are concerned, with zero background detection of the deep ultraviolet (DUV, 200-280 nm) signal under sunlight or room lighting, the DUV (SBDUV) PD solar curtains are an excellent tool for applications such as ozone detection, space communications, missile plume detection, flame monitoring, biological medical analysis, and offshore oil inspection [2, 89]. Ultraviolet (UV) nanostructure PDs are reviewed in many articles [2, 55, 85, 89]. Except for ref. [2], which reviews ZnSe, all these reviews do not take into account II-VI PDs. Wang et al. [87] reviewed the zinc-containing nanocrystals for energy conversion applications covering the ZnS and ZnSe nanocrystals. Type I and type II core-shell quantum dot (QD) synthesis is reviewed well in the chapter of Drofs et al. [25].

Charge carriers in QD-based PDs are transferred in three dimensions resulting in great trapping states on the surface. Therefore, there are some difficulties in the process of electron transfer between QDs. While colloidal QDs can effectively create electron-hole pairs, all the photogenerated charge carriers are not transferred to the corresponding electrodes, as is required for excellent device performance. They travel through the adjacent QDs and are trapped before reaching the electrodes. This problem is solved by bringing the QDs closer enough using short-chain ligands (such as trioctylphosphine oxide (TOPO), TOP, Oleic acid, Oleylamine) to boost the performance through the best inter-dot charge transport. In this case the electron wave function is strongly delocalized between the neighboring QDs, enhancing the photocurrent.

17.2   II-VI Semiconductor Quantum Dots-Based PDs and Their Fabrication Methods

The widest band gap II-VI semiconductor, ZnS, has a large bandgap of 3.72 eV (2.5 nm Bohr radius) for cubic zinc-blende and 3.77 eV for hexagonal wurtzite crystals, making it suitable for visible-blende UV devices [93]. The quantum size effect can tune the bandgap beyond 4.43 eV by reducing the QD size below the Bohr radius.

A large-scale, cost-effective method is vital for QDs production (read the Chaps. 12 and 13, Vol. 1). Liquid phase epitaxy, hot injection, sputtering, thermal evaporation, chemical vapour deposition (CVD) are expensive due to their equipment, limiting large scale production, or require high temperature and pressure [54]. Wet chemical methods are simpler and cheaper methods for the synthesis of II-VI compounds. It is a low temperature, inexpensive method suitable for mass production. This liquid-phase synthesis of QDs results in a good dispersion and stability in a solvent, easy preparation, pot reaction, low-cost, and controlling particle size [54]. The most important factor in this method is the choice of the appropriate composition of the solution for synthesis, which, in addition to the precursors of the components of II-VI compounds, may include surfactants and polyelectrolytes. For example, for the synthesis of ZnS QDs by this method, Li et al. [54] used sodium dodecyl sulfate as a passivating agent, controlling their nucleation growth rate and confirming its stability. The resulted PD is highly suitable for UV detection by its wide bandgap and high electron mobility (600 cm^2/V.s). He et al. [36] introduced the 3-mercaptopropionic acid as a stabilizing reagent and a source of sulfur in the synthesis of Cu:ZnS QDs assisted by microwave irradiation. Compared with the conventional method, the microwave-assisted wet chemical method accelerates the synthesis of Cu:ZnS QDs, while no phase transformation occurs, and the crystal quality is improved. The QD tunable range is possible by controlling the reaction time.

Large bandgap ZnS QDs have high environmental stability and can be used for photocatalysts, solar energy conversion, phosphor, and photodetectots. Xia et al. [95] fabricated a solar-blind, deep-UV (SBDUV) ZnS QDs PD by adopting ZnS QDs via solution process with a bandgap tuned beyond 4.68 eV and emitting a sharp exciton peak at 256 nm with a fast and stable response. Premkumar et al. [72] demonstrated visible blind UV detection using ZnS QDs grown by a simple reflux technique. The ZnS QDs UV sensor exhibits the best photocurrent response and high responsitivity. However, their long recombination lifetime slows the device response, limiting their applications.

ZnSe QDs also have a wide-direct bandgap (the bulk bandgap is 2.67 eV), enabling UV detection. Their spectrum cover 300–450 nm wavelength (A and B UV regions), broader than the UV spectrum covered by GaN. A slight mismatch between ZnSe and GaAs can be corrected by growing ternary or quaternary ZnSe structures on the available GaAs substrates [2]. ZnSe is the least toxic and high photoluminescence QD material. However, the ZnSe microfluidic platform still remains a challenge due to the low solubility of most Zn precursors and the low reactivity of Zn-Se [33, 39, 79]. Bratskaya et al. [11] reported on the green synthesis of ZnSe QDs on an aqueous solution of polyampholyte chitosan derivatives-N-(2-carboxyethyl) chitosan, producing a high intensity of photoluminescence. ZnSe devices are environmentally friendly and used for biological imaging, DNA analysis, and therapy diagnosis. Their high optical nonlinearity makes them adequate for optical limiting and ultrafast switch devices [15, 65]. Fang et al. [27] demonstrated the first ZnSe nanobelts sensor by chemical vapor deposition (CVD) method that had an ultralow dark current (below 10^{-14} A, the detection limit of the current meter) and response time ($\tau < 0.3$ s) less than conventional ZnSe PD sensors.

Mishra et al. [64] synthesized QDs of CdS, another wide band gap II-VI semiconductor, by a simple co-precipitation method. Absorption in the UV-visible region has an absorption peak at 427 nm where it is blue-shifted by 0.48 eV compared to bulk CdS. In [32] CdS QDs were synthesized by liquid phase technique at room temperature and atmospheric pressure using the sodium alkyl sulfonate as capping agent. Magic size CdS QDs were synthesized by Dickon and Hu [24] by a no-injection reaction using 1-octadecene as a solvent, a cadmium salt as a source of cadmium, and the organic sulfur 1-dodecanethiol as a source of sulfur. The acid acted as a capping ligand on the surface and gently controlled the solubility. As a result, single-sized CdS QDs with a fixed position of the emission wavelength were formed due to homogeneous nucleation and size uniformity. Husham et al. [38] propose a simple microwave-assisted chemical bath deposition (CBD) to synthesize a self-powered PDs from nanocrystalline CdS.

Jiang et al. [40] also developed CdS QDs-based photodetector. They modified the CdS QDs by Au nanoparticles to increase the performance via surface plasmon resonance (SPR) and then used MXene as an electrode. $Ti_3C_2T_x$ MXene, a 2D structure, exhibits Ohmic contacting with CdS at a non considerable Schottky barrier of 0.07 eV height. First, the prepared CdS core-Au nanoparticles were dispersed in ethanol. Then the uniform solution above was dropped onto fluorophlogopite substrate (1 cm × 1 cm) and natural air-drying at room temperature to fabricate the device. The MXene colloidal solution was dropped on both sides of CdS core-Au nanoparticles as electrodes. Under illumination, hot electrons are injected from to CdS QDs from two sides: due to SPR from Au nanoparticles and due to its photothermal property from MXene. The photoresponse polarity depends on the higher electrons transfer to CdS. Jiang et al. [40] have found that a broad-spectrum, beginning from deep UV to near IR can be detected at high response by developed PD. The specific detectivity at 405 nm was equaled ~$1.3 \cdot 10^{11}$ Jones [40].

However, the use of CdS QDs is prohibited in Europe from an environmental point of view. Therefore, when using CdS QDs, it is necessary to form protective coatings on its surface. In this case, we get a core-shell structure, where CdS performs the functions of the core, and the less toxic material forms the shell. One such example is the quantum dots of the CdTeSe/ZnS core with a surface modified with trioctylphosphine oxide (TOPO) [12].

17.3 QD Core/Shell Structures

The nanoscale QDs have a high surface-to-volume ratio. A significant fraction of atoms are located on the surface, and some atoms are not fully coordinated. As a result, we have a surface with increased reactivity, prone to chemical interaction with the surrounding atmosphere, primarily with oxygen and water vapor. Dangling bonds in QD surface defects also create nonradiative recombination centers, which worsen their optical and transport properties. To eliminate unwanted interactions that are the main cause of instability of QDs parameters, the QDs surface needs to be passivated by surface ligands. The loss of ligands that occurs during film formation, leads to the formation of dangling bonds and defects on the surface, which can be evidenced by low PL QY. Coating the QD with an inorganic shell material with a larger band gap can reduce the reactivity of the QD surface with oxygen and water. Encapsulation of QD cores with shells effectively passivates anionic and cationic surface areas, eliminates harmful effects, prevents heavy metals from entering the environment, improves bothe the photostability against photooxidation, and the photoluminescence (PL) quantum efficiency. It is important to note that the match of the crystal structure and the lattice constant between the core and the shell are important to avoid lattice strain at the interface. With a big mismatch and when the shell becomes thick, the strain-induced defects increase, resulting in the decrease of PL QY. It was found that alloying and graded shells relax strain with increasing shell thickness [29, 33, 52].

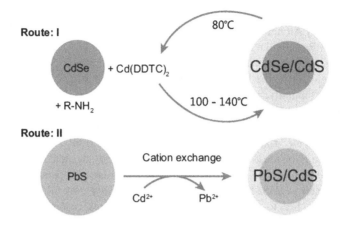

Fig. 17.1 (**a**) Schematic illustration of shell growth. Route I: layer-by-layer growth or CBD of CdSe–CdS core–shell QDs. Route II: The cation-exchange reaction used to convert core-only PbS QDs into core–shell PbS–CdS QDs. (Reprinted from Ref. [80]. Published 2017 by Oxford University Press as open access)

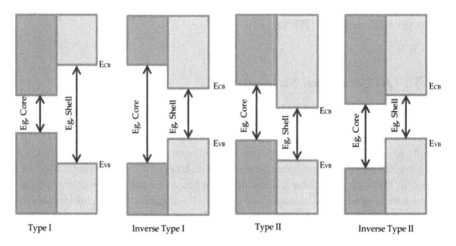

Fig. 17.2 Types of core-shell QDs based on band alignment. (Reprinted with npermission from Ref. [86]. Copyright 2015: Elsevier)

In most cases, core-shell structures are formed using the two paths shown in Fig. 17.1. As a result, depending on the properties of the materials used, two types of core/shell structures can be formed: type-I and type-II. Figure 17.2 shows the band gap alignments for each of these types. Table 17.1 compares QDs classification based on their band alignment, the location of carriers, quantum yields, stokes shift, and the average absorption and emission ranges [1]). It also provides examples of core/shell QDs based on II-VI semiconductors and lists some of their limitations.

Table 17.1 Types of core/shell QDs based on band alignment

Parameter	Type I	Inverse Type I	Type II	Inverse Type II
Band gap	The band gap of the core is smaller than the band gap of the shell, as well as the band gap of the core falls within the band gap of the shell	The band gap of the core is greater than the band gap of the shell, as well as the band gap of the shell falls within band gap of the core	Valence band edge of the core is within the band gap of the shell or conduction band edge of the shell is within the band gap of the core	Conduction band edge of the core is within the band gap of the shell or valence band edge of the shell is within the band gap of the core
Excited electrons/ holes positions	Excited electrons and holes are completely confined in the core region	The excited electrons and holes are completely or partially confined in the shell based on the thickness of the shell.	One charge carrier either excited electron or hole is confined to the core, while the other is mostly confined to the shell	One of the excited electrons or the holes are delocalized in the core/shell structure, and the other one is confined within the core.
Quantum yield (QY)	Higher QY and long-term stability	Lower QY and poor stability	Lower QY and poor stability	Relatively higher QY and fair stability
Stokes shift	Small	Significantly large	Large	Large and tunable via controlling the size of the core and thickness of the shell
Average absorption range	(400–500) nm	(400–500) nm	(600–800) nm	(300–1600) nm
Average emission range	(430–600) nm	(400–700) nm	(700–1000) nm	(700–1000) nm
Limitations	The shell can trap charge carriers which leads to reduced fluorescence QY	Both the excited electrons and holes may leak to the surface	One of the excited electrons or hole leak to the surface	The excited electron or hole can be absorbed leading to reduced excited decay time one carrier is mostly confined to the core, while the other is mostly confined to the shell
Construct/ materials	CdSe/ZnS CdSe/CdS CdS/ZnS	CdS/HgS CdS/CdSe ZnSe/CdSe	CdTe/CdSe CdSe/ZnTe CdSe/ZnSe	InP/CdS PbS/CdS

Source: Reprinted from Ref. [1]. Published 2019 by MDPI as open access. Table 1

Table 17.1 shows that when going from Type-I to Type-II, the difference in the bandgap between the core and shell decreases, and the excited electrons are spatially depart from each other. The stokes shift increased while the quantum yield and stability decreased with increasing absorption wavelength. As a result, Type-I is more suitable as concentrators compared to Type-II because they show no leakage of excited electrons, show better stability and also offer higher QY [1]. It is necessary to note that in comparison with organic passivation, core-shell passivation provides higher quantum yield, extraordinary photoluminescence, enhanced optical properties, increased half life time, improved stability towards photo-oxidation and finally better structural properties. As a result, the core/shell confinement results in a higher power conversion efficiency than the only core QD structures [41].

CdS/ZnS core/shell nanoparticles were successfully synthesized using a two-step chemical synthesis method. Compared to CdS and ZnS nanoparticles, CdS/ZnS core/shell QDs exhibit a photocatalytic enhancement due to the reduced recombination of photogenerated electron-hole pairs and increasing the CdS/ZnS core/shell photocatalytic properties [74]. Increasing cadmium concentration in the shell for $ZnS/Cd_xZn_{1-x}S$ core/shell QDs decreases the interface optical phonon frequencies due to the reduced mass of the CdZnS shell [37].

CdSe/ZnS, CdSe/CdS, and CdSe/ZnS/CdS nanocrystals are of great importance because their luminescence wavelength overlaps the full vision wavelength and high photoluminescence [6, 96]. Coating CdSe core with CdS shell (or CdS/ZnS multishell) exhibits redshift and boost absorption and PL spectra. Therefore, the semiconductor CdSe/CdS/ZnS core-shell QDs are an alternative to fluorescent nanoparticles for data storage equipment, sensors, optoelectronic devices, and solar cells. CdSe/ZnS core-shell can also be synthesized via the successive ion layer adsorption and reaction (SILAR) method. This conventional hot injection synthesis method injects Zn and S precursors into a solution containing CdSe QDs. The few structural differences and the similarity of internal energies result in zinc blende and wurtzite structures for CdSe QD morphology. CdSe/ZnS QDs of zinc blend crystal structure has the best optoelectronic properties compared to the wurtzite type. The element of high average atomic number (Z) has high intensity. The average Z value of the CdSe core ($Z_{Cd} = 48$ and $Z_{Se} = 34$) is twice that of the ZnS shell ($Z_{Zn} = 30$ and $Z_S = 16$) [29].

ZnSe has numerous advantages such as high stability and better doping effects. However, it takes some time before this becomes decisive for its application due to problems in realizing high-quality ZnSe QDs. ZnSe QDs adhere to any nanoparticle surface, creating a nano-hybrid system with enhanced lifetime and high performance. Passivating the ZnSe QD core with a broader bandgap (like ZnS) shell reduces the recombination and allows better stability [2, 61, 76, 83]. It is proved that ZnS-based PDs have a lower leakage current. A colloidal ZnTe/ZnSe core/shell QDs of the type-II structure extends the PL wavelength from blue to amber [9]. This structure has an energy conversion efficiency 11 times higher than ZnSe QDs for photovoltaic devices. This improvement is a result of the properties of the Type-II core shell structure that increase carrier extraction and extend wavelength [9]. When the shell thickness increases, the ZnSe/CdSe core/shell QDs are tuned from Type-I to Type-II and then back to Type-I, see Fig. 17.2, when the shell thickness is increased [13]. ZnSe is used as a shell for CdSe QD optical amplifier with a high emission coefficient and low noise [7, 13]. ZnS is reliable as a shell for type-I alignment with ZnSe QDs for high fluorescence. The growing mode of the ZnS shell is modified by tuning the precursor from kinetic (fast) to thermodynamic (slow) regimes. In the thermodynamic regime, the quantum yield increases, and on-off blinking reduces, which prevents the growth of traps at the core/shell interface [43].

A ZnO/ZnS core-shell nanorod array as a photoanode is synthesized as a self-powered photoelectrochemical cell, using facile chemical solution method such as SILAR [57]. This PD showed an improvement in both photosensitivity and sensitivity. This improvement is a result of excess light collection and rapid separation of photo-generated electron-hole pairs due to ZnS coating. The highest photoresponsivity was obtained at seven SILAR cycles for ZnS deposition. Further increment ZnS SILAR cycles reduces the photoresponsivity of ZnO/ZnS array UV PDs, where a thick layer of ZnS nanoparticles may hinder the hole transport and increase the recombination of the photon-generated electron-hole pairs. Lin et al. [57] also found that this PD had the best wavelength selectivity making it suitable for the UV-A range [57].

17.4 Doping Influence on QDs Properties

In addition to the controllability of the size and shape of QDs, doping with impurities is vital for the developing electric and optical properties. The energy levels of the dopant lie in the bandgap of the host crystal and change the electronic and optical properties of the host crystal. In particular, the emission and absorption spectra vary depending on the dopant and its concentration (see Fig. 17.3). II-VI semiconductor QDs can be doped with transition metals and lanthanide ions such as Ni^+, Mn^{2+}, Ag^{2+}, Cr^{3+}, Eu^{+3}, Sm^{3+}, and Tb^{3+} [42, 60]. For example, Mn^{2+} doped ZnS QD shows luminescence at 413 and 585 nm wavelengths. ZnS:Mn QDs were synthesized by employing a simple aqueous method using starting materials: zinc sulfate, sodium sulfide, manganese acetate tetrahydrate, a complex agent, and a solvent [16]. On the other hand, defects generated at a high dopant concentration quench the PL. Precisely this effect was observed at a high Mn^{2+} concentration; in addition to PL quenching, a redshift of the orange (impurity) luminescence wavelength and a blueshift of the edge luminescence of the initial ZnS QDs were also observed [16]. This means that this effect must be taken into account when doping QDs.

It should be noted that Mn^{2+} ions are the common dopant used for ZnS QDs due to their low cost and ability to change the electronic and optical properties of ZnS QDs. The most important advantages of Mn-doped ZnS QDs are low toxicity, longer excited-state lifetimes (on millisecond), significant Stokes shift between excitation and emission wavelengths which avoids self-absorption, the absorption range of doped QDs is extended, insignificant interference from associated compounds, no light scattering, increased thermal and environmental stability. Mn^{2+} dopant is also exhibiting a hyperfine splitting in the paramagnetic resonance. The excitation

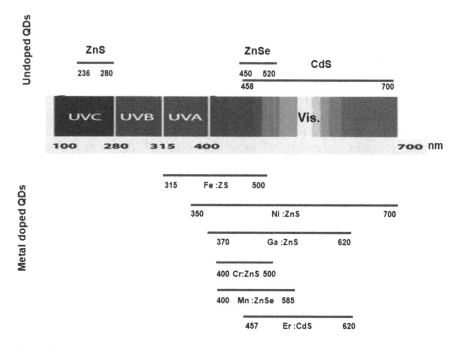

Fig. 17.3 Emission spectra of undoped and doped ZnS, ZnSe, and CdS QDs

occurs first through the host material; then, energy is transferred to the impurity. The Mn^{2+} doped ZnS QDs range is limited from yellow to orange [16, 67]. Lee et al. [53] also observed that QDs from an Mn^{2+} doped ZnS core with a ZnS shell show improved PL intensity, but electric current decreases due to the core–shell electron confinement effect. Al-Haddad et al. [3] prepared Mn^{2+} and Ce^{3+} doped ZnS nanocrystals by simple microwave irradiation and found that responsivity, detectivity, and quantum efficiency for Ce^{3+} doped ZnS are higher than that for Mn^{2+} doped ZnS nanocrystals.

Conversely, Cu dopants existing as either Cu^{+} or Cu^{2+} state, offer more possibilities to develop optical properties and extend the visible range of ZnS QDs. The Cu:ZnS QDs have a tuned range between 500–595 nm wavelength [19, 36]. The samples of Cu- and Fe-codoped ZnS QDs exhibit a UV peak at 315 nm wavelength in addition to a visible one at 500 nm for the co-doped sample. These samples were prepared using a hot injection method [31]. According to Grandhi et al. [31] midgap energy states occur due to doping with transition metals and a charge transfer between the energy band of the host material and the level of doping in the band gap. Cui et al. [19] suggest that doped copper atoms locate at the vacancy of the Zn atom of the QDs in the Cu^{2+} ion form. The defect levels are the shallow accepter levels reducing the n-type photovoltaic characteristic in the Cu-doped QDs.

Cr-doped ZnS QDs, prepared by co-precipitation method, have two absorption peaks in the visible range at 425 nm and 595 nm, related to Cr^{3+}, in addition to the UV peak below 370 nm, characterized the intrinsic ZnS bandgap absorption [97].

Study has also found that due to the dopant intrinsic property, Cr-doped ZnS QDs exhibit weak ferromagnetism or super-paramagnetism at room temperature. The Cr-doped ZnS QDs demonstrate a sharp electron paramagnetic resonance, which indicates an increase in the number of unpaired electrons upon doping. The fourfold spin degeneracy of the ground state of the Cr^{3+} is removed by a subsequent low symmetric field and splits into Kramers doublets, Ms. = ±3/2 and ± 1/2. Hence, Cr:ZnS QDs are promising for spintronic device applications [71].

Ni-doped ZnS QDs have also recognized ferromagnetic properties compared to paramagnetism with bare ZnS QDs [78]. It was found that Ni-doped ZnS QDs exhibit size-reduction attributed to the small Ni-ion radius than Zn-ion. The ionic radii of Ni^{2+} and Zn^{2+} are 0.07 nm and 0.074 nm, respectively. Structural characterization of ZnS:Ni has shown that Ni during doping incorporates in the ZnS matrix at interstitial or substitutional sites. This difference between ion radius diminishes the bond length that connects S-anion and Ni-cation, vice versa to Zn-S bond length, causes a lattice strain and leads to stress in the ZnS crystal. This stress shifts the PL peak position of the Ni-doped ZnS QDs [49]. The sample color is also changed from light green to dark green at high Ni-ion concentrations.

Ag:ZnS QDs were prepared by wet chemical co-precipitation. These samples were found to show improved PL [77]. A blue PL emission peak with a short radiative lifetime was found. Sahai et al. [77] believe that Ag:ZnS QDs are the best choice for display devices due to undelayed image transformation.

As for the nature of the emerging photoluminescence peaks, the studies performed in [35] for ZnS QDs synthesized via the vapor-liquid-solid technique and dopped with metal ions emited in the visible band, have shown that blue emission is ascribed to S vacancies, green emission is due to Zn vacancies, and yellow-orange and red colors are responsible for impurity defects in the host ZnS. Defect states alert the balance of participating states in the recombination processes and substantially modify the radiative recombination kinetics in the host matrix [35].

Similar studies were carried out for ZnSe QDs. It was found that ZnSe QDs doped with transition metals, such as Cu:ZnSe, Mn:ZnSe, and Fe:ZnSe QDs, still have the same cubic zinc blende crystal type after doping. However, the photoluminescence intensity of Mn:ZnSe QDs and Fe:ZnSe QDs is lower than that of Cd:ZnSe and Cu:ZnSe QDs, which indicates a greater number of structural defects in these QDs compared to ZnSe QDs doped with Cd and Cu [99]. In addition, for these samples, due to the increase in size after doping, a red shift in photoluminescence is observed for ZnSe QDs doped with transition metal ions.

An analysis of the optical properties of doped ZnS and ZnSe quantum dots shows that doping with rare earth elements is promising in applications such as high-performance multicolor photodetectors and nonlinear optics. A multicolor light-detecting PD was also fabricated using Er^{3+}-doped CdS nanoribbons (Er-CdS NRs) [22]. Dedong et al. [22] reported that this PD can detect blue, red, and near-infrared light. This wide range of detection results from the energy band structure of Er-CdS NRs and especially the transitions of Er^{3+} ions energy levels.

Some parameters of photodetectors developed on the basis of quantum dots of wide band gap II-VI compounds are listed in Table 17.2.

Table 17.2 Photoresponsivity and the on-off ratio of photodetectors based on ZnS, ZnSe, and CdS nanostructures

Photodetectors	Responsivity (R) (A/W)	λ (nm)	On-off ratio	Ref.
ZnS nanobelts	0.12	320		[27]
ZnS QDs	0.31	390	413	[72]
ZnS QDs	5.83	365		[54]
CdS QDs	0.3 μA/W	454		[32]
CdS nanorods	1.23×10^4	450		[100]
Er:CdS nanorods	$9.19 \times 10^3 - 3.46 \times 10^4$	457–955		[22]
Pd/Ce doped ZnS/Pd	5.59	300		[3]
ZnS QDs	0.1 mA/W	254		[95]
ZnO/ZnS core-shell nanorod	0.056	340		[57]
ZnO/ZnS MC	177	320–380		[10]
ZnSe QD-CdTe nanoneedle	–			[76]
ZnSe-PbSe nanowire	40	350–510		[23]
1D ZnS/CdS	–	300–550	1.2×10^5	[59]
CdS/CdSSe axial nanowire	1.18×10^2	500	10^5	[34]
CdS/CdS:SnS$_2$ superlattice nanowire	2.5×10^3	350–650	1.5×10^5	[30]
CdS QD-Au/MXene	86 mA/W	405		[40]

17.5 1D Structures and QDs in Heterostructure-Based Photodetectors

The fabrication of heterostructures based on 1D structures and QDs is another approach to fabrication of photodetectors. For the hybrid or heterostructure-based PDs, the enhanced photoresponse is due to large nanostructure surface-to-volume ratio allowing more oxygen molecules to adsorb on the surface, giving a thicker charge depletion layer and reducing the dark current. The multiple scattering light among the nanostructure branches can further increase the light absorption, increasing the photocurrent. The energy diagram of heterostructures based on QDs of various materials can be represented as shown in Fig. 17.2. Electrons generated under illumination are transferred into the lower bandgap side while holes are into the higher side. The separation of the photogenerated electron-hole pairs reduces the rate of charge carriers recombination and significantly increases the photoresponse [30, 59].

Currently, there are reports on the formation of heterostructures using a wide variety of materials. For example, Deng et al. [23] fabricated PbSe-ZnSe nanowire PDs as flexible PDs. The PbSe-ZnSe nanowire has type II alignment where electrons and holes are separated at the interface. This PD covers UV-visible-NIR spectral regions. Lou et al. [59] proposed to use one-dimensional ZnS/CdS heterostructures synthesized by thermal evaporation for the development of such photodetectors. For these PDs, an ultrahigh on-off ratio (10^5) was obtained, which is several orders of magnitude higher than for devices with pure ZnS or CdS nanostructures. At a wavelength of 300–550 nm, the sensitivity was high due to the coupling between parts of the ZnS/CdS heterostructure, and the response time was short. It was also found that the integration of these ZnS and CdS II-VI semiconductors into the ZnS/CdS heterostructure is beneficial for broadband photodetection in the UV-visible range (Lou et al. [59]. ZnS works in the UV range, while CdS works in the visible range. Such photodetectors are required for light image processing, memory storage or switches with high on-off ratio, significant stability and high response speed. Lou et al. [59] also studied PDs based on one-dimensional ZnS/CdS heterostructures fabricated on flexible PET substrates. Under bending conditions, stability, photoresponse to visible light, and photocurrent were also proved to be excellent for this flexible PDs, promising portable and flexible devices [59].

Yadav et al. [89] also fabricated flexible PDs, using ZnO/CdS core/shell micro/nanowire synthesized by the facile hydrothermal method. They have found that the created piezoelectric potential inside this structure enhances charge carrier generation and separation in the metal-semiconductor contact or p-n junction. The broad polarization creates a piezoelectric potential under stress in the wurtzite crystal of the ZnO nanowire. Yadav et al. [89] determined that this PD is sensitive both to green light (548 nm), and UV light (372 nm). The sensitivity of ZnO/CdS core/shell micro/nanowire was three orders higher than CdS nanoribbon and six times higher than CdS nanowire. In addition, the responsivity was significantly increasing by more than ten times compared to the unstrained nanowire. Yadav et al. [89] believe that the improvement in the characteristics of a photonic sensor was due to coupling of piezoelectric, electrical, and optical properties in the developed semiconductor device. The internal piezoelectric field in ZnO controls the charge transfer through carrier separation, optimizing the photoresponse, and the piezoelectric potential itself acts as a "gate" voltage [99].

Benyahia et al. [10] elaborated an eco-friendly ZnO/ZnS composite-based UV-visible PDs with high photoresponse. PDs exhibited a very few dark-current (5 nA) and high on-off ratio (78 dB). In addition, prepared ZnO/ZnS PDs had low optical losses and high absorption at UV-visible, guaranteeing 270% enhancement than the conventional ZnS thin-film PDs. The thermal treatment evaporates the ZnS layer and creates cavities. It was found that the forming such ZnO/ZnS structures with voided regions increases the light-trapping due to multiple reflections acting as optical confinement regions, making this structure a promising high photoresponse multispectral photodetector [10].

CdS/CdS:SnS$_2$ superlattice nanowires synthesized using co-evaporation technology with local environmental control have also been used to develop UV photodetectors [30]. Its photo-responsivity was about seven times higher than pure CdS PD. However, it was found that the photocurrent of superlattice devices increased linearly rapidly at low light intensity and then slowly increased at high light intensity. This means that the photosensitivity decreases at high light intensity. According to Gou et al. [30], this effect is associated with a slow increase in the photocurrent upon saturation of photogenerated carriers.

Guo et al. [34] have shown that high-quality CdS/CdS$_x$Se$_{1-x}$ axial heterostructure nanowires, grown using a moving-source improved CVD method, allowing a large-scale synthesis, are also promising for development of UV-visible PDs. For these photodetectors, the spectral range of 350–650 nm, response time (~68 μs), and on-off ratio of 10^5 were determined. This performance was higher compared to PD developed on CdS and CdS_xSe_{1-x} nanostructures separately (Guo et al. [34].

It should be noted that 1D structures and quantum dots can also form heterostructures. It is known that the fast capture of photogenerated electrons at the interface is still a challenge for efficient light-harvesting from photonic devices using ZnSe QDs. Given this problem, it has been suggested that hybridizing QDs with other nanostructures can contribute to more efficient separation of the photogenerated electron-hole pairs and enhance light harvesting. It is this approach that Saeed et al. [76] used when developing a photodetector based on ZnSe QDs. ZnSe QDs, synthesized by the wet chemical method, were adhered to one-dimensional (1D) nanostructures. Said et al. (2020) showed that the attachment of QDs to one-dimensional (1D) nanostructures indeed promotes efficient electron transfer from the QD surface to the 1D nanostructure. It was found that for the ZnSe QD-CdTe nanoneedle heterostructure, the PL peak is redshifted, which is related to nanohybrid interfaces. In addition, the PL peak of the nanohybrid structure becomes more intense, which is associated with the rapid transfer of excitation energy from the QD as it is the higher energy side (ZnSe band gap ~2.7 eV) to the nanoneedle with a small band gap (CdTe band gap ~1.2 eV). Estimates have shown that the charge-energy transfer from the QD to the nanoneedle occurs in <800 ps. All these results promise nanohybrid photovoltaic devices with better performance [76].

Quantum dot-quantum dot is another form of heterostructure. In particular, UV PDs with such configuration were developed by Peng et al. [69]. ZnO and ZnS spherical nanoshells connected as hollow-sphere bilayer nanofilm were constructed by the self-assembly method. The developed bilayer QDs-based UV PDs demonstrated higher sensitivity, excellent stability, and short response times compared to monolayer devices. The ZnO/ZnS QDs and ZnS/ZnO QDs bilayer nanofilms have type II staggered heterostructure. The electron affinity of ZnO is higher than that of ZnS. According to Peng et al. [69], holes are transported to the ZnS layer and electrons to the ZnO layer. The photocurrent enhancement is correlated to the electron-hole separation by the internal electric field in the nanofilm bilayer, reducing their recombination and increasing photocurrent compared to the monolayer devices.

The experiment showed that heterostructures for the manufacture of UV photodetectors can be formed not only between nanostructured chalcogenides. In particular, Zhao et al. [100] fabricated UV PDs based on CdS–Si heterojunctions. CdS NRs were grown on SiO_2/Si substrate over a large scale using CVD. The PD developeted had a high responsivity due to best crystallinity and a large surface of CdS NR [100]. While indirect bandgap of Si is an obstacle for its UV-visible detection, porous silicon (PSi) as a nano-Si structure with direct bandgap is a good counterpart with limitations on UV detection. Electrochemical etching is usually used to form a porous structure in silicon [47]. It has been found that wider PSi bandgap, controlled as a nanostructure, is favored for good band alignment between QDs and PSi. It facilitates the transfer of photo-induced electrons from the QDs to PSi. In particular, good band alignment between QDs (1.91 eV) and PSi (1.77 eV) was achieved by Chou et al. [17], which contributed to the improvement of optoelectronic device performance. CdSe/CdS/ZnS QDs in (Chou et al. [17] were synthesized by chemical reaction and incorporated into PSi layer, prepared by anodic etching of a p-type silicon substrate. Das et al. [21] used the same approach to fabricate UV-visible PDs based on a ZnS-PSi:p-Si hybrid heterostructure.

17.6 QDs-Polymer Hybrid Strcutures

QDs-polymer hybrid strcutures are a kind of analogue of inorganic heterojunctions, in which one of the components is organic. At present, in the development of photodetectors based on QDs-polymer hybrid strcutures, and as an organic semiconductor, such polymers as poly(3,4-ethylenedioxythiophene): poly(styrene sulfonate) (PEDOT: PSS) [51, 90], cellulose [73], PQT-12 polymer [50], 6,13-Bis(triisopropy lsilylethynyl)- pentacene (TIPS-pentacene) [91], and poly-(3-hexylthiophene-2,5-diyl) (P3HT) [51] have been tested. Quantum dots of CdS, ZnS, CdSe and CdSe/CdS, CdS/ZnS and PbS/CdS core-shell QDs were used. When developing solar cells with different polymers and quantum dots, even more II-VI compounds were tested.

The formation of hybrid heterostructures was usually carried out by layer-by-layer deposition of polymers and QDs using a low-cost spin coating method. It is important to keep in mind that the manufacture of a multilayer structure in this way requires the use of orthogonal solvents to ensure the integrity of the underlying layer when applying the upper layers. A typical structure of photodetectors based on QDs-polymer hybrid strcutures is shown in Fig. 17.4.

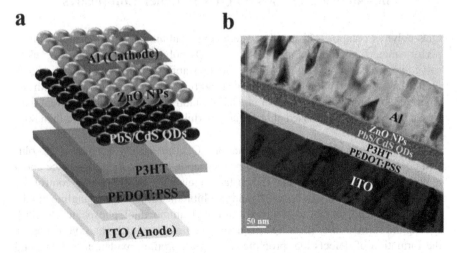

Fig. 17.4 Device structure and TEM images of the all solution-processed SWIR photodetectors. a Schematic device structure, b cross-sectional (scale bar, 50 nm). (Reprinted from Ref. [51]. Published 2020 by Springer as open access)

In such structures, metal oxides such as an indium tin oxide (ITO), TiO_2 or ZnO are used as transparent conductive contacts and electron transport layers, QDs of II-VI semiconductors act as an absorbing photoactive layer, and the polymer provides separation of photogenerated carriers and acts as the hole transport layer. Metal oxides also serve as a kind of protective coating that improves the stability of photodetectors.

Using the considered approach, photodetectors were developed both for SWIR [51] and for the visible spectral region [50, 90, 91]. At the same time, the developed photodetectors have acceptable characteristics. For example, the solution-processable PbS/CdS core-shell QD-based SWIR photodetector exhibited a high on/of (light/dark) ratio of 11, high detectivity of 4.0×10^{12} Jones, fast response (110 ms) and fall (133 ms) times [51]. CdSe QDs-PQT-12 hybrid structure-based PD developed by [50] showed band-pass response over the visible spectrum with a sharp cutoff for higher wavelengths at $\sim$610 nm. The maximum responsivity and the detectivity of the self-powered photodetector was $\sim$3.3 mA/W and 5.4×10^9 $cmHz^{1/2} W^{-1}$, respectively, at a wavelength of $\sim$420 nm under the optical power density of $\sim$130 $\mu W/cm^2$. The rise time and fall time of the device were found to be $\sim$12 and $\sim$15 ms, respectively. Yang et al. [91] reported that the pentacene/CdSe@ZnS QDs composite device exhibited excellent electrical and optical properties with current switch ratio I_{on}/I_{off} of 10^4 (λ = 513 nm), photosensitivity (P) of 10^5, responsivity R of 0.33 mA/W and detectivity (D) of 1.5×10^{11} Jones at drain voltage of $\sim$35 V and light intensity of 1.6 mW/cm^2. Yang et al. [91] believe that the pentacene/CdSe@ZnS QDs hybrid strusture could be one of the most promising candidates for channel transport layer materials for photodetectors.

17.7 Photodetectors Based on QDs-Polymer Composites

QDs-polymer composites is another interesting combination used for improvement performance of photodetectors. The use of QDs-polymer composites allows us to solve two problems. The first is to improve the stability of QDs. As is known, QDs do not have stable parameters and their properties change during contact with the environment. For example, Xiao et al. [101] established that the photocurrent density of the pure CdS QD-based photodetector exhibited a big drop of more than 80% in 24 hours.

QDs in solution are quite stable formations. QDs are usually synthesized in solution containing surface ligands with a headgroup tethered to the QD surface and a hydrocarbon tail directed away. The surface ligands act as stoppers to control the QD growth and surfacepassivating ligands prohibit the formation of dangling bonds. Commonly used ligand head groups are carboxyl, amino, thiol, phosphate, etc. [80]. During the formation of films, QDs usually lose ligands, which is accompanied by the formation of defects that promote oxygen adsorption, oxidation of QDs, and affect the transport of carriers between them. Shell growth, i.e., the formation of core-shell structures, which was discussed in the previous sections, is an effective strategy to mitigate the effects of ligand loss. Polymers can also have the same function. To improve film stability and conductivity, QDs are typically dispersed in a conductive organic matrix that acts as a medium for charge carrier transport. In addition, organic molecules grafted onto the surface of QDs can protect them from aggregation and enhance solubility in non-polar solvents, facilitating the fabrication of uniform and stable QD-based films.

The second task is to improve the photoelectric characteristics of polymer optoelectronic devices [82]. In recent years, flexible, low-power, and low-cost polymer-based devices have been increasingly used [84]. But, polymers do not belong to materials with high absorption coefficient and high quantum efficiency of radiation. At the same time, the QDs of II-VI compounds have these parameters. Therefore, the combination of these materials in one device is quite logical.

QDs-polymer composites can be prepared using various approaches. To incorporate CdS QDs into the polypyrrole (PPy) polymer network, the synthesized CdS QDs were added during the polymerization of PPy. The electrospray method was used to deposit the PPy/CdS QD composite onto the substrate [8]. ZnS quantum dots were incorporated in polyacrylamide by adding an aqueous suspension of ZnS sol in anacrylamide:bisacrylamide copolymer with following polymerasation [63]. To prepare the CdTe/CdS core-shell QD-polymethyl methacrylate (PMMA) composite, PMMA powder was dissolved in chloroform, then QDs were added to the PMMA solution and sonicated for 1 hour. The mixture was then heated to evaporate the solvents [28]. Films were prepared using spin-coating method. The CdSe/PVA nanocomposite thin films were deposited on chemically clean glass substrate by reacting Cd^{2+} dispersed PVA with sodium selenosulphite [70]. The final solution containing Cd^{2+} Cd^{2+} and Se^{2-} ions in the polymeric matrix was coated onto chemically clean glass substrate by dip coating technique and then subjected to thermolysis at 300° C. For incorporation of CdSe/ZnS QDs in PMMA, the PMMA was stirred with QDs in a chloroform solution to form polymer. Then the polymer was paved on a quartz substrate, and kept drying for 72 hours to obtain a uniform film with hundred micrometers thickness [94]. To form a P3HT-CdTe QDs composite, a mixed solution of P3HT and CdTe QDs was spin-coated onto a hole transport layer coated with indium tin oxide (ITO) glass [88]. Subsequently, the films were annealed in a solvent in an atmosphere of 1,2-dichlorobenzene (DCB) for 8 hours. However, Stiff-Roberts et al. [82] noted that one of the biggest problems with these methods of forming QDs-polymer composites is the inability of many solution-based methods to control the internal QDs morphology in the polymer bulk film. For example, they observed that drop-cast and spin-cast CdSe/polymer hybrid nanocomposite films demonstrated the aggregation of CdSe CQDs into micrometre scale clusters inside the polymer bulk. Furthermore, the CdSe CQD concentration demonstrated a strong spatial dependence on the distance from the film centre. Stiff-Roberts et al. [82] believe that resonant IR matrix-assisted pulsed laser evaporation (RIR-MAPLE) has more potential to control the internal and surface morphologies of conjugated polymers, as well as the distribution of QDs in hybrid nanocomposites. However, this method does not belong to the simple and cheap methods of film formation, which drastically limits the possibilities of its use.

Experiment has shown that polymer-QDs composites can find the widest application in the development of various detectors. Phukan and Saikia [70] found that a thin film of CdSe/PVA nanocomposite is suitable for use as a window layer in the fabrication of a CdSe/CdTe solar cell. NIR PDs with good responsivity at 850 nm based on PPy doped with CdS QDs were fabricated by Amiri and Alizadeh [8]. The PDs were prepared with a simple and cost-effective method and had planner structure. The PPy/CdS QDs (22 ppm) PD exhibited a high overall performance with a photoresponsivity of 3.8 mA/W, on/off ratio of 120, external quantum effi ciency (EQE) of 560% and specific detectivity of up to 2.1×10^{16} Jones. According to Amiri and Alizadeh [8], the increase mobility of charge carriers in PPy/CdS QDs was indicated as one of the most important factors for enhance sensitivity of PDs. Wei et al. [88] have shown that by selective passivating deep traps on CdTe QDs surfaces with P3HT, the photodetectors maintain a high gain of 50 but have a 25,000 times shorter response time of 2 µs compared to photodetectors based on CdTe QDs. The high specific detectivity of approaching 10^{13} Jones in the UV–vis spectral range makes it possible to directly detect weak light with an intensity of less that 1 pW cm^{-2}. The Fig. 17.5 shows the PD structure and the EQE spectra of UV-visible photodetectors based on P3HT-CdTe QDs composite developed by Wei et al. [88].

Feizi and Zare [28] developed a new gamma sensor based on CdTe/CdS-PMMA composite. The results showed that the dose rate response of the prepared sensor was linear in the dose rate range of 50–145 mGy/min (see Fig. 17.6a). Yang et al. [92] suggested using CdSe/ZnS core–shell QDs-polyaniline nanocomposites to develop a broadband photodetector based on a single nanowire (NW) (see Fig. 17.6b). The photodetector showed excellent light response in the spectral range of 350 nm to 700 nm. Moreover, the spectral range could be tuned via the size change of QDs. The external quantum efficiency value reaches up to 10^6, the responsivity value reaches up to 10^5 A/W, and the response time value is down to 8 ms.

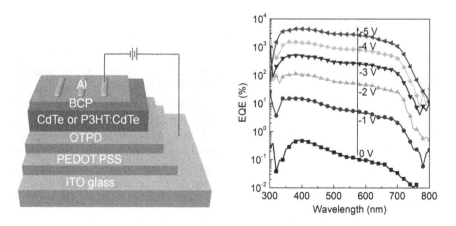

Fig. 17.5 (a) The device structure of P3HT:CdTe nanocomposite photodetectors; (b) The EQE spectra at different reverse bias for the P3HT:CdTe device. (Reprinted with permission form Ref. [88]. Copyright 2015: Wiley)

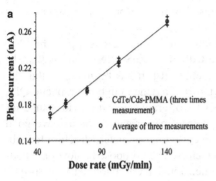
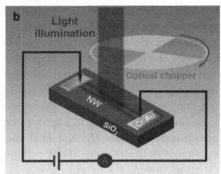

Fig. 17.6 (**a**) Dose rate–response curve of CdTe/CdS-PMMA nanocomposite-based gamma detector. Reprinted with permission from [28]. Copyright 2018: Springer. Fig. 5; (**b**) Schematic diagram of the CdTe/CdS-PMMA NW-based photodetecror. (Reprinted with permission from Ref. [92]. Copyright 2016: RSC)

The presented results indicate that QDs-polymer composites are indeed of interest for the development of photodetectors. However, it must be recognized that the problem of long-term stability of the parameters of QDs-polymer composites detectors during operation, thermal exposure and UV irradiation has not been completely resolved.

17.8   PDs Based on 0D-2D Hybrid Structures

In the last decade, great hopes in the development of electronics and optoelectronics are associated with the use of 2D materials. The term 2D materials refers to crystalline solids consisting of a single layer of atoms. Of all known 2D materials, graphene is the most studied. While the low-cost, low toxicity, biocompatibility, chemical inertness, stretching, wearable flexibility, high conductivity, high carrier transport characteristics, and broad absorption bandwidth make graphene PDs at the top of materials used for optoelectronics and photodetection applications, their absorption rate is only 2.3%, originates from its gapless property limiting their response and then reducing graphene applications [56, 83].

The experiment and calculations showed that the most effective way to improve the parameters of graphene-based photodetectors is the hybridization of graphene with quantum dots such as QDs of II-VI semiconductors. In these graphene-QDs hybrid devices: graphene offers high carrier transport conductive channels, while the QDs of II-VI semiconductors work as a light-harvesting part, where photoelectric conversion occurs by providing high absorption. When a QD absorbs a photon, the created electron-hole pair separates at the graphene-QD interface under the built-in field resulting from their different work functions. This contributes to a decrease in the resistance of graphene and the appearance of a photoresponse, which is much higher than the photoresponse caused by the absorption of radiation by graphene itself [6, 56, 83].

Graphene-QDs hybrid structures can be fabricated using various approaches. For example, the hybrid structure of graphene-CdSe/CdS/ZnS QDs (core/multishell) was fabricated by the Langmuir-Blodgett method [6]. It has been shown that the transfer of photogenerated carriers from QDs to graphene in hybrid graphene-QDs structure indeed improves the optical properties of graphene and provides superior light absorption, helping to improve photocurrent performance and achieve good detection performance [6]. Kan et al. [44] also used hybrid heterostructure for fabrication of deep ultraviolet (DUV) photodetectors. They proposed a novel methodology to construct a hybrid zero−/two-dimensional DUV photodetector (p-type graphene/ZnS QDs/4H-SiC) with photovoltaic characteristics. The schematic illustration of the preparation process is shown in Fig. 17.7. Here, the single-layer p-type graphene (p-Gr), ZnS QDs film and 4H-SiC single crystal substrate are constructed into a PIN junction. The device exhibited excellent selectivity for the DUV light and has an ultrafast response speed (rise time: 28 µs and decay time: 0.75 ms), which are much better than those reported for conventional photoconductive photodetectors.

The same effect was achieved after the fabrication of the hybrid graphene-CuInS$_2$/ZnS core/shell QDs structure [68]. In hybrid graphene-CuInS$_2$/ZnS QDs PD graphene was fabricated by CVD [55, 56], and CuInS$_2$/ZnS QDs were synthesized via hot injection. The direct semiconductor CuInS$_2$ QDs emits at 680 nm but is chemically unstable. Introducing ZnS shell solves the stability problem of CuInS$_2$ QDs and improves the PL. The emission wavelength for this core/shell QD structure is extended to 610–760 nm. In addition, this ternary structure has a long PL lifetime. These merits make CuInS$_2$/ZnS QDs adequate for application in PDs. According to [56, 68], the high response of hybrid graphene-CuInS$_2$/ZnS QDs is achieved due to the long lifetime of carriers of CuInS$_2$/ZnS QDs and high mobility due to single-layer graphene.

There are also reports of a flexible, friendly environment, large area, and cost-effective UV PDs with a high response based on hybrid structures such as graphene-ZnSe/ZnS core/shell QDs [83] and hybrid graphene-CdS QDs [14, 81]. It is believed that in the hybrid graphene-nanostructure PDs, the high photoconductive gain results from the ratio of the nanocrystal (long) carrier lifetime to the (short) graphene transit time. Nevertheless, the slow recombination limits devices-speed below

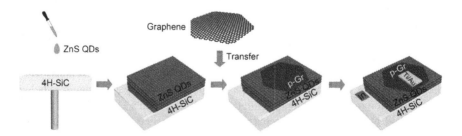

Fig. 17.7 Fabrication procedure of the DUV photodetector (p-type graphene/ZnSQDs/4H-SiC. (Reprinted with permission from Ref. [44]. Copyright 2019: ACS)

a 100 Hz. However, Spirito et al. [81] showed that it is possible to synthesize hybrid graphene-CdS QDs UV-PDs with a fast photocurrent decay time, which make it possible to detect repetition rates above the required limit, for example, during video recording.

It is important to note that the same technique can be used to improve the parameters of photodetectors based on other 2D nanomaterials such as ReS_2 [66] and antimonene []. For example, Qin et al. [66] suggested using this approach when developing photodetectors based on 2D transition metal dichalcogenide ReS_2. ReS_2 is considered as a promising candidate for optoelectronic applications due to its direct bandgap character and optical/electrical anisotropy. However, the narrow spectrum and the low absorption of ReS_2 limits its optoelectronic applications. At the same time it is known that CdSe/CdS/ZnS multi-core/shell QDs are stable and have high absorption in UV-visible spectral range. Qin et al. [66] fabricated the hybrid CdSe/CdS/ZnS multi-core/shell QD-ReS_2 film heterostructure using process shown in Fig. 17.7. The spin-coated process fabricates the QDs while the thick ReS_2 films are prepared by CVD and then exfoliated into a monolayer using a novel ultrasonic-assisted exfoliation approach. It was established that under 589 nm laser irradiation, the responsivity of ReS_2 phototransistor decorated with II-VI quantum dots could be enhanced by more than 25 times (up to ~654 A/W), and the rising and recovery time can be also reduced to 3.2 and 2.8 s, respectively. According to Qin et al. [66], the excellent optoelectronic performance of developed PD is originated from the coupling effect of quantum dots light absorber and cross-linker ligands 1,2-ethanedithiol. Photoexited electron-hole pairs in quantum dots can separate and transfer efficiently due to the type-II band alignment and charge exchange process at the interface. Thus, the built-in field at the interface prevents the photogenerated carriers in QDs from recombination.

The responsivity and the on-off ratio of some hybrid-based PDs are listed in the Table 17.3.

Table 17.3 Photoresponsivity and the on-off ratio of hybrid-based ZnS, ZnSe, and CdS photodetectors

Photodetectors	Responsivity (R) (A/W)	λ (nm)	Ref.
Hybrid ZnS-PSi:p-Si	0.9–1.1	365, 400	[21]
Hybrid CdSe/CdS/ZnS QDs-PSi MSM	0.24	~900	[17]
Graphene-CdS QDs	40	450	[14]
Graphene-CdS nanocrystals	3.4×10^4	349	[81]
Graphene-CdSe/CdS/ZnS QDs	46	365	[6]
Graphene-ZnSe/ZnS core/shell QD	2×10^3	405	[83]
Graphene-CuInS$_2$/ZnS core/shell QDs	2.5×10^5	650	[68]
Graphene-CuInS$_2$/ZnS QDs	35	660	[56]
p-type graphene/ZnS QDs/4H-SiC	0.29 mA/W	250	[44]
ReS$_2$-CdSe/CdS/ZnS multicore/shell QDs	654	589	[66]
Antimonene-CdS QDs	10 μA/W	700	[97]

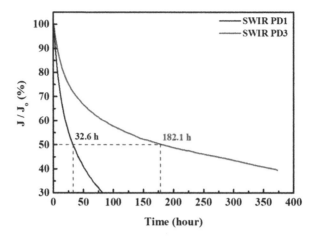

Fig. 17.8 Lifetime characteristics of SWIR PD1 (PbS QDs-polymer) and SWIR PD3 (PbS/CdS QDs-polymer) without encapsulation, under constant voltage operation −1 V at room temperature. (Reprinted from Ref. [51]. Published 2020 by Springer as open access)

17.9   Outlook

This chapter discusses various photodetectors based on QDs of wide band gap II-VI compounds. It is shown that these QDs are indeed a promising materials for creating efficient photodetectors for various purposes. No doubts, there are problems that require additional research and hinder the widespread use of QD-based PDs. The main problem is the low parameter stability of QDs-based photodetectors. For example, Kwon et al. [51] studied unencapsulated SWIR PDs, developed using PbS QDs (PD1) and PbS/CdS core-shell QDs-based hybrid structures (see Fig. 17.1), and established that the lifetime of PbS QDs-based photodetectors (PD1) is only 32 h (Fig. 17.8). The use of PbS/CdS core-shell structures (PD3) improves the stability of photodetectors. The increase in device stability was attributed to the passivating characteristics of the thick CdS shell, which served as a physical barrier to oxygen and moisture penetration. But even in this case lifetime was only 182 h. Of course, this is a very short lifetime of detectors intended for the market. But there is hope that this problem will be solved.

Acknowledgments   G. Korotcenkov is grateful to the State Program of the Republic of Moldova, project 20.80009.5007.02, for supporting his research.

References

1. Abou Elhamd AR, Al-Sallal KA, Hassan A (2019) Review of core/shell quantum dots technology integrated into building's glazing ew of core/shell quantum dots technology integrated into building's glazing. Energies 12:1058

2. Alaie Z, Nejad SM, Yousefi MH (2015) Recent advances in ultraviolet photodetectors. Mater Sci Semicond Proces 29:16–55
3. Al-Haddad RM, Ali IM, Ibrahim IM, Ahmed NM (2015) Photoconductivity and performance of Mn^{2+} and Ce^{3+} doped ZnS quantum dot detectors. Eng Tech J B 33:1608–1618
4. Al-Khursan AH (2006) Gain of excited states in the quantum-dots. Phys E 35:6–8
5. Al-Khursan AH (2005) Intensity noise characteristics in quantum-dot lasers: four-level rate equations analysis. J Luminescen 113:129–136
6. Al-Alwani AJK, Chumakov AS, Shinkarenko OA, Gorbachev IA, Pozharov MV, Venig S, Glukhovskoy EG (2017) Formation and optoelectronic properties of graphene sheets with CdSe/CdS/ZnS quantum dots monolayer formed by langmuir-schaefer hybrid method. Appl Surf Sci 424:222–227
7. Al-Mossawi MA, Al-Shatravi AG, Al-Khursan AH (2012) CdSe/ZnSe quantum-dot semiconductor optical amplifiers. Insciences J 2:52–62
8. Amiri M, Alizadeh N (2020) Highly photosensitive near infrared photodetector based on polypyrrole nanoparticle incorporated with CdS quantum dots. Mater Sci Semicond Proces 111:104964
9. Bang J, Park J, Lee JH, Won N, Nam J, Lim J et al (2010) ZnTe/ZnSe (Core/Shell) type-II quantum dots: their optical and photovoltaic properties. Chem Mater 22:233–240
10. Benyahia K, Djeffal F, Ferhati H, Bendjerad A, Benhaya A, Saidi A (2021) Self-powered photodetector with improved and broadband multispectral photoresponsivity based on ZnO-ZnS composite. J Alloys Comp 859:158242
11. Bratskaya S, Sergeeva K, Konovalova M, Modind E, Svirshchevskaya E, Sergeev A et al (2019) Ligand-assisted synthesis and cytotoxicity of ZnSe quantum dots stabilized by N-(2-carboxyethyl) chitosans. Colloids Surf B: Biointerfaces 182:10342
12. Bursa B, Rytel K, Skrzypiec M, Prochaska K, Wróbel D (2018) Thin film of CdTeSe/ZnS quantum dots on water subphase: thermodynamics and morphology studies. Dyes Pigments 155:36–41
13. Cao J, Jiang Z-J (2020) Thickness dependent shell homogeneity of ZnSe/CdSe core/shell nanocrystals and their spectroscopic and electron and hole transfer dynamics properties. J Phys Chem C 124:12049–12064
14. Chan Y, Dahua Z, Jun Y (2020) Fabrication of hybrid graphene/CdS quantum dots film with the flexible photo-detecting performance. Phys E 124:114216
15. Chen C, Zhu B, Zhang X, Gao Y, Wang G, Gu Y (2018) Synthesis and enhanced third-order nonlinear optical effect of ZnSe/graphene composites. AIP Adv 8:065025
16. Chen X, Liu W, Zhang G, Wu N, Shi L, Pan S (2015) Efficient photoluminescence of Mn^{2+}-doped ZnS quantum dots sensitized by hypocrellin A. Adv Mater Sci Eng 2015:412476
17. Chou C, Cho H, Hsiao VKS, Yong KT, Law WC (2012) Quantum dot-doped porous silicon metal–semiconductor metal photodetector. Nanoscale Res Lett 7:291
18. Chuang SL (2009) Physics of photonic devices. Wiley, New York
19. Cui JY, Li KY, Ren L, Zhao J, Shen TD (2016) Photogenerated carriers enhancement in Cu-doped ZnSe/ZnS/L-cys self-assembled core-shell quantum dots. J Appl Phys 120:184302
20. Dakhil T, Abdulalmuhsin SM, Al-Khursan AH (2018) Quantum efficiency of CdS quantum dot photodetectors. Micro Nano Lett 13:1185–1187
21. Das M, Sarmah S, Sarkar D (2021) Distinct and enhanced ultraviolet to visible ZnS-porous silicon (PS):p-Si hybrid metal-semiconductor-metal (MSM) photodetector. Mater Today Proc 46:247–6252
22. Dedong H, Ying-Kai L, Yu DP (2015) Multicolor photodetector of a single Er3+-doped CdS nanoribbon. Nanoscale Res Lett 10:285
23. Deng J, Lv W, Zhang P, Huang W (2022) Large-scale preparation of ultra-long ZnSe-PbSe heterojunction nanowires for flexible broadband photodetectors. J Sci Adv Mater Dev 7:100396
24. Dickson RE, Hu MZ (2015) Chemical synthesis and optical characterization of regular and magic-sized CdS quantum dot nanocrystals using 1-dodecanethiol. J Mater Res 30:890–895

25. Dorfs D, Hickey SG, Eychmüller A (2010) Type-I and type-II core-shell quantum dots: synthesis and characterization. In: Kumar CSSR (ed) Semiconductor nanomaterials. Wiley-VCH, Weinheim, pp 331–366
26. El-Toni AM, Habila MA, Labis JP, ALOthman ZA, Alhoshan M, Elzatahry AA, Zhang F (2016) Design, synthesis and applications of core–shell, hollow core, and nanorattle multi-functional nanostructures. Nanoscale 8:2510–2531
27. Fang X, Bando Y, Liao M, Gautam UK, Zhi C, Dierre B et al (2009) Single-crystalline ZnS nanobelts as ultraviolet-light sensors. Adv Mater 21:2034–2039
28. Feizi S, Zare H (2018) Masoumeh hoseinpour investigation of dosimetric characteristics of a core–shell quantum dots nano composite (CdTe/CdS/PMMA): fabrication of a new gamma sensor. Appl Phys A Mater Sci Process 124:420
29. Fernández-Delgado N, Herrera M, Tavabi AH, Luysberg M, Borkowski RED, Cantóc PJR et al (2018) Structural and chemical characterization of CdSe-ZnS core-shell quantum dots. Appl Surf Sci 457:93–97
30. Gou G, Dai G, Qian C, Liuc Y, Fu Y, Tianb Z et al (2016) High-performance ultraviolet pho-todetectors based on CdS/CdS:SnS$_2$ superlattice nanowires. Nanoscale 8(30):14580–14586
31. Grandhi GK, Krishna M, Mondal P, Viswanatha R (2020) Cation co-doping into ZnS quan-tum dots: towards visible light sensing applications. Bull Mater Sci 43:301
32. Gu Y, Tang L, Guo X (2019) Preparation and photoelectric properties of cadmium sulfide quantum dots. Chinese Phys B 28:047803
33. Guidelli EJ, Lignos I, Yoo JJ, Lusardi M, Bawendi MG, Baffa O, Jensen KF (2018) Mechanistic insights and controlled synthesis of radioluminescent ZnSe quantum dots using a microfluidic reactor. Chem Mater 30:8562–8570
34. Guo P, Xu J, Gong K (2016) On-nanowire axial heterojunction design for high-performance photodetectors. ACS Nano 10:8474–8848
35. Hafeez M, Al-Asbahi BA, Hj Jumali MH, Yahaya M, Inam F, Bhopal MF, Bhatti AS (2020) Critical role of defect states on visible luminescence from ZnS nanostructures doped with Au, Mn and Ga. Mater Sci Semicond Proces 117:105193
36. He L, Yang L, Liu B, Zhang J, Zhang C, Liu S et al (2019) One-pot synthesis of color-tunable copper doped zinc sulfide quantum dots for solid-state lighting devices. J Alloys Comp 787:537–542
37. Huang W, Yuan Z, Ren Y, Zhou K, Cai Z, Yang C (2019) Interface optical phonons and its ternary mixed effects in ZnS/Cd$_x$Zn$_{1-x}$S quantum dots. Phys E 108:60–67
38. Husham M, Hassan Z, Selman AM (2016) Synthesis and characterization of nanocrystalline CdS thin films for highly photosensitive self-powered photodetector. Eur Phys J Appl Phys 74:10101
39. Ippen C, Greco T, Kim Y, Kim J, Oh MS, Han CJ et al (2014) ZnSe/ZnS quantum dots as emitting material in blue QD-LEDs with narrow emission peak and wavelength tenability. Organic Electron 15:126–131
40. Jiang T, Huang Y, Meng X (2020) CdS core-Au/MXene-based photodetectors: positive deep-UV photoresponse and negative UV–Vis-NIR photoresponse. Appl Surf Sci 513:145813
41. Jamshidi A, Yuan C, Chmyrov V, Widengren J, Sun L, Ågren H (2015) Efficiency enhanced colloidal Mn-doped type II Core/Shell ZnSe/CdS quantum dot sensitized hybrid solar cells. J Nanomater 2015:921903
42. Jbara AS, Abood HI, Al-Khursan AH (2012) Effect of doping and in-composition on gain of long wavelength III-nitride QDs. J Opt 41:214–223
43. Ji B, Koley S, Slobodkin I (2020) ZnSe/ZnS core/shell quantum dots with superior optical properties through thermodynamic shell growth. Nano Lett 20:2387–2395
44. Kan H, Zheng W, Lin R, Li M, Fu C, Sun H et al (2019) Ultra-fast photovoltaic-type deep-ultraviolet photodetector using hybrid zero−/two dimensional hetero junctions. ACS Appl Mater Interfaces 11(8):8412–8418
45. Kaushik S, Singh R (2021) 2D layered materials for ultraviolet photodetection: a review. Adv Optic Mater 9(11):2002214

46. Khani O, Rajabi HR, Yousefi MH, Khosravi AA, Jannesari M, Shamsipur M (2011) Synthesis and characterizations of ultra-small ZnS and $Zn_{(1-x)}Fe_xS$ quantum dots in aqueous media and spectroscopic study of their interactions with bovine serum albumin. Spectrochim Acta A 79:361–369

47. Korotcenkov G (ed) (2015) Porous silicon: from formation to application. Vol. 1: formation and properties. Taylor and Fracis Group/CRC Press, Boca Raton

48. Korotcenkov G (2014) Handbook of gas sensor materials properties, advantages and short-comings for applications, New trends and technologies, vol 2. Springer, New York

49. Kumar V, Rawal I, Kumar V, Goyal PK (2019) Efficient UV photodetectors based on Ni-doped ZnS nanoparticles prepared by facial chemical reduction method. Physica B 575:411690

50. Kumar H, Kumar Y, Rawat G, Kumar C, Mukherjee B, Pal BN, Jit S (2017) Colloidal CdSe quantum dots and PQT-12-based low-temperature self-powered hybrid photodetector. IEEE Photon Technol Lett 29(20):1715–1718

51. Kwon J-B, Kim S-W, Kang B-H, Yeom S-H, Lee W-H, Kwon D-H et al (2020) Air-stable and ultrasensitive solution-cast SWIR photodetectors utilizing modifed core/shell colloidal quantum dots. Nano Convergence 7:28

52. Lad AD, Rajesh C, Khan M, Ali N, Gopalakrishnan IK, Kulshreshtha SK, Mahamuni S (2007) Magnetic behavior of manganese-doped ZnSe quantum dots. J Appl Phys 101:103906

53. Lee Y-J, Cha J-M, Yoon C-B, Lee S-E (2018) Study on UV opto-electric properties of ZnS:Mn/ZnS core-shell QD. J Korean Ceram Soc 55:55–60

54. Li R, Tang L, Zhao Q, Teng KS, Lau SP (2020) Facile synthesis of ZnS quantum dots at room temperature for ultra-violet photodetector applications. Chem Phys Lett 742:137127

55. Li Z, Xu K, Wei F (2018a) Recent progress in photodetectors based on low-dimensional nanomaterials. Nanotechnol Rev 7:393–411

56. Li F, Guo C, Pan R, Zhu Y, You L, Wang J et al (2018b) Integration of green $CuInS_2$/ZnS quantum dots for high-efficiency light-emitting diodes and high-responsivity photodetectors. Opt Mater Express 8:314–323

57. Lin H, Wei L, Wu C, Chen Y, Yan S, Mei L, Jiao J (2016) High-performance self-powered photodetectors based on ZnO/ZnS core-shell nanorod arrays. Nanoscale Res Lett 11:420

58. Liu H, Zhong H, Zheng F, Xie Y, Li D, Wu D, Zhou Z, Sun XW, Wang K (2019) Near-infrared lead chalcogenide quantum dots: synthesis and applications in light emitting diodes. Chinese Phys B 28:128504

59. Lou Z, Li L, Shen G (2016) Ultraviolet/visible photodetectors with ultrafast, high photosensitivity based on 1D ZnS/CdS heterostructures. Nanoscale 8:5219–5225

60. Maity P, Singh SV, Biring S, Pal BN, Ghosh AK (2019) Selective near-infrared (NIR) photodetectors fabricated with colloidal CdS: Co quantum dots. J Mater Chem C 7:7725

61. Memon UB, Chatterjee U, Gandhi MN, Tiwari S, Duttagupta SP (2014) Synthesis of ZnSe quantum dots with stoichiometric ratio difference and study of its optoelectronic property. Procedia Mater Sci 5:1027–1033

62. Miao S, Cho Y (2021) Toward green optoelectronics: environmental-friendly colloidal quantum dots photodetectors. Front Energy Res 9:666534

63. Mir FA (2010) Structural and optical properties of ZnS nanocrystals embedded in polyacrylamide. J Optoelectron Biomed Mater 2(2):79–84

64. Mishra SK, Srivastava RK, Prakash SG, Prakash SG, Yadav RS, Panday AC (2011) Structural, photoconductivity and photoluminescence characterization of cadmium sulfide quantum dots prepared by a co-precipitation method. Electron Mater Lett 7:31–38

65. Nikesh VV, Lad AD, Kimura S, Nozak S, Mahamuni S (2006) Electron energy levels in ZnSe quantum dots. J Appl Phys 100:113520

66. Qin JK, Ren DD, Shao WZ, Li Y, Miao P, Sun ZY et al (2017) Photoresponse enhancement in monolayer ReS_2 phototransistor decorated with CdSe-CdS-ZnS quantum dots. ACS Appl Mater Interfaces 9:45

67. Pacheco ME, Castells CB, Bruzzone L (2017) Mn-doped ZnS phosphorescent quantum dots: coumarins optical sensors. Sensors Actuators B Chem 238:660–666

68. Pan R, Wang J (2019) Ultra-high responsivity graphene-CIS/ZnS QDs hybrid photodetector. Proc SPIE 10843:108431B
69. Peng L, Han S, Hu X (2014) Photocurrent enhancement mechanisms in bilayer nanofilm-based ultraviolet photodetectors made from ZnO and ZnS spherical nanoshells. Nanoscale Res Lett 9:388
70. Phukan P, Saikia D (2013) Optical and structural investigation of CdSe quantum dots dispersed in PVA matrix and photovoltaic applications. Intern J Photoenergy 2013:728280
71. Poornaprakash B, Kumar KN, Chalapathi U, Reddeppa M, Poojitha PT, Park SH (2016) Chromium doped ZnS nanoparticles: chemical, structural, luminescence and magnetic studies. J Mater Sci Mater Electron 27:6474–6479
72. Premkumar S, Nataraj D, Bharathi G, Ramya S, Thangadurai TD (2019) Highly responsive ultraviolet sensor based on ZnS quantum dot solid with enhanced photocurrent. Sci Rep 9:18704
73. Ranjan PS, Bhuyan RK (2021) Organic polymer and perovskite CdSe–CdS QDs hybrid thin film: a new model for the direct detection of light elements. J Mater Sci Mater Electron 32:5538–5547
74. Reddy CV, Shim J, Cho M (2017) Synthesis, structural, optical and photocatalytic properties of CdS/ZnS core/shell nanoparticles. J Phys Chem Solids 103:209–217
75. Rosencher E, Vinter B (2002) Optoelectronics. Cambridge University Press, Cambridge
76. Saeed S, Iqbal A, Iqbal A (2020) Photo-induced charge carrier dynamics in a ZnSe quantum dot-attached CdTe system. Proc Royal Soc A 476:20190616
77. Sahai S, Husain M, Shanker V, Singh N, Haranath D (2011) Facile synthesis and step by step enhancement of blue photoluminescence from Ag-doped ZnS quantum dots. J Colloid Interface Sci 357:379–383
78. Sahin O, Horoz S (2018) Synthesis of Ni: ZnS quantum dots and investigation of their properties. J Mater Sci Mater Electron 29:16775–16781
79. Senthilkumar K, Kalaivani T, Kanagesan S, Balasubramanian V (2012) Synthesis and characterization studies of ZnSe quantum dots. J Mater Sci Mater Electron 23:2048–2052
80. Shang Y, Ning Z (2017) Colloidal quantum-dots surface and device structure engineering for high-performance light-emitting diodes. National Sci Rev 4:170–183
81. Spirito D, Kudera S, Miseikis V, Giansante C, Coletti C, Krahne R (2015) UV light detection from CdS nanocrystal sensitized graphene photodetectors at kHz frequencies. J Phys Chem C 119:23859–23864
82. Stiff-Roberts AD, Lantz KR, Pate R (2009) Room-temperature, mid-infrared photodetection in colloidal quantum dot/conjugated polymer hybrid nanocomposites: a new approach to quantum dot infrared photodetectors. J Phys D Appl Phys 42:234004
83. Sun YL, Xie D, Sun MX, Teng CJ, Qian L, Chen RS et al (2018) Hybrid graphene/cadmium free ZnSe/ZnS quantum dots phototransistors for UV detection. Sci Rep 8:5107
84. Tavasli A, Gurunlu B, Gunturkun D, Isci R, Faraji S (2022) A review on solution-processed organic phototransistors and their recent developments. Electronics 11:316
85. Tian W, Li L, Lu H (2015) Nanoscale ultraviolet photodetectors based on one dimensional metal oxide nanostructures. Nano Res 8:382–405
86. Vasudevan D, Gaddam RR, Trinchi A, Cole I (2015) Core-shell quantum dots: properties and applications. J Alloys Comp 636:395–404
87. Wang A, Buntine MA, Jia G (2019) Recent advances in zinc-containing colloidal semiconductor nanocrystals for optoelectronic and energy conversion applications. Chem Electro Chem 6:1–17
88. Wei H, Fang Y, Yuan Y, Shen L, Huang J (2015) Trap engineering of CdTe nanoparticle for high gain, fast response, and low noise P3HT:CdTe nanocomposite photodetectors. Adv Mater 27(34):4975–4981
89. Yadav PVK, Ajitha B, Reddy YAK, Sreedhar A (2021) Recent advances in development of nanostructured photodetectors from ultraviolet to infrared region: a review. Chemosphere 279:130473

90. Yan Y, Wu X, Chen Q, Liu Y, Chen H, Guo T (2019) High-performance low-voltage flexible photodetector arrays based on all-solid-state organic electrochemical transistors for photosensing and imaging. ACS Appl Mater Interf 11:20214–20224

91. Yang Z, Lin S, Liu J, Zheng K, Lu G, Ye B et al (2020) High performance phototransistors with organic/quantum dot composite materials channels. Org Electron 78:105565

92. Yang X, Liu Y, Lei H, Li B (2016) An organic–inorganic broadband photodetector based on a single polyaniline nanowire doped with quantum dots. Nanoscale 8:15529–15537

93. Yaraki MT, Tayebi M, Ahmadieh M, Tahriri M, Vashaee D, Tayebi L (2017) Synthesis and optical properties of cysteamine-capped ZnS quantum dots for aflatoxin quantification. J Alloys Comp 690:749–758

94. Yingming S, Pan H, Chu H, Mamat M, Abudurexiti A, Li D (2021) Core-shell CdSe/ZnS quantum dots polymer composite as the saturable absorber at 1.3 μm: influence of the doping concentration. Phys Lett A 400:127307

95. Xia Y, Zhai G, Zheng Z, Lian L, Liu H, Zhang D, Gao J, Zhai T, Zhang J (2018) Solution-processed solar-blind deep ultraviolet photodetectors based on strongly quantum confined ZnS quantum dots. J Mater Chem C 6:11266–11271

96. Xia X, Liu Z, Du G, Li Y, Ma M (2010) Wurtzite and zinc-blende CdSe based core/shell semiconductor nanocrystals: structure, morphology and photoluminescence. J Luminescence 130:1285–1291

97. Zeng X, Zhang J, Huang F (2012) Optical and magnetic properties of Cr-doped ZnS nano-crystallites. J Appl Phys 111:123525

98. Zhang F, Ding Y, Zhang Y, Zhang X, Wang ZL (2012a) Piezo-phototronic effect enhanced visible and ultraviolet photodetection using a ZnO-CdS core-shell micro/nanowire. ACSNano 6:9229–9236

99. Zhang Y, Shen Y, Wang X, Zhu L, Han B, Ge L, Tao Y, Xie A (2012b) Enhancement of blue fluorescence on the ZnSe quantum dots doped with transition metal ions. Mater Lett 78:35–38

100. Zhao W, Liu L, Xu M (2017) Single CdS nanorod for high responsivity UV–visible photodetector. Adv Optic Mater 5:1700159

101. Xiao Q, Hu C-X, Wu H-R , Ren Y-Y, Li X-Y, Yang Q-Q, et al. (2020) Antimonene-based flexible photodetector. Nanoscale Horizons 5(1):124–130

Chapter 18
Solution-Processed Photodetectors

**Shaikh Khaled Mostaque, Abdul Kuddus, Md. Ferdous Rahman,
Ghenadii Korotcenkov, and Jaker Hossain**

18.1 Introduction

The current technological revolution has been spurred by design of novel materials that can be systematically adjusted to gain distinctive features for usage in a number of applications. Till now, variety of researches have been carried out with II-VI semiconductor materials as they generally encompass the entire spectrum from ultraviolet (UV) to far infrared (IR). Besides, their electron energy band gap corresponds to visible region and can be tuned with adjusting the ternary or quaternary compounds providing the facility to fabricate optoelectronic devices for various wavelengths [20]. Additional information one can find in the Chaps. 1, 2, 3, 4, 5, and 6 (Vol. 1).

II-VI semiconductors have been extensively investigated over a long period, since the observation of phosphorescence from ZnS crystals in 1866, especially in

S. K. Mostaque · J. Hossain (✉)
Solar Energy Laboratory, Department of Electrical and Electronic Engineering, University of Rajshahi, Rajshahi, Bangladesh
e-mail: jak_apee@ru.ac.bd

A. Kuddus
Solar Energy Laboratory, Department of Electrical and Electronic Engineering, University of Rajshahi, Rajshahi, Bangladesh

Graduate School of Science and Engineering, Saitama University, Saitama, Japan

M. F. Rahman
Department of Electrical and Electronic Engineering, Begum Rokeya University, Rangpur, Bangladesh

G. Korotcenkov
Department of Physics and Engineering, Moldova State University, Chisinau, Moldova

© The Author(s), under exclusive license to Springer Nature Switzerland AG 2023
G. Korotcenkov (ed.), *Handbook of II-VI Semiconductor-Based Sensors and Radiation Detectors*, https://doi.org/10.1007/978-3-031-20510-1_18

the areas of light emitters (e.g. phosphor for television CRT, electro-luminescent cells for flat display, light emitting diodes (LEDs), Schottky contact diodes, lasers), photo−/radiation detectors (e.g. photoconductive detectors, solar cells, thermal/far-infrared detectors, radiation and particle detectors), optics (e.g electro-optics, magneto-optics, optical switching and bistability), electronics (e.g. thin film transistors, integrated optoelectronics, charge couple devices, varistors, passivation and blocking layer), sensors, data storage and so on [20]. Thus far, research involving II-VI band gap semiconductors have obtained significant achievements in various applications via formation of different p-n junctions and patterns while controlling the doping, applying suitable contacts, and performing various etching and passivation methods. Despite the fact that current formation techniques such as ion beam sputtering, magnetron sputtering, ion beam assisted deposition, plasma source assisted deposition, and electron beam deposition enable high performance applications, neither single coating technique or model is better compared across all circumstances [34].

Currently, there are two main directions in the development of electronic and optoelectronic devices. The first direction is microminiaturization, aimed at the development of monolithic integrated circuits, and the second is the search for technological approaches that make it possible to manufacture cheap devices for wide application. Solution-processed technology refers to the second direction in the development of semiconductor optoelectronic devices such as solar cells and photodetectors. Low temperature solution processed semiconductors are an emerging class of photoactive materials that can be processed using wet chemical methods. They are attractive from a technological point of view for several reasons [10, 8, 12, 41, 42]: they can be deposited from a solution over large areas using readily available cheap manufacturing techniques. This simple manufacturing technology is processed at low temperatures and under ambient conditions with minimal material consumption. In addition, these low cost deposition techniques are compatible with (i) a wide range of versatile and uncommon substrates, including flexible substrates, (ii) various types of auxiliary materials, which can be used for contacts and transport layers, and (iii) contemporary sensing systems.

Modern photodetectors mainly use photodiodes based on crystalline and single-crystal inorganic semiconductors, such as silicon or III–V compounds. As a rule, for the manufacture of such devices, epitaxial methods are used, which require expensive equipment. Therefore, photodetectors made from solution-processed semiconductors have in recent years been considered as candidates for the next generation of light detectors [10, 8, 12].

II-VI compounds fully meet the requirements for materials that can be used in solution-processed technology. First, nanocrystals and colloidal quantum dots of II-VI compounds used in this technology can be synthesized by various chemical methods. Secondly, II-VI compounds have specific optical and electrical properties. In addition to the manufacturing benefits offered by solution-processed semiconductors, II-VI semiconductors have the advantage that their optoelectronic properties can be tailored. Especially appealing for photodetection is the ability to control the semiconducting optical band gap. Compositional tuning of the II-VI semiconductor alloys allows the modification of the semiconductor bandgap from the visible through to the near-infrared.

In this chapter, we intend to discuss about the state of the art performance of solution processed techniques for II-VI semiconductor compounds, especially in the photodetector applications. Here, the focus has been given on non-oxide II-VI wide band compounds.

18.2 Solution Processed II-VI Semiconductor-Based Solar Cells and Photodetectors

When developing solution processed photodetectors, two approaches are used. The first approach involves the formation of photosensitive layers of II-VI compounds on substrates directly during their synthesis [4, 37]. These methods include chemical bath deposition (CBD), electrochemical deposition, spray pyrolysis, successive ionic layer adsorption and reaction (SILAR) or successive ionic layer deposition (SILD) [24, 38, 40, 49, 54]. A description of these methods for II-VI compounds can be found in the Chap. 10 (Vol. 1). This is a fairly simple method for forming films of II-VI compounds. But if it is necessary to deposit semiconductor films over large areas, problems arise with their uniformity and reproducibility of their parameters. In addition, the properties of the films strongly depend on the deposition parameters.

The second approach is based on the synthesis of nanocrystallites (NCs) or colloidal particles, their separation from the solution, and subsequent use for deposition on the substrate surface by various methods [12, 41, 42]. Typically, nanocrystals and colloidal particles are synthesized using methods such as sol-gel, hydrothermal, solvothermal, hot injection, and so on [16]. These methods are described in sufficient detail in the Chap. 11 (Vol. 1). From the synthesized nanoparticles, a paste or solution is formed, which are used to form layers of II-VI compounds. The advantage of this approach is the ability to control the parameters of nanoparticles at the stage of synthesis. The deposition of films of II-VI compounds in this case is carried out using methods such as spin coating, roll-to-roll printing, spraying, ink jet printing, drop coating, dip coating, and doctor blading (Fig. 18.1). A brief description of these methods is presented in Table 18.1.

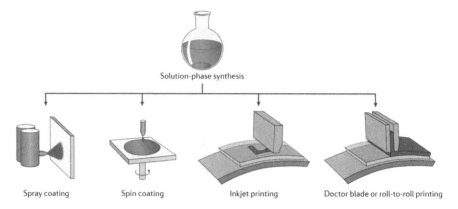

Fig. 18.1 Solution-processed materials are synthesized in the form of colloidal semiconductor inks. These can be deposited and assembled in solid films using spray coating or spin coating, or manufacturing techniques such as inkjet printing, doctor blading or roll-to-roll printing. (Reprinted with permission from Ref. [12]. Copyright 2017: Springer Nature)

Table 18.1 Methods usually used for preparing films in the frame of solution-processed technology

Method	Description
Casting	In this process the material to be deposited is dissolved in liquid form in a solvent. The material can be applied to the substrate by spraying or spinning. Once the solvent is evaporated, a thin film of the material remains on the substrate. The thicknesses that can be cast on a substrate range all the way from a tens of nanometers to tens of micrometers. The control on film thickness depends on exact conditions, but can be sustained within +/−10% in a wide range. Delamination and cracking can occur if the film is too thick. This method gives a more uniform and a more reproducible membrane than dip coating.
Spin coating	In the *spin coating* process, the substrate spins around an axis which should be perpendicular to the coating area. The quality of the coating depends on the rotation velocity, rheological parameters of the coating liquid and surrounding atmosphere. The coating thickness varies between several hundreds of nanometers and up to 10 micrometers. Desired thickness obtained by precursor dilution, spin speed and number of layers. Equipment similar to that of spin-coat tracks used for photoresist deposition.
Spray coating	Precursor is atomized to form fine aerosol which then is deposited on a wafer. Deposition enhanced by electrostatic charging of aerosol. Desired thickness is controlled by adjusting deposition time and number of layers. The coating step is suitable for establishing an in-line process.
Slip casting	*Slip casting* is a technique in which a suspension (slip) is poured into a porous mold. The mold's pores absorb the liquid, and particles are compacted on the mold walls by capillary forces, i.e. solidify, producing parts of uniform thickness. Once dried to the leather-hard stage, the molds are opened and the cast object removed to dry completely before firing.

(continued)

Table 18.1 (continued)

Method	Description
Tape casting	The *tape casting* is a technique for continuous production of the films according to the "Doctor-Blade-principle". In this process a suspension of solid state particles in an organic solvent or water, mixed together with strengthening plasticizers and/or binders can be used. The slip is cast onto a precisely machined stone plate, on which the carrier film is moved smoothly and without perturbations. By the doctor blade the slurry is spread homogeneously on the surface of the tape. Drying and firing are final stages of the actual tape forming.
Dip coating	*Dip coating* techniques can be described as a process where the substrate to be coated is immersed in a liquid and then withdrawn with a well-defined withdrawal speed under controlled temperature and atmospheric conditions. The coating thickness is mainly defined by the withdrawal speed, by the solid content and the viscosity of the liquid. The applied coating may remain wet for several minutes until the solvent evaporates. This process can be accelerated by heated drying. In addition, the coating be exposed to various thermal, UV, or IR treatments for stabilization.
Screen printing	*Screen printing* is a printing technique that uses a woven mesh to support an ink-blocking stencil. The paste (ink) used is a mixture of the material of interest, an organic binder, and a solvent. The attached stencil forms open areas of mesh that transfer ink or other printable materials which can be pressed through the mesh as a sharp-edged image onto a substrate. A roller or squeegee is moved across the screen stencil, forcing or pumping ink past the threads of the woven mesh in the open areas. After printing, the wet films are allowed to settle for 15–30 min to flatten the surface and are dried. This removes the solvents from the paste. Subsequent firing burns off the organic binder and metallic particles are reduced or oxidized and glass particles are sintered. It can be used to print on a wide variety of substrates, including paper, paperboard, plastics, glass, metals, fabrics, and many other materials.
Inkjet printing	*Inkjet technologies*, which are based on the 2D printer technique of using a jet to deposit tiny drops of ink onto substrate, are perfectly suited to controllably dispense small and precise amounts of "liquid" to precise locations. The available inkjet technologies include: (1) Continuous inkjet; (2) Drop on demand inkjet; (3) Thermal inkjet; and (4) Piezo inkjet. The "liquid" materials can encompass low to high viscosity fluids, colloidal suspensions, frits, metallic suspensions, and almost any other material that can be dispersed in a liquid carrier material. The carrier material can be aqueous or non-aqueous based solvent material. When printed, liquid drops of these materials instantly cool and solidify to form a layer of the part. Usually inkjet printing is accompanied by thermal treatment.

Source: Data extracted from Ref. [23]

18.3 Solution-Processed Photodetectors with Direct Wet Chemical Synthesis of Photosensitive Layers

As an example, let us consider two methods for the synthesis of II-VI semiconductor films, which are the most commonly used in the manufacture of solution-treated photodetectors.

18.3.1   Chemical Bath Deposition (CBD)

Chemical bath deposition is an old method for fabrication of photo detectors and a still a popular method for thin film deposition because of simplicity, scalability, low temperature, and less energy consumption [38, 40, 54]. Hou et al. [16] believe that thin film of II-VI compounds deposited by soft chemical solution method can be useful in production of large area thin films with low cost. In CBD method, the substrate is dipped into the solution consisting chalcogen source material, a metal ion, an extra base and a complexing agent. The method is based on the delayed release of chalcogenide ions into an alkaline mixture, with occurrence of film formation during the surpassing of ionic product (IP) to the solubility product (SP).

Chemical bath deposition method is acceptable for all II-VI compounds. In this process ZnS can be synthesized using zinc sulfate heptahydrate ($ZnSO_4 \cdot 7H_2O$) as the targeted ion source for Zn, and thiourea ($CS(NH_2)_2$) served as an ion source of S [38, 40]. 0.03 M $ZnSO_4 \cdot 7H_2O$ and 0.5 M ($CS(NH_2)_2$) with 2.5 M of ammonia were continuously being stirred at room temperature and kept in a water bath. The substrates were dipped into solution and the temperature was then raised to 60 °C with continuous stirring to ensure pH at around 9.8. Keeping the substrates for 45 min resulted deposition of films, which were finally washed with distilled water to get rid of porous ZnS. Consecutive drying resulted uniform ZnS thin film which is ready to be annealed. The annealed films were more homogeneous and came out with less cracks.

To deposit thin films of ZnTe, solid tellurium oxide, TeO_2, and zinc chloride, $ZnCl_2$, can be used as a source of Te and Zn, respectively [54]. While $ZnCl_2$ was melted with distilled water, both HCl and distilled water took part in melting TeO_2. The precipitation solution was prepared with mixing all of these compounds with a magnetic stirrer at 70 °C. With mixing the following reactions occurred:

$$TeO_2 + 4HCl \leftrightarrow TeCl_4 + 2H_2O \qquad (18.1)$$

and

$$ZnCl_2 + TeCl_4 \rightarrow ZnTe + Cl_2 \uparrow + H_2O \uparrow \qquad (18.2)$$

After the deposition of films, the precipitation was done for different times and at different temperatures. Though the amount of time did not affect the film performance, the film prepared with 85 °C exhibited the lowest energy gap.

Likewise, CdSe film, specially studied for optoelectronic and photodetecting applications, can be prepared with chemical bath deposition technique [14, 17]. Example of such process is shown in Fig. 18.2.

During deposition time, temperature, pH of solution and concentration of the metal ions played the key role in above method. The proper optimization was possible by controlling the emancipation of Cd^{2+} and Se^{2-} ions. For regulation of

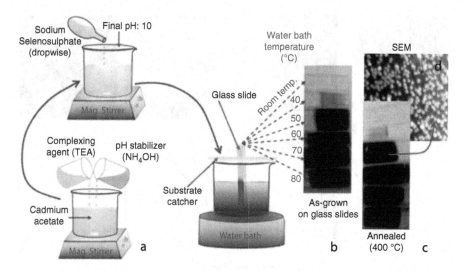

Fig. 18.2 (**a**) The procedure for making precipitation solution (**b**) Samples after deposition on glass slides at different constant precipitation temperature (**c**) CdSe film after annealing at 400 °C and (**d**) SEM image. (Adapted from Ref. [17]. Published 2021 by Frontiers Media as open access)

cadmium hydrolysis, triethanolamine (TEA) worked as the complexing agent here. The following reactions took place for the preparation of precipitation solution.

$$Cd^{2+} + TEA \rightarrow \left[Cd\left(TEA\right) \right]^{2+} \tag{18.3}$$

$$NH_3 + H_2O \rightleftharpoons NH^{4+} + OH^- \tag{18.4}$$

$$\left[Cd\left(TEA\right) \right]^{2+} + Na_2SeSO_3 + 2OH^- \rightarrow CdSe + Na_2SO_4 + TEA + H_2O \tag{18.5}$$

In this process, smaller peaks at XRD tends to get larger with deposition temperature and the best result arrived between 75 and 80 °C, exhibited in Fig. 18.3a. Though XRD small peaks started exhibiting at 25 °C annealing, an elevated (111) peak for the particular case arrived at 450 °C because of reduced crystallite size with annealing temperature. The transmittance data in Fig. 18.3b reveals that with higher deposition and annealing temperatures, the transmittance falls. Accordingly, the band gap values get decreased and it has been found that the band gap values for 50 °C, 70 °C and 80 °C falls in the region of visible spectrum. Thus, the film can be suitably applied for photodetection in the visible region of solar spectrum.

Chemical bath deposition for the formation of high photosensitive CdSe films in Ag/CdSe/ITO-based photodetectors was also used in [17]. Figure 18.4a represents the setup for chemical bath deposition where ITO coated glass substrate has been emerged. The 50 mL bath has been filled with 10 mL 0.1 M CdCl$_2$·2H$_2$O, 3.5 mL of 30% NH$_3$ aqueous, 10 mL solution of Na$_2$SeSO$_3$ and distilled water for laboratory purpose [39, 40]. Figure 18.4b represents the experimental setup for measuring

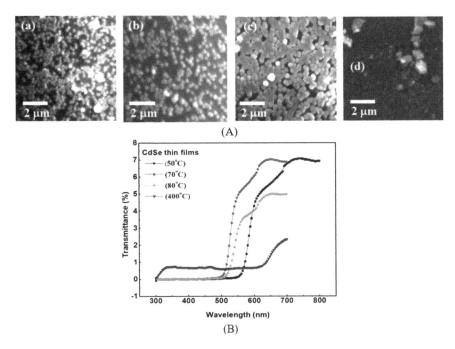

(A)

(B)

Fig. 18.3 (**a**) Characterization of CdTe thin films deposited at (a) 50 °C, (b) 70 °C, (c) 80 °C and (d) 50 °C (annealed at 400 °C), (**b**) Transmittance plot. (Adapted from Ref. [17]. Published 2021 by Frontiers Media as open access)

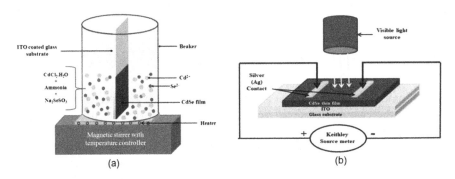

Fig. 18.4 (**a**) Deposition of CdSe films in chemical bath; (**b**) Experimental setup to measure photoresponse. (Reprinted from Ref. [39]. Published by SAMRIDDHI as open access)

metal-semiconductor-metal, i.e. Ag/CdSe/Ag structure photodetector. Owing to suitable direct bandgap, high absorption coefficient and high photosensitivity, CdSe can take part in optical activity in infrared and visible part of the spectrum [17]. The I-V characteristics of Ag/CdSe/Ag metal-semiconductor-metal structure confirm the ohmic nature of the fabricated device and it exhibited a good photosensitivity and photo-responsivity under the illumination of visible-light.

Another interesting option for manufacturing a solution processed photodetector was proposed by Chen et al. [3]. In this process, chemical bath deposition of ZnSe nanoparticles on Si wire (SiW) arrays makes fabricating well-aligned 3D hetero-structured ZnSe NP/SiW arrays with a high on/off ratio photocurrent and photocata-lytic activity. ZnSe/SiW 3D heterostructures demonstrated excellent performance UV photocurrent response due to a large surface-to-volume ratio and active area, as well as a fast conductive pathway [3]. The coating of ZnSe nanowires on Si wafer using CBD method provides excellent photodetection with immediate decay greater than 99.85%, on/off ratio greater than 700 and fast photoresponse speed less than 0.4 μs. Under the exposure of UV radiation, the formation of electron-hole pair occurs in ZnSe. The electron goes to conduction band and hole stays in the valance band. The recombination was possible to avoid because of the ready formation of 3D heterostructure that has the advantage for migration of photogenerated electrons to the surface.

18.3.2   Successive Ionic Layer Adsorption and Reaction Technique (SILAR)

Adsorption, which is the base for SILAR method, is a surface activity that takes place between ions and the substrate's surface and is caused by the cohesive force, Van der Walls force, chemical attractive force etc. between the ions in the solution and the substrate's surface. Here, the substrate particles are held in place by the imbalanced or residual force of the atoms or molecules on the substrate surface. The approach entails two steps (i) adsorption and rinsing, and (ii) reaction and rinsing. The cations in the precursor solution are adsorbed on the surface of the substrate in this initial stage, forming the Helmholtz electric double layer [39]. For instance, a self-powered photoelectrochemical cell-type UV detector with a ZnO/ZnS core-shell nanorod array as the active photoanode and deionized water as the electrolyte demonstrates a quick photoresponse time of around 0.04 s and outstanding spec-trum selectivity [3]. Here, the shell consist of ZnS nanoparticles was grown through the steps of SILAR method. First, a ZnO nanorod array substrate was immersed into 0.1 M zinc nitrate aqueous solution and rinsed with deionized water. On the second step it was dipped into 0.1 M Na_2S aqueous solution and again rinsed with deion-ized water. This process was repeated for 2, 7, 10, and 15 cycles to generate differ-ent thicknesses of ZnS nanoparticles.

Figure 18.5c shows that ZnS nanoparticles have been uniformly deposited on ZnO nanorod in SILAR process and have come out with a rough surface which is favorable for light scattering. The light scattering causes photo-anodes to exhibit lower transmittance than bare FTO. The transmittance gets lowered with increase of SILAR cycles as visible in Fig. 18.5a. From Fig. 18.5b, it is clear that the photosen-sitivity gets improved with more than 0.3 A/W just after adding ZnS with only two SILAR cycles. The sample with ZnS seven SILAR cycles at 340 nm has the best

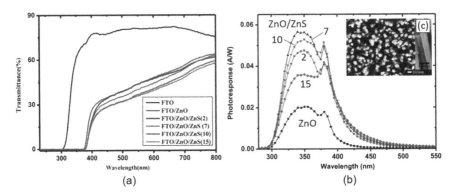

Fig. 18.5 (**a**) UV-visible transmission spectrum, (**b**) spectral sensitivity characteristics of UV photodetectors at different SILAR cycles, (**c**) FESEM view of ZnO/ZnS core-shell nanorod array with high magnification of nanorod at inset. (Adapted from Ref. [31]. Published 2016 by Springer as open access)

photoresponsivity of 0.056 A/W which is 180% enhancement in comparison with bare ZnO. The new band energy alignment of the ZnO/ZnS contact is responsible for accelerating the separation of e-h pairs with this regard. The research further suggests that the improved photo responsivity has been evolved from the application of ZnS with SILAR method [31].

18.4 Solution Processed Photodetector Fabricated Using Methods of Thick Film Technology

As we indicated earlier, the second approach used in the manufacture of solution processed photodetectors involves the separation of the processes of nanoparticle synthesis and their deposition on a substrate to form a photosensitive layer. As an example of this approach to the development of photodetectors, consider several technological processes used in the manufacture of solution processed photodetectors based on II-VI semiconductors. For example, Saeed et al. [44] fabricated ZnS:Mn-based metal-semiconductor-metal junction UV photodetector based on the nanorod networks using this approach. Mn^{2+}-doped ZnS nanorods were synthesized by a facile hydrothermal method. The ZnS:Mn NRs were dispersed into DI water to make a suspension. The typical concentration was 0.2 mg/ml. A droplet of the ZnS:Mn NRs suspension was drop-cast on Ag electrodes and allowed to dry at room temperature. After soaking the substrate in a 7:3 mixture of butylamine and acetonitrile, the substrate was annealed for 3 min at 80 °C to prepare the films. The device exhibited visible blindness, superior ultraviolet photodetection with a responsivity of 1.62 A/W, and significantly fast photodetection response with the rise and decay times of 12 and 25 ms, respectively. The shift in sensitivity to the UV region is due to the fact that when the particle size of ZnS is minimized far below the Bohr diameter (5 nm) depending on strong quantum confinement, it becomes conceivable to adjust the band gap of ZnS quantum dot above 4.43 eV for photodetection in the ultraviolet region.

The same approach was used by Kuang et al. [25] and Xia et al. [53] in developing solar-blind deep ultraviolet (DUV) photodetectors (PDs). Xia et al. [53] fabricated PDs using colloidal ZnS quantum dots (QDs) via spin coating and ligand exchange. The ZnS QD solar-blind DUV PDs showed a fast response ($t_r = 0.35$ s, $t_d = 0.07$ s) and good responsivity (~0.1 mA/W). Colloidal ZnS QDs with an exciton peak at 265 nm, corresponding to a band gap of 4.68 eV, were synthesized by a hot injection method. Kuang et al. [25] used cubic ZnS QDs with a particle size of ~2.29 nm synthesized by wet chemical method to fabricate UV photodetectors. The layer of ZnS QDs on the top of interdigital Au electrodes was formed using the drop coating method. The photodetector showed a cutoff at 300 nm, a photosensitivity of 8, responsivity of 1.60 mA/W, and detectivity of 5.51×10^9 cm Hz$^{1/2}$ W^{-1} at 254 nm, when operated at a bias voltage of 15 V.

Mei et al. [35] showed that the spin-casting method can also be used to form multilayer structures, such as TiO$_2$/CdS/CdTe-based solar cells. A simple solution process under ambient conditions developed by Mei et al. [35] is shown in Fig. 18.6. A TiO$_2$ film with a thickness of 40 nm was prepared by depositing a Ti^{2+} precursor onto the FTO substrate and spin-casted at 2500 rpm for 15 s, then the substrate was annealed at 500 °C for 1 h to eliminate any organic solvent and form a compact TiO$_2$ thin film. Several drops of the CdS NC solution with different concentrations (5 mg/mL, 10 mg/mL, 15 mg/mL, and 20 mg/mL) were then deposited onto the FTO/TiO$_2$ and spin-casted at 3000 rpm for 20 s. Following this, the substrate was transferred to a hot plate and annealed at 150 °C for 10 min, then transferred to another hot plate and annealed at 380 °C for 30 min. One wash with isopropanol was used to remove any impurities. The CdTe NCs were then deposited layer by layer onto the FTO/

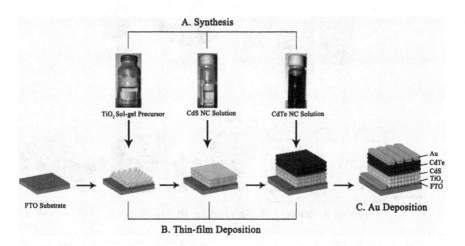

Fig. 18.6 A schematic of the fabrication process of the NC solar cells. (Reprinted from Ref. [35]. Published 2018 by MDPI as open access)

TiO₂/CdS substrate with a process described previously in [40]. Finally, several drops of saturated CdCl₂/methanol were put onto the FTO/TiO₂/CdS/CdTe substrate and spin-casted at 1100 rpm for 20 s, then transferred onto a hot plate at 330–420 °C for 15 min. Sixty nanometers of Au was deposited via thermal evaporation through a shadow mask with an active area of 0.16 cm² to make the electrode contact. The introduction of a thin layer of CdS NC film (~5 nm) between the CdTe and TiO₂ resulted in optimized band alignments and reduces the interface defects. As a result, solar cells manufactured using the described technology showed a power conversion efficiency (PCE) of 5.16% [35].

For the manufacture of photodetectors by this method, commercially available powders of II-VI semiconductors as well can be used. This approach was used by Rahman et al. [41, 42]. They showed that the deposition of CdS thin films by spin coating method using thiol-amine co-solvents has good potential in the fabrication of low cost films for high efficiency applications. The process involves commercially available CdS powder (99.9999%), ethylene-di-amine, 1,2 ethanedi-thiol, and Triton X-100 for the preparation of CdS precursor solution. The simple process first takes a mixture of ethylene-di-amine and 1,2 ethane-di-thiol in 10:1 ratio. CdS powder of 0.3 wt% is then dissolved with the mixture. This can be done with a magnetic stirrer with keeping the solution more than the room temperature. The process may take around 15 h at 50 °C to completely dissolve the CdS particles. The process is illustrated in Fig. 18.7a. Addition of small amount of surfactant (e.g. Triton X-100 with 0.1 wt %) has been found to produce consistent and high quality films [41, 42].

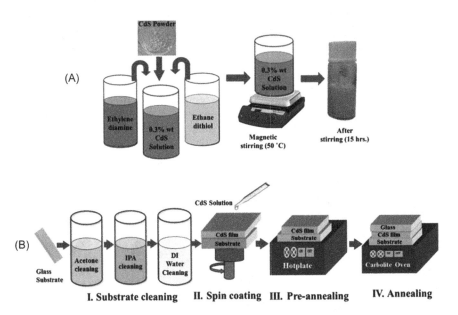

Fig. 18.7 (**a**) Processing steps for preparation of CdS precursor solution; (**b**) Processing steps for deposition of CdS film in simple spin coating method. (Reprinted with permission from Ref. [41]. Copyright 2020: Springer)

The deposition can be completed on glass substrate with simple spin coating technique. Rahman et al. [41] coated one time on perfectly cleaned glass substrate at speed more than 1000 rpm for 45 s. To remove the remaining solvents, the coated films were pre-annealed at 90 °C. The high temperature annealing requires a glass protector to get rid from oxidation. The process is described in Fig. 18.7b. In this process, the method offers good cyrstallinity at annealing temperature of 300 °C.

Photodetectors based on II-VI semiconductor-polymer composites can also be fabricated using the principles of solution processed technology. Huynh et al. [18] fabricated a CdSe/P3HT-based photodetector using this approach. CdSe/P3HT 200 nm layer between an aluminum electrode and a transparent conducting electrode of PEDOT:PSS was spin-cast from a solution containing 90% wt % CdSe nanorods in P3HT and pyridinechloroform as solvent. The active area of the device was 1.5 mm by 2.0 mm. Adjusting the band gap by changing the radius of the nanorod allowed Huynh et al. [18] to match the absorption spectrum of the photodetector with the spectrum of solar radiation. As a result, a photovoltaic device consisting of CdSe nanorods 7 nm in diameter and 60 nm long and a conjugated poly-3(hexylthiophene) polymer demonstrated an external quantum efficiency of more than 54%, and a monochromatic power conversion efficiency of 6.9% under illumination of 0.1 mW/cm^2 at 515 nm. It is important that with a decrease in the length of nanorods to 7 nm, the external quantum efficiency decreased to 20%.

18.5 Combined Approach to the Fabrication of Solution-Processed Photodetectors

Solution-processed photodetectors can also be produced by combining the above methods. This approach is mainly used in the formation of multilayer structures. For example, Sekhar Reddy et al. [45] in the manufacture of NiO/CdS heterojunction based photodetectors (see Fig. 18.8), the CdS layers upon ITO coated PET substrate were deposited by photochemical deposition technology at ambient condition, and the NiO layers were deposited from NiO nanoparticles by the spin-coating method. The solution for CdS deposition was prepared by mixing 7 ml of 0.05 M CdCl$_2$ with 6.8 ml of 0.25 M sodium citrate at maintained pH 5, then 2.5 ml of KOH of a pH ten followed by 5 ml of buffer to reach the pH 12, and finally 27.5 ml of deionized water. The total solution was mixed 5 min; then the ITO-PET substrate was immersed in this reaction solution and put in a sealed box illumined by UV source (313 nm) in a room temperature setting. After 1.5 hrs, the CdS-coated substrate was taken out, washed with deionized water, and dried up with nitrogen. NiO nanoparticle powders were dispersed in ethanol and subjected to spinning at 3000 rpm for 30 s. After deposition, the p-n heterojunction was heated at 60 °C for 30 mins. The final device was completed by depositing the top aluminium metal contacts (120 nm) on the heterojunction using a shadow mask.

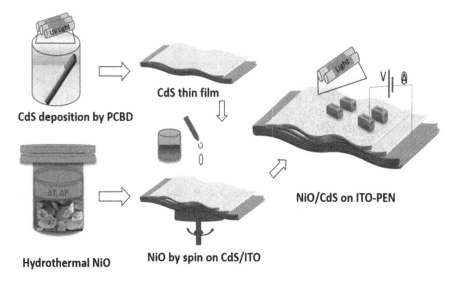

Fig. 18.8 Scheme of the solution manufacturing process for CdS, NiO and the photodetector final device construction. (Reprinted with permission from Ref. [45]. Copyright 2021: Elsevier)

The same approach was taken by Li et al. [30] in the fabrication of organic-inorganic metal halide perovkite solar cells (PKSCs) (see Fig. 18.8). ZnSe layer was synthesizes using chemical bath deposition method, while the TiO_2 and metal halide perovkite were spin coated. Li et al. [30] have found that low-temperature solution-processed ZnSe can be used as a potential electron transportation layer (ETL) for PKSCs. Optimized device with ZnSe ETL has achieved a high power conversion efficiency (PCE) of 17.78% with negligible hysteresis, compared with the TiO_2 based cell (13.76%). Li et al. [30] believe that this enhanced photovoltaic performance is attributed to the suitable band alignment, high electron mobility, and reduced charge accumulation at the interface of ETL/perovskite. Encouraging results were obtained when the thin layer of ZnSe cooperated with TiO_2. It shows that the device based on the TiO_2/ZnSe ETL with cascade conduction band level can effectively reduce the interfacial charge recombination and promote carrier transfer with the champion PCE of 18.57%. In addition, the ZnSe-based device exhibits a better photostability than the control device due to the greater ultraviolet (UV) light harvesting of the ZnSe layer, which can efficiently prevent the perovskite film from intense UV-light exposure to avoid associated degradation (see Fig. 18.9b).

Rose et al. [43] for the fabrication of solution-processed CdTe/CdS photodetectors (see Fig. 18.10) proposed another way. If the CdS layer was grown by chemical-bath deposition (CBD), then the CdTe layer was formed by close-spaced sublimation. The I-V results (average and standard deviation) with AM1.5 illuminations are shown in the Table 18.2. It is seen that photodetectors made in this way showed an average efficiency of ~12.6%.

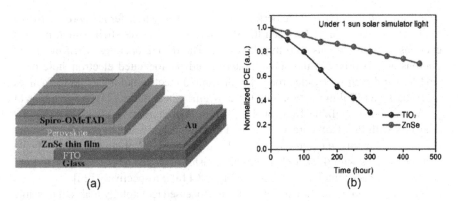

(a)   (b)

Fig. 18.9 (**a**) Schematic view of the typical cell architecture of organic–inorganic metal halide perovskite solar cells (PKSCs). (**b**) The normalized power conversion efficiency (PCE) decay of devices based on ZnSe and TiO_2 ETLs as a function of storage time upon 1 sun irradiation. (Reprinted with permission from Ref. [30]. Copyright 2018: ACS)

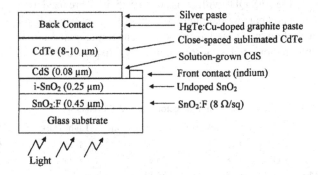

Fig. 18.10 The structure of CdTe/CdS-based solar cell. The CdS layer was grown by chemical-bath deposition (CBD). (Reprinted with permission from Ref. [43]. Copyright 1999: Wiley)

Table 18.2 AM1.5 I-V results for 38 solar cell baseline set

Parameter	Efficiency	V_{oc} (mV)	J_{sc} (mA/cm^2)	FF (%)	Area (cm^2)
Average	12.6%	820	21.8	70.6	0.86
Standard deviation	0.5%	8	0.4	2.0	0.39

Source: Reprinted with permission from Ref. [43]. Copyright 1999: Wiley

A combined approach to the manufacture of solution-processed photodetectors can also offer such a way when wet chemical methods are used to synthesize core shell structures, and the photosensitive layer of the solution-processed photodetector is formed by the spin coating method. In particular, Kwon et al. [26] proposed such an approach for the development of a PbS/CdS core-shell-based photodetector. While PbS can be a cost-effective option due to its tunable bandgap, photosensitivity, and, most crucially, solution-processability, findings suggest that depositing inorganic CdS onto the PbS core can significantly increase photo and thermal stabilities, and therefore device performance. The charge transfer is improved by the gradient interfacial layer between the PbS core and the CdS shell, which permits excitons to partially leak into the shell [26]. The device operates when incoming photons are absorbed in the active material and photoexcited electron–hole pairs (EHPs) are driven to the electrodes by an applied electric field. The performances under dark and IP illumination with 0.1 mW/cm^2 power density and reverse bias of −1 V are shown in Table 18.3 for three different combinations. It is seen that the detectors with PbS/CdS core shell structure show a significant improvement in on/ off ratio, detectivity and photo response. In addition, the IR photo response indicate that the devices with thicker shell exhibit a faster response with -1 V bias and fall time performance which is displayed in Fig. 18.11a, b respectively [26].

Zhou et al. [56] have shown that solution-processed technology makes it possible to implement more complex structures. For example, Zhou et al. [56], using the principles of solution-processed technology, fabricated upconversion photodetectors, the structure of which is shown in Fig. 18.12a. The infrared photodetector converted incident light of 1600 nm into visible emitted light (see Fig. 18.12b). Luminescent CdSe/ZnS QDs with emission at 525 nm were used as the active material for the LED, and narrow-bandgap PbS QDs (300–1600 nm), synthesized via the hot injection method, were used as the active sensitizing material in the photodetector. The PbS nanoparticles were uniformly dispersed with a diameter of

Table 18.3 Performance of SWIR photodetector with IR illumination

Device	J_{dark} (mA/cm^2)	J_{light} (mA/cm^2)	Light/Dark ratio	D* (Jones)
PbS-QD	9.32	12.514	1.34	6.16×10^{11}
Thin shell PbS/CdS QD	5.884	35.3	5.99	1.35×10^{12}
Thick shell PbS/CdS QD	5.0526	56.856	11.25	7.14×10^{12}

Source: Reprinted from Ref. [26]. Copyright 2020: Springer Open. Open access

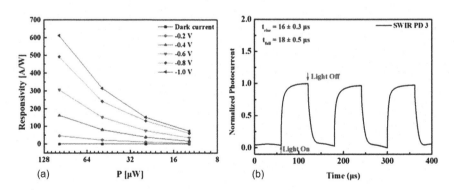

Fig. 18.11 (a) IR photo response for the thicker PbS/CdS QD under −1 V bias and (b) Transient photo response of the shell under -1 V bias. The light intensity of 0.1 mW/cm^2. (Reprinted from Ref. [26]. Published 2020 by Springer Open as open access)

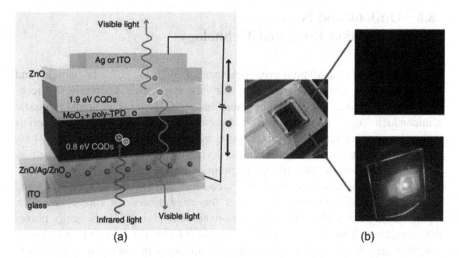

Fig. 18.12 (a) Structure and composition of upconversion devices herein. (b) Photograph of a sample clamped in the measurement box (left), and images of a device (area of 0.05 cm²) with (bottom) and without (top) infrared light (940 nm, 10 mW cm⁻²). (Reprinted with permission from Ref. [56]. Copyright 2020: Springer Nature)

approximately 6 nm. Zinc oxide (ZnO) with embedded Ag nanoparticles was used as the ETL and poly-(N,N-bis(4-butylphenyl)-N,N-bis(phenyl) benzidine) (poly-TPD) as the hole transport layer. The PbS layer was deposited via layer-by-layer spin coating. For each layer, two drops of PbS solution were spin cast on to the ZnO substrate at 2500 rpm for 10s. CdSe/ZnS core/shell QDs dissolved in octane (10 mg/ml) were spin coated at 2000 rpm for 30 s, followed by heat treatment at 80 °C for 10 min. The ZnO film was spin cast at 2000 rpm for 30 s, followed by heat treatment at 80 °C for 10 min. Finally, 100 nm Ag or 140 nm ITO were deposited using an Angstrom Engineering deposition system.

The sensitivity of the resulting photodetectors corresponds to the absorption of QDs and exhibits a definite exciton peak at 1500 nm. When reverse biased, the devices show gain with a peak sensitivity of >20 AW⁻¹ in the shortwave infrared and >60 AW⁻¹ in the visible range. However, devices without Ag nanoparticles show no gain and provide sensitivity below 0.3 AW⁻¹. The devices show typical diode characteristics in the dark. The value of the dark current at a bias of −1 V does not exceed 13 μAcm⁻². The resulting noise equivalent power (NEP) reached $3 \times 10{-}14$ W/Hz$^{1/2}$, resulting in a D* of 6×10^{12} Jones at 180 Hz. Zhou et al. [56] consider this to be in line with the best previously published PbS-based infrared photodetectors with a similar bandgap. According to Zhou et al. [56], this top-emitting upconversion device can be used for bioimaging.

18.6   Outlook and Perspectives of Solution-Processed Technology

There is no doubt that solution-processed technology, using colloidal particles and nanocrystals, is a promising technology for wide application. Currently, this technology is widely used for applying organic materials. However, colloidal inorganic semiconductor nanocrystals share the important advantages of organic materials such as low-temperature solution-processing and controllable synthesis. Meanwhile, compared to organics, nanocrystals-based devices with proper particle surface passivation exhibit broader absorption spectrum and more efficient charge transport. The QD wavelength selectivity can be tuned by adjusting the QD materials and dimensions, which eliminates the need for color filters for image detectors. Optoelectronic applications based on these nanomaterials include solar cells, photodetectors, phosphors, and light-emitting diodes (LEDs). But it must be recognized that the traditional methods of colloidal nanoparticles deposition, which are widely used at the moment, are good for research purposes or for forming layers over large areas used in solar cells (see Fig. 18.13). In the manufacture of photodetectors intended for the market, the use of such methods as spin coating, dip coating or roll-to-roll technique may not be effective. In this regard, inkjet-printing techniques [2, 15, 28, 47, 57] as well as a number of other printing technologies [11, 21], which have been intensively developed over the past two decades, seem very promising.

Currently, inkjet printing has begun to be used in the development of ZnS:Mn, CdS, CdSe, and CdSe/ZnS light emitted diodes, phosphors [19, 22] and color conversion element for full-color LED displays [50]. An example of the

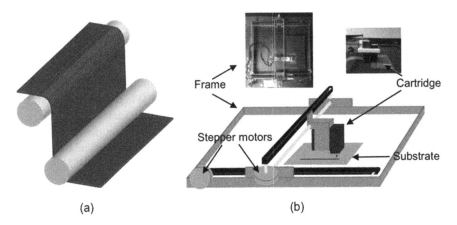

(a)   (b)

Fig. 18.13 (**a**) Schematic diagram of roll-to-roll technique. Using a roll-to-roll technique to manufacture solution-processed solar cells reduces the manufacturing, deployment and energy costs involved. (Reprinted with permission from Ref. [13]. Copyright 2012: Springer Nature). (**b**) Schematic diagram of inkjet printing system. The inkjet printer consists of a XY moving stage, a print head with cartridge, and a control board. (Reprinted from Ref. [7]. Published 2020 by Springer as open access)

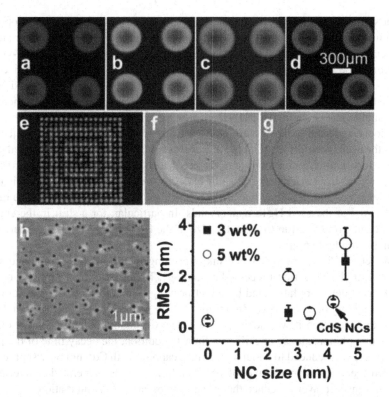

Fig. 18.14 Fluorescence microscope images of (**a–d**) single-colour and (**e**) multi-colour (1×1 cm²) arrays of CdS (**a**) and CdSe@ZnS NCs (**b–e**) in 5 wt% PS. In (**f–g**) optical profiler 3D images of pixels formed of (**f**) CdS and (**g**) CdSe@ZnS NCs (3.4 nm) in 5 wt% PS, respectively. In (**h**) 2D view of the AFM image of CdS NCs in 5 wt% PS. In the graph RMS roughness values of pixels with CdS and CdSe@ZnS NCs in 3 wt% and 5 wt% PS. The value corresponding to 0 nm size is referred to bare PS pixels. (Reprinted with permission from Ref. [19]. Copyright 2009: Elsevier)

implementation of luminescent structures based on colloidal CdS nanoparticles and differently sized CdSe@ZnS nanocrystals (NCs) using the inkjet printing technique is shown in Fig. 18.14. Figure 18.14 shows that regular shape disk-like pixels, luminescent from blue to red, were dispensed with a diameter of 400–600 µm and with no satellites. The microstructures based on CdSe@ZnS NCs present a slight concavity as evidenced by the slight change in shading, with no pronounced "coffee-staining effect" (Fig. 18.14f), while multiple rings are in the pixels formed of CdS NCs (Fig. 18.14e). Ingrosso et al. [19] believe that the manufactured highly luminescent and non-bleachable microstructures can be integrated in polymer displays and coloured wall papers. The reported approach can be extended to functionalize a variety of polymers with different functional colloidal NCs to fabricate polymer based micro- and nano-electronic components.

But experiment shows that inkjet printing technology can be easily transferred to the production of photodetectors [1, 9, 13, 46]. For example, there are already reports of fabrication using this method of HgTe NCs based photodetectots operating up to 3 μm wavelengths [1]. Detectors operating in this spectral region are of particular importance for biological applications, remote sensing and night-vision imaging. The hydrophobic HgTe nanocrystals (NC) were initially synthesized in aqueous solution at room temperature via a reaction between $Hg(ClO_4)_2$ and H_2Te gas in the presence of short-chain hydrophilic thiols such as thioglycerol or mercaptoethanolamine as stabilizer. For inkjet-printing a 2 wt% HgTe NC/chlorobenzen solution was found to have a suitable viscosity and surface tension for ejection from the used piezo-driven print head. Boberl et al. [1] reported that fabricated photodetectors demonstrated room temperature detectivities up to $D* = 3.2 \times 10^{10}$ cm $Hz^{1/2}$ W^{-1} at a wavelength of 1.4 μm close to the important telecommunication wavelength region. The long-wavelength cut-off of the photoresponse can be tuned by the size of the used HgTe nanocrystals. In particular, for a shift in the cut-off wavelength from 1–2 μm to 3 μm the size of the HgTe nanocrystals was increased from 3–4 nm up to 6 nm.

Wu et al. [51] showed that the inkjet-printing method can be combined with other methods of solution-processed technology. To optimize the parameters of UV ZnO-based detectors fabricated by inkjet printing, Wu et al. [51] suggested to use the dip coating of as-prepared ZnO films by CdS nanoparticles. They reported that using this approach it was achieved a 3 orders of magnitude enhancement of the ultraviolet photoresponse of ZnO thin film. In addition, the decay time of the photoresponse was reduced to about 4 ms. Thus, capping with CdS not only suppressed the detrimental passivation layer of ZnO thin films, but also generated an interfacial carrier transport layer to reduce the probability of carrier recombination.

But we are sure that the main advantage of the inkjet-printing technique is that it allows to combine microminiaturization and solution processed technology. And this means that a cheap technology for manufacturing pixelated photodetectors is emerging. Currently, the minimum nozzle diameter is ~100 nm (https://www.fujifilm.com). This means that if a drop with a diameter of 100 nm is formed, it can form a round film with a diameter of ~200 nm [48]. However, the best devices for inkjet printing actually used provide a resolution of ~2 μm [27], while the resolution of conventional devices is in the range of 20–50 μm [21]. Therefore, in its current incarnation, inkjet printing is not particularly well-suited to the production of extremely high-resolution electronics on the order of those currently produced by conventional means. However, by delivering droplets that are a handful of micrometres in diameter, reasonably well-resolved ordinary optoelectronic and photoelectronic devices may be produced.

In addition, the use of the inkjet printing technique makes it possible to produce multicolor photodetectors without much difficulty. In particular, the possibility of manufacturing such photodetectors were demonstrated by Cook et al. [5]. Schematic diagram of printing of ZnO QDs (black), PbS QDs (blue), and FeS_2 (red) NCs on graphene channels between two nearest-neighbor Au electrodes on SiO_2 (500 nm)/ Si substrates is shown in Fig. 18.15. Three different nozzles containing ZnO QDs, PbS QDs, and FeS_2 NCs were employed for printing each of them on the different places. The above example and the presence of a large number of methods for the

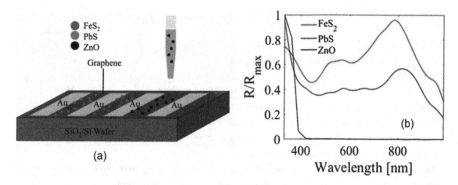

Fig. 18.15 (a) Schematic diagram of printing of ZnO QDs (black), PbS QDs (blue), and FeS₂ (red) NCs on graphene channels defined between two nearest-neighbor Au electrodes on SiO₂ (500 nm)/Si substrates. (b) Normalized spectral photosensitivity of elements forming a multicolor photodetector. (Reprinted with permission from Ref. [5]. Copyright 2019: ACS)

Table 18.4 Commercially available conductive inks for inject printing

Name	Material	Manufacturer	
NanoGold	Au nanoparticle	Sigma Aldrich	https://www.sigmaaldrich.com
NanoSilver	Ag nanoparticle	Sun Chemical	https://www.sunchemical.com
Clevios ™PH 1000	PEDOT:PSS	Heraeus	https://www.heraeus.com
Polyaniline (Emeraldine salt)	PANi nanoparticles	Sigma Aldrich	https://www.sigmaaldrich.com

PEDOT:PSS poly(3,4-ethylenedioxythiophene) polystyrene sulfonate

synthesis of colloidal particles and nanocrystals of II-VI compounds (Chaps. 11, 12, and 13, Vol. 1) indicates that the use of inkjet-printing technology in the manufacture of photodetectors based on II-VI compounds should not cause any particular difficulties. Ink-jet printing of colloidal nanocrystals is material effective and highly reproducible, and can be applied not only for the photosensitive materials but also for fabrication of the electrodes [52], opening up prospects for low cost, all ink-jet printed photodetector devices. Commercially available conductive inks for inject printing are listed in Table 18.4. With an appropriate selection of QDs of II-VI compounds, it is possible to manufacture multicolor selective photodetectors simultaneously sensitive in the IR, visible and UV spectral regions.

The inkjet-printing technique can also be used to form the heterostructures [33] needed to fabricate tandem photodetectors. It is also possible to simultaneously deposit nanoparticles of different II-VI semiconductors. In particular, Miethe et al. [36] thus formed a CdSe/CdS gel-network, which may be a prototype of the quasi-type-II superlattice structure. The use of inkjet-printing techniques also facilitates the integration of II-VI semiconductor-based photodetectors with signal processing silicon integrated circuits. Inkjet printing offers unique advantages in mass

Table 18.5 Typical inkjet ink composition

Component	Function	Loading (w/w%)
II-VI semiconductor colloidal NPs or NCs	Key component	0.1–10
Solvent	Dispersion/dissolution medium	50–90
Co-solvent(s)	Controls drying ("coffee-ring")	0–50
Viscosity modification	Surface tension modification	0–50
Surfactant	Modifies surface tension improves wetting	0–5
Viscosity modifier (dissolved)	Generally, increases viscosity	<1
Humectant	Low volatility, prevents ink drying in nozzles	0–20
Other	pH buffer; biocide; fungicide; dispersant; defoamer; binder (polymer)	<1

Source: Data extracted from Refs. [2, 6, 28, 32, 48, 57]

scalability, cost reduction, low waste, and direct deposition on targeted regions. This means that inkjet printing makes it possible to embed photonic components based on II-VI semiconductor functional nanomaterials and quantum nanostructures into highly crystalline structures of desired morphology with CMOS read-out circuits without the need for any additional chemical treatments that affect the properties of these circuits.

It is understood that the ink preparation process can be very complex depending on the material used, its compatibility with carrier solvents and other ink components, and its ease of dissolution or dispersion. Therefore, ink formulation becomes an absolutely critical aspect of functional material application and device fabrication. Although inks may be formulated in many different ways, the composition of the ink is generally similar to that given in Table 18.5. Ink preparation requires careful control of fluid properties, wetting behavior, drying behavior, interaction with a given substrate, maintaining dispersion and, above all, maintaining functionality. We have to admit that at the moment, there is a big gap between a simple suspension or solution and real inks suitable for creating optoelectronic structures [32]. But there is no doubt that these problems in relation to II-VI compounds will be successfully solved, which will contribute to the rapid development of inkjet-printing technology for manufacturing II-VI semiconductor-based photodetectors.

18.7 Conclusion

This chapter has reviewed some modern methods of II-VI semiconductor synthesis as well as approaches to photodetectors fabrication. The analysis showed that the development of photodetectors based on solution-processed II-VI semiconductors has advanced significantly over the past decade due to advances in materials science and device development. This has led to the development of photodetectors that combine desired performance with manufacturing advantages based on state-of-the-art technology. The achieved results indicate that solution-processed devices can indeed be considered as promising candidates for the commercialization of low-cost photodetectors.

Since II-VI semiconductors and their quantum dots have a wide range of controllable band gaps, the solution- processed technology allows to tune this value to detect radiation in any part of the solar spectrum. However, in all cases, the solution process often requires proper optimization of its parameters in order to get the best possible result from the intended device. Solution- processed photovoltaic materials typically contain a significant degree of disorder, often due to the boundaries that define randomly oriented nanocrystals. Therefore, transport within these domains as well as at their boundaries (including transport-limiting mechanisms such as charge carrier capture) requires attention and optimization. For example, studies carried out by Zhang et al. [55] have shown that manipulating the lifetime and transport domains of majority and minority carriers, one can significantly increase the sensitivity of CdTe photoconductors and achieve a detectivity of up to 5×10^{17} Jones in the visible and near infrared ranges. Likewise, the best reported gain (G) $\times$ bandwidths (BW) products (above 6×10^6 Hz) was achieved for visible CdS photoconductors after improvement of carrier mobility [29].

Acknowledgments G. Korotcenkov is grateful to the State Program of the Republic of Moldova, project 20.80009.5007.02, for supporting his research.

References

1. Boberl M, Kovalenko MV, Gamerith S, List EJW, Heiss W (2007) Inkjet-printed nanocrystal photodetectors operating up to 3 μm wavelengths. Adv Mater 19:3574–3578
2. Calvert P (2001) Inkjet printing for materials and devices. Chem Mater 13:329
3. Chen Y-H, Li W-S, Liu C-Y, Wang C-Y, Chang Y-C, Chen L-J (2013) Three-dimensional heterostructured ZnSe nanoparticles/Si wire arrays with enhanced photodetection and photocatalytic performances. J Mater Chem C 1:1345–1351
4. Chesman ASR, Duffy NW, Martucci A, De Oliveira Tozi L, Singh TB, Jasieniak JJ (2014) Solution-processed CdS thin films from a single source precursor. J Mater Chem C 2(2014):3247–3253
5. Cook B, Gong M, Ewing D, Casper M, Stramel A, Elliot A, Wu J (2019) Inkjet printing multicolor pixelated quantum dots on graphene for broadband photodetection. ACS Appl Nano Mater 2:3246–3252
6. Croucher M, Hair M (1989) Design criteria and future directions in inkjet ink technology. Ind Eng Chem Res 28(11):1712
7. Da Costa TH, Choi J-W (2020) Low-cost and customizable inkjet printing for microelectrodes fabrication. Micro Nano Syst Lett 8:2
8. Das S, Dhara S (eds) (2021) Chemical solution synthesis for materials design and thin film device applications. Elsevier, Amsterdam
9. Dong Y, Zou Y, Song J, Li J, Han B, Shan Q et al (2017) All-inkjet-printed flexible UV photodetector. Nanoscale 9:8580
10. Eslamian M (2017) Inorganic and organic solution-processed thin film devices. Nano-Micro Lett 9:3

11. Fukuda K, Someya T (2017) Recent progress in the development of printed thin-film transistors and circuits with high-resolution printing technology. Adv Mater 29(25):1602736
12. García de Arquer FP, Armin A, Meredith P, Sargent EH (2017) Solution-processed semiconductors for next-generation photodetectors. Nat Rev Mater 2:16100
13. Graetzel M, Janssen RAJ, Mitzi DB, Sargent EH (2012) Materials interface engineering for solution-processed photovoltaics. Nature 488:304–312
14. Hankare PP, Bhuse VM, Garadkar KM, Delekar SD, Mulla IS (2004) Chemical deposition of cubic CdSe and HgSe thin films and their characterization. Semicond. Sci. Technol. 19:70.
15. Haverinen H, Myllyla R, Jabbour G (2009) Inkjet printing of light-emitting quantum dots. Appl Phys Lett 94:073108
16. Hou X, Aitola K, Lund PD (2021) TiO_2 nanotubes for dye-sensitized solar cells—A review. Energy Sci Eng 9:921–937
17. Hussain S, Iqbal M, Khan AA, Khan MN, Mehboob G, Ajmal S et al (2021) Fabrication of nanostructured cadmium selenide thin films for optoelectronics applications. Front Chem 9:661723
18. Huynh WU, Dittmer JJ, Alivisatos AP (2002) Hybrid nanorod-polymer solar cells. Science 295:2425–2427
19. Ingrosso C, Kim JY, Binetti E, Fakhfouri V, Striccoli M, Agostiano A et al (2009) Drop-on-demand inkjet printing of highly luminescent CdS and CdSe@ZnS nanocrystal based nanocomposites. Microelectron Eng 86:1124–1126
20. Jain M (1993) II-VI semiconductor compounds. World Scientific, Singapore
21. Khan S, Lorenzelli L, Dahiya RS (2015) Technologies for printing sensors and electronics over large flexible substrates: a review. IEEE Sensors J 15:3164–3185
22. Kim JY, Ingrosso C, Fakhfouri V, Striccoli M, Agostiano A, Curri ML, Brugger J (2009) Inkjet-printed multicolor arrays of highly luminescent nanocrystal-based nanocomposits. Small 5(9):1051–1057
23. Korotcenkov G (2014) Handbook of gas sensor materials, Vol. 2: new trends and technologies. Springer, New York
24. Korotcenkov G, Tolstoy V, Schwank J (2006) Successive ionic layer deposition (SILD) as a new sensor technology: synthesis and modification of metal oxides. Meas Sci Technol 17:1861–1869
25. Kuang W-J, Liu X, Li Q, Liu Y-Z, Su J, Harm TH (2018) Solution-processed solar-blind ultraviolet photodetectors based on ZnS quantum-dots. IEEE Photon Technol Lett 30(15):1384–1387
26. Kwon J-B, Kim S-W, Kang B-H, Yeom S-H, Lee W-H, Dae-Hyuk Kwon D-H et al (2020) Air-stable and ultrasensitive solution-cast SWIR photodetectors utilizing modified core/shell colloidal quantum dots. Nano Converg 7:28
27. Kwon HJ, Chung S, Jang J, Grigoropoulos CP (2016) Laser direct writing and inkjet printing for a sub-2 μm channel length MoS_2 transistor with high-resolution electrodes. Nanotechnology 27:405301
28. Le H (1998) Progress and trends in ink-jet printing technology. J Imaging Sci Technol 42(1):49
29. Lee J-S, Kovalenko MV, Huang J, Chung DS, Talapin DV (2011) Band-like transport, high electron mobility and high photoconductivity in all-inorganic nanocrystal arrays. Nat Nanotechnol 6:348–352
30. Li X, Yang J, Jiang Q, Lai H, Li S, Xin J, Chu W, Hou J (2018) Low-temperature solution-processed ZnSe electron transport layer for efficient planar perovskite solar cells with negligible hysteresis and improved photostability. ACS Nano 12:5605–5614
31. Lin H, Wei L, Wu C, Chen Y, Yan S, Mei L, Jiao J (2016) High-performance self-powered photodetectors based on ZnO/ZnS core-shell nanorod arrays. Nanoscale Res Lett 11(1):420
32. Magdassi S (ed) (2010) The chemistry of inkjet inks. World Scientific Publishing, Singapore
33. Marjanovic N, Hammerschmidt J, Perelaer J, Farnsworth S, Rawson I, Kus M et al (2011) Inkjet printing and low temperature sintering of CuO and CdS as functional electronic layers and Schottky diodes. J Mater Chem 21:13634

34. Mbam SO, Nwonu SE, Orelaja OA, Nwigwe US, Gou XF (2019) Thin-film coating; historical evolution, conventional deposition technologies, stress-state micro/nano-level measurement/ models and prospects projection: a critical review. Mater Res Express 6(12):122001

35. Mei X, Wu B, Guo X, Liu X, Rong Z, Liu S et al (2018) Efficient CdTe nanocrystal/TiO$_2$ hetero-junction solar cells with open circuit voltage breaking 0.8 V by incorporating a thin layer of CdS nanocrystal. Nano 8:614

36. Miethe JF, Luebkemann F, Schlosser A, Dorfs D, Bigall NC (2020) Revealing the correlation of the electrochemical properties and the hydration of inkjet-printed CdSe/CdS semiconductor gels. Langmuir 36:4757–4765

37. Miskin CK, Dubois-Camacho A, Reese MO, Agrawal R (2016) A direct solution deposition approach to CdTe thin films. J Mater Chem C 4(2016):9167–9171

38. Mohammed RY (2021) Annealing effect on the structure and optical properties of CBD-ZnS thin films for windscreen coating. Materials 14:6748

39. Nikam CP, Gosavi NM, Gosavi SR (2020) Low-cost visible-light photodetector based on Ag/ CdSe Schottky diode fabricated using soft chemical solution method. SAMRIDDHI: J Phys Sci Eng Technol 12(2):62–67

40. Pawar SM, Pawar BS, Kim JH, Joo O-S, Lokhande CD (2011) Recent status of chemical bath deposited metal chalcogenide and metal oxide thin films. Curr Appl Phys 11:117–161

41. Rahman MF, Hossain J, Kuddus A, Tabassum S, Rubel MHK, Shirai H, Ismail ABM (2020a) A novel synthesis and characterization of transparent CdS thin films for CdTe/CdS solar cells. Appl Phys A Mater Sci Process 126:145

42. Rahman MF, Hossain J, Kuddus A, Tabassum S, Rubel MHK, Rahman MM et al (2020b) A novel CdTe ink-assisted direct synthesis of CdTe thin films for the solution-processed CdTe solar cells. J Mater Sci 55:7715–7730

43. Rose DH, Hasoon FS, Dhere RG, Albin DS, Ribelin RM, Li XS, Mahathongdy Y (1999) Fabrication procedures and process sensitivities for CdS/CdTe solar cells. Prog Photovolt Res Appl 7:331–340

44. Saeed S, Dai R, Janjua RA, Huang D, Wang H, Wang Z, Ding Z, Zhang Z (2021) Fast-response metal−semiconductor−metal junction ultraviolet photodetector based on ZnS:Mn nanorod networks via a cost effective method. ACS Omega 6(48):32930–32937

45. Sekhar Reddy KC, Selamneni V, Syamala Rao MG, Meza-Arroyo J, Sahatiya P, Ramirez-Bon R (2021) All solution processed flexible p-NiO/n-CdS rectifying junction: applications towards broadband photodetector and human breath monitoring. Appl Surf Sci 568:150944

46. Sliz R, Lejay M, Fan JZ, Choi M-J, Kinge S, Hoogland S et al (2019) Stable colloidal quantum dot inks enable inkjet-printed high-sensitivity infrared photodetectors. ACS Nano 13(10):11988–11995

47. Taylor R, Church K, Sluch M (2007) Red light emission from hybrid organic/inorganic quantum dot AC light emitting displays. Displays 28:92

48. Tekin E, Smith P, Schubert U (2008) Inkjet printing as a deposition and patterning tool for polymers and inorganic particles. Soft Matter 4:703

49. Tolstoi VP (2009) New routes for the synthesis of nanocomposite layers of inorganic compounds by the Layer-by-Layer scheme. Russ J Gen Chem 79:2578–2583

50. Wang X, Yuan M, Qin M (2016) Surface energy-modulated inkjet printing of semiconductors. In: Yun I (ed) Printed electronics – current trends and applications. INTECH, pp 5–24

51. Wu Y, Tamaki T, Volotinen T, Belova L, Rao KV (2010) Enhanced photoresponse of inkjet-printed ZnO thin films capped with CdS nanoparticles. J Phys Chem Lett 1:89–92

52. Wu Y, Li Y, Ong BS, Liu P, Gardner S, Chiang B (2005) High-performance organic thin-film transistors with solution-printed gold contacts. Adv Mater 17:184–187

53. Xia Y, Zhai G, Zheng Z, Lian L, Liu H, Zhang D et al (2018) Solution-processed solar-blind deep ultraviolet photodetectors based on strongly quantum confined ZnS quantum dots. J Mater Chem C 6:11266–11271

54. Younus IA, Ezzar AM, Uonis MM (2020) Preparation of ZnTe thin films using chemical bath deposition technique. Nanocomposites, 6:4:165–172

55. Zhang Y, Hellebusch DJ, Bronstein ND, Ko C, Ogletree DF, Salmeron M, Alivisatos AP (2016) Ultrasensitive photodetectors exploiting electrostatic trapping and percolation transport. Nat Commun 7:11924
56. Zhou W, Shang Y, de Arquer PG, Xu K, Wang R, Luo S et al (2020) Solution-processed upconversion photodetectors based on quantum dots. Nat Electron 3:251–258
57. Zhouping Y, Yongan H, Ningbin B, Xiaomei W, Youlun X (2010) Inkjet printing for flexible electronics: materials, processes, and equipment. Chin Sci Bull 55(30):3383

Chapter 19
Multicolor Photodetectors

Paweł Madejczyk

19.1 General

The requirement for improved target recognition and temperature estimation has led to an increasing interest in infrared (IR) detection in more than one band. Multicolor detectors have the ability to detect a number of different infrared bands or a number of different wavelengths in the same band separately and independently [40].

The goal of multiband IR visualization is to overcome the contrast limitations that exist with single band imagery [37]. For typical display of single band infrared imagery, a monochrome display is used and the limitation is that a human viewer has an instantaneous dynamic range equivalent to only 100 shades-of-gray, a 1% contrast difference. By using two of more infrared bands to create a color space, an associated composite color image can be created using visible color bands for display to human viewers. In a color image people can discriminate millions of colors defined by varying hue, saturation, and brightness. Multicolor detectors can determine the absolute temperature of objects in the scene. They play many important roles in Earth and planetary remote sensing, medicine, astronomy, and many others. Multicolor detection is the basic requirement for third generation IR devices [35].

One of the earlier concepts for infrared dual band (DB) vision was a system consisted of two single-color focal plane arrays and a beam splitter [27]. This worked, but there was considerable difficulty in optical alignment to a precision such that the exact same image feature could be accurately compared on the two focal planes at the pixel level. It also had the drawbacks of dual vacuum enclosures and cooling systems. Recent advances in material, electronic, and optical technologies have led to the development of novel types of electronically tunable filters, including so-called adaptive or tunable Focal Plane Arrays (FPAs) [10].

P. Madejczyk (✉)
Military University of Technology, Warsaw, Poland
e-mail: pawel.madejczyk@wat.edu.pl

© The Author(s), under exclusive license to Springer Nature Switzerland AG 2023
G. Korotcenkov (ed.), *Handbook of II-VI Semiconductor-Based Sensors and Radiation Detectors*, https://doi.org/10.1007/978-3-031-20510-1_19

There are currently several types of detector technologies which offer multicolor capability over a wide IR spectral range. In the wavelength regions of interest, such as short-wavelength IR (SWIR: 1–3 µm), medium-wavelength IR (MWIR: 3–5 µm), and long-wavelength (LWIR: 5–14 µm), there are four technologies that are developing multispectral detection: HgCdTe [34], quantum well IR photodetectors(QWIP) [13], antimonide-based type-II superlattices [31], and quantum dot IR photodetectors (QDIPs) [21].

In this chapter we focus on multicolor detectors from II-VI materials. Among them only Mercury Cadmium Telluride (MCT, $Hg_{1-x}Cd_xTe$, HgCdTe) offers multicolor capability.

19.2 HgCdTe Multicolor Detectors

HgCdTe is a pseudo-binary alloy semiconductor that crystallizes in the zinc blende structure. It has a large optical coefficient that enable a high quantum efficiency. Since firstly synthesized over 60 years ago by British scientists [22] HgCdTe provides an unprecedented degree of freedom in IR detector design. Because of its bandgap tunability with composition x, $Hg_{1-x}Cd_xTe$ has become the most versatile material for detector applications over the entire IR range [19, 28]. For these reasons the HgCdTe alloy is also an unique material for designing multicolor IR structures [36].

Historically, multicolor detectors using HgCdTe photoconductors have been demonstrated in the early 1970s [14]. Worth to mention is a concept of Italian scientists who designed very first photovoltaic multicolor HgCdTe detector integrating the physically separated single-color samples of different wavelengths into a three-band multispectral detector [11]. However, this concept did not contribute to the development of multicolor focal plane arrays and was only limited to the multicolor detector as a single element.

New integrated multicolor (multiple cutoffs) detectors have been fashioned from multiple layer structures, where two or more color detectors are integrated into a single pixel. Figure 19.1 presents the general idea of a three-color detector pixel. Shorter wavelengths of infrared flux are absorbed in the upper layer (Absorber 3), while longer wavelengths are transmitted to lower placed layers and are absorbed there, successively. Therefore, the energy gap E_g of individual absorbers should be selected in such a way that the top layer (through which the stream incidences firstly) has the widest energy gap, that is: $E_{g1} < E_{g2} < E_{g3}$. The barrier layers separate absorbers preventing or minimizing the photocurrent leaking into the adjacent band.

19.2.1 Dual-Band HgCdTe Detectors

Considering the operation mode, there are two basic kinds of multicolor detectors: (1) Sequential, called also a bias – selectable detector where the detector bias is changed to activate one of the colors at a time and (2) Simultaneous, where all colors are detected simultaneously [6].

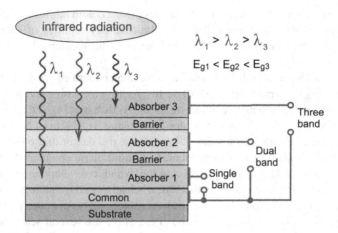

Fig. 19.1 Structure of a three-color detector pixel. Shorter wavelengths of infrared flux are absorbed in the upper layer (Absorber 3), while longer wavelengths are transmitted successively to lower placed layers

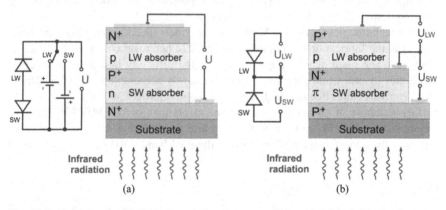

Fig. 19.2 The operating modes of two-color detectors: (**a**) sequential and (**b**) simultaneous. (Pictures present both simplified electrical diagrams and exemplary structures)

Figure 19.2 presents simplified electrical diagrams and exemplary structures of different operating modes of two-color detectors. Sign plus "+" in layers description denotes high doping and the capital letter denotes wider bandgap.

The sequential-mode detector has a single indium bump per unit cell that permits sequential bias selectivity of the spectral bands associated with operating back-to-back photodiodes. Alternatively, simultaneous detector typically contains multiple electrical contact per unit cell and are grounded through the substrate on which detector is grown. Photosignal is extracted directly from one of contacts, while the second contact, common to both spectral bands, provides the sum of the two photo-currents. While this architecture provides the benefit of real-time imaging, the geometry of the pixel reduces the optical fill factor and reduces the maximum achievable quantum efficiency. The challenges involved with fabricating arrays having multiple contacts per unit cell further limit the development of simultaneous detectors. For the above reason sequential detectors have emerged as the favored two-color technology.

Back-to-back photodiode two-color detector as an integrated pixel was first demonstrated using quaternary III-V alloy ($Ga_xIn_{1-x}As_yP_{1-y}$) absorbing layers in a lattice-matched InP structure sensitive to two different SWIR bands [4]. A decade later the original back-to-back concept was designed using HgCdTe at Santa Barbara Research Center [5] and Rockwell [3]. First successful demonstration of two-color 128 × 128 elements FPA was implemented in liquid phase epitaxial (LPE)-grown HgCdTe devices [42]. The real progress in the dynamic development of two-color detectors has taken place since the mid-90s, when advanced technologies were used to obtain complex HgCdTe heterostructures: molecular beam epitaxy (MBE) and metal-organic vapor phase epitaxy (MOVPE). There are several scientific centers that reported dual-band (DB) HgCdTe detectors using epitaxial techniques. MBE was used at: Rockwell (today Teledyne) [2, 39], Chinese Academy of Sciences (ChAs) [16], Sofradir-CEA-Leti [7, 32], Raytheon [38, 41] and AIM [9, 43]. In turn, DB HgCdTe detectors using MOVPE were reported by: Lockheed Martin (today BAE Systems) [26, 33] and British workers [1, 12, 25, 29].

Both technologies (MBE and MOVPE) reveal their specific capabilities and limitations in the fabrication of sophisticated HgCdTe heterostructures for multi-color detectors. The main disadvantage of MBE is the difficulty in activating acceptor impurities in situ, while abrupt interfaces are hardly achievable in MOVPE deposited heterostructures due to a high temperature growth (over 300 °C).

One example of earlier promising implementations of DB MW/LW HgCdTe detectors was MOVPE grown P-n-N-P structure shown in Fig. 19.3. The cross section and the energy band profiles present the dual-band (DB) detector composed

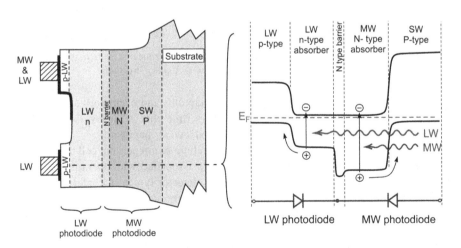

Fig. 19.3 Cross section and energy band profiles for the simultaneous P-n-N-P dual-band (DB) HgCdTe detector. (Adapted with permission from Ref. [33]. Copyright 1998: SPIE)

electrically of two back-to-back HgCdTe photodiodes. It was designed for simultaneous operation mode. The LW photodiode is a P-on-n heterojunction, grown directly on top of the MW photodiode, which is an n-on-P heterojunction. A thin n-type compositional barrier layer is placed between the MW and LW absorber layers. This barrier layer forms isotype n-N heterojunctions at the interfaces, which prevent MW photocarriers from diffusing into the LW absorber layer and prevents LW photocarriers from diffusing into the MW absorber layer. Two indium bump interconnects in each unit cell provide independent electrical access to the back-to-back MW and LW photodiodes, and allow the MW and LW photocurrents to be separate and independent. Lockheed Martin demonstrated MOVPE grown 64 × 64 FPA with 75 μm pixel pitch reported by [26]. Simultaneous MW/LW DB HgCdTe detectors were fabricated from a P-n-N-P films grown *in situ* by interdiffused multilayer process (IMP) MOVPE onto lattice-matched CdZnTe substrates. Detectivity at λ_{PEAK} (f/2.9) of 4.8×10^{11} cmHz$^{1/2}$/W for MWIR (4.3 μm cutoff wavelength) and of 7.1×10^{10} cmHz$^{1/2}$/W for LWIR (10.1 μm cutoff wavelength) were obtained at 78 K. Simultaneity, separate and independent MW and LW integration within each unit cell, and full stare efficiency were maintained as key features. Improved R_0A for the HgCdTe detectors and lower operating temperatures allowed near-BLIP performance at low backgrounds. There were promising plans to advance simultaneous DB HgCdTe FPA technology to larger array sizes, smaller unit cells, and higher performance but unfortunately this technology was abruptly suspended from unknown reasons at the end of last century.

In turn Raytheon reported MBE-grown HgCdTe M/LWIR DB 649 × 480 FPA with 20 μm pixel pitch operating in the sequential mode [41]. Triple-layer n-P-n heterojunction (TLHJ) device structures were deposited on 100-mm (211)Si substrates. The wafers showed low macrodefect densities (<300 cm^2). Typical 81 K cutoff wavelengths of 5.1 μm for MWIR and 9.6 μm for LWIR were obtained. The FPAs exhibited high pixel operabilities in each band with NETD operabilities up to 99.98% for the MWIR band and 98.7% for the LWIR band at 81 K, at f/3 background.

Figure 19.4 presents scanning electron microscopy (SEM) images of chosen DB HgCdTe detectors designed at different technologies and operating in different modes. The achievements of German technology in the field of two-color detectors with HgCdTe are also worth emphasizing. Figure 19.4a presents a fragment of MBE-grown 640 × 512 FPA with 20 μm pixel pitch reported by AIM [43]. Sequential MW/LW DB HgCdTe detectors were fabricated from a n-p-P-P-N films deposited on CdZnTe substrates. NEDT was lower than 18 mK and 25 mK for MWIR and LWIR ranges, respectively. ROIC technology based on analog CMOS let to operate with frame rate equal 100 Hz at eight outputs. Authors emphasize, that one key technology for processing narrow and deep mesa trenches an etching technique, which satisfies, among others, several key requirements: anisotropy, high selectivity, stoichiometricity, and freedom from damage. Anisotropy and high selectivity are preconditions to achieve a high aspect ratio and by this a high fill factor. A requirement just as important for the processing of DB arrays is a technique providing damage-free etching to avoid layer damage and/or Hg-diffusion. A technique which meets all these demands is dry etching by inductively coupled plasma (ICP).

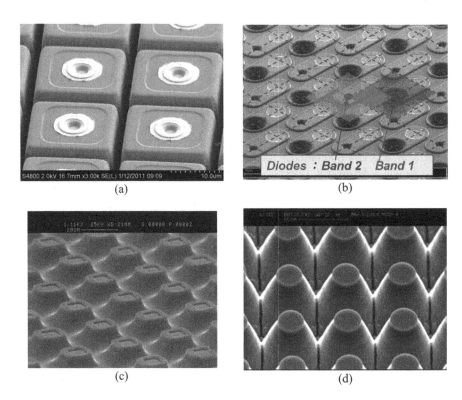

Fig. 19.4 Scanning electron microscopy (SEM) images of different dual-band (DB) HgCdTe detector designs: (**a**) MBE-grown with 20 μm pixel pitch. (Reprinted with permission from Ref. [43]. Copyright 2011: SPIE), (**b**) MBE grown with 20 μm pixel pitch. (Reprinted with permission from Ref. [32]. Copyright 2011: SPIE), (**c**) MOVPE grown with 24 μm pixel pitch. (Reprinted with permission from Ref. [29]. Copyright 2008: SPIE), (**d**) MOVPE grown with 12 μm pixel pitch. (Reprinted with permission from Ref. [25] Copyright 2019: SPIE)

Another example of MBE possibilities for DB detectors are French scientists' achievements. Figure 19.4b shows a fragment of MBE-grown 640 × 512 FPA with 24 μm pixel pitch reported by Sofradir-CEA-Leti [32]. Simultaneous or sequential MW/LW DB HgCdTe detectors were fabricated from a n-p-P-P-N or n-p-P-N films deposited on substrates issued from homemade 90-mm diameter Cadmium Telluride (CdTe) ingots sliced into crystal-oriented rectangular wafers. The typical performances in each band of the DB FPA are close to what is routinely obtained in single detectors of the same cut-off wavelength. NEDT was lower than 20 mK and 25 mK for MWIR and LWIR ranges, respectively. ROIC technology based on CMOS let to operate with frame rate equal 90 Hz at two analog outputs per band. Spectral crosstalk was lower than 1% and the operability higher than 99.5% was achieved.

Many original solutions of British scientists made a significant contribution to the development of DB detectors. Figure 19.4c, d present the micrographs of MOVPE grown FPAs with 24 μm and 12 μm pixel pitches, respectively reported by Leonardo (formerly Selex ES). Sequential MW/LW DB HgCdTe detectors were fabricated from a n-P-N structures deposited on low cost gallium arsenide (GaAs) substrates. 640 × 512 FPAs were demonstrated with NEDT parameter lower than 25 mK and the spectral crosstalk lower than 0.2% due to the selective filters application. The pixel operability of >99% in each waveband has been achieved. Constant progress in pixel sizes reduction was reported from 30 μm at initial research to 12 μm at current implementation.

MCT multicolor heterostructures are extremely challenging material, which presents many different problems to each growth technique. Epitaxial techniques not only remove many of the hazards associated with the high pressure processes required for conventional bulk growth but also improve diode performance router by offering increased area and improved crystalline quality, alloy composition, and dopant and thickness control, and by enabling the use of semiconductor manufacturing processing methods. The ability to control composition, thickness, and dopants within the growth facilitates device engineering opportunities such that multicolor heterostructure designs can be realized [8, 24]. MBE and MOVPE technologies compete against each other providing a high quality HgCdTe material for the production of multi-color detectors. Today, both technologies can offer HgCdTe heterostructures for DB FPA with comparable final parameters: NEDT <20 mK, pixel size <20 μm, array size 640 × 512, crosstalk <1% covering IR spectral ranges required by the industry.

Figure 19.5 shows examples of current – voltage characteristics of DB MCT detectors. The current-voltage characteristics shown in Fig. 19.5a were measured in a cryogenic prober on back-to-back test diodes to sample the complete DB diode stack by changing the bias polarity of the diode from positive to negative. They

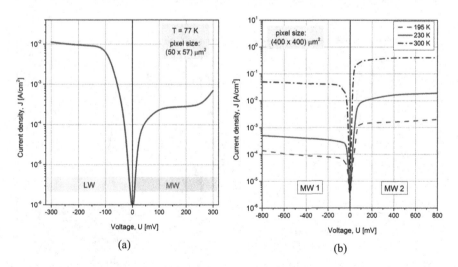

Fig. 19.5 Examples of current – voltage characteristics of DB HgCdTe detectors: (**a**) (50 × 57) μm^2 structures measured at 77 K. (Adapted with permission from Ref. [43]. Copyright 2011: SPIE), and (**b**) (400 × 400) μm^2 measured at elevated temperatures. (Adapted from Ref. [20]. Published 2019 by Springer as open access)

concern (50×57) μm^2 MBE grown-n-p-P-P-N structures. In Fig. 19.5a left-hand side, the reverse biased part of the LWIR curve is graphed and on the right-hand side, the reverse biased part of the MWIR diode. These combined current-voltage curve represents the well-known shape of back-to-back diodes. The LWIR diode show an expected small spread in reverse bias, but nevertheless demonstrates reasonable diode characteristics. The performance of the LWIR diodes will be further improved by appropriate measures concerning layer growth and array processing, mainly in terms of a further reduction in dislocation density. The I-V curve of the MWIR diode exhibits a broad plateau in reverse bias demonstrating low leakage currents and by this an excellent diode performance. Figure 19.5b shows current – voltage characteristics of (400×400) μm^2 mesa structures measured at elevated temperatures [20]. In left-hand side, the reverse biased part of the MW1 curves is graphed and on the right-hand side, the reverse biased part of the MW2 diode. The advanced calculations found, that I-V characteristics were shaped with different generation- recombination mechanisms occurring through trap states related to metal (mercury) vacancies and the dislocations. The diffusion current strongly influences I-V characteristics at elevated temperatures.

Figure 19.6 illustrates the examples of normalized spectral responses of DB HgCdTe detectors from MBE grown structures. Figure 19.6a shows example of spectral responses of MW/LW detector [43]. The cut-off wavelengths are 5.4 μm for the MWIR and 9.1 μm for the LWIR band at 60 K. As can be seen, the cross-talk between the two bands is very low. By exclusion of the wavelength range between 5 and 7 μm, which is of no interest for applications due to atmospheric absorption, the cross-talk of the two spectral bands is less than 1%. Figure 19.6b presents the spectral response for DB HgCdTe detector operating at the temperature 230 K. The

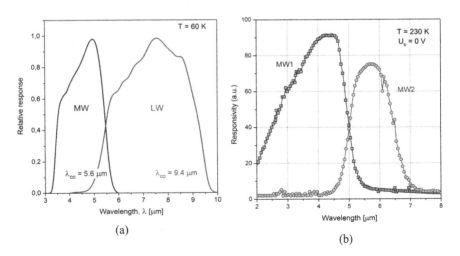

Fig. 19.6 Examples of normalized spectral responses of DB HgCdTe detectors: (**a**) MW/LW type. (Adapted with permission from Ref. [43]. Copyright 2011: SPIE), and (**b**) MW1/MW2 type. (Reprinted from Ref. [23]. Published 2020 by Springer as open access)

measurements were performed without the bias voltage applying (U_b = 0 V). The MW1 pins were connected to the spectrophotometer, and the spectral response characteristic was obtained with the $\lambda_{50\%CO}$ = 5 μm and next the connections were switched to the MW2 pins and the spectral response was taken with the $\lambda_{50\%CO}$ = 6.5 μm. The MW2 responsivity is of several percentages lower than that MW1, because the MW2 absorber is too thin and failed to capture all photons. Moreover, there is evident gradient of the composition x in the MW2, which disorders the absorption process. There is also the substantial gradient of the arsenic concentration within the whole area of the MW2 absorber what can contribute to higher recombination of the minority carriers that mutes the signal [23].

Table 19.1 presents the comparison of DB HgCdTe FPA parameters published over last 20 years. Results of major scientific centers from USA, UK, France, Germany and China are given. Worth to note is FPA's operating temperature oscillating around 77 K.

Raytheon Vision Systems (RVS) developed a 640 × 480 two-color ROIC type SB-275 based on time division multiplexed integration (TDMI). A photograph of an SB-275 developed on the basis of MBE grown DB MCT 640 × 480 FPA is shown in Fig. 19.7. In that solution, the bias polarity on the detector is switched many times within a single frame period. This allows interlacing of the integration times of the two spectral bands. Fast subframe switching of less than 1 msec is typically employed. This achieves good simultaneity between the two spectral bands with minimal spectral latency. The MWIR band is integrated by summing the charge collected from the individual subframe integration periods. The LWIR band is integrated by averaging the charge collected from the individual subframe integrations. This approach optimizes the effective charge storage capacities of the two bands. The SB-275 operates at a frame rate of 30 Hz.

Equally advanced DB IR system was elaborated by British scientists on the basis of MOVPE grown HgCdTe 640 × 512 FPA [29]. A new silicon readout integrated circuit has been designed specifically for CONDOR II arrays at Leonardo (Fig. 19.8). Two capacitors are incorporated into each pixel to allow integration and readout of both bands in a single frame offering quasi-simultaneous operation. A 4:1 capacitance ratio between the LW and MW bands enables equivalent thermal sensitivity to be achieved at the same well fill. Several different modes of operation can be used which make the ROIC highly versatile and suitable for many different applications. Figure 19.9 presents imaging from CONDOR bispectral detector (F2, 4 ms/0.4 ms integration time).

These different modes come under four main categories: MW band only, single stare per frame; LW band only, single stare per frame; alternate wavebands between frame or frames, one waveband per frame; and interleaved mode where MW and LW bands are imaged and output within the same detector frame, in any order. Performance in the two single band modes can be further improved by a feature which integrates the charge on both capacitors. This feature provides increased charge storage, improved NETDs and allows some matching to different optical systems. Interleaved mode offers the greatest flexibility with a choice of integrating a combination of wavebands several times within a single frame. This capability

Table 19.1 Comparison of DB HgCdTe FPA parameters

Parameter	Center						
	Lockheed Martin	Leonardo	Rockwell	ChAS	Sofradir/CEA-Leti	Raytheon	AIM
Growth	MOVPE		MBE				
Structure	p-n-N-N-P	n-p-N	p-n-N-P-N	n-p-P-P-N	n-p-P-P-N/n-p-P-N	n-p-n	n-p-P-P-N
Mode	Simultaneous	Sequential	Simultaneous	Simultaneous	Simultaneous/sequential	Sequential	Sequential
Spectral range (Band 1) (µm)	3–4.3	3–6.5	3–3.9	2–4.8	3–5	3–5.1	3–5
Spectral range (Band 2) (µm)	4.3–10	6–11	3.9–5	5–9.7	8–9.5	5.1–9.6	8–9.5
Array	64 × 64	640 × 512	128 × 128	128 × 128	640 × 512	649 × 480	640 × 512
Pixel pitch (µm)	75	12	40	50	24	20	20
Operating temperature	78 K	80 K	78 K	78 K	77 K	78 K	77 K
NEDT (mK) Band 1	12–20	25	9.3	10[a]	15–20	39	25
NEDT (mK) Band 2	6.2–7.5	25	13.3	10[a]	20–25	26	18
Cross talk	10%	0.2%	3–6%	0.7–1.25	<1%	1–10%	<1%
Reference	[33]	[25]	[39]	[16]	[32]	[41]	[43]

[a]Estimated values on the basis of the Detectivity

Fig. 19.7 MBE grown DB HgCdTe 640 × 480 FPA type SB-275 in which the detector bias polarity is alternated many times within a single frame period. (Reprinted with permission from Ref. [30]. Copyright 2005: SPIE)

Fig. 19.8 CONDOR II
dual waveband MW/LW
integrated detector cooler
assembly. MOVPE grown
HgCdTe 640 × 512 array is
inside the assembly.
(Reprinted with permission
from Ref. [29]. Copyright
2008: SPIE)

Fig. 19.9 Imaging from CONDOR bispectral detector (F2, 4 ms/0.4 ms integration time). (Reprinted with permission from Ref. [24]. Copyright 2020: John Wiley & Sons Ltd)

provides a mechanism to improve temporal alignment between the integration times as MW integration times are invariably longer than LW. The system configures and controls the detector operation and function dynamically through a high speed digital serial interface. The increased amount of signal information from the array is output from the detector using eight buffered outputs to enable higher frame rates of up to 120 Hz in single waveband modes and 60 Hz in dual waveband modes to be achieved. The ROIC also provides other read out modes, windowing and with configurable frame size, scan direction and comprehensive windowing functions.

19.2.2 Three-Color HgCdTe Detectors

A natural path in the evolution of multi-color detectors is the development of three color detectors. Collection of signals from more than two infrared bands provides enhanced target discrimination and identification. So far, there are only a few works devoted to the practical implementation of triple band detectors with HgCdTe presented by British researchers [15, 17]. The bias dependent cut-off is achieved by employing three absorbers in an n-p-n structure with low p-doped electronic barriers at the junctions, see Fig. 19.10. The first n-region (in direction of radiation) defines the shorter wave (SW) side; the p-region, the intermediate wave (IW) response; and the top n-layer the longer wave (LW) response. At low biases either the SW or LW response would dominate, depending on which junction is reverse biased, in a similar way to a 2-color detector. In this bias range, the barriers prevent electron flow from the IW region from both the photogenerated carriers from IW absorption, and by direct injection from the forward biased junction. As the barrier region is low doped, any applied bias will predominately fall on this side of the junction. Increasing reverse bias will, therefore, reduce the barrier until eventually energetic electrons generated by IW photons can cross the junction.

Figure 19.11 shows the obtained spectral response of a three-color MCT detector at various biases. The cut-on wavelength at 2.33 μm is given by a coated Ge window. In positive bias the LW/IW junction is in reverse bias and a bias independent

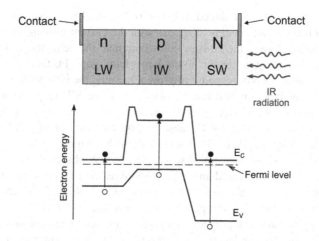

Fig. 19.10 Three-color concept and associated zero-bias energy band diagram. (Adapted with permission from Ref. [15]. Copyright 2006: SPIE)

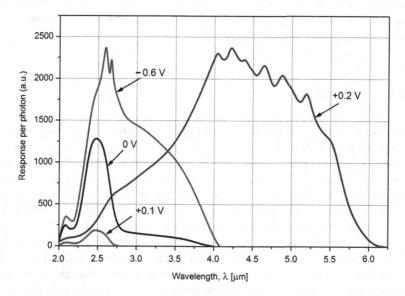

Fig. 19.11 Spectral response of a three-color MCT detector at various biases. (Adapted with permission from Ref. [15]. Copyright 2006: SPIE)

LW spectrum is obtained above 0.2 V. The doping levels chosen result in no barrier lowering at the LW/IW junction at these applied biases. The response below 4.2 μm is due to incomplete absorption in the IW absorber resulting in carrier generation in the LW absorber at these wavelengths. Carriers generated in the IW absorber have insufficient energy to surmount the LW barrier.

As the positive bias is reduced to below 0.2 V, the LW signal collapses and a signal from the SW side begins to appear with the current flowing in the opposite direction. In this regime the built-in fields dominate the behavior, and the largest field is at the SW/IW junction due to the larger band-gaps. Further reduction in the bias to 0 V causes the SW response to grow. Changing the bias polarity to negative puts the SW/IW junction into reverse bias, giving the SW response with cut-off 2.8 μm. Increasing the bias magnitude lowers the electron barrier at this junction and allows a response from the IW absorber, moving the cut-off out to 4.2 μm as shown at −0.6 V in Fig. 19.11. The increase in the SW signal with increasing negative bias is due to incomplete SW absorption in the SW absorber.

Because of the complicated and expensive fabrication process, numerical simulation has become a critical tool for the development of HgCdTe bandgap-engineered devices. It was shown, that the performance of a three-color detector is critically dependent on the barrier doping level and position in relation to the junction [18]. A small shift of the barrier location and doping level causes significant changes in spectral responsivity. This serious disadvantage of the considered three-color detector should be eliminated in further investigation. To ensure adequate barrier lowering, the barrier must be low doped and of the same polarity as the intermediate absorber, and be located adjacent to the junction. The MOVPE growth process is shown to readily achieve such control.

19.3 Conclusions

The presented analysis of the literature on infrared detectors shows a huge progress in improving the parameters of multi-color detectors with HgCdTe over the last 30 years. Significant progress has been observed in the crosstalk reducing, increasing the size of arrays, reducing pixel sizes and increasing the signal-to-noise ratio as defined by the NEDT parameter. The dynamic development of multi-color detectors is related to the progress in advanced epitaxial techniques: MBE and MOVPE. The improvements in processing techniques: the photolithography, dry etching and passivation, as well as the readout electronics (ROIC) and the compound image analysis tools achievements are equally important. Multi-color detectors developed so far operate basically at cryogenic temperatures what is their general disadvantage related to the cost, weight, size and reliability. So, future studies should focus on the operation temperature increasing. The ultimate performance potential of MCT will ensure that it is the material of choice for all high performance infrared systems including multispectral detectors.

References

1. Abbott P, Pillans L, Knowles P, McEwen RK (2010) Advances in dual-band IRFPAs made from HgCdTe grown by MOVPE. Proc SPIE 7660:766035-1-11

2. Almeida LA, Thomas M, Larsen W, Spariosu K, Edwall DD, Benson JD, Mason W, Stoltz AJ, Dian JH (2002) Development and fabrication of two-color mid- and short-wavelength infrared simultaneous unipolar multispectral integrated technology focal-plane arrays. J Electron Mater 30(7):669–676

3. Blazejewski ER, Arias JM, Williams GM, McLevige W, Zandian M, Pasko J (1992) Bias-switchable dual-band HgCdTe infrared photodetector. J Vac Sci Technol B 10:1626

4. Campbell JC, Dentai AG, Lee TP, Burrus CA (1980) Improved two-wavelength demultiplexing InGaAsP photodetector. IEEE J Quantum Electron 16:601

5. Casselman TN, Walsh DT, Myrosznyk JM, Kosai K, Radford WA, Schultz EF, Wu OK (1990) An integrated multispectral IR detector structure. In: Extended abstracts of the U.S. workshop on the physics and chemistry of Mercury Cadmium Telluride, San Francisco, 2–4 October 1990

6. Casselman TN (1997) State of infrared photodetectors and materials. Proc SPIE 2999:1–10

7. Destefanis G, Baylet J, Ballet P, Castelein P, Rothan F, Gravrand O, Rothman J, Chamonal JP, Million A (2007) Status of HgCdTe bicolor and dual-band infrared focal plane arrays at LETI. J Electron Mater 36(8):1031–1044

8. Dvoretsky SA, Mikhailova NN, Remesnik VG, Sidorov YG, Shvets VA, Ikusov DG et al (2019) MBE-grown MCT hetero- and nanostructures for IR and THz detectors. Opto-Electron Rev 27(3):282–290

9. Eich D, Ames C, Breiter R, Figgemeier H, Hanna S, Lutz H, Mahlein KM, Schallenberg T, Sieck A, Wenisch J (2019) MCT-based high performance bispectral detectors by AIM. J Electron Mater 48(10):931–936

10. Faraone L (2005) MEMS for tunable multi-spectral infrared sensor arrays. Proc SPIE 5957:59570F

11. Fiorito G, Gasparrini G, Svelto F (1976) Multispectral $Hg_{1-x}Cd_xTe$ photovoltaic detectors. Infrared Phys 16:531–534

12. Gordon NT, Abbott P, Giess J, Graham A, Halis JE, Hall DJ, Hipwood L, Jones CL, Maxey CD, Price J (2007) Design and assessment of metal-organic vapour phase epitaxy-grown dual wavelength infrared detectors. J Electron Mater 36(8):931–936

13. Gunapala SD, Bandara SV, Liu JK, Mumolo JM, Ting DZ, Hill CJ, Nguyen J, Rafol SB (2010) Demonstration of 1024x1024 pixel dual-band QWIP focal plane array. Proc SPIE 7660:76603L

14. Halpert H, Musicant BI (1972) N-color (Hg, Cd)Te photodetectors. Appl Opt 11:2157–2161

15. Hipwood LG, Jones CL, Maxey CD, Lau HW, Fitzmaurice J, Catchpole RA, Ordish M (2006) Three-color MOVPE MCT diodes. Proc SPIE 6206:620612

16. Hu W, Ye Z, Liao L, Chen H, Chen L, Ding R, He L, Chen X, Lu W (2014) 128×128 long-wavelength/mid-wavelength two-color HgCdTe infrared focal plane array detector with ultralow spectral cross talk. Opt Lett 39(17):5184–5187

17. Jones CL, Hipwood LG, Price J, Shaw CJ, Abbott P, Maxey CD, Lau HW, Catchpole RA, Ordish M, Knowles P (2007) Multi-colour IRFPAs made from HgCdTe grown by MOVPE. Proc SPIE 6542:654210

18. Jóźwikowski K, Rogalski A (2007) Numerical analysis of three-colour HgCdTe detectors. Opto-Electron Rev 15:215–222

19. Kinch MA (2014) State-of-the-art infrared detector technology. SPIE Press, Bellingham

20. Kopytko M, Gawron W, Kębłowski A, Stępień D, Martyniuk P, Jóźwikowski K (2019) Numerical analysis of HgCdTe dual-band infrared detector. Opt Quant Electron 51:62

21. Krishna S, Forman D, Annamalai S, Dowd P, Varangis P, Tumolillo T, Gray A, Zilko J, Sun K, Liu M, Campbell J, Carothers D (2006) Two-color focal plane arrays based on self-assembled quantum dots in a well heterostructure. Phys Status Solidi (c) 3(3):439–443

22. Lawson WD, Nielson S, Putley EH, Young AS (1959) Preparation and properties of HgTe and mixed crystals of HgTe-CdTe. J Phys Chem Solids 9(3–4):325–329

23. Madejczyk P, Gawron W, Kębłowski A, Mlynarczyk K, Stępień D, Martyniuk P, Rogalski A, Rutkowski J, Piotrowski J (2020) Higher operating temperature IR detectors of the MOCVD grown HgCdTe heterostructures. J Electron Mater 49:6908–6917

24. Maxey CD, Capper P, Baker IM (2020) MOVPE growth of cadmium mercury telluride and applications. In: Irvine S, Capper P (eds) Metalorganic Vapor Phase Epitaxy (MOVPE): growth, materials properties, and applications. Wiley, Hoboken, pp 293–324

25. McEwen KR, Hipwood L, Bains S, Owton D, Maxey C (2019) Dual waveband infrared detectors using MOVPE grown MCT. Proc SPIE 11002:1100218-1–1100218-6
26. Mitra P, Barnes SL, Case FC, Reine MB, O'Dette P, Starr R, Hairston A, Kuhler K, Weiler MH, Musicant BL (1997) MOCVD of bandgap – engineered HgCdTe p-n-N-P dual – band infrared detector arrays. J Electron Mater 26(6):482–487
27. Norton P (2002) HgCdTe infrared detectors. Opto-Electron Rev 10(3):159–174
28. Piotrowski J, Rogalski A (2007) High-operating temperature infrared photodetectors. SPIE Press, Bellingham
29. Price JPG, Jones CL, Hipwood LG, Shaw CJ, Abbott P, Maxey CD et al (2008) Dual-band MW/LW IRFPAs made from HgCdTe grown by MOVPE. Proc SPIE 6940:69402S
30. Radford WA, Patten EA, King DF, Pierce GK, Vodicka J, Goetz P et al (2005) Third generation FPA development status at Raytheon vision systems. Proc SPIE 5783:331–339
31. Razeghi M, Dehzangi A, Li J (2021) Multi-band SWIR-MWIR-LWIR Type-II superlattice based infrared photodetector. Results Opt 2:100054
32. Reibel Y, Chabuel F, Vaz C, Billon-Lanfrey D, Baylet J, Gravrand O, Ballet P, Destefanis G (2011) Infrared dual band detectors for next generation. Proc SPIE 8012:801238-1-13
33. Reine MB, Hairston A, O'Dette P, Tobin SP, Smith FTJ, Musicant BL, Mitra P, Case FC (1998) Simultaneous MW/LW dual-band MOCVD HgCdTe 64 × 64 FPAs. Proc SPIE 3379:200–212
34. Rogalski A (2000) Dual-band infrared detectors. Proc SPIE 3948:17–30
35. Rogalski A (2019) Infrared and terahertz detectors, 3rd edn. CRC Press, Boca Raton
36. Rutkowski J, Madejczyk P, Piotrowski A, Gawron W, Jóźwikowski K, Rogalski A (2008) Two-colour HgCdTe infrared detectors operating above 200K. Opto-Electron Rev 16:321–327
37. Scribner D, Schuler J, Warren P, Satyshur M, Kruer M (1998) Infrared color vision: separating object from backgrounds. Proc SPIE 3379:2–13
38. Smith E, Venzor G, Gallagher A, Reddy M, Peterson J, Lofgreen D, Randolph J (2011) Large-format. J Electron Mater 40(8):1630–1636
39. Tennant WE, Thomas M, Kozlowski LJ, McLevige WV, Edwall DD, Zandian M et al (2001) A novel simultaneous unipolar multispectral integrated technology approach for HgCdTe IR detectors and focal plane arrays. J Electron Mater 30(6):590–594
40. Vallone M, Goano M, Tibaldi A, Hanna S, Eich D, Sieck A, Figgemeier H, Ghione G, Bertazzi F (2020) Challenges in multiphysics modeling of dual-band HgCdTe infrared detectors. Appl Opt 59(19):5656–5663
41. Vilela MF, Olsson KR, Norton EM, Peterson JM, Rybnicek K, Rhiger DR et al (2013) High-performance M/LWIR dual-band HgCdTe/Si focal-plane arrays. J Electron Mater 42(11):3231–3238
42. Wilson JA, Patten EA, Chapman GR, Kosai K, Baumgratz B, Goetz P et al (1994) Integrated two-color detection for advanced FPA applications. Proc SPIE 2274:117–125
43. Ziegler J, Eich D, Mahlein M, Schallenberg T, Scheibner R, Wendler J et al (2011) The development of 3rd gen IR detectors at AIM. Proc SPIE 8012:801237-1-13

Chapter 20
Flexible Photodetectors Based on II-VI Semiconductors

Mingfa Peng and Xuhui Sun

20.1 Introduction

Photodetector as an important component that can invert incident light into electric signals have attracted intensive research interest and exhibited various promising applications, including optical communication, imaging technique, environment monitoring, remote sensing, *etc*. [1–4]. Conventional photodetectors are usually fabricated on rigid substrates or wafers using inorganic semiconductor materials as functional materials, such as three-dimensional (3D) bulk crystalline Si, InGaAs, HgCdTe and related heterostructures in high-performance visible and infrared photodetectors [5, 6]. However, to obtain high-performance photodetectors, these functional materials should be fabricated with large thickness due to their low light absorption coefficients, and thus resulting in complex and expensive fabrication process and hindering their application in flexible electronics. Compared to photodetectors on rigid substrates, flexible photodetectors exhibit many advantages such as a good bendability, foldability and even stretchability as well as weight light, which have triggered a widely concerned in wearable electronics including wearable monitoring, wearable image sensing, self-powered integrated electronics, *etc*. [7–10]. For flexible photodetectors, not only the "5S" key parameters including photoresponse sensitivity, signal-to-noise ratio, speed, selectivity, and stability, but

M. Peng
School of Electronic and Information Engineering, Jiangsu Province Key Laboratory of Advanced Functional Materials, Changshu Institute of Technology, Changshu, Jiangsu, People's Republic of China

X. Sun (✉)
Institute of Functional Nano & Soft Materials (FUNSOM), and Jiangsu Key Laboratory for Carbon-based Functional Materials and Devices, Soochow University, Suzhou, Jiangsu, People's Republic of China
e-mail: xhsun@suda.edu.cn

© The Author(s), under exclusive license to Springer Nature Switzerland AG 2023 469
G. Korotcenkov (ed.), *Handbook of II-VI Semiconductor-Based Sensors and Radiation Detectors*, https://doi.org/10.1007/978-3-031-20510-1_20

also the mechanical flexibility should be taken into account to evaluate the performance of the devices [11, 12]. Therefore, the flexible substrates, device structure and functional materials are three important factors for flexible photodetectors. Up to now, several flexible substrates have been employed to construct flexible photodetectors, for instance, paper, polyethylene terephthalate (PET), polyethylene naphthalate (PEN) and polyimide (PI), *etc.* [13–16]. The stability and flexibility of these substrates should be paid more attention. In addition, the flexible Si based photodetectors can also be fabricated by reducing the thickness of Si film to tens to hundreds of nanometres, leading to the low light absorption [7]. Moreover, different device structures including planar photoconductive and vertical photodiode types have been demonstrated in high-performance flexible photodetectors [15, 17].

Apart from the flexible substrates and device structures, the functional materials as a building blocks play a key role in determining the photoelectronic performance of the flexible photodetectors. In recent years, various functional materials have been exploited and demonstrated in flexible photodetectors, including organic semiconductors, metal halide perovskites, two-dimensional (2D) nanostructures, *etc.* [7, 9, 18]. As an alternative, low dimensional II-VI semiconductors which can be divided into zinc chalcogens (*e.g.* ZnO and ZnS, *etc.*) and cadmium chalcogens (*e.g.* CdS, CdSe, and CdS_xSe_{1-x}, *etc.*) exhibit promising applications in optoelectronic devices due to their unique characteristics, such as direct bandgap, excellent optical and electric properties, and high quantum efficiency [19, 20]. For example, the optical properties and bandgap structure of the 0D II-VI nanostructures could be easily tuned by adjusting the size and shape of the nanostructures [21, 22]. 2D II-VI nanostructures in the nature with covalent bonds in all three dimensions exhibit many distinct properties different from 2D layered materials, such as abundant surface dangling bands, high-activity and high-energy surface [23, 24]. So far, different dimensional II-VI semiconductors, such as 0 D nanocrystals (NCs) and quantum dots (QDs) [25, 26], 1D nanowires (NWs) and nanoribbons (NRs) [27, 28], 2D nanosheets and related heterostructures [29], have been successfully exploited and demonstrated in flexible photodetectors due to their unique properties. However, the photoresponse of the single II-VI nanostructures based flexible photodetectors just only focus on UV (*e.g.* ZnS) or visible (*e.g.* CdS and CdSe) light region due to their intrinsic optical bandgap resulting in narrow wavelength range of light absorption, which will limit their potential application in some specific field, such as wide spectrum light detection, NIR light imaging, and night vision technique, *etc.* Furthermore, the II-VI semiconductors hybrid structure or heterostructure can be easily fabricated on flexible substrate, which can be employed as functional materials for flexible photodetectors [30, 31], thus improving the photoelectric performance and realizing broadband or NIR photoresponse.

With the rapid development of the Internet of Things (IOTs), flexible photodetectors exhibit a promising application in wearable electronics. However, the progress review of flexible photodetectors based on II-VI semiconductors are very few. In this chapter, we introduce the most recent progress on low dimensional II-VI semiconductors (0D, 1D, 2D and their related heterostructures) based flexible photodetectors and their application in wearable electronic. Firstly, the sensing mechanisms and key figures of merits for photodetectors have been summarized and discussed, which will help us to better understand and study various photodetectors. Then, the typical flexible photodetectors based on different dimensional II-VI semiconductor nanostructures have been introduced, in which the functional materials synthesis method have also been discussed. We mainly focus on the device structure and fabrication process of the II-VI semiconductors based flexible photodetector, as well as the mechanical flexibility and electric stability. In this review, we also discuss the design of various hybrid nanostructure or heterostructure for the flexible photodetectors to improve the photoelectric performance and realize broadband or NIR photoresponse. Furthermore, various applications of the II-VI semiconductors based flexible photodetectors have been summarized, including wearable monitoring sensors, image sensors, and self-powered integrated wearable electronics. At last, we summarize the review and discuss the challenges and perspectives of the II-VI semiconductors based flexible photodetectors.

20.2 Device Structure and Substrate Materials of Flexible Photodetectors

It has witnessed many important progresses in architecture design and fabrication of flexible photodetectors in the past decades. Generally, photodetectors can be classified as photoconductors, photodiodes and phototransistors according to their different working mechanisms. Through architectural design, various featured photodetectors could be obtained. In addition, the mechanical flexibility is a significant factor to evaluate the performance of the flexible photodetectors. Therefore, the device fabrication and the selection of flexible substrate materials are also very important for flexible photodetectors. In this section, we will mainly introduce the device structure and flexible substrate materials.

20.2.1 Device Structure

Photodetectors can be divided into lateral architecture and vertical architecture in terms of their different spatial layouts between electrodes and photoactive materials. Compared to the lateral structure photodetectors, the vertical structure devices exhibit shorter carrier transport lengths due to the smaller electrodes spacing. Photodiode is a typical vertical configuration structure device (Fig. 20.1), in which the photoactive materials were designed and fabricated between the bottom and top electrodes. Therefore, the photodiodes usually exhibit a faster response speed, an improved detectivity and high sensitivity [32]. However, the flexible transparent conductive materials should be employed as electrodes for this vertical configuration device and thus bring about complex processes for the flexible photodiode fabrication.

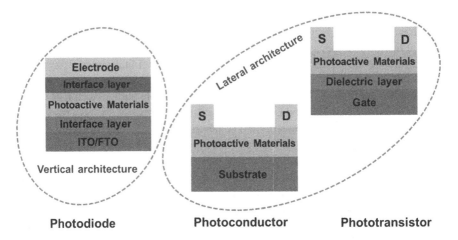

Fig. 20.1 Architectures of the typical flexible photodetectors

Compared with photodiode, the device structure of the photoconductor with lateral architecture is simple and easy to fabricate. The device architecture of the photoconductor is composed with photoactive materials as channel and two symmetrical electrodes as source and drain. Therefore, the flexible photodetector could be easily fabricated by using various metal electrodes rather than only transparent electrodes. In addition, photoconductors could obtain high photocurrent and gain but relatively low detectivity due to the high dark current resulting from the formation of ohmic contact and their unbalance transport process of the photogenerated charge carriers [18].

Phototransistor is similar with the configuration of the field-effect-transistors (FETs) and can be regarded as a special case of the photoconductor [33], which is also a lateral architecture device with a dielectric layer and gate electrode as floating gate to modulate the conductivity of the semiconductors effectively. Therefore, the lifetime of the photo-carriers of phototransistors has been prolonged and thus lead to a higher gain but slower response speed than photoconductors.

20.2.2 Substrate Materials

In order to obtain high performance flexible photodetectors, substrate materials, electrodes and photoactive materials should be taken into account and designed. Recently, various materials including polymer, paper and fiber have been employed as substrates for flexible photodetectors. In addition, silicon can be exploited as flexible substrates by reducing the thickness to tens to hundreds of nanometres. Furthermore, mica can also be used as substrate materials to construct flexible photodetector. In this part, we will mainly introduce the substrate materials in flexible photodetectors.

Polymer materials including polyethylene terephthalate (PET), polyimide (PI), polyethylene naphthalate (PEN), and polydimethylsiloxane (PDMS) have been

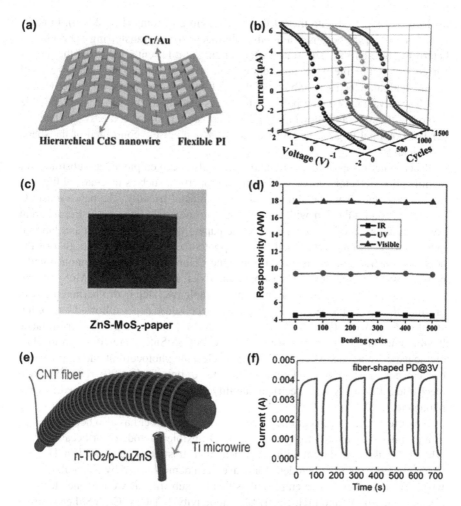

Fig. 20.2 (**a, b**) CdS nanowires based flexible photodetector on PI substrate. ((**a, b**) Reprinted with permission from Ref. [34]. Copyright 2015: American Chemical Society). (**c, d**) ZnS-MoS$_2$ heterostructure based flexible photodetector on paper substrate. ((**c, d**) Reprinted with permission from Ref. [36]. Copyright 2017: Wiley-VCH). (**e, f**) n-TiO$_2$/p-CuZnS heterostructure based flexible photodetector on Ti fiber substrate. ((**e, f**) Reprinted with permission from Ref. [38]. Copyright 2018: Wiley-VCH)

widely utilized as substrates for flexible photodetectors due to their intrinsic flexibility and mechanical stability. The photoactive materials can be directly placed on the polymer substrates by various strategies including spin-coating, transferring, magnetron sputtering, and printing, *etc*. For instance, Li et al. [34] directly transferred the synthesized 1D CdS NWs on PI substrates by the contact printing method to fabricate flexible photodetector (Fig. 20.2a). After bending 1500 cycles, the performance of the PI substrate based flexible photodetectors remained almost unchanged compared with that of the device before bending (Fig. 20.2b). In addition, Lou et al. [35] proposed a flexible photodetector by transferring 1D ZnS/CdS heterostructure on PET substrate. After bending at different conditions, the photocurrent of the flexible photodetectors exhibit no clear change indicating the good mechanical and electrical stability of the PET substrate based flexible photodetectors. The advantages of the polymer materials usually as the substrate of flexible photodetectors are as following: (1) The intrinsic flexibility and easy availability, (2) The potential large-scale fabrication, (3) The precisely controlled of the shape and thickness.

Paper materials as another alternative have also been employed as substrates for flexible photodetectors because of intrinsic properties, such as mechanical flexibility, low cost and easy availability, and environmental-friendly. The paper substrate exhibits a good adhesion with the photoactive materials because of the inherent porous structure and hygroscopicity of the paper, therefore the design and fabrication of the paper substrate based flexible photodetectors are similar to that of the polymer substrates based devices. For instance, Gomathi et al. [36] demonstrated a paper substrate based flexible broadband photodetector with the ZnS/MoS$_2$ heterostructure as photoactive material by using a simple two-step hydrothermal method, in which the layered MoS$_2$ was firstly grown on cellulose paper followed by synthesis of ZnS on MoS$_2$ (Fig. 20.2c, d). Gou et al. [37] proposed a high-performance flexible photodetector by depositing P3HT/CdS/CdS:SnS$_2$ nanowires hybrid films on the paper substrate. After bending 200 cycles, the photocurrent and responsivity of the paper substrate based flexible photodetector only decline 9%, revealing good mechanical flexibility and electrical stability of the paper substrate based flexible photodetector.

Apart from the aforementioned substrate materials, fiber has also been exploited as a promising candidate for constructing 1D flexible photodetector because of its inherent flexibility and wearability. A typical fiber substrate based flexible n-TiO$_2$/p-CuZnS heterostructure photodetectors have been demonstrated by Xu et al. [38], in which Ti microfiber can be employed as flexible substrate as well as one electrode due to its flexibility and high electrical conductivity, n-TiO$_2$/p-CuZnS/ heterostructure as photoactive materials and CNT fiber as another electrode (Fig. 20.2e, f). After bending 200 cycles, optoelectronic properties of the flexible device remained stable owing to the bendable properties of the fiber substrate based flexible photodetector. Furthermore, a double-twisted broadband perovskite flexible photodetector based on fiber substrate was also proposed by Sun et al. [39]. The double-twisted flexible photodetector exhibited excellent flexibility and bending stability even after bending 60 cycles at the bending angle of 90.

20.3 Flexible Photodetectors Based on II-VI Semiconductors

Various II-VI nanostructures have been widely employed in photodetectors as photoactive materials due to their intrinsic direct band-gap characteristics and excellent optical and electrical properties. In the following section, we will discuss the flexible photodetectors based on low-dimensional (0D, 1D and 2D) II-VI semiconductors and address their photoelectric performances.

20.3.1 0D II-VI Nanostructures Based Flexible Photodetectors

0D nanostructures including nanocrystals (NCs) and quantum dots (QDs) were usually synthesized by solution routing. The optical properties and bandgap structure of the 0D nanostructures could be easily tuned by adjusting the size and shape of the nanostructures during the synthesis process. For example, the PL lifetimes and their absorption and emission spectra of the core-shell CdSe/CdS QDs could be successfully adjusted by altering the thickness of the CdS shells [40]. Therefore, the solution processed 0D II-VI nanostructures exhibits many advantages in the fabrication of flexible photodetectors, including low-temperature synthesis process, large-scale production, and compatibility with flexible substrates, *etc*. Recently, various 0D II-VI nanostructures have been exploited as sensing materials for the flexible photodetectors, such as ZnS NPs, CdS QDs, CdSe QDs, *etc*. For instance, Yang et al. [41] proposed a flexible photodetector based on a hybrid film of graphene/CdS quantum dots by using chemical bath deposition method. The fabricated visible photodetectors exhibit high performance due to excellent interface between the graphene and the in suit growth of CdS QDs. The photodetectors exhibit good photoelectric properties with the responsivity up to 40 A/W at the wavelength of 450 nm. In addition, some universal method such as spin-coating and dip-coating have been explored to fabricate the flexible photodetector [42, 43]. For example, Hsiao et al. [44] fabricated a flexible photodetector on PEN substrate with s with CdSe/ZnS core-shell QDs coated on ZnO nanorod arrays as photoactive materials. The responsivity of the heterostructure photodetectors increased from 3.2 to 73.2 A/W with increasing the cover of the core-shell QDs from 0 to 1.00% at the wavelength of 380 nm (Fig. 20.3c). The significant enhancement of the optoelectronic performance can be attributed to the high surface-to-volume ratios of the ZnO/QDs nanostructure easily providing a directed path for electronic transport. Xia et al. [45] proposed solar-blind deep ultraviolet photodetectors by spin-coating colloidal ZnS quantum dots (QDs) solution on the electrode device. The photodetectors exhibit fast response ($t_r = 0.35$ s, $t_d = 0.07$ s) and good responsivity of ~0.1 mA/W at 265 nm light illumination. The further study indicated that the response of the ZnS QD photodetectors can be attributed to the drift of the photogenerated carriers due to the sensitivity of the ZnS QDs thin film to their surface states. However, all these process based flexible photodetectors will need expensive and time-consuming process, such as photolithography and lift-off technique.

Motivated by these problem, various novel techniques including all inject printing [44], screen printing [46], roll-to-roll printing [47], three-dimensional (3D) printing [48], and laser direct writing method [49] as alternative routes have been employed in flexible electronics and exhibit many advantages, such as rapid prototyping, high precision, and energy efficiency, *etc*. For instance, Lin et al. [49]

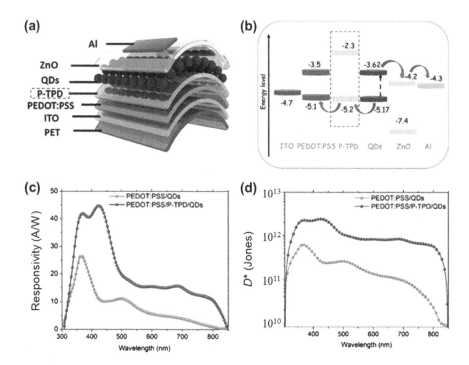

Fig. 20.3 (a) Schematic illustration of the device with double hole transport layer (HTL). (b) Energy band diagram of the photodetector with double HTL (c) Responsivity and (d) detectivity of the single HTL and double HTL devices as a function of wavelength. (Reprinted with permission from Ref. [50]. Copyright 2019: Royal Society of Chemistry)

demonstrated flexible ZnS/SnO_2 ultraviolet photodetectors by using a mask-free laser direct writing method with ZnS/SnO_2 heterostructure as photoactive materials, PI as flexible substrate, and highly conductive graphene obtained from the underneath PI by CO_2 laser irradiation as the lateral electrodes. The devices exhibited high optoelectronic performance resulted from the good interfaces between the graphene electrodes and the photoactive materials of ZnS/SnO_2 by this in suit growth method. The photocurrent of the devices has no obvious degradation after 500 bending cycles, presenting the excellent flexibility due to the reduced thickness of the device by lateral electrode structure. Subsequently, Shen et al. [50] proposed a vertical architecture flexible self-powered $CdSe_xTe_{1-x}$ QD photodetector with a double hole transport layer (HTL) of PEDOT:PSS/P-TPD (Fig. 20.3a, b). The photodetection capacity was extended significantly from UV to NIR region due to the well matched energy lever between P-PPD layer and the PEDOT:PSS/QD layers. The flexible QDs based photodetectors exhibit a low dark current density of 1.03×10^{-6} mA/cm^2, and a large specific detectivity of ~2.6×10^{12} Jones at 450 nm light illumination (Fig. 20.3c, d). More importantly, the flexible photodetectors exhibited excellent mechanical and electrical stability even after 150 cycles bending at different bending angles.

20.3.2 1D II-VI Semiconductors Based Flexible Photodetectors

1D semiconductor nanostructures, including nanowires (NWs) [51], nanoribbons (NRs) [52] and nanotubes (NTs) [53], have drawn much attention as functional building blocks in nanoscale electronic and optoelectronic devices due to their unique optical and electrical properties. Compared with other nanostructure, 1D nanostructures have a simple dimensional characteristic, which make it easy to realize quantitative analysis of some important parameters for optoelectronics. In addition, 1D nanostructures have been beneficial to the charge carrier transportation and thus exhibit excellent photoelectronic performance because of their large surface to volume ratio and high quality crystal structure as well. Furthermore, 1D nanostructures have an intrinsic mechanical flexibility and easy to fabricate flexible photodetectors due to their above characteristics. Up to now, various 1D II-VI semiconductors have been employed as functional materials in flexible photodetectors, such as ZnS NWs [36, 49, 54], CdS NWs [55, 56], CdSe NRs [57, 58], CdS_xSe_{1-x} NWs or NRs [27, 59], and their hybrid nanostructures [60–62]. For instance, Zhang et al. [63] proposed core-shell ZnS/InP nanowires based ultraviolet flexible photodetectors with a high photoconductive gain of 4.36×10^2, a high detectivity of 5.10×10^{12} Jones, and fast response speed of 0.74 s/0.38 s, respectively. The excellent optoelectronic performance can be attributed to the formation of core-shell heterostructures. Furthermore, the results also indicated that the flexible photodetectors have excellent mechanical flexibility and electrical stability by bending the device even for 1200 times at different bending angles.

Recently, various 1D visible light absorption sensor materials of CdS, CdSe or ternary CdS_xSe_{1-x} have also been widely investigated in photodetectors. For instance, Shen et al. employed hierarchical CdS nanowires as building block to fabricate flexible photodetector with ultrahigh sensitivity [34]. The results indicate that the CdS NWs based flexible photodetectors exhibit a high current on/off ratio larger than 2500, the response time of 0.2 s at 407 nm, as well as a super mechanical flexibility and electric stability even after 1500 cycles of bending. Sun et al. synthesized 1D ternary CdS_xSe_{1-x} NRs and demonstrated them in flexible visible photodetector with an ultrahigh current on/off ratio of 4×10^6, a high responsivity of 1.24×10^3 A/W, and excellent mechanical stability as well [27]. Moreover, organic-inorganic hybrid nanostructure has also been demonstrated in flexible photodetectors by Shen's group. They proposed a hybrid flexible photodetectors by introducing high hole-transport rate and strong absorption materials of P3HT into high electrical conductivity materials of CdSe nanowires. The hybrid device exhibits an extremely high on/off switching ratio larger than 500, fast response time of 10 ms, and excellent mechanical flexibility and stability as well. Recently, Lin et al. [64] demonstrated 1D CdS NRs/2D WSe_2 heterostructure with piezo-phototronic effect for a performance enhanced flexible photodetector. Figure 20.4a shows the optical image of the flexible 1D CdS/2D WSe_2 heterostructure photodetector. The heterostructure device fabrication processes are shown in Fig. 20.4b, c. Briefly, 2D WSe_2 was precisely transferred on one-terminal of the pre-dispersed CdS NW by using an accurate transfer platform equipped with micromanipulators, and then the metal electrodes were patterned on the surface of the NW and nanosheet with electron beam lithography and lift-off process. Figure 20.4d shows the schematic illustration of experimental setup for the applied strain and the piezo- phototronic process, in which the strain is applied through a two-point bending apparatus. The results exhibit that the photocurrent increase with increasing the compressive strain, while the photocurrent increased from 0.32 to 0.65 nA with the light intensity of 16.9 μW/cm² at the compressive strain of −0.73%. The optimized photoresponsivity of the device is of 33.4 A/W (Fig. 20.4e).

In addition, many other efforts have been devoted to realize the broadband photodetector based on 1D II-VI semiconductors by integrating semiconductors of 1D ZnTe, ZnS, or CdS semiconductors with strong light-absorbing materials such as colloidal PbX (X = S, Se) quantum dots, perovskite or even other 2D materials [30, 35, 65–67]. For instance, Liu et al. [68] demonstrated flexible visible-light photodetector based on high-quality ZnTe nanowires on polymer substrate. The photocurrent of the flexible photodetectors maintained at 30 nA + 1.2 nA under different bending radius. More importantly, the photocurrent of the flexible device remained

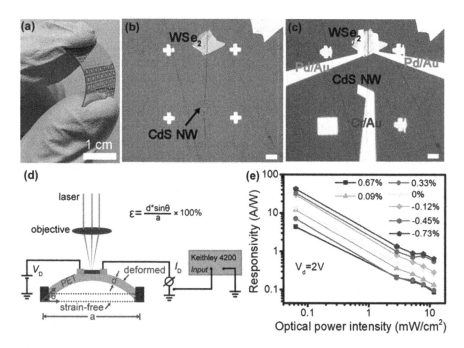

Fig. 20.4 (a) Optical image flexible 2D WSe₂ nanosheet/1D CdS nanowire heterojunction photodetector. SEM image of (**b**) 2D WSe₂/1D CdS heterojunction and (**c**) heterojunction based device. (**d**) Schematic illustration of experimental setup for the applied strain and the piezo-photoelectronic process. (**e**) Strain dependence of the photoresponsivity of the flexible photodetector at a bias of 2 V. (Reprinted with permission from Ref. [64]. Copyright 2018: Royal Society of Chemistry)

unchanged even after 250 cycles bending. Subsequently, Wang et al. [60, 61] demonstrated a piezo-phototronic effect enhanced UV and visible photodetector by using ZnO/CdS core/shell nanowire or carbon-fiber/ZnO-CdS double shell nanowire as sensor materials. For the flexible photodetector, CdS NWs were firstly synthesized on the surface of ZnO NWs to form core/shell structure, and then the obtained single CdS/ZnO NW was bonded on a polymer substrate to fabricate flexible photodetector, which shows excellent photoresponse from UV to visible (372–574 nm) with the responsivity of 11 A/W at 548 nm [61]. The photocurrent and sensitivity of the core/shell structure photodetector is 1000 times higher than that of 1D CdS NRs. Liang et al. [69] have also proposed flexible resistivity-type UV-visible photodetectors based on CdS NWs with Ag NWs as electrode by using a non-transfer process. The flexible photodetector exhibits a UV-visible light sensitivity with the maximum sensitivity of 120 and the response time of 6 ms.

20.3.3   2D II-VI Semiconductors Based Flexible Photodetectors

Since the discovery of graphene [70], various layered two-dimensional (2D) nanostructures, such as transition metal dichalcogenides (TMDs) (e.g. MoS_2 and WS_2) [71, 72], Indium selenide (InSe) [71, 73], black phosphorus (BP) have been explored as building blocks in electronics and optoelectronics due to their unique dimensional dependent properties. For these layered structures, the atoms are usually bonded with strong covalent in-plane but with weak van der Waals interactions out-of-plane direction [74]. This characteristic allows that high-quality multilayer or even single layer 2D semiconductors can be easily obtained by simple mechanical exfoliation from bulk crystals. In contrast, there are larger number of non-layered semiconductors in the nature with covalent bonds in all three dimensions, and thus make it difficult to obtain 2D non-layered semiconductors. However, these non-layered materials exhibit many their distinct properties different from 2D layered materials, such as abundant surface dangling bands, high-activity and high-energy surface [23, 75, 76]. Inspired by these unique features, various 2D non-layered semiconductors including 2D II-VI groups semiconductors have been attracted considerable interest and have been synthesized by wet-chemical synthesis, such as CdS, CdSe, CdS_xSe_{1-x}, and ZnO *etc* [77–83]. However, the lateral sizes of the semiconductors are small than 300 nm, which limit their practical application in optoelectronics. Thus, many efforts have been devoted to explore larger lateral size 2D II-VI semiconductors by chemical vapor deposition (CVD) method and demonstrated them in photodetectors.

For instance, Hu et al. [84] fabricated 2D non-layered CdS on MoS_2 substrate by epitaxial growth technique with the uniform thickness of 50 nm and the lateral size larger than 1 μm. The formed vertical heterostructure of 2D CdS/MoS_2 have been demonstrated in photodetectors and the results exhibited a broad wavelength photoresponse (larger than 680 nm) and the responsivity enhance over 50 time (70.8 mA/W at 610 nm light illumination), which can be attributed to the epitaxial growth of 2D CdS and photoinduced electrons transfer between the energy levels of CdS and MoS_2. Subsequently, Meng et al. [29, 85] synthesized 2D CdSe and CdS_xSe_{1-x} on mica substrate via van der Waals epitaxy technique. Briefly, $CdCl_2$ located at the centre of tube and mixture S and Se powder located at the upstream of the $CdCl_2$ were employed as powder source to synthesize 2D CdS_xSe_{1-x} by using a vapor deposition method. SEM image shows that the shape of the as-synthesized 2D CdS_xSe_{1-x} flakes is hexagon with isolated domains and the thickness is around 76 nm (Fig. 20.5a). Figure 20.5b represents the schematic illustration of the flexible photodetectors, and the linear I-V curves of the photodetector indicate that a good ohmic contact has been formed between 2D CdS_xSe_{1-x} flakes and metal electrodes. The photocurrent of the 2D CdS_xSe_{1-x} flakes based photodetector increase with

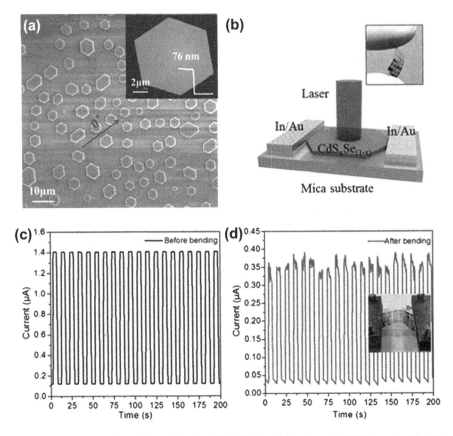

Fig. 20.5 (**a**) SEM image of the synthesized 2D CdS_xSe_{1-x} flakes. Inset is AFM image and height profile of a representative 2D CdS_xSe_{1-x}. (**b**) Schematic illustration of 2D CdS_xSe_{1-x} based flexible photodetector. Inset is the optical image of the flexible photodetector. I-$Time$ characteristics of the flexible photodetector (**c**) before and (**d**) after 50 times bending cycles. (Reprinted with permission from Ref. [29]. Copyright 2017: Royal Society of Chemistry)

increasing the incident light intensities from 0.56 to 5.73 mW/cm^2 under 450 nm light illumination. The demonstrated 2D CdS$_x$Se$_{1-x}$ flexible photodetectors exhibit a high responsivity of 703 A/W, specific detectivity of 3.41 × 10^{10} Jones, external quantum efficiency (EQE) of 1.94 × 10^3, and a good mechanical stability after repeated bending, respectively. Although the photocurrent shows a slightly decrease after bending, the flexible photodetector still exhibits a good mechanical stability with clear on/off photoresponse (Fig. 20.5c, d).

Recently, Sun et al. presented 2D non-layered CdS$_x$Se$_{1-x}$ nanosheets with lateral size of 10 μm by one-step physical evaporation method. By introducing the strong light absorption material of PbS QDs or perovskite [20, 86], the photoelectric performance of the 2D CdS$_x$Se$_{1-x}$ based heterostructure photodetectors was improved significantly compared with that of pristine 2D CdS$_x$Se$_{1-x}$ nanosheet. In addition, Zhai et al. [24] proposed self-limited epitaxial growth of 2D non-layered CdS flakes on mica substrate with ultrathin thickness of 6 nm and large lateral size of 40 μm by using In$_2$S$_3$ as surface passivation agent. The growth mechanism of ultrathin 2D CdS was also studied by the experiments and theoretical calculation. The 2D CdS flakes based flexible photodetector was fabricated by a photolithography and lift-off process and then was directly attached to the flexible PET film, exhibiting the responsivity and detectivity of 0.05 A/W and 7.81 × 10^8 Jones, respectively, at the light intensity of 26.6 mW/cm^2 under 400 nm light illumination. The flexible photodetector also showed excellent mechanical stability with the current on/off ratio of ~10^4 even after bending 200 cycles, which can be attributed to the ultrathin 2D CdS flakes with outstanding deformation tolerance.

20.4   Applications of Flexible Photodetector

The above mentioned II-VI semiconductors based flexible photodetectors are individual device structure, which are not suitable for the practical application in monitoring [87], imaging [88], and optical communication [89] because their operation are based on the integrated high-density device arrays. In this section, we will mainly introduce the recent applications of II-VI semiconductors flexible photodetectors, including wearable monitoring sensors, image sensors, self-powered integrated electronics.

20.4.1   Wearable Monitoring Sensors

The wearable monitoring sensors usually demand a precise detection of output signals due to the weak incident light intensity or low light intensity variation in many practical applications [11]. Thus, the detectivity and responsivity of the photodetector should be larger enough to make sure that the monitoring systems work normally. Furthermore, the integrated photodetectors should be flexible, stable and skin feeling for the practical application of wearable monitoring sensors. In order to continuous monitor human blood waves, Kim et al. [90] successfully presented stretchable optoelectronic sensors by integrated QDs based LEDs and QDs based photodetectors with elastomeric substrates (Fig. 20.6). The optoelectronic properties of the flexible sensors were stable under different deformations due to the using of graphene electrodes ensuring the excellent bendability of the devices. The

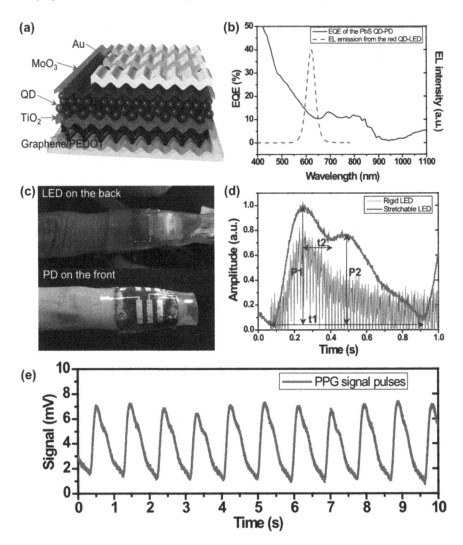

Fig. 20.6 (**a**) Schematic diagram of QDs based flexible photodetector. (**b**) The external quantum efficiency (EQE) spectra of the QD photodetector. (**c**) Optical image of skin-mounted optoelectronic sensor composed of QD-LEDs and QD photodetectors wrapped around the finger. (**d**) Real-time optoelectronic signal pulse from a stretchable QD photodetector. (**e**) Real-time record over several pulse periods under illumination from the red QD-LED. (Reprinted with permission from Ref. [90]. Copyright 2017: American Chemical Society)

emission of the LEDs could be efficiently controlled by adjusting the composition of the QD, such as CdSe/CdS/ZnS (core/shell/shell) QDs for the red-emissive while CdSe/ZnS QDs for the green and blue-emissive. The schematic diagram of QDs based flexible photodetector is shown in Fig. 20.6a, in which the QDs and graphene were used as photoactive materials and electrode by assembling them on elastomeric PEN substrates via transfer printing method. MoO_3 and TiO_2 layers were used as hole and electron transport layers, respectively. The external quantum efficiency (EQE) of the photodetectors can reach 12% at a wavelength of 618 nm (Fig. 20.6b). To demonstrate the application of the flexible devices, the fabricated QDs based LEDs and photodetectors had been stretched around the tip of a forefinger (Fig. 20.6c). The pressure pulse signals of the sensors could be detected by monitoring the variation of light absorption resulted from illumination in the skin. One pulse wave from a stretchable QD photodetector could be obtained by using a stretchable QDs based LED as light source under a luminous intensity of 520 cd m^{-2} (Fig. 20.6d), revealing that the flexible sensor showed two clearly distinguishable peaks. Figure 20.6e showed the real-time record over several pulse periods under illumination from the red QD-LED. The results indicated that the proposed stretchable optoelectronic sensors exhibited excellent optoelectronic properties as well as health-monitoring capability.

Similarly, Yan et al. [91] proposed novel flexible photodetector arrays based on all-solid-state organic electrochemical transistors (OECT) for photosensing and imaging. The OECT based phototransistor with CdSe/ZnS quantum dots as photoactive materials exhibit excellent responsivity and detectivity of 6.9×10^5 A/W and 5.8×10^{12} Jones under 355 nm UV light illumination, respectively. Furthermore, the flexible image sensors have been fabricated by integrating the flexible OECTs based phototransistor into 10×10 pixel on PET substrate. The demonstrated flexible image sensors have demonstrated excellent detection ability for broadband and thus can clearly distinguish the target images composed of red and white light even under bending states. Most recently, Lee et al. [92] reported a stretchable wearable electronics based on CdSe/ZnS quantum-dots for visual display of body movement and skin temperature signals from skin-attached sensors. Firstly, an array of CdSe/ZnS QD-LEDs was fabricated on top of the pre-formed NOA63 islands array and then liquid metal Galinstan was spray-coated on the top of the QDs-LED array as conductive electrode. Finally, the array of QD-LEDs on the islands was transferred onto a stretchable Eco-PDMS elastomer substrate. The fabricated QDs based wearable electronics exhibited stable operation under 50% uniaxial and 30% biaxial strains due to the architectural design minimizing the strain applied to the QD-LEDs by transforming the strain onto the stretchable elastomer substrate and the liquid metal electrode. More importantly, the stretchable QD-LED array were demonstrated in the visual pattern display by monitoring the extent of knee bending and the changes of human skin temperature.

20.4.2 Image Sensors

The other important application of the flexible photodetector is image sensing, which can detect environments and objects and have a specific application according to their different sensing wavelength range, such as UV imaging for skin health monitoring and remote sensing, visible light imaging for face recognition system, and infrared imaging for medical diagnosis and night vision, *etc.* [11, 93, 94] For instance, Li et al. [95] presented ZnO quantum dot decorated Zn_2SnO_4 nanowire heterojunction photodetectors for flexible ultraviolet image sensors. To investigate the image sensing, 10×10 pixel flexible heterojunction photodetectors were integrated to recognize the letters under the bending states (Fig. 20.7). Figure 20.7a shows the optical image of the flexible photodetector arrays on a PET substrate. The flexible photodetector shows excellent stability with almost identical *I-V* characteristics even after bending 2000 cycles. According to the energy band diagram (Fig. 20.7b), the photogenerated electron-hole pairs will be efficiently separated and the holes migrate from NWs to QDs to neutralize the absorbed O_2^- and then increase the electron concentration, thus enhancing the photoresponse performance of the heterostructure photodetectors. Figure 20.7c, d presents the schematic illustrations of using a 10×10 flexible photodetector array to sense the letters "F" and "E" under different bending directions, in which the letter patterns were placed on the back side of the devices and the incident light was also illuminated from the back side. It is obviously that the corresponding output images of letters "F" and "E" can be clearly observed under bending conditions (Fig. 20.7e, f). The results indicate that the QDs based heterostructure has a promising application in flexible UV image sensing.

In addition, a rectifying diode-type flexible photodetectors based on II-VI semiconductors have also been demonstrated in image sensing. This type device is usually fabricated with a vertical structure, the top and bottom electrodes were used to extracted electrons and holes, while the photoactive materials located in middle layer and closely contact with electron and hole transporting layers, respectively. Brabec et al. [96] proposed the solution-processed flexible photodetector in visible and near-infrared imaging with P3HT/O-IDTBR as the photoactive and with ZnO as electron transporting layer. The photodiode shows a record responsivity of 0.42 A/W and external quantum efficiency (EQE) of 69% at 755 nm NIR light, respectively. More importantly, the photodiodes based image sensor can realize objects detection under visible and NIR light conditions with excellent imaging performance. Similarly, Kim et al. [97] reported a matrix-type multiplexed photodiode array with perovskite film as photoactive materials and with spin-coated ZnO NPs as electron transporting layer by using a novel spin-on-patterning (SoP) process. By controlling the spin-coating rate and the temperature of the precursor solution, the optimized pattern yield can be obtained in SoP process. Subsequently, a 10×10 multiplexed ultrathin and flexible image sensor array was fabricated to realize a high-resolution imaging of dot array patterns. The device exhibited the responsivity and detectivity of 0.109 A/W and 1.35×10^{12} Jones at 530 nm, respectively, which can be compared with the commercial Si photodetector of 0.01 A/W and 4×10^{12} Jones.

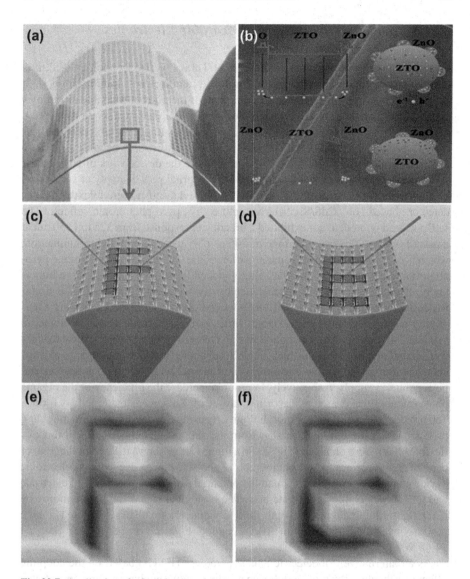

Fig. 20.7 Application of a flexible photodetector array as UV image sensor. (**a**) Optical images of as-fabricated flexible photodetector arrays on a PET substrate. (**b**) Energy band diagrams of carrier separation mechanism. (**c, d**) Schematic illustrations of using a 10 × 10 flexible photodetector array to sense the letters "F" and "E" under different bending directions. (**e, f**) Corresponding output images of letters "F" and "E" by the flexible photodetector array. (Reprinted with permission from Ref. [95]. Copyright 2017: American Chemical Society)

20.4.3 Self-Powered Integrated Wearable Electronics

With the rapid development of the Internet of Things (IOTs) technology [98], it is more and more urgent for wearable electronic systems towards minimization, high integration and multifunctional. However, most of the current wearable electronics usually works by the external power source supply including battery and solar cell, which will inevitably increase the system size and thus limit their practical application. Consequently, various self-powered photodetectors have been developed to integrate into wearable electronics, which can drive the device work directly by the self-powered function without the demands of external power supply or series connecting wires. For instance, Dai et al. [31] proposed a self-powered flexible photodetector based on CdS/Si heterostructure by pyro-phototronic effect. The self-powered photodetector exhibited a broadband response from 325 to 1550 nm at a bias of 0 V and fast response time of 245 μs /277 μs at 325 nm light illumination, which is over the optical bandgap lamination of the CdS and Si because of the pyro-phototronic effect in CdS and thus photoinduced heating or cooling. More importantly, the responsivity and specific detectivity were also improved 67.8 times. Recently, Park et al. [99] reported self-powered ultra-flexible electronics via nano-grating-patterned organic photovoltaics which can realize the detection of biometric signals with high SNR when applied to skin tissue. The flexible biomedical devices were integrated into a 1 μm thick flexible substrate in which the organic electrochemical transistors and organic photovoltaic (Fig. 20.8a) were used as sensors and power sources supply, respectively. In order to increase the efficiency of the organic photovoltaics, the charge transporting layers of ZnO were patterned to form a 760 nm periodic nano-grating morphologies by using a room-temperature moulding process. When the device separated from the supporting substrate, 3 μm thick of flexible OPVs were obtained (Fig. 20.8b). Figure 20.8c shows the circuit diagram of cardiac signal recording and photograph of the self-powered integrated electronic device attached to a finger, in which the potential difference between Gel electrode and OECT was employed as gate bias. The biological signal can be monitored by LED light illumination. The optimized organic photovoltaics exhibited a high power-conversion efficiency of 10.5% and a high power-per-weight ratio of 11.46 W/g.

As an alternative, triboelectric nanogenerators (TENGs) based on the coupling effect of triboelectrification and the electrostatic induction have been demonstrated in self-powered flexible electronics by harvesting external mechanical energy and converting it into electric power to driver the devices [100]. Recently, TENGs as a power supply have attracted intensive interest in self-powered gas sensing [98], flexible UV light detection [101], self-powered weighting [102], *etc.* For instance, Zhang et al. [101] reported a self-powered wearable real-time UV Photodetector by coupling effects of triboelectric and photoelectric. The schematic illustration of the flexible real-time UV light detection system is show in Fig. 20.8d, and the inset is the optical image of flexible self-powered photodetector on human body. The flexible photodetection system is consist of flexible-film-based TENG working in

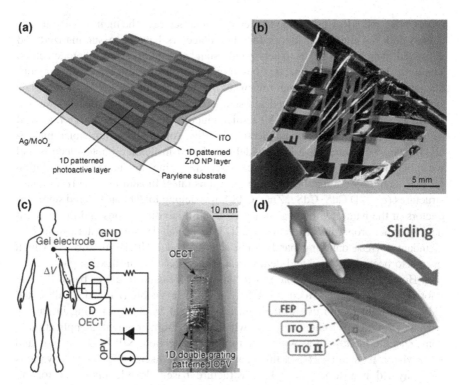

Fig. 20.8 (**a**) Structure diagram of the organic photovoltaic (OPV) device. (**b**) Photography of the OVP device was wrapped around an object. (**c**) The circuit diagram of cardiac signal recording and photograph of the self-powered integrated electronic device attached to a finger. ((**a**–**c**) Reprinted with permission from Ref. [99]. Copyright 2018: Springer Nature). (**d**) Schematic diagram of the flexible photodetector in tapping-mode. ((**d**) Reprinted with permission from Ref. [101]. Copyright 2020: American Chemical Society)

tapping mode as energy harvester, UV photodetector as sensor, and a series of commercial LEDs as alarm. The output voltages of the constant resistance changes with increasing the UV light intensities, which can be reflected directly through series LEDs on/off display in circuit. The working mechanism can be attributed to impedance matching effect between NPs-based UV photodetector and TENG. The proposed self-powered flexible UV photodetector may pave a new way in self-powered wearable electronics.

20.5   Conclusion and Outlook

In this chapter, the most recent progress on different dimensional II-VI semiconductors based flexible photodetectors and their application in wearable electronics have been introduced, including 0D nanostructures (QDs and NCs), 1D nanostructures

(NWs, NBs and NRs) and 2D non-layered nanostructures. The main sensing mechanisms and key figures of merits for photodetectors have been summarized and discussed, which will help us to better understand and study various photodetectors. The typical flexible photodetectors based on different dimensional II-VI semiconductor nanostructures have been reviewed, in which the functional II-VI materials synthesis method have also been discussed. We mainly focus on the device structure and fabrication process of the II-VI semiconductors flexible photodetector, as well as the mechanical flexibility and electric stability. Among these functional materials, 0D nanostructures have been widely exploited in flexible photodetectors and wearable electronics for their large scale manufacturing, low temperature solution process, and easy compatible with flexible substrates. In addition, 2D II-VI nanostructures (*e.g.* 2D CdS, CdSe, ZnO, CdS_xSe_{1-x}) belong to 2D non-layered semiconductors in the nature with covalent bonds in all three dimensions and thus exhibit many distinct properties different from 2D layered van der Waals materials, such as abundant surface dangling bands, high-activity and high-energy surface, which have also been developed and demonstrated most recently in flexible photodetectors. However, the photoresponse of the single II-VI nanostructures based flexible photodetectors just only focus on UV (*e.g.* ZnS) or visible (*e.g.* CdS, CdSe, and CdS_xSe_{1-x}) light region due to their intrinsic optical bandgap resulting in narrow wavelength range of light absorption, which will limit their potential application in some specific filed, such as wide spectrum light detection, NIR light imaging, and night vision technique, *etc.* In this chapter, we also discuss the designed various II-VI hybrid nanostructure or heterostructure based flexible photodetectors to improve the photoelectric performance and realize broadband or NIR photoresponse. Finally, various application of the II-VI nanostructures based flexible photodetectors have been summarized, including wearable monitoring sensors, image sensors, and self-powered integrated wearable electronics.

Although significant research progresses on II-VI semiconductors based flexible photodetectors have been achieved, there are still many challenges and opportunities in future studies, such as optimizing synthesis process to obtain more uniformly and high quality crystalline nanostructure, further improving the adhesion properties between flexible substrate, functional materials and metal electrodes, and thus improving the mechanical flexibility and electric stability of the flexible photodetectors. (1) For flexible photodetectors, various II-VI nanostructures have been mainly synthesized by using solution process, chemical vapor deposition, or various printing techniques, while different synthesis process have their unique characteristics but also have their disadvantages. For instance, large lateral size (>1 μm) and high-quality crystalline 2D non-layered II-VI nanostructures have been mainly synthesize by CVD routing, but it will limit their application in large scale integrated flexible photodetector arrays due to the difficult large scale synthesis by this routing. The solution process synthesis of 0D II-VI nanostructure exhibits potential large scale manufacturing characteristics and can be easily spin-coated on flexible substrate, but the post ligands-exchange step is further needed to improve the conductivity of the NPs or QDs by organic solvent and thus resulting in the degradation of the flexible substrate and damage the photoelectric performance of the flexible photodetectors. Thus, the functional materials synthesis strategies should be optimized to realize the penitential application for large scale and high-performance flexible photodetectors. (2) The mechanical flexibility and electronic stability of the flexible photodetector can be slightly affected after several hundred times bending or twisting and resulting in the deterioration of contact property between the functional materials and metal electrodes and the flexible substrate. Thus, it is urgently to need further study and more endeavours to improve the adhesion property between the functional materials and electrodes and substrate, thus ensuring the flexibility and stability of the flexible photodetectors.

The II-VI nanostructures based flexible photodetectors exhibit promising and inspiring application in wearable electronic including wearable monitoring sensors, image sensors and self-powered integrated wearable electronic in Sect. 20.4. For wearable monitoring sensors, flexible photodetector integrated into human body to monitor the environment UV signals for skin disease prevention according to the resistance variation caused by UV light illumination and thus lighting the LEDs. Thus, the integrated flexible photodetectors with ultra-flexibility, high sensitivity and skin like should be paid more attention for the practical application of wearable monitoring sensors. For image sensors, few works including 10 × 10 pixel flexible heterojunction photodetectors based on ZnO have been integrated and demonstrated in UV image sensing system to sense the letters with bending states. However, the visible and NIR image sensors are also very important for the practical application in wearable electronics, such as visible light imaging for face recognition system, and infrared imaging for medical diagnosis and night vision, *etc*. Therefore, we should pay more attention not only to the fabrication of wearable image sensors in different sensing wavelength range but also the precise and complicated imaging recognition system for special application environments. For self-powered wearable integrated electronics, various self-powered photodetectors have been developed to integrate into wearable electronics to efficiently drive the photodetector by integrating p-n junction, solar cells or organic photovoltaics. These strategies have been proved to be the effective ways while the complicated fabrication process will limit their potential extensive application. Thanks to the development of TENG technique, TENG based flexible photodetector serve as another choice for the self-powered wearable integrated electronics. However, this strategy is still in its infancy, there are many problems to solve in future research, such as the output electric signals stability in mechanical energy harvesting process, the integrated stability of the whole systems. All in all, several significant progresses about the flexible photodetector integrated wearable electronic have been exploited in most recently, while there are still tremendous challenges for the integration of wearable electronic in real-time data monitoring, processing, and outputting.

References

1. Luo P, Zhuge F, Wang F, Lian L, Liu K, Zhang J et al (2019) PbSe quantum dots sensitized high-mobility Bi_2O_2Se nanosheets for high-performance and broadband photodetection beyond 2 μm. ACS Nano 13(8):9028–9037

2. Zhu T, Yang Y, Zheng L, Liu L, Becker ML, Gong X (2020) Solution-processed flexible broadband photodetectors with solution-processed transparent polymeric electrode. Adv Funct Mater 30(15):1909487
3. Xu Y, Lin Q (2020) Photodetectors based on solution-processable semiconductors: recent advances and perspectives. Appl Phys Rev 7(1):011315
4. Konstantatos G (2018) Current status and technological prospect of photodetectors based on two-dimensional materials. Nat Commun 9(1):1–3
5. Xie C, Yan F (2017) Flexible photodetectors based on novel functional materials. Small 13(43):1701822
6. Buscema M, Island JO, Groenendijk DJ, Blanter SI, Steele GA, van der Zant HS et al (2015) Photocurrent generation with two-dimensional van der Waals semiconductors. Chem Soc Rev 44(11):3691–3718
7. Chow PCY, Someya T (2019) Organic photodetectors for next-generation wearable electronics. Adv Mater 32(15):1902045
8. Long M, Wang P, Fang H, Hu W (2018) Progress, challenges, and opportunities for 2D material based photodetectors. Adv Funct Mater 29(19):1803807
9. Dong T, Simões J, Yang Z (2020) Flexible photodetector based on 2D materials: processing, architectures, and applications. Adv Mater Interfaces 7(4):1901657
10. Yokota T, Fukuda K, Someya T (2021) Recent progress of flexible image sensors for biomedical applications. Adv Mater 33(19):2004416
11. Cai S, Xu X, Yang W, Chen J, Fang X (2019) Materials and designs for wearable photodetectors. Adv Mater 31(18):1808138
12. Zheng L, Deng X, Wang Y, Chen J, Fang X, Wang L et al (2020) Self-powered flexible TiO_2 fibrous photodetectors: heterojunction with P3HT and boosted responsivity and selectivity by Au nanoparticles. Adv Funct Mater 30(24):2001604
13. Selamneni V, Raghavan H, Hazra A, Sahatiya P (2021) MoS_2/paper decorated with metal nanoparticles (Au, Pt, and Pd) based plasmonic-enhanced broadband (visible-NIR) flexible photodetectors. Adv Mater Interfaces 8(6):2001988
14. Zhang Y, Huang P, Guo J, Shi R, Huang W, Shi Z et al (2020) Graphdiyne-based flexible photodetectors with high responsivity and detectivity. Adv Mater 32(23):2001082
15. Wang M, Tian W, Cao F, Wang M, Li L (2020) Flexible and self-powered lateral photodetector based on inorganic perovskite $CsPbI_3$–$CsPbBr_3$ heterojunction nanowire array. Adv Funct Mater 30(16):1909771
16. Schneider DS, Grundmann A, Bablich A, Passi V, Kataria S, Kalisch H et al (2020) Highly responsive flexible photodetectors based on MOVPE grown uniform few-layer MoS_2. ACS Photonics 7(6):1388–1395
17. Fuentes-Hernandez C, Chou W-F, Khan TM, Diniz L, Lukens J, Larrain FA et al (2020) Large-area low-noise flexible organic photodiodes for detecting faint visible light. Science 370:698–701
18. Hao D, Zou J, Huang J (2019) Recent developments in flexible photodetectors based on metal halide perovskite. InfoMat 2(1):139–169
19. Majithia RY (2013) Microwave-assisted synthesis of II-VI semiconductor micro-and nanoparticles towards sensor applications, vol 3572238. Texas A&M University. ProQuest Dissertations Publishing
20. Peng M, Xie X, Zheng H, Wang Y, Zhuo QQ, Yuan G et al (2018) PbS quantum dots/2D non-layered CdSxSe1-x nanosheets hybrid nanostructure for high-performance broadband photodetectors. ACS Appl Mater Interfaces 10(50):43887–43895
21. Jang Y, Shapiro A, Isarov M, Rubin-Brusilovski A, Safran A, Budniak AK et al (2017) Interface control of electronic and optical properties in IV–VI and II–VI core/shell colloidal quantum dots: a review. Chem Commun 53(6):1002–1024
22. Kagan CR (2019) Flexible colloidal nanocrystal electronics. Chem Soc Rev 48(6):1626–1641
23. Zhou N, Yang R, Zhai T (2019) Two-dimensional non-layered materials. Mater Today Nano 8:100051

24. Jin B, Huang P, Zhang Q, Zhou X, Zhang X, Li L et al (2018) Self-limited epitaxial growth of ultrathin nonlayered CdS flakes for high-performance photodetectors. Adv Funct Mater 28(20):1800181
25. Mitra S, Aravindh A, Das G, Pak Y, Ajia I, Loganathan K et al (2018) High-performance solar-blind flexible deep-UV photodetectors based on quantum dots synthesized by femtosecond-laser ablation. Nano Energy 48:551–559
26. Peng M, Wang Y, Shen Q, Xie X, Zheng H, Ma W et al (2019) High-performance flexible and broadband photodetectors based on PbS quantum dots/ZnO nanoparticles heterostructure. Sci China Mater 62(2):225–235
27. Peng M, Wen Z, Shao M, Sun X (2017) One-dimensional CdS_xSe_{1-x} nanoribbons for high-performance rigid and flexible photodetectors. J Mater Chem C 5(30):7521–7526
28. Park J, Lee J, Noh Y, Shin K-H, Lee D (2016) Flexible ultraviolet photodetectors with ZnO nanowire networks fabricated by large area controlled roll-to-roll processing. J Mater Chem C 4(34):7948–7958
29. Xia J, Zhao YX, Wang L, Li XZ, Gu YY, Cheng HQ et al (2017) van der Waals epitaxial two-dimensional $CdS_xSe_{(1-x)}$ semiconductor alloys with tunable-composition and application to flexible optoelectronics. Nanoscale 9(36):13786–13793
30. Wang S, Zhu Z, Zou Y, Dong Y, Liu S, Xue J et al (2019) A low-dimension structure strategy for flexible photodetectors based on perovskite nanosheets/ZnO nanowires with broadband photoresponse. Sci China Mater 63(1):100–109
31. Dai Y, Wang X, Peng W, Xu C, Wu C, Dong K et al (2018) Self-powered Si/CdS flexible photodetector with broadband response from 325 to 1550 nm based on pyro-phototronic effect: an approach for photosensing below bandgap energy. Adv Mater 30(9):1705893
32. Saran R, Curry RJ (2016) Lead sulphide nanocrystal photodetector technologies. Nat Photonics 10(2):81–92
33. Liu CH, Chang YC, Norris TB, Zhong Z (2014) Graphene photodetectors with ultra-broadband and high responsivity at room temperature. Nat Nanotechnol 9(4):273–278
34. Li L, Lou Z, Shen G (2015) Hierarchical CdS nanowires based rigid and flexible photodetectors with ultrahigh sensitivity. ACS Appl Mater Interfaces 7(42):23507–23514
35. Lou Z, Li L, Shen G (2016) Ultraviolet/visible photodetectors with ultrafast, high photosensitivity based on 1D ZnS/CdS heterostructures. Nanoscale 8(9):5219–5225
36. Gomathi PT, Sahatiya P, Badhulika S (2017) Large-area, flexible broadband photodetector based on ZnS-MoS_2 hybrid on paper substrate. Adv Funct Mater 27(31):1701611
37. Gou G, Dai G, Wang X, Chen Y, Qian C, Kong L et al (2017) High-performance and flexible photodetectors based on P3HT/CdS/CdS: SnS_2 superlattice nanowires hybrid films. Appl Phys A Mater Sci Process 123(12):1–8
38. Xu X, Chen J, Cai S, Long Z, Zhang Y, Su L et al (2018) A real-time wearable UV-radiation monitor based on a high-performance p-CuZnS/n-TiO_2 photodetector. Adv Mater 30(43):1803165
39. Sun H, Tian W, Cao F, Xiong J, Li L (2018) Ultrahigh-performance self-powered flexible double-twisted fibrous broadband perovskite photodetector. Adv Mater 30(21):1706986
40. Bae WK, Padilha LA, Park YS, McDaniel H, Robel I, Pietryga JM et al (2013) Controlled alloying of the core-shell interface in CdSe/CdS quantum dots for suppression of auger recombination. ACS Nano 7:3411–3419
41. Chan Y, Dahua Z, Jun Y, Linlong T, Chongqian L, Jun S (2020) Fabrication of hybrid graphene/CdS quantum dots film with the flexible photo-detecting performance. Physica E Low Dimens Syst Nanostruct 124:114216
42. Zheng Z, Zhuge F, Wang Y, Zhang J, Gan L, Zhou X et al (2017) Decorating perovskite quantum dots in TiO_2 nanotubes array for broadband response photodetector. Adv Funct Mater 27(43):1703115
43. Shen T, Yuan J, Zhong X, Tian J (2019) Dip-coated colloidal quantum-dot films for high-performance broadband photodetectors. J Mater Chem C 7(21):6266–6272

44. Hsiao Y-J, Ji L-W, Lu H-Y, Fang T-H, Hsiao K-H (2016) High sensitivity ZnO nanorod-based flexible photodetectors enhanced by CdSe/ZnS core-shell quantum. IEEE Sensors J 17(12):3710–3713

45. Xia YX, Zhai G, Zheng Z, Lian L, Liu H, Zhang D et al (2018) Solution-processed solar-blind deep ultraviolet photodetectors based on strongly quantum confined ZnS quantum dots. J Mater Chem C 6(42):11266–11271

46. Chu L, Hu R, Liu W, Ma Y, Zhang R, Yang J et al (2018) Screen printing large-area organo-metal halide perovskite thin films for efficient photodetectors. Mater Res Bull 98:322–327

47. Tong S, Yuan J, Zhang C, Wang C, Liu B, Shen J et al (2018) Large-scale roll-to-roll printed, flexible and stable organic bulk heterojunction photodetector. Npj Flex Electron 2(1):1–8

48. Park SH, Su R, Jeong J, Guo SZ, Qiu K, Joung D et al (2018) 3D printed polymer photodetectors. Adv Mater 30(40):1803980

49. Zhang C, Xie Y, Deng H, Tumlin T, Zhang C, Su JW et al (2017) Monolithic and flexible ZnS/SnO$_2$ ultraviolet photodetectors with lateral graphene electrodes. Small 13(18):1604197

50. Shen T, Binks D, Yuan J, Cao G, Tian J (2019) Enhanced-performance of self-powered flexible quantum dot photodetectors by a double hole transport layer structure. Nanoscale 11(19):9626–9632

51. Cui Y, Zhong Z, Wang D, Wang W, Lieber CM (2003) High performance silicon nanowire field effect transistors. Nano Lett 3:149–152

52. Duan X, Niu C, Zaanen J, Sahi V, Chen J, Parce JW et al (2003) High-performance thin-film transistors using semiconductor nanowires and nanoribbons. Nature 425:274–278

53. Cho N, Roy Choudhury K, Thapa RB, Sahoo Y, Ohulchanskyy T, Cartwright AN et al (2007) Efficient photodetection at IR wavelengths by incorporation of PbSe–carbon-nanotube conjugates in a polymeric nanocomposite. Adv Mater 19(2):232–236

54. Tian W, Zhang C, Zhai T, Li SL, Wang X, Liu J et al (2014) Flexible ultraviolet photodetectors with broad photoresponse based on branched ZnS-ZnO heterostructure nanofilms. Adv Mater 26(19):3088–3093

55. Jie JS, Zhang WJ, Jiang Y, Li YQ, Lee ST (2006) Photoconductive characteristics of single-crystal CdS nanoribbons. Nano Lett 6:1887–1891

56. Xing X, Zhang Q, Huang Z, Lu Z, Zhang J, Li H et al (2016) Strain driven spectral broadening of Pb ion exchanged CdS nanowires. Small 12(7):874–881

57. Jiang Y, Zhang WJ, Jie JS, Meng XM, Fan X, Lee ST (2007) Photoresponse properties of CdSe single-nanoribbon photodetectors. Adv Funct Mater 17(11):1795–1800

58. Wu P, Dai Y, Sun T, Ye Y, Meng H, Fang X et al (2011) Impurity-dependent photoresponse properties in single CdSe nanobelt photodetectors. ACS Appl Mater Interfaces 3(6):1859–1864

59. Takahashi T, Nichols P, Takei K, Ford AC, Jamshidi A, Wu MC et al (2012) Contact printing of compositionally graded CdS$_{(x)}$Se$_{(1-x)}$ nanowire parallel arrays for tunable photodetectors. Nanotechnology 23(4):045201

60. Zhang F, Niu S, Guo W, Zhu G, Liu Y, Zhang X et al (2013) Piezo-phototronic effect enhanced visible/UV photodetector of a carbon-Fiber/ZnO-CdS double-shell microwire. ACS Nano 7:4537–4544

61. Zhang F, Ding Y, Zhang Y, Zhang X, Wang ZL (2012) Piezo-phototronic effect enhanced visible and ultraviolet photodetection using a ZnO-CdS core-shell micro/nanowire. ACS Nano 6:9229–9236

62. Rai SC, Wang K, Ding Y, Marmon JK, Bhatt M, Zhang Y et al (2015) Piezo-phototronic effect enhanced UV/visible photodetector based on fully wide band gap type-II ZnO/ZnS core/shell nanowire array. ACS Nano 9(6):6419–6427

63. Zhang K, Ding J, Lou Z, Chai R, Zhong M, Shen G (2017) Heterostructured ZnS/InP nanowires for rigid/flexible ultraviolet photodetectors with enhanced performance. Nanoscale 9(40):15416–15422

64. Lin P, Zhu L, Li D, Xu L, Wang ZL (2018) Tunable WSe$_2$–CdS mixed-dimensional van der Waals heterojunction with a piezo-phototronic effect for an enhanced flexible photodetector. Nanoscale 10(30):14472–14479

65. Hu L, Yang J, Liao M, Xiang H, Gong X, Zhang L et al (2012) An optimized ultraviolet-a light photodetector with wide-range photoresponse based on ZnS/ZnO biaxial nanobelt. Adv Mater 24(17):2305–2309
66. Zheng Z, Gan L, Zhang J, Zhuge F, Zhai T (2017) An enhanced UV-vis-NIR and flexible photodetector based on electrospun ZnO nanowire array/PbS quantum dots film heterostructure. Adv Sci 4(3):1600316
67. Gao T, Zhang Q, Chen J, Xiong X, Zhai T (2017) Performance-enhancing broadband and flexible photodetectors based on perovskite/ZnO-nanowire hybrid structures. Adv Opt Mater 5(12):1700206
68. Liu Z, Chen G, Liang B, Yu G, Huang H, Chen D et al (2013) Fabrication of high-quality ZnTe nanowires toward high-performance rigid/flexible visible-light photodetectors. Opt Express 21(6):7799–7810
69. Pei Y, Pei R, Liang X, Wang Y, Liu L, Chen H et al (2016) CdS-nanowires flexible photodetector with ag-nanowires electrode based on non-transfer process. Sci Rep 6(1):21551
70. Xia F, Mueller T, Lin YM, Valdes-Garcia A, Avouris P (2009) Ultrafast graphene photodetector. Nat Nanotechnol 4(12):839–843
71. Tamalampudi SR, Lu YY, Kumar UR, Sankar R, Liao CD, Moorthy BK et al (2014) High performance and bendable few-layered InSe photodetectors with broad spectral response. Nano Lett 14(5):2800–2806
72. Chhowalla M, Shin HS, Eda G, Li LJ, Loh KP, Zhang H (2013) The chemistry of two-dimensional layered transition metal dichalcogenide nanosheets. Nat Chem 5(4):263–275
73. Feng W, Zheng W, Cao W, Hu P (2014) Back gated multilayer InSe transistors with enhanced carrier mobilities via the suppression of carrier scattering from a dielectric interface. Adv Mater 26(38):6587–6593
74. Koppens FH, Mueller T, Avouris P, Ferrari AC, Vitiello MS, Polini M (2014) Photodetectors based on graphene, other two-dimensional materials and hybrid systems. Nat Nanotechnol 9(10):780–793
75. Xie Z, Xing C, Huang W, Fan T, Li Z, Zhao J et al (2018) Ultrathin 2D nonlayered tellurium nanosheets: facile liquid-phase exfoliation, characterization, and photoresponse with high performance and enhanced stability. Adv Funct Mater 28(16):1705833
76. Wang F, Wang Z, Yin L, Cheng R, Wang J, Wen Y et al (2018) 2D library beyond graphene and transition metal dichalcogenides: a focus on photodetection. Chem Soc Rev 47(16):6296–6341
77. Ithurria S, Dubertret B (2008) Quasi 2D colloidal CdSe platelets with thicknesses controlled at the atomic level. J Am Chem Soc 130(49):16504–16505
78. Ithurria S, Bousquet G, Dubertret B (2011) Continuous transition from 3D to 1D confinement observed during the formation of CdSe nanoplatelets. J Am Chem Soc 133(9):3070–3077
79. Schlenskaya NN, Yao Y, Mano T, Kuroda T, Garshev AV, Kozlovskii VF et al (2017) Scroll-like alloyed CdS_xSe_{1-x} nanoplatelets: facile synthesis and detailed analysis of tunable optical properties. Chem Mater 29(2):579–586
80. Kelestemur Y, Dede D, Gungor K, Usanmaz CF, Erdem O, Demir HV (2017) Alloyed heterostructures of $CdSe_xS_{1-x}$ nanoplatelets with highly tunable optical gain performance. Chem Mater 29(11):4857–4865
81. Sun Z, Liao T, Dou Y, Hwang SM, Park M-S, Jiang L et al (2014) Generalized self-assembly of scalable two-dimensional transition metal oxide nanosheets. Nat Commun 5(1):1–9
82. Xu Y, Zhao W, Xu R, Shi Y, Zhang B (2013) Synthesis of ultrathin CdS nanosheets as efficient visible-light-driven water splitting photocatalysts for hydrogen evolution. Chem Commun 49(84):9803–9805
83. Maiti PS, Meir N, Houben L, Bar-Sadan M (2014) Solution phase synthesis of homogeneously alloyed ultrathin CdS_xSe_{1-x} nanosheets. RSC Adv 4(91):49842–49845
84. Zheng W, Feng W, Zhang X, Chen X, Liu G, Qiu Y et al (2016) Anisotropic growth of nonlayered CdS on MoS_2 Monolayer for functional vertical heterostructures. Adv Funct Mater 26(16):2648–2654

85. Zhu DD, Xia J, Wang L, Li XZ, Tian LF, Meng XM (2016) van der Waals epitaxy and photo-response of two-dimensional CdSe plates. Nanoscale 8(22):11375–11379

86. Peng M, Ma Y, Zhang L, Cong S, Hong X, Gu Y et al (2021) All-inorganic CsPbBr$_3$ perovskite nanocrystals/2D non-layered cadmium sulfde selenide for high-performance photodetectors by energy band alignment engineering. Adv Funct Mater 31:2105051

87. An J, Le T-SD, Lim CHJ, Tran VT, Zhan Z, Gao Y et al (2018) Single-step selective laser writing of flexible photodetectors for wearable optoelectronics. Adv Sci 5(8):1800496

88. Lee W, Liu Y, Lee Y, Sharma BK, Shinde SM, Kim SD et al (2018) Two-dimensional materials in functional three-dimensional architectures with applications in photodetection and imaging. Nat Commun 9(1):1–9

89. Rein M, Favrod VD, Hou C, Khudiyev T, Stolyarov A, Cox J et al (2018) Diode fibres for fabric-based optical communications. Nature 560:214–218

90. Kim TH, Lee CS, Kim S, Hur J, Lee S, Shin KW et al (2017) Fully stretchable optoelectronic sensors based on colloidal quantum dots for sensing photoplethysmographic signals. ACS Nano 11(6):5992–6003

91. Yan Y, Wu X, Chen Q, Liu Y, Chen H, Guo T (2019) High-performance low-voltage flexible photodetector arrays based on all-solid-state organic electrochemical transistors for photo-sensing and imaging. ACS Appl Mater Interfaces 11(22):20214–20224

92. Lee Y, Kim DS, Jin SW, Lee H, Jeong YR, You I et al (2022) Stretchable array of CdSe/ZnS quantum-dot light emitting diodes for visual display of bio-signals. Chem Eng J 427:130858

93. Chen X, Shehzad K, Gao L, Long M, Guo H, Qin S et al (2019) Graphene hybrid structures for integrated and flexible optoelectronics. Adv Mater 32(27):1902039

94. Yip S, Shen L, Ho JC (2019) Recent advances in flexible photodetectors based on 1D nanostructures. J Semicond 40(11):111602

95. Li L, Gu L, Lou Z, Fan Z, Shen G (2017) ZnO quantum dot decorated Zn$_2$SnO$_4$ nanowire heterojunction photodetectors with drastic performance enhancement and flexible ultraviolet image sensors. ACS Nano 11(4):4067–4076

96. Gasparini N, Gregori A, Salvador M, Biele M, Wadsworth A, Tedde S et al (2018) Visible and near-infrared imaging with nonfullerene-based photodetectors. Adv Mater Technol 3(7):1800104

97. Lee W, Lee J, Yun H, Kim J, Park J, Choi C et al (2017) High-resolution spin-on-patterning of perovskite thin films for a multiplexed image sensor array. Adv Mater 29(40):1702902

98. Shen Q, Xie X, Peng M, Sun N, Shao H, Zheng H et al (2018) Self-powered vehicle emission testing system based on coupling of triboelectric and chemoresistive effects. Adv Funct Mater 28(10):1703420

99. Park S, Heo SW, Lee W, Inoue D, Jiang Z, Yu K et al (2018) Self-powered ultra-flexible electronics via nano-grating-patterned organic photovoltaics. Nature 561:516–521

100. Fan F-R, Tian Z-Q, Lin Wang Z (2012) Flexible triboelectric generator. Nano Energy 1(2):328–334

101. Zhang Y, Peng M, Liu Y, Zhang T, Zhu Q, Lei H et al (2020) Flexible self-powered real-time ultraviolet photodetector by coupling triboelectric and photoelectric effects. ACS Appl Mater Interfaces 12(17):19384–19392

102. Xie X, Wen Z, Shen Q, Chen C, Peng M, Yang Y et al (2018) Impedance matching effect between a triboelectric nanogenerator and a piezoresistive pressure sensor induced self-powered weighing. Adv Mater Technol 3(6):1800054

Chapter 21
Self-Powered Photodetector

Hemant Kumar and Satyabrata Jit

21.1 Introduction

The development in the very-large-scale-integration (VLSI) technology has enabled the Internet of Things (IoT) networks to find applications in remotely controlled monitoring of the forest, civil works, agricultural lands, wireless capsule endoscopy, telepresence robots etc. [1]. These monitoring devices can capture and stream data containing high-resolution images. It consists of processing and transmitting a large volume of data wirelessly through IoT devices. A large number of various sensors are required for converting various physical phenomena into electrical signals in the IoT devices and systems. Photodetector may be viewed as an optical sensor which is an important component of the IoT based optical image processing networks [2]. In general, photodetector is a device used to convert optical signals of some desired wavelengths into electrical signals in the form of a current or a voltage. In addition to IoT applications, photodetectors are used in the field of cameras, non-destructive testing, mimicking artificial eye, optical computing systems, biological and environmental applications [3].

The very basic form of the photodetectors is simply a pn junction operated under reverse bias condition by using an externally connected power supply. When optical signal is incident on the device, excess electron-hole pairs are generated in the device due to the photoelectric effect. The applied reverse bias creates a large

H. Kumar
Department of Electronics and Communication Engineering, Jaypee Institute of Information Technology, Noida, India
e-mail: hemant.kumar@jiit.ac.in

S. Jit (✉)
Department of Electronics Engineering, Indian Institute of Technology (Banaras Hindu University), Varanasi, India
e-mail: sjit.ece@iitbhu.ac.in

© The Author(s), under exclusive license to Springer Nature Switzerland AG 2023
G. Korotcenkov (ed.), *Handbook of II-VI Semiconductor-Based Sensors and Radiation Detectors*, https://doi.org/10.1007/978-3-031-20510-1_21

electric field in the depletion region of the pn junction which helps in drifting out these excess photogenerated electrons and holes in opposite directions at a very high speed. The collection of the carriers results in an illumination-dependent current component, called the photocurrent, in the photodetectors. The external power supply in the form of a battery is thus an integral part of the conventional photodetectors for their operation [4]. However, self-powered photodetectors under discussion do not require any external power supply in the form of a battery for their operation. This class of photodetectors senses the incident photons and generates sufficient power internally by photovoltaic effect for their operation. The self-powered photodetectors are very much suitable for future generation IoT devices to be operated with optimal or minimal power requirements.

The working of self-powered photodetectors is closely related to that of the photovoltaic devices (i.e. solar cells) operated under short-circuited mode or open-circuited mode [5]. The self-powered photodetectors can be classified into two types. The first type of self-powered photodetectors independently generates sufficient power on their own and they do not require any external power supply or a separate energy harvesting unit for their operation [6]. The second type of self-powered photodetectors is the integration of the conventional photodetectors with a separate energy harvesting unit which supplies energy required for the operation of the photodetectors connected to it [7]. The comparative illustration between self-powered photodetector and traditional photodetector is shown in Fig. 21.1. The minimal circuit requirements with ideal conditions have been used to demonstrate the difference between a traditional photodiode (which requires an external power

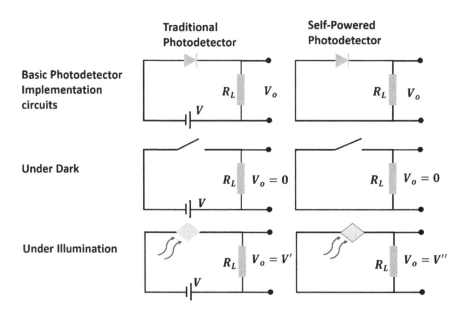

Fig. 21.1 Differentiation between the working of traditional and self-powered photodetector under minimal conditions

supply for its operation) and a self-powered detector (which requires no power supply for its operation). Both the devices under dark conditions behave as an open circuited element due to zero reverse bias current/dark current. Under illumination, both the conventional and self-powered photodetectors act as an illumination-dependent current source as shown in the Fig. 21.1.

The tremendous development in the synthesis and fabrication of low-dimensional materials with a variety of new electrical and optical properties have enabled the researchers to explore the nanostructured materials for the fabrication of self-powered photodetectors. Different nanostructures such as nanobelts (NBs), nanoribbons (NRs), and other one-dimensional nanostructures of high carrier mobility and optical absorption coefficient have been used for the fabrication of self-powered photodetectors [8, 9]. However, the requirement of sophisticated and expensive equipment for the fabrication of the inorganic nanostructures puts major restriction in achieving low-cost inorganic nanomaterials based self-powered photodetectors. To fabricate the low-cost photodetectors, researchers have explored simple and cost-effective solution-processed quantum dots (QDs) for photodetectors [6, 7], and solar cells [10]. Various organic semiconducting materials have also been used for the fabrication of low-cost self-powered photodetectors [5]. The major limitations of the organic materials are their low carrier mobility and poor stability. That is why, the organic photodetectors suffer from poor performance over inorganic photodetectors. Therefore, some researchers have investigated hybrid type self-powered photodetectors using both the organic and inorganic semiconductors in same the device [5, 11, 12].

This chapter presents various types of self-powered photodetectors of different kind of materials and device structures. The discussion is restricted to the Schottky-based self-powered photodetectors and heterojunction-based self-powered photodetectors. Finally, spectrum selective self-powered photodetectors have also been introduced.

21.2   II–VI Materials Based Self-Powered Schottky Photodetectors

The Schottky junction or rectifying junction is basically a junction between a metal and a n-type (p-type) semiconductor with the work function of the metal is higher (lower) than the n-type (p-type) semiconductor. Otherwise, the metal-semiconductor junction is known as the ohmic junction. The Schottky junction is considered to be equivalent to that of a p^+-n (n^+-p) junction if the semiconductor on which the contact is formed is a n-type (p-type) material. In other words, the Schottky junction behaves similarly as a p-n junction diode which conducts current only in the forward biased condition. On the other hand, a metal-semiconductor ohmic junction simply behave as a resistor which transports current in both directions irrespective of its biasing condition. Both types of junctions are used in both the conventional and self-powered photodetector devices. Two major photodetector structures are described below:

1. **Photoconductor Type Photodetectors:** This type of photodetectors consists of a semiconductor with two ohmic contacts on it [13]. This structure requires a continuous high potential between the two ohmic contacts made on the semiconductor. If an optical illumination is applied on the device, excess electron-hole pairs are generated in the semiconductor which changes the conductivity of the material thereby increasing the current in the external circuit. Since both contacts are of ohmic type, polarity of biasing does not affect the device characteristics. The device's response is expressed in terms of the variation of the resistivity under illumination. These structures provide very high photoconductive gain and high responsivity.

2. **Schottky Junction Photodiodes:** This type of photodetectors use one Schottky contact and one ohmic contact on the semiconductor working as the photoactive material in the device [14]. This kind of structure will produce rectifying behavior due to the Schottky junction. The Schottky photodiodes are operated under a reverse bias condition. Thus, a very small current flows in the external circuit under dark condition. When the device is exposed to illumination, excess electron-hole pairs are generated due to photoelectric effect in the device. These photo-generated excess electrons and holes are drifted in the opposite directions by the high electric field in the depletion region of the Schottky junction resulted due to the applied reverse bias voltage. The extraction of photogenerated electrons and holes results in a significant current in the circuit proportional to the intensity of the incident illumination.

In general, the Schottky junction photodiodes are faster and more sensitive to incident photons than the photoconductor type photodetectors. However, the selection of both the metals and photoactive materials plays crucial roles in determining the performance of the Schottky photodetectors. Some important works on the Schottky contact based self-powered photodetectors using II-VI group materials are discussed in the following sub-sections.

21.2.1 Graphene or Carbon Based Self-Powered Schottky Photodetectors

Graphene is one of the most versatile 2D materials used for making the Schottky contacts on CdSe and ZnO materials [8, 9, 15]. Jin et al. [8] fabricated a graphene-based Schottky junction on a CdSe nanobelt (NB) at $1000\ °C$ using the chemical vapor deposition (CVD) method shown in Fig. 21.2a. They achieved a very high photosensitivity of 3.5×10^5 and the transient behavior of $82\ \mu s$ rise time (t_r) and $179\ \mu s$ of decay time (t_d) shown in Fig. 21.2c with the corresponding experimental schematic shown in Fig. 21.2b. The CdSe NB/graphene-based Schottky device

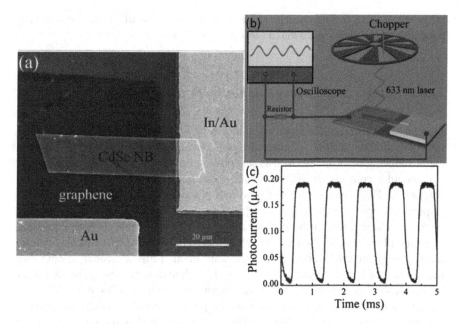

Fig. 21.2 (a) FESEM image of the photodetector showing CdSe NB with Graphene; (b) Schematic diagram for obtaining a transient response using chopper; (c) Photodetector's transient response under 633 nm laser at 1000 Hz light switching frequency. (Reprinted with permission from [8]. Copyright 2012: The Royal Society of Chemistry)

showed a very high photoconductive gain of 28 at 633 nm laser light. Gao et al. [15] fabricated a similar structure on PET flexible substrates and reported an improved transient response with 70 μs rise-time and 137 μs fall-time. Though they did not achieve any significant improvement in the transient response, but the realization of the CdSe based self-powered photodetectors on the PET flexible substrates could be viewed as a great achievement.

Graphene is also used for Schottky contact formation on the ZnO for fabricating the fast-responsive self-powered UV photodetectors. Boruah et al. [9] sandwiched a layer of ZnO nanorods (NR) between two graphene layers to demonstrate a very fast photodetector with the rise-time of $t_r = 3\ s$ and fall-time of $t_d = 0.47\ s$ under UV illumination. The significant improvement in the response speed encouraged other researchers to work on ZnO/graphene based photodetectors. Chen et al. [16] fabricated a ZnO/graphene Schottky contact based self-power photodetector by treating the ZnO with H_2O_2 to improve the barrier height at the ZnO/Graphene interface. They observed a significant improvement in the transient characteristics of the device with a rise-time of only 32 ms. Duan et al. [17] used the graphene-based Schottky contact on a thick film of Al-doped ZnO nanorods to achieve a significant improvement in the UV-to-Visible rejection ratio of 1×10^2 with a responsivity of 0.039 A/W. A self-powered Schottky photodiode based on carbon dot-decorated ZnO Nanorods on a graphite-coated paper substrate was fabricated by Sinha et al.

[18] for UV Detection applications. The transient response of the device was very poor with a rise-time $t_r = 2$ s and a fall-time $t_d = 3.2$ s.

21.2.2 Other Self-Powered Schottky Photodetectors

Being a noble metal, Gold (Au)-based electrode is frequently used for the fabrication of Schottky junction photodetectors. Chen et al. [19] used an asymmetric metal-semiconductor-metal (MSM) self-powered photodetector structure with two Au Schottky contacts of different widths on the same semiconductive material (MgZnO). The asymmetric self-powered photodetector showed an increased responsivity upto 20 mA/W. They [19] also demonstrated the impact of asymmetric Au contact structures on the performance of the device in the UV region. Their proposed photodetector showed the self-powered characteristics when asymmetric ratio was maintained above 20:1. Benyahia et al. [20] fabricated a ZnO-ZnS microstructure-based broadband self-powered photodetector using Au Schottky contact. The device showed a wideband response ranging from 300 nm to 900 nm.

Au Schottky electrode is also used on composite materials for the fabrication of self-powered photodetectors. Ebrahimi et al. [21] fabricated an Au Schottky contact on $Sn_xZn_{1-x}S$ composite material prepared by using the low-cost solvothermal process. They observed a reduction in the bandgap of the materials with increased Sn content in the composite material. Schematic illustrations of the device fabrication and variation of bandgap with Urbach energy (for varying concentration) are shown in Fig. 21.3a, and b, respectively. The device was shown to possess the self-powered feature in the UV-B region. The I-V characteristics of pure ZnS based device and band-alignment diagram are shown in Fig. 21.3c, and d, respectively. In another work [22], the same group investigated the transient response characteristics with rise-time of $t_r = 1.9$ ms and fall-time of $t_d = 2.6$ ms of the same device structure for varying bandgaps of $Sn_xZn_{1-x}S$ obtained by varying the Sn concentration in the active material.

Besides the gold, silver (Ag) with a work function of 4.2 eV [23] is also used for the fabrication of the Schottky junctions with ZnO for self-powered photodetectors. Purushotaman et al. [23] reported Ag/ZnO nanorods (NRs) Schottky junction photodetector shown in Fig. 21.4a. They developed floral ZnO NRs over the flexible PVDF substrates and explored the piezotronics property of the ZnO to obtain the self-powered feature of the photodetectors. The SEM image in Fig. 21.4b shows the surface morphology of the PVDF film while the SEM images of Fig. 21.4c–e show the morphology of the film containing the floral ZnO NRs. The flexible nature of the self-powered photodetector is illustrated in Fig. 21.4f.

ZnO-based self-powered photodetectors are primarily designed to operate in the UV region. For the self-powered photodetectors operating in the visible or NIR region, researchers either use doped ZnO or use different materials for the active region of the device. Selenium-based microrods were used to achieve wideband photo-response ranging from UV to the entire visible region [24] as shown in

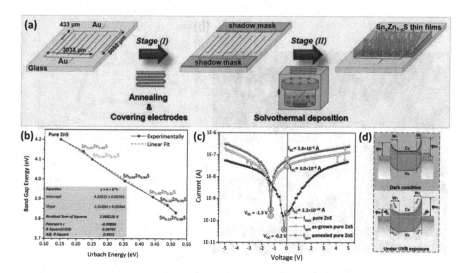

Fig. 21.3 (a) Schematic indicating the fabrication process for obtaining $Sn_xZn_{1-x}S$ thin film based Self-Powered photodetector; (b) Variation in band-gap against Urbrach energy for varying concentration of Sn and Zn; (c) I-V curves under the dark and UV illumination for the MSM UV detector (as-grown and annealed) based on Au/pure ZnS thin films/Au planar structure; (d) Band-Alignment of MSM device under dark and under UV illumination. (Reprinted with permission from [21]. Copyright 2020: Elsevier)

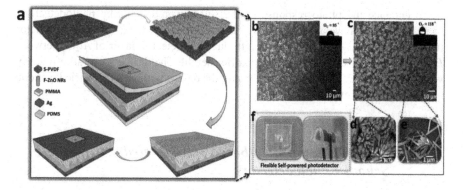

Fig. 21.4 (a) Schematic for the fabrication of flexible self-powered photodetector; (b) Indicates the untreated PVDF film; (c) SEM image of floral ZnO NRs; (d, e) magnified results, and (f) Fabricated flexible self-powered photodetector. (Reprinted with permission from [23]. Copyright 2018: American Chemical Society)

Fig. 21.5a, b. A Schottky junction between the Se and Ga-In alloy was created to obtain a very low dark current of 200 fA. The fabricated device showed a very high responsivity and detectivity of 408 mA/W and 1.30×10^{13} J, respectively, with $t_r = 124$ μs and $t_d = 146$ μs as shown in Fig. 21.5c.

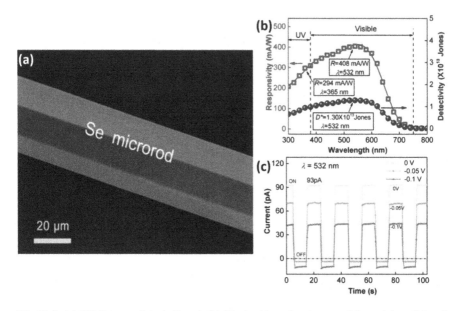

Fig. 21.5 (a) SEM image of single Se rod; (b) Obtained broadband responsivity and detectivity of the self-powered photodetector, and (c) Transient response of the device at 532 nm wavelength under different bias. (Reprinted with permission from [24]. Copyright 2019: American Chemical Society)

To enhance the photo response of the self-powered Schottky photodetectors, some researchers have used a dielectric layer between the metal and semiconductor in the device [25]. Zhang et al. [25] introduced a dielectric layer of Al_2O_3 between Pt and ZnO to increase the Schottky barrier height. The device's overall efficiency was improved by 2.77 times as compared to the device without a dielectric layer due to the Piezotronic effect.

21.2.3 II–VI Quantum Dots-Based Self-Powered Schottky Photodetectors

A very limited amount of research is reported on the quantum dots (QDs) based self-powered photodetectors. Kumar et al. [6] fabricated a self-powered Schottky junction photodiode using Au on CdSe QDs deposited on ZnO QDs based electron transport layer as shown in Fig. 21.6a, b. The device showed an optical response over 350–750 nm (i.e. covering the complete visible region) with an ultrafast response of $t_r = 18$ ms and $t_d = 17.9$ ms. It was shown that the inclusion of the ZnO QDs layer improved the photoresponse of the fabricated device by nearly 17 times as shown in Fig. 21.6c. The same group [26] also fabricated ITO/ZnO QDs/CdSe QDs/(Pd, Au) based self-powered photodetectors and compared the photoresponse

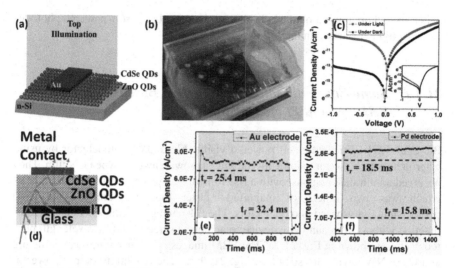

Fig. 21.6 (a) Schematic of QD-based self-powered photodetector; (b) Optical image of the fabricated device; (c) I-V characteristics under dark and illumination, and inset showing I-V characteristics without charge transport layer. Reprinted with permission from [6]. Copyright 2017: IEEE; (d) Schematic of the device, and (e, f) Transient behavior device with Au and Pd electrode respectively. Reprinted with permission from [26]. Copyright 2019: IEEE

of the Au/CdSe QDs and Pd/CdSe QDs Schottky contact photodetectors as illustrated in Fig. 21.6d. They observed that the metal electrode greatly affected the spectral coverage of the device. The Pd based electrode increased the quantum efficiency by 2.1 times by reducing the FWHM (full width at the half maximum) from 190 nm to 61 nm. Further, the Pd based device showed better transient response of $t_r = 18.5$ ms and $t_d = 15.8$ ms over the Au based device. Comparative results are shown in Fig. 21.6e, f. Moreover, the spectral coverage of the device reported in [26] was observed to be reduced (380–600 nm) as compared to the device reported in [6]. In case of [6], light was incident from the top side of the device whereas the light was allowed to enter into the device from the backside of the device considered in [26]. The improved performance in [26] was attributed to the filtering action of the ZnO QDs in addition to its normal electron transportation from the active region of photodetector.

21.3 Heterojunction Based Self-Powered Photodetectors

A heterojunction is formed between two different types of semiconducting materials of different bandgap energies. In this section, we will discuss some state-of-the-art heterojunction based self-powered photodetectors. We will first consider some conventional heterojunction photodetectors. Then some special self-powered photodetectors using the 2D transition metal dichalcogenide (TMD) heterostructure

materials and organic/inorganic hybrid heterojunction based self-powered photodetectors will be discussed in the following.

21.3.1   Some Conventional Heterojunction Based Self-Powered Photodetectors

Bie et al. [27] fabricated a self-powered visible-blind UV photodetector by using
heterojunctions between a single n-type ZnO nanowire and a p-type GaN film. They
observed an ultrafast response with a rise time of ~20 μs and a fall-time of ~219 μs
which was two orders of magnitude faster than the conventional ZnO
photoconductivity-based photodetectors. They also integrated the ZnO/GaN heterojunction with a CdSe nanowire device to achieve a selective multiwavelength photodetector as shown in Fig. 21.7a. The SEM images of the n-ZnO/p-GaN structure
and CdSe NW device are shown in Fig. 21.7b, c. The maximum output power of
1.1 μW was obtained from the self-powered photodetector device [27].

Yamada et al. [28] fabricated a transparent self-powered photodetector using
p-CuI and n-InGaZnO heterojunction. The device worked as a self-powered device

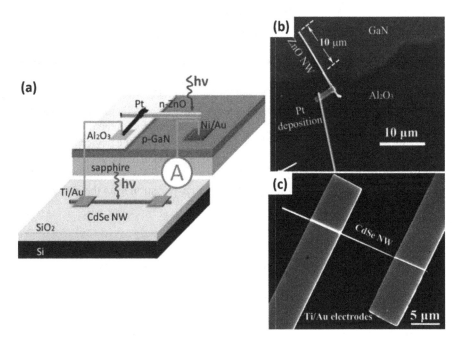

Fig. 21.7 (a) Schematic of n-ZnO/p-GaN with derived CdSe NW photodetector, (b) SEM image
of ZnO NW and GaN, and (c) SEM image of CdSe NW. (Reprinted with permission from [27].
Copyright 2010: John Wiley and Sons)

under photovoltaic mode and showed a spectral sensitivity under UV region only. Rana et al. [29] fabricated a transparent self-powered photodetector using a p-NiO and n-ZnO heterojunction. Pyro-phototronic effect and photovoltaic mode were explored to improve the responsivity and detectivity by 5460% and 6063% in the UV region, respectively. The proposed device showed a a t_r = 3.92 μs and t_d = 8.90 μs. Authors also tried to increase the spectral coverage of the detector by integrating narrowband semiconductors.

Huang et al. [30] reported Si NW/InGaZnO self-powered photodetector to achieve the spectral coverage in UV-NIR region by integrating the wide bandgap material InGaZnO with the lower bandgap material Si. The device showed a peak at ~400 nm with a response time of 0.2 ms. Similarly, a self-powered photodetector was fabricated by Hasan et al. [31] by integrating n-ZnO NRs (wide bandgap) with p-Si (low bandgap). They obtained a rise-time t_r = 25 ms and fall-time t_d = 22 ms.

Heterojunctions between the carbon and II-VI group semiconductors are also reported for self-powered photodetectors. Huag et al. [32] reported a self-powered UV–visible photodetector based on ZnO/graphene/CdS/electrolyte heterojunctions. The device showed major response in the UV region with some tails extended to the blue part of the visible spectrum. Though the authors considered the detector under self-powered category, the results were not encouraging enough for such a claim. Hatch et al. [33] reported a n-ZnO/p-Copper Thiocynate (CuSCN) heterojunction based self-powered photodetector for operating in the UV region. The device showed an ultrafast response of t_r = 500 ns and t_d = 6.7 μs. In another work, the same group attempted to improve the response speed by reducing t_r from 500 ns to 25 ns of the device in the same device structure by just modifying the ZnO annealing temperature. Ghamgosar et al. [34] fabricated a self-powered photodetector using n-ZnO and $p − Co_3O_4$ in core-shell structure with an Al_2O_3 buffer layer. The buffer layer was shown to improve the photoresponsivity by six times of the device. However, the results were not encouraging enough to consider the device under the self-bias category. Guo et al. [35] fabricated a ZIF-8@H:ZnO core-shell NRs array/ Si based self-powered broadband photodetector. This device also showed a very poor self-biased nature. Researchers have also tried to deplete the active layer by introducing various heterostructures to improve the photoresponse characteristics of the photodetectors. Mishra et al. [36] used ZnO/GaN heterostructure to improve the dark to photocurrent ratio. They analyzed a variety of structures, though the devices did not offer any promising results under self-biased conditions. Effect of doping on the performance characteristics of the self-powered photodetectors have also been investigated by some researchers. Ga doping in ZnO is one of the most common techniques employed for expanding the photoresponse of the ZnO based photode-tectors. In this direction, Saha et al. [37] prepared Ga doped ZnO NW coated with Ag and deposited the same over the p-Si substrate as shown in Fig. 21.8a, b. The fabricated Ga-ZnO/p-Si heterojunction offered a spectral coverage from 320 nm to 400 nm. A varying photo-response with red-shifted characteristics was observed with increased Ga doping as shown in Fig. 21.8c.

Zhou et al. [38] fabricated a Perovskite/Ga-doped ZnO (GZO) NRs heterojunction-based self-powered photodetector with a broad spectrum covering the entire visible

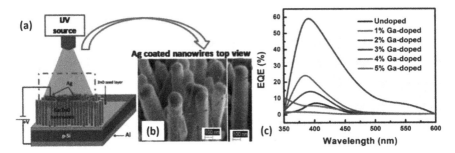

Fig. 21.8 (**a, b**) Schematic of the fabricated device using Ga-doped ZnO NW with Ag coating also indicating with SEM image, and (**c**) EQE of different Ga-doped ZnO devices. (Reprinted with permission from [37]. Copyright 2020: Elsevier)

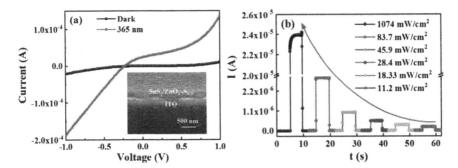

Fig. 21.9 (**a**) I–V characteristics under dark and illumination, and vertical SEM image indicating device structure is shown in inset and (**b**) Transient response under self-bias for the different illumination power densities. (Reprinted with permission from [40]. Copyright 2020: John Wiley and Sons)

band (400–800 nm). The device did not exhibit any variation in the transient response for varying Ga doping concentrations. Chen et al. [39] fabricated a self-powered photodetector using $n - Mg_xZn_{1-x}O$ alloy/p-Si heterojunctions. The photoresponse characteristics of the device were optimized by modifying the pizeopotential $Mg_xZn_{1-x}O$ by controlling its Mg content. Jiang et al. [40] fabricated a $SnS_2/ZnO_{1-x}S_x$ heterojunction based self-powered photodetector shown in the inset of Fig. 21.9a with its current-voltage characteristics under dark and under illumination (365 nm) shown in the Fig. 21.9a. The fabricated self-powered photodetector exhibited a broadband detection capability over 365 nm to 850 nm. The transient response of the detector is shown in Fig. 21.9b with varying power intensity. The device showed $t_r = 49.51$ *ms* and $t_d = 25.93$ *ms*.

21.3.2   2D Semiconducting Transition Metal Dichalcogenide (TMD) Heterostructure Materials Based Self-Powered Photodetectors

Researchers have also explored 2D semiconducting transition metal dichalcogenide (TMD)-based van der Waals (vdW) heterostructures for self-powered operations [41, 42]. Zou et al. reported a vertical GaSe/MoS$_2$ pn heterojunction self-powered photodetector grown by liquid gallium (Ga)-assisted chemical vapor deposition method [41]. The device exhibited a sizeable open-circuit voltage (0.61 V) with a broadband detection capability over 375–633 nm. Under self-powered operation, the device showed a high responsivity of 900 mA/W and a fast response speed of 5 ms. A self-powered broadband photodetector based on vertically stacked WSe$_2$/Bi$_2$Te$_3$ $p–n$ heterojunctions was reported by Liu et al. [42]. The device exhibited a maximum short-circuited current of 18 nA and an open circuit voltage of 0.25 V at 633 nm (with an incident power density of 26.4 mW/cm^2). The device's detection range was extended from the visible to near-infrared (375–1550 nm). The photodetector showed a fast response time ($\sim$210 μs) and high responsivity (20.5 A/W at 633 nm and 27 mA/W at 1550 nm) under zero bias voltage [42]. A flexible self-powered photodiode using MoS$_2$/WSe$_2$ vertically stacked vdW heterostructures was reported by Lin et al. [43]. Though they failed to measure the device's response time accurately, the detector was expected to provide a fast response speed. Ahn et al. [44] fabricated a MoTe$_2$/MoS$_2$ semi-vertically stacked vdW heterojunction based self-powered photodetector with a photo detection capability ranging from the violet (405 nm) to near-infrared (1310 nm) [44]. A WSe$_2$/Bi$_2$O$_2$Se 2D/2D vdW heterostructure-based self-powered photodetector with a broadband spectrum over 365 to 2000 was reported by Luo et al. [45]. The schematic of the fabricated self-powered photodetector showing vdW based heterostructure between WSe$_2$ and Bi$_2$O$_2$Se is shown in Fig. 21.10a. The transient response of the detector under 532 nm wavelength (20 kHz) at $V_{ds} = -5V$ gave the rise time of 2.4 μs and fall time of 2.6 μs (Fig. 21.10b). The electrical power generated by the WSe$_2$/Bi$_2$O$_2$Se heterostructure under different illumination intensities is shown in Fig. 21.10c as a function of bias voltage. Recently, Zheng et al. [46] demonstrated the growth of quasi-van der Waals epitaxial (QvdWE) based vertically aligned one-dimensional (1D) GaN nanorod arrays (NRAs) on TMDs/Si substrates for high-performance self-powered photodetection applications. A competitive photovoltaic photoresponsivity over 10 AW^{-1} under a weak detectable light signal was demonstrated under self-bias conditions.

21.3.3   Inorganic-Organic Hybrid Heterostructures Based Self-Powered Photodetectors

Organic semiconductors are suitable for the flexible electronics devices. But the major drawbacks of the organic semiconductors are their poor carrier mobility, low optical absorption coefficient and poor stability over longer period. That is why,

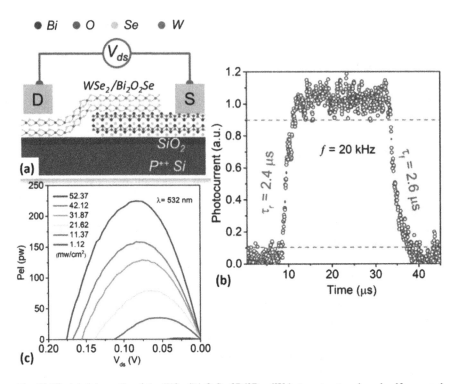

Fig. 21.10 (**a**) Schematic of the WSe₂/Bi₂O₂Se 2D/2D vdW heterostructure-based self-powered photodetector, (**b**) Transient response of the device showing rise time of 2.4 μs and fall time of 2.6 μs under 532 nm (20 KHz) at $V_{ds} = -5$ V, and (**c**) Generated electrical power by the heterostructure as a function of V_{ds}. (Reprinted with permission from [45]. Copyright 2020: John Wiley and Sons)

many researchers have explored the hybrid heterojunctions formed between an organic semiconductor and an inorganic semiconductor for the self-powered photodetectors. Sarkar et al. [47] fabricated the PEDOT:PSS/ZnO@CdS NR core-shell heterojunction based self-powered photodetectors. The photo-response spectrum covered the UV and visible region with a transient response of 20 ms. Zhan et al. [48] fabricated a rGO/ZnO hybrid heterojunction based self-powered photodetector. Game et al. [12] explored the hybrid heterojunction between n-type Nitrogen-doped ZnO and p-type Spiro-MeOTAD for fabricating self-powered visible photodetectors with rise time $t_r = 200$ μs and fall-time or decay time $t_d = 950$ μs. The self-biasing characteristic of the detector is shown in Fig. 21.11a, b. The experimental setup and transient characteristics of the device are shown in Fig. 21.11c, d, respectively. Kumar et al. [49] fabricated CdSe QDs/PQT-12 based hybrid self-powered photodetector shown in Fig. 21.12a. The photovoltaic effect in the fabricated detector is shown in Fig. 21.12b. The device exhibited a rise-time $t_r = 15.32$ ms and a fall-time $t_d = 12.01$ ms as shown in Fig. 21.12c.

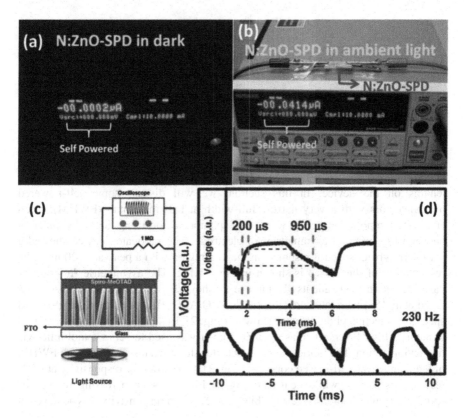

Fig. 21.11 Current N: ZnO-SPD self-powered photodetector (a) under dark, (b) under illumination, (c) Schematic of the experimental setup for obtaining a transient response, and (d) transient behavior under self-bias condition. (Reprinted with permission from [12]. Copyright 2014: The Royal Society of Chemistry)

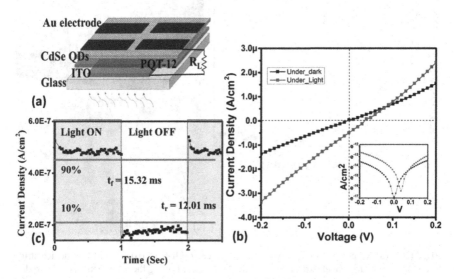

Fig. 21.12 (a) Schematic of fabricated CdSe QD/PQT-12 self-powered photodetector; (b) I-V characteristics of hybrid photodetector, and (c) Transient response of the self-powered hybrid photodetector. (Reprinted with permission from [49]. Copyright 2017: IEEE)

21.4   Spectrum Selective Self-Powered Photodetectors

Spectrum selective photodetectors with high detectivity, high signal-to-noise ratio, and fast response speed are important for various applications such as fluorescent-based biomedical imaging [50] and mimicking the human eye [51]. Spectrum selectivity of the photodetectors is defined as its ability to detect the incident photons of a single desired wavelength (or photon in a very narrowband of wavelengths centered about the desired wavelength) from the incident photons of different wavelengths on the device. In this section, we will discuss some self-powered photodetectors with a very narrow full-width at half maximum (FWHM). Chen et al. [52] reported a self-powered UV photodetector using Schottky junction between MgZnO and asymmetric Au electrode. The fabricated device showed a very narrowband spectral response in the UV region with a peak at ~320 nm. The responsivity of the device is shown in Fig. 21.13a. The asymmetric Au contact-based device structure and its SEM image are shown Fig. 21.13b, and c, respectively.

Ni et al. [53] fabricated an Au/i-ZnO/n-ZnO/Al$_2$O$_3$ structure based self-powered spectrum selective photodetector shown in Fig. 21.14a. The n-ZnO layer was used to act as a filter layer, whereas the i-ZnO layer was used to act as a multiplier via impact ionization. The fabricated photodetector showed a responsivity with FWHM of 9 nm under the self-biased condition. Fig. 21.14b shows the responsivity at −5 V reverse bias voltage, while the inset of Fig. 21.14b shows the increasing trend of the responsivity with increased reverse bias voltage. High responsivity was achieved at

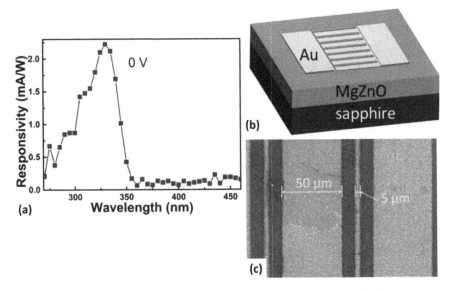

Fig. 21.13 (**a**) Responsivity of the Au/MgZnO device at self-bias, (**b**) schematic of device structure with asymmetric Au electrode, and (**c**) SEM image confirming the asymmetric Au electrodes. (Reprinted with permission from [52]. Copyright 2019: IOP Publishing)

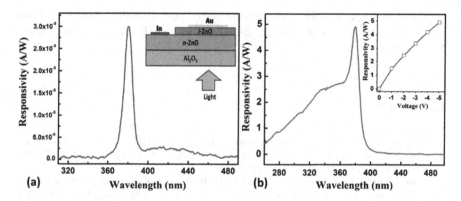

Fig. 21.14 (a) Responsivity of the i-ZnO/n-ZnO device at self-bias with inset showing device structure, and (b) Improved responsivity of the fabricated device under −5 V indicating reduced FWHM with a variation of responsivity against applied reverse bias voltage shown in inset. (Reprinted with permission from [53]. Copyright 2012: American Chemical Society)

high reverse bias voltage at the cost of reduced spectrum selectivity, as shown in Fig. 21.14b. Shen et al. [54] fabricated a homojunction between n-ZnO and p-ZnO for UV detection applications. They deposited both n-ZnO and p-ZnO on a sapphire substrate by using the molecular beam epitaxy method. Li, and N dopants were used to achieve the p-ZnO to act a filter layer, whereas the n-ZnO acted as the active layer. The combined effect of the filter layer and the active layer provided a very narrow spectral width of 9 nm. The detection capability of the detector was extended from the visible to the NIR region. Li et al. [55] fabricated CdSe/$p - Sb_2(S_{1-x}Se_x)_3$ heterojunction based self-powered narrowband photodetectors without using any filter layer to achieve a FWHM of ~35 nm. The device was capable to detect light in 650–900 nm. It was shown that the FWHM of the device could be controlled by varying the concentration of Se. Wei et al. [56] fabricated a p-NiO/Al-doped n-ZnO heterojunction based self-powered photodetector using the low-cost chemical bath deposition technique. The device was shown to possess a narrow FWHM at ~400 nm. Kumar et al. [57] fabricated a TiO_2/Si-based self-powered photodetector demonstrates its application in digital communication by sending optical pulses.

21.5 Conclusion

Advances in different types of self-powered photodetectors using Schottky junctions, heterojunctions, 2D heterostructure materials, colloidal quantum dots, and hybrid organic/inorganic heterojunctions have been discussed in this chapter. Among various II-VI group materials, most of the researchers have mainly explored ZnO and CdSe for the fabrication of self-powered photodetectors to primarily cover the UV and visible regions. It is observed that only a limited amount of work on

QD-based self-powered photodetectors has been reported. Particular emphasis must be given to developing self-powered photodetectors with detection capability in the NIR and IR regions.

References

1. Ayaz M, Ammad-Uddin M, Sharif Z, Mansour A, Aggoune E-HM (2019) Internet-of-Things (IoT)-based smart agriculture: toward making the fields talk. IEEE Access 7:129551–129583. https://doi.org/10.1109/ACCESS.2019.2932609
2. Masoud M, Jaradat Y, Manasrah A, Jannoud I (2019) Sensors of smart devices in the Internet of Everything (IoE) era: big opportunities and massive doubts. J Sens 2019:1–26. https://doi.org/10.1155/2019/6514520
3. Jansen-van Vuuren RD, Armin A, Pandey AK, Burn PL, Meredith P (2016) Organic photo-diodes: the future of full color detection and image sensing. Adv Mater 28(24):4766–4802. https://doi.org/10.1002/adma.201505405
4. Rawat G, Somvanshi D, Kumar Y, Kumar H, Kumar C, Jit S (2016) Electrical and ultraviolet-a detection properties of E-beam evaporated n-TiO$_2$ capped p-Si nanowires heterojunction pho-todiodes. IEEE Trans Nanotechnol:1–1. https://doi.org/10.1109/TNANO.2016.2626795
5. Bera A, Das Mahapatra A, Mondal S, Basak D (2016) Sb$_2$S$_3$/Spiro-OMeTAD inorganic–organic hybrid p–n junction diode for high performance self-powered photodetector. ACS Appl Mater Interfaces 8(50):34506–34512. https://doi.org/10.1021/acsami.6b09943
6. Kumar H, Kumar Y, Mukherjee B, Rawat G, Kumar C, Pal BN et al (2017) Electrical and optical characteristics of self-powered colloidal CdSe quantum dot-based photodiode. IEEE J Quantum Electron 53(3):1–8. https://doi.org/10.1109/JQE.2017.2696487
7. Lu H, Tian W, Cao F, Ma Y, Gu B, Li L (2016) A self-powered and stable all-perovskite photodetector-solar cell nanosystem. Adv Funct Mater 26(8):1296–1302. https://doi.org/10.1002/adfm.201504477
8. Jin W, Ye Y, Gan L, Yu B, Wu P, Dai Y et al (2012) Self-powered high performance photodetec-tors based on CdSe nanobelt/graphene Schottky junctions. J Mater Chem 22(7):2863. https://doi.org/10.1039/c2jm15913a
9. Boruah BD, Mukherjee A, Misra A (2016) Sandwiched assembly of ZnO nanowires between graphene layers for a self-powered and fast responsive ultraviolet photodetector. Nanotechnology 27(9):095205. https://doi.org/10.1088/0957-4484/27/9/095205
10. Sargent EH (2012) Colloidal quantum dot solar cells. Nat Photonics 6(3):133–135. https://doi.org/10.1038/nphoton.2012.33
11. Xie Y, Wei L, Li Q, Wei G, Wang D, Chen Y et al (2013) Self-powered solid-state photodetec-tor based on TiO$_2$ nanorod/spiro-MeOTAD heterojunction. Appl Phys Lett 103(26):261109. https://doi.org/10.1063/1.4858390
12. Game O, Singh U, Kumari T, Banpurkar A, Ogale S (2014) ZnO(N)–Spiro-MeOTAD hybrid photodiode: an efficient self-powered fast-response UV (visible) photosensor. Nanoscale 6(1):503–513. https://doi.org/10.1039/C3NR04727J
13. Kumar Y, Kumar H, Rawat G, Kumar C, Jit S (2018) Spectrum selectivity and respon-sivity of ZnO nanoparticles coated Ag/ZnO QDs/Ag UV photodetectors. IEEE Photon Technol Lett 30(12):1147–1150. https://doi.org/10.1109/LPT.2018.2836978
14. Kumar Y, Kumar H, Mukherjee B, Rawat G, Kumar C, Pal BN et al (2017) Visible-blind Au/ZnO quantum dots-based highly sensitive and spectrum selective Schottky photodiode. IEEE Trans Electron Devices 64(7):2874–2880. https://doi.org/10.1109/TED.2017.2705067
15. Gao Z, Jin W, Zhou Y, Dai Y, Yu B, Liu C et al (2013) Self-powered flexible and transpar-ent photovoltaic detectors based on CdSe nanobelt/graphene Schottky junctions. Nanoscale 5(12):5576. https://doi.org/10.1039/c3nr34335a

16. Chen D, Xin Y, Lu B, Pan X, Huang J, He H et al (2020) Self-powered ultraviolet photovoltaic photodetector based on graphene/ZnO heterostructure. Appl Surf Sci 529:147087. https://doi.org/10.1016/j.apsusc.2020.147087

17. Duan L, He F, Tian Y, Sun B, Fan J, Yu X et al (2017) Fabrication of self-powered fast-response ultraviolet photodetectors based on graphene/ZnO:Al nanorod-array-film structure with stable Schottky barrier. ACS Appl Mater Interfaces 9(9):8161–8168. https://doi.org/10.1021/acsami.6b14305

18. Sinha R, Roy N, Mandal TK (2020) Growth of carbon dot-decorated ZnO Nanorods on a graphite-coated paper substrate to fabricate a flexible and self-powered Schottky diode for UV detection. ACS Appl Mater Interfaces 12(29):33428–33438. https://doi.org/10.1021/acsami.0c10484

19. Chen H-Y, Liu K-W, Chen X, Zhang Z-Z, Fan M-M, Jiang M-M et al (2014) Realization of a self-powered ZnO MSM UV photodetector with high responsivity using an asymmetric pair of Au electrodes. J Mater Chem C 2(45):9689–9694. https://doi.org/10.1039/C4TC01839G

20. Benyahia K, Djeffal F, Ferhati H, Bendjerad A, Benhaya A, Saidi A (2021) Self-powered photodetector with improved and broadband multispectral photoresponsivity based on ZnO-ZnS composite. J Alloys Compd 859:158242. https://doi.org/10.1016/j.jallcom.2020.158242

21. Ebrahimi S, Yarmand B (2020) Solvothermal growth of aligned SnxZn1-xS thin films for tunable and highly response self-powered UV detectors. J Alloys Compd 827:154246. https://doi.org/10.1016/j.jallcom.2020.154246

22. Ebrahimi S, Yarmand B (2021) Tunable and high-performance self-powered ultraviolet detectors using leaf-like nanostructural arrays in ternary tin zinc sulfide system. Microelectron J 116:105237. https://doi.org/10.1016/j.mejo.2021.105237

23. Purusothaman Y, Alluri NR, Chandrasekhar A, Vivekananthan V, Kim S-J (2018) Direct in situ hybridized interfacial quantification to stimulate highly flexile self-powered photodetector. J Phys Chem C 122(23):12177–12184. https://doi.org/10.1021/acs.jpcc.8b02604

24. Chang Y, Chen L, Wang J, Tian W, Zhai W, Wei B (2019) Self-powered broadband Schottky junction photodetector based on a single selenium microrod. J Phys Chem C 123(34):21244–21251. https://doi.org/10.1021/acs.jpcc.9b04260

25. Zhang Z, Liao Q, Yu Y, Wang X, Zhang Y (2014) Enhanced photoresponse of ZnO nanorods-based self-powered photodetector by piezotronic interface engineering. Nano Energy 9:237–244. https://doi.org/10.1016/j.nanoen.2014.07.019

26. Kumar H, Kumar Y, Mukherjee B, Rawat G, Kumar C, Pal BN et al (2019) Effects of optical resonance on the performance of metal (Pd, Au)/CdSe quantum dots (QDs)/ZnO QDs optical cavity based Spectrum selective photodiodes. IEEE Trans Nanotechnol 18:365–373. https://doi.org/10.1109/TNANO.2019.2907529

27. Bie Y-Q, Liao Z-M, Zhang H-Z, Li G-R, Ye Y, Zhou Y-B et al (2011) Self-powered, ultrafast, visible-blind UV detection and optical logical operation based on ZnO/GaN nanoscale p-n junctions. Adv Mater 23(5):649–653. https://doi.org/10.1002/adma.201003156

28. Yamada N, Kondo Y, Cao X, Nakano Y (2019) Visible-blind wide-dynamic-range fast-response self-powered ultraviolet photodetector based on CuI/In-Ga-Zn-O heterojunction. Appl Mater Today 15:153–162. https://doi.org/10.1016/j.apmt.2019.01.007

29. Rana AK, Kumar M, Ban D-K, Wong C-P, Yi J, Kim J (2019) Enhancement in performance of transparent p-NiO/n-ZnO heterojunction ultrafast self-powered photodetector via Pyro-Phototronic effect. Adv Electron Mater 5(8):1900438. https://doi.org/10.1002/aelm.201900438

30. Huang C-Y, Huang C-P, Chen H, Pai S-W, Wang P-J, He X-R et al (2020) A self-powered ultraviolet photodiode using an amorphous InGaZnO/p-silicon nanowire heterojunction. Vacuum 180:109619. https://doi.org/10.1016/j.vacuum.2020.109619

31. Hassan JJ, Mahdi MA, Kasim SJ, Ahmed NM, Abu Hassan H, Hassan Z (2012) High sensitivity and fast response and recovery times in a ZnO nanorod array/p-Si self-powered ultraviolet detector. Appl Phys Lett 101(26):261108. https://doi.org/10.1063/1.4773245

32. Huang G, Zhang P, Bai Z (2019) Self-powered UV–visible photodetectors based on ZnO/graphene/CdS/electrolyte heterojunctions. J Alloys Compd 776:346–352. https://doi.org/10.1016/j.jallcom.2018.10.225

33. Hatch SM, Briscoe J, Dunn S (2013) A self-powered ZnO-Nanorod/CuSCN UV photodetector exhibiting rapid response. Adv Mater 25(6):867–871. https://doi.org/10.1002/adma.201204488

34. Ghamgosar P, Rigoni F, Kohan MG, You S, Morales EA, Mazzaro R et al (2019) Self-powered photodetectors based on Core–Shell ZnO–Co$_3$O$_4$ nanowire heterojunctions. ACS Appl Mater Interfaces 11(26):23454–23462. https://doi.org/10.1021/acsami.9b04838

35. Guo T, Ling C, Li X, Qiao X, Li X, Yin Y et al (2019) A ZIF-8@H:ZnO core–shell nanorod arrays/Si heterojunction self-powered photodetector with ultrahigh performance. J Mater Chem C 7(17):5172–5183. https://doi.org/10.1039/C9TC00290A

36. Mishra M, Gundimeda A, Garg T, Dash A, Das S, Vandana et al (2019) ZnO/GaN heterojunction based self-powered photodetectors: influence of interfacial states on UV sensing. Appl Surf Sci 478:1081–1089. https://doi.org/10.1016/j.apsusc.2019.01.192

37. Saha R, Karmakar A, Chattopadhyay S (2020) Enhanced self-powered ultraviolet photoresponse of ZnO nanowires/p-Si heterojunction by selective in-situ Ga doping. Opt Mater 105:109928. https://doi.org/10.1016/j.optmat.2020.109928

38. Zhou H, Yang L, Gui P, Grice CR, Song Z, Wang H et al (2019) Ga-doped ZnO nanorod scaffold for high-performance, hole-transport-layer-free, self-powered CH3NH3PbI3 perovskite photodetectors. Sol Energy Mater Sol Cells 193:246–252. https://doi.org/10.1016/j.solmat.2019.01.020

39. Chen Y-Y, Wang C-H, Chen G-S, Li Y-C, Liu C-P (2015) Self-powered n-MgZnO/p-Si photodetector improved by alloying-enhanced piezopotential through piezo-phototronic effect. Nano Energy 11:533–539. https://doi.org/10.1016/j.nanoen.2014.09.037

40. Jiang J, Huang J, Ye Z, Ruan S, Zeng Y (2020) Self-powered and broadband photodetector based on SnS$_2$/ZnO$_{1-x}$S$_x$ heterojunction. Adv Mater Interfaces 7(20):2000882. https://doi.org/10.1002/admi.202000882

41. Zou Z, Liang J, Zhang X, Ma C, Xu P, Yang X et al (2021) Liquid-metal-assisted growth of vertical GaSe/MoS$_2$ p–n heterojunctions for sensitive self-driven photodetectors. ACS Nano 15(6):10039–10047. https://doi.org/10.1021/acsnano.1c01643

42. Liu H, Zhu X, Sun X, Zhu C, Huang W, Zhang X et al (2019) Self-powered broad-band photodetectors based on vertically stacked WSe$_2$/Bi$_2$Te$_3$ *p–n* heterojunctions. ACS Nano 13(11):13573–13580. https://doi.org/10.1021/acsnano.9b07563

43. Lin P, Zhu L, Li D, Xu L, Pan C, Wang Z (2018) Piezo-Phototronic effect for enhanced flexible MoS$_2$/WSe$_2$ van der Waals photodiodes. Adv Funct Mater 28(35):1802849. https://doi.org/10.1002/adfm.201802849

44. Ahn J, Kang J-H, Kyhm J, Choi HT, Kim M, Ahn D-H et al (2020) Self-powered visible–invisible multiband detection and imaging achieved using high-performance 2D MoTe2/MoS2 Semivertical heterojunction photodiodes. ACS Appl Mater Interfaces 12(9):10858–10866. https://doi.org/10.1021/acsami.9b22288

45. Luo P, Wang F, Qu J, Liu K, Hu X, Liu K et al (2021) Self-driven WSe$_2$/Bi$_2$O$_2$Se Van der Waals Heterostructure photodetectors with high light on/off ratio and fast response. Adv Funct Mater 31(8):2008351. https://doi.org/10.1002/adfm.202008351

46. Zheng Y, Cao B, Tang X, Wu Q, Wang W, Li G (2022) Vertical 1D/2D heterojunction architectures for self-powered Photodetection application: GaN Nanorods grown on transition metal Dichalcogenides. ACS Nano 16(2):2798–2810. https://doi.org/10.1021/acsnano.1c09791

47. Sarkar S, Basak D (2015) Self powered highly enhanced dual wavelength ZnO@CdS Core–Shell Nanorod arrays photodetector: an intelligent pair. ACS Appl Mater Interfaces 7(30):16322–16329. https://doi.org/10.1021/acsami.5b03184

48. Zhan Z, Zheng L, Pan Y, Sun G, Li L (2012) Self-powered, visible-light photodetector based on thermally reduced graphene oxide–ZnO (rGO–ZnO) hybrid nanostructure. J Mater Chem 22(6):2589–2595. https://doi.org/10.1039/C1JM13920G

49. Kumar H, Kumar Y, Rawat G, Kumar C, Mukherjee B, Pal BN et al (2017) Colloidal CdSe quantum dots and PQT-12-based low-temperature self-powered hybrid photodetector. IEEE Photon Technol Lett 29(20):1715–1718. https://doi.org/10.1109/LPT.2017.2746664

50. Guo Z, Park S, Yoon J, Shin I (2014) Recent progress in the development of near-infrared fluorescent probes for bioimaging applications. Chem Soc Rev 43(1):16–29. https://doi.org/10.1039/C3CS60271K

51. Jansen-van Vuuren RD, Pivrikas A, Pandey AK, Burn PL (2013) Colour selective organic photodetectors utilizing ketocyanine-cored dendrimers. J Mater Chem C 1(22):3532. https://doi.
org/10.1039/c3tc30472h

52. Chen H, Sun X, Yao D, Xie X, Ling FCC, Su S (2019) Back-to-back asymmetric Schottky-
type self-powered UV photodetector based on ternary alloy MgZnO. J Phys Appl Phys
52(50):505112. https://doi.org/10.1088/1361-6463/ab452e

53. Ni P-N, Shan C-X, Wang S-P, Li B-H, Zhang Z-Z, Zhao D-X et al (2012) Enhanced responsivity of highly Spectrum-selective ultraviolet photodetectors. J Phys Chem C 116(1):1350–1353.
https://doi.org/10.1021/jp210994t

54. Shen H, Shan CX, Li BH, Xuan B, Shen DZ (2013) Reliable self-powered highly spectrum-
selective ZnO ultraviolet photodetectors. Appl Phys Lett 103(23):232112. https://doi.
org/10.1063/1.4839495

55. Li K, Lu Y, Yang X, Fu L, He J, Lin X et al (2021) Filter-free self-power CdSe/Sb$_2$(S$_{1-x}$,Se$_x$)$_3$
near infrared narrowband detection and imaging. InfoMat 3(10):1145–1153. https://doi.
org/10.1002/inf2.12237

56. Wei C, Xu J, Shi S, Bu Y, Cao R, Chen J et al (2020) The improved photoresponse properties of
self-powered NiO/ZnO heterojunction arrays UV photodetectors with designed tunable Fermi
level of ZnO. J Colloid Interface Sci 577:279–289. https://doi.org/10.1016/j.jcis.2020.05.077

57. Kumar M, Park J-Y, Seo H (2021) High-performance and self-powered alternating current ultraviolet photodetector for digital communication. ACS Appl Mater Interfaces
13(10):12241–12249. https://doi.org/10.1021/acsami.1c00698

Index

© The Editor(s) (if applicable) and The Author(s), under exclusive license to   517
Springer Nature Switzerland AG 2023
G. Korotcenkov (ed.), *Handbook of II-VI Semiconductor-Based Sensors
and Radiation Detectors*, https://doi.org/10.1007/978-3-031-20510-1

Printed in the United States
by Baker & Taylor Publisher Services

Printed in the United States
by Baker & Taylor Publisher Services